HAZARDOUS WASTE

LAND TREATMENT

Edited by

KIRK W. BROWN
GORDON B. EVANS, JR.
BETH D. FRENTRUP

Written by

J. ADAMS, D.C. ANDERSON, K.W. BROWN,
K.C. DONNELLY, G.B. EVANS, JR.,
B.D. FRENTRUP, S.G. JONES, D.C. KISSOCK,
C. SMITH, J.C. THOMAS

BUTTERWORTH PUBLISHERS
Boston•London
Sydney•Wellington•Durban•Toronto

An Ann Arbor Science Book

Ann Arbor Science is an imprint of Butterworth Publishers.

Library of Congress Catalog Card Number 83-72485
ISBN 0-250-40636-5

10 9 8 7 6 5 4 3 2 1

Butterworth Publishers
10 Tower Office Park
Woburn, MA 01801

Printed in the United States of America

Kirk W. Brown is the Principal of K. W. Brown and Associates, Inc., an environmental research and consulting firm, and is a Professor in the Soil and Crop Science Department at Texas A&M University in College Station, Texas. Dr. Brown has conducted extensive research programs for government agencies and provided consulting services to government and private clients. His research has covered the fate and movement of metals, salts, plant nutrients, and pesticides in soils, the environmental impact of land disposing municipal and industrial waste, and the effect of chemicals on the integrity of clay liners. Dr. Brown founded a consulting firm in 1980 to bring together a broad range of technical expertise to assist clients in developng state-of-the-art solutions to the environmental problems associated with waste disposal activities. K. W. Brown and Associates, Inc. has designed land treatment facilities for both hazardous and nonhazardous wastes and has developed management, operating, monitoring, and closure/post-closure plans for these facilities. Dr. Brown has over 100 technical publications and reports to his credit covering many areas of research and development.

ACKNOWLEDGEMENTS

This book is a synthesis of the thought, research and experience of many individuals involved in government, industry, academia, and consulting. An exhaustive list of these people would be impossible to compile, yet a number of people who collaborated with the authors during preparation and editing of the manuscript deserve mention. Particular thanks go to the following:

Carlton C. Wiles, U.S. EPA Solid and Hazardous Waste Research
    Division, Cincinatti, Ohio
Michael P. Flynn, U.S. EPA Office of Solid Waste, Washington, D.C.
Lloyd E. Deuel, Jr., Research Scientist, Soil & Crop Sciences
    Department, Texas A&M University
J. David Zabcik, Senior Associate, K. W. Brown & Associates, Inc.
Ronald L. Shiver, K. W. Brown & Associates, Inc.
Eric T. Clarke
David L. Guth
Keith Honey
Jean R. Richardson
Brandon D. Smith

PREFACE

In the waste management field, recent years have shown us both the environmental damage caused by inattention to good practices and the economic infeasibility of some of the most advanced management schemes. Land treatment, which uses the surface soil as the treatment medium, has become an attractive alternative for various waste types. For municipal sewage effluents and sludges, land treatment has become generally accepted and practiced as a relatively reliable and inexpensive alternative during the post-advanced wastewater treatment era. Since the 1950's, the relatively low hazard and ease of treatment of most petroleum refining wastes led some parts of the oil industry to begin using the practice. Today, companies involved in petroleum refining and related operations represent a substantial portion of the land treatment practitioners. However, over fifty industrial categories presently use land treatment. Land treatment is now being considered for increasingly complex wastes, hazardous wastes being a special subset. Because of this, there is an ever greater need to understand the behavior and fate of waste constituents in a given soil environment, particularly their mobility and toxicity.

Land treatment, also called landfarming, soil incorporation, land application, and others, is commonly known but often little understood. In general, in the soil environment, organic wastes may be effectively degraded and inorganic constituents immobilized while associated environmental liabilities remain low. To maintain these positive qualities, one should recognize that, in contrast with other types of waste disposal, land treatment is a knowledge intensive and dynamic process. Not only should careful planning be emphasized, but management should be responsive to varying treatment conditions. A good analogy can be made with industrial process design, where bench scale studies and pilot plants precede final design, construction, and operation.

This book explains the factors involved in designing, operating, and monitoring land treatment units from initial site selection and waste characterization through facility design, operation, and closure. Although the book concentrates on hazardous wastes, the same scientific and engineering principles apply regardless of the degree of hazard associated with the waste. The important differences between land treating hazardous and nonhazardous wastes are the rigor with which certain aspects of design, management, and monitoring are followed, based on the perceived degree of hazard, and the level of government regulatory controls imposed on the facility. Land treatment units designed in accordance with the procedures discussed in this book should meet or exceed the current requirements of the U.S. Environmental Protection Agency for hazardous waste.

Much of the information and many of the concepts presented herein are a compilation of the research and experience of the authors. Original thought is interwoven with references to the work of others in this and related fields. As such, the book is designed to be a practical guide for anyone involved in the design or operation of land treatment units. It may also prove valuable to those who will review the land treatment designs of others and to those who wish to further the study of land treatment.

To the uninitiated, the contents will serve to define and describe the system components and processes. The practitioner will find this book to be a convenient source of pertinent data and references. Indeed, we hope the approach and concepts presented here may expand, reinforce and even reshape the reader's understanding of land treatment technology.

February 1983   K.W.B.
                                                                G.B.E.
                                                                B.D.F.

TABLE OF CONTENTS

TABLE OF CONTENTS (continued)

LIST OF TABLES

LIST OF TABLES (continued)

LIST OF TABLES (continued)

LIST OF TABLES (continued)

LIST OF FIGURES

25

LIST OF FIGURES (continued)

LIST OF FIGURES (continued)

LIST OF FIGURES (continued)

INTRODUCTION 1

Gordon B. Evans, Jr.
K. W. Brown

The elimination of vast and increasing quantities of waste is an important issue facing any growing, industrialized society. Waste products, the inevitable consequence of production and consumption processes, require proper handling to minimize public health and environmental hazards. Historically, instances of poor disposal technology or management have caused extensive environmental damage and human suffering. In the United States in recent years, problems related to waste disposal began to surface whose real and potential consequences led to federal legislative action in the passage of the Resource Conservation and Recovery Act of 1976 to regulate the management of hazardous waste. Due to the tighter controls placed on waste management, the limitations of many of the disposal technologies used in the past are becoming apparent to representatives of industry; federal, state and local governments; and the general public. Along with these realizations has come a reassessment of the waste factor when evaluating the technical and economic feasibility of any industrial process.

Development of the best, most cost-effective technologies for handling municipal and industrial wastes is essential. Ideally, a method of treatment and disposal results in the degradation of any decomposable materials and the transformation and/or immobilization of the remaining constituents at no unacceptable risk to human health or the environment and at little cost. All techniques will fall short of this ideal, but some methods will prove more effective than others.

Land treatment is one such alternative for handling waste that can simultaneously constitute treatment and final disposal. Land treatment is the controlled application of waste onto or into the surface horizon of the soil accompanied by continued monitoring and management, to degrade, transform or immobilize waste constituents. Properly designed and managed land treatment facilities should be able to accomplish disposal without contaminating runoff water, leachate water, or the atmosphere. Additionally, some systems might be managed so that the land used for disposal is free of undesirable concentrations of residual materials that would limit the use of the land for other purposes in the future.

The emphasis herein is placed on hazardous wastes, a special subset of wastes having properties likely to cause environmental harm or affect human health when mismanaged. However, no attempt is made here to define or specify the meaning of the term "hazardous." In general, one who seeks a specified set of criteria leading to a list of wastes that are hazardous should regard definitions such as that developed by the U.S. EPA (Title 40 Code of Federal Regulations, Part 261). Ultimately, the concept of hazard or the degree of hazard is closely associated with the combined effects of waste characteristics, site description, and facility design and operation. Regardless of the definition chosen, the same principles apply to land treating both hazardous and nonhazardous wastes. The primary differences are the rigor with which design guidelines are followed and the degree of regulatory constraints placed on design and operation.

Land treatment is already widely practiced by numerous industries for handling hazardous industrial waste. Although many facilities have successfully used land treatment for their waste, the lack of systematic studies or monitoring of most facilities has limited the amount of knowledge available on important parameters and waste-site interactions. Additionally, many potentially land treatable wastes have not been tested or have been examined under only a limited range of conditions. To design or evaluate a hazardous waste land treatment (HWLT) unit, information is needed on site and waste characteristics, soil and climatic conditions, application rates and scheduling, decomposition products, and contingency plans to avert environmental contamination. In addition, the facility design should minimize potential problems such as the accumulation of toxic inorganic and recalcitrant organic waste constituents in the soil, as well as surface and groundwater pollution and unacceptable atmospheric emissions. Given these many concerns, land treatment design should be approached with interdisciplinary expertise having a ready source of current information on land treatment performance and practice.

The information presented in this document is to be used in assessing the technical aspects of hazardous waste land treatment. Where values are given in subsequent chapters for the parameters important to land treatment (e.g., application rates), the intent is to provide a guide to reasonable ranges, as gathered from the best available sources. Because the actual range for a given parameter will be largely site-specific, design and operating parameters may frequently fall outside of the ranges presented. Instances where parameters fall outside of these ranges signal that further information is needed or that the waste or site may not be suitable for land treatment.

The objectives of this book are to describe current land treatment knowledge and technology and to provide methods to design or evaluate the potential performance of a proposed or existing HWLT unit based on information about waste and site, operation and maintenance plans, monitoring needs, and closure objectives. This document takes a comprehensive decision-making approach to land treatment, from initial site selection through closure and post-closure activities. Additional information sources are referenced liberally to help provide state-of-the-art answers to the multitude of design considerations. The guidance presented is consistent with the current EPA regulations which are briefly summarized in Section 1.4 of this chapter.

1.1   THE ROLE OF LAND TREATMENT

An understanding of the potential usefulness and associated environmental risks of the various disposal options helps to place land treatment in perspective as a sound means of waste treatment and disposal. Hazardous waste disposal options are narrowing due to increasing environmental constraints, soaring energy costs, widespread capital shortages, and a desire to decrease potentially high long-term liabilities. Compared to other disposal options, properly designed and managed land treatment units carry low combined short and long-term liabilities. In the short-term, the land treated wastes are present at or near the land surface so that monitoring can rapidly detect any developing problems and management adjustments can be made in a preventive fashion. Also by virtue of using surface soils for waste treatment, management activities can exert direct and immediate control on the treatment/disposal process. Since most organic wastes undergo relatively rapid and near complete degradation, and hazardous metals are practically immobilized in an aerobic soil environment, long-term monitoring, maintenance and potential cleanup liabilities are potentially lower than with other waste disposal options. Many wastes are well suited to land treatment and because of the potentially lower liabilities associated with this method of waste disposal and the relatively low initial and operating costs, this option is becoming increasingly attractive to industry.

In a nationwide survey of HWLT, 197 facilities disposing of more than 2.45×10^6 metric tons of waste per year were identified. Over half of these were associated with petroleum refining and production (K. W. Brown and Associates, Inc., 1981). In a study of the waste disposal practices of petroleum refiners, 1973 records were compared with projections for 1983 and a general trend toward the increasing use of land treatment was evident (Rosenberg et al., 1976). Approximately 15% of the HWLT units were associated with chemical production. Industries providing electric, gas and sanitary services and producing fabricated metal items were the next largest users of HWLT, each having approximately 7% of the total number of units (K. W. Brown and Associates, Inc., 1981). Table 1.1 shows the numbers of land treatment units classed according to industry, using the standard industrial classification (SIC) codes for major industrial groups. Geographically, land treatment units are concentrated in the Southeastern United States from Texas to the Carolinas with a few scattered in the Great Plains and Far West regions (Figure 1.1). Most are found in areas having

TABLE 1.1 LAND TREATMENT USAGE BY MAJOR INDUSTRY GROUP*

SIC Code[†]	Description	Number of Units
29	Petroleum refining and related industries	105
28	Chemicals and allied products	30
49	Electric, gas, and sanitary services	16
34	Fabricated metal products, except machinery and transportation equipment	12
97	National security and international affairs	9
24	Lumber and wood products, except furniture	7
36	Electrical and electronic machinery, equipment, and supplies	5
20	Food and kindred products	4
22	Textile mill products	4
39	Miscellaneous manufacturing industries	3
35	Machinery, except electrical	3
26	Paper and allied products	3
13	Oil and gas extraction	2
44	Water transportation	2
76	Miscellaneous repair services	2
02	Agricultural production – livestock	1
30	Rubber and miscellaneous plastics products	1
33	Primary metal industries	1
37	Transportation equipment	1
51	Wholesale trade – nondurable goods	1
82	Educational services	1

* K. W. Brown and Associates, Inc. (1981).

† A listing of HWLT units by more specific SIC codes appears in Appendix A.

intensive petrochemical refining and processing activities and moderate climates. Of the 182 facilites for which areas are reported, the facility sizes range from 0.005 to 1668 acres; however, the median size is only 13.5 acres. Therefore, although there are a few very large facilities, the distribution is strongly skewed toward the small facilities, as illustrated by a bar graph (Figure 1.2).

Ten to fifteen percent of all industrial wastes (roughly 30-40 billion kg annually) are considered to be hazardous (EPA, 1980). Many wastes currently being disposed by other methods without treatment could be treated and rendered less hazardous by land treatment, often at lower cost. Of the six main groups of hazardous materials which have been found to migrate from sites to cause environmental damage (Table 1.2), three are prime candidates for land treatment. These three are (1) solvents (halogenated solvents may benefit from some form of pretreatment to enhance their biodegradability), (2) pesticides, and (3) oils (EPA, 1980). Land treatment is not, however, limited to these classes of wastes and may be broadly applicable to a large variety of wastes. The design principles and management practices for land treatment of waste discussed in this document are directed to the treatment and disposal of hazardous industrial waste.

TABLE 1.2   LAND TREATABILITY OF THE SIX MAIN GROUPS OF HAZARDOUS MATERIALS MIGRATING FROM DISPOSAL SITES*[†]

Hazardous Material Group	Land Treatability
(1) Solvents and related organics such as trichloroethylene, chloroform and toluene	High
(2) PCBs and PBBs	Limited
(3) Pesticides	High
(4) Inorganic chemicals such as ammonia, cyanide, acids and bases	Limited
(5) Heavy metals	Limited
(6) Waste oils and greases	High

* EPA (1980).

[†] High land treatability does not infer immunity from environmental damage. Only through proper design and management of a land treatment unit can the desired level of treatment be obtained and the migration of hazardous materials be prevented.

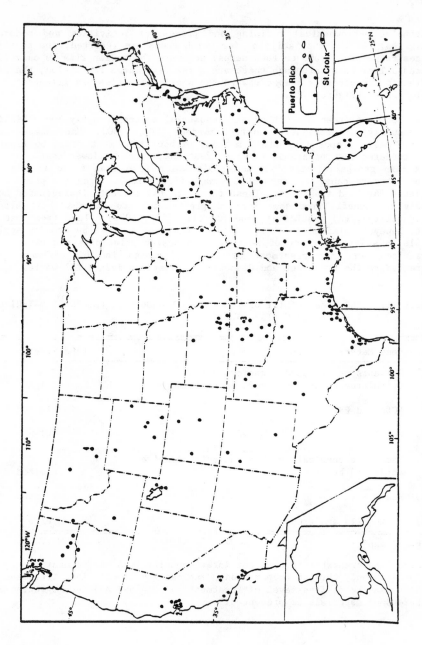

Figure 1.1. Areal distribution of land treatment units.

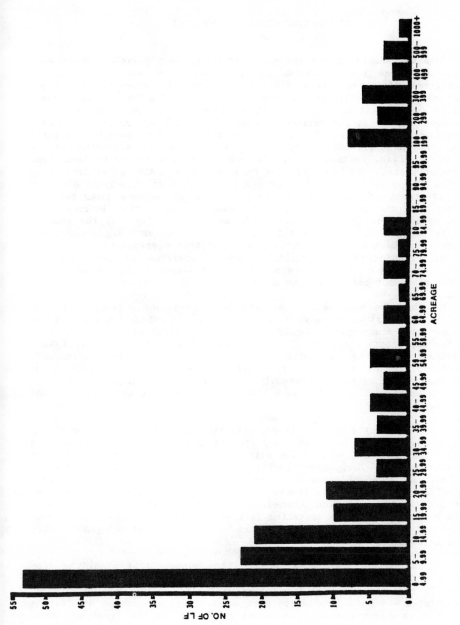

Figure 1.2. Size distribution of land treatment units.

In a well designed and operated HWLT unit, most hazardous waste constituents become less hazardous as they degrade or are transformed or immobilized within the soil matrix. In addition, the long-term maintenance and monitoring liabilities and the concomitant risk of costly cleanup efforts are minimized. However, it is important to remember that land treatment activities use unlined surface soils which are subject to direct contaminant losses via air, water or food chain; consequently, facility management has a tremendous impact on both the treatment effectiveness and the potential for contamination. If improperly designed or managed, land treatment units could cause various types of human health or environmental damage. The potential for such problems has not been closely studied for land treatment of hazardous wastes, but, it is evident from research conducted on the land treatment of nonhazardous waste that damages sometimes occur. For land treatment to be an effective system, the process must be managed to operate within given ranges for various design parameters. Frequent or consistent violation of these parameters could cause the inadvertant release of pollutants to the environment. The following brief discussion of the various means of contaminant migration emphasizes the importance of careful design and management.

Probably the most obvious pathway for contaminant migration at HWLT units is runoff since waste materials are often exposed on the soil surface or mixed into a nonvegetated soil surface. If control structures for runoff are improperly constructed or maintained, high concentrations of suspended and soluble waste constituents could be released to the environment. Therefore, control structures that are adequate to prevent release of untreated runoff water are obviously essential parts of a good design and the management plan should ensure that these structures are inspected and repaired, when necessary.

Since HWLT units are not lined, attention must be given to the potential for leaching of hazardous constituents to groundwater. Interactions between the waste and soil at the site may either increase or decrease the leaching hazard. Management practices, which can affect the biological, physical and chemical state of waste constituents in the treatment zone, can be designed to minimize leaching if the mobility of the waste constituents and their degradation products is carefully evaluated before operations begin. During the operating life of the facility, unsaturated zone monitoring provides information that can be used to adjust management practices to control leaching.

Release to the atmosphere is the third pathway that should be controlled. Emissions of volatile organic constituents can be reduced by carefully choosing the method and time of waste application. Wind-blown particulates can be controlled by management practices such as maintaining a vegetative cover and/or optimal water content in the treatment zone. Odors, another cause for concern, can also generally be controlled through management practices.

Migration of contaminants to the food chain must be prevented. If food chain crops are grown during the active life of the HWLT unit, the crop must be free of contamination before it is harvested and used for either animal or human food. In addition, waste constituents should not be allowed to accumulate in surface soils to levels that would cause a food chain hazard if food chain crops are likely to be grown.

Sites for HWLT units should be selected considering the potential pathways for contamination. Testing methods that can be used to predict waste-site interactions and the potential for contamination by each of these pathways are presented in this document. Facility design and management to minimize operational problems during the active life and at closure are also discussed.

1.3   SOURCES OF TECHNICAL INFORMATION

This document is not intended to encompass a thorough review of all the literature pertinent to the topic of land treatment of waste. Instead, information is provided which is specifically pertinent to the land treatment of hazardous waste. For many considerations, specific information and examples are sparingly few in the literature; therefore, it was necessary to draw on professional experience, the available published information on land treatment of municipal effluents and sludges, and associated literature concerning the fate of chemicals applied to soils. There are a number of sources from which the reader may obtain additional information on the principles and procedures of land treatment of waste. Some of the available books dealing with various aspects of this topic are listed in Table 1.2.

1.4   OVERVIEW OF REGULATIONS

Standards for all hazardous waste land disposal facilities regulated under the Resource Conservation and Recovery Act became effective on January 26, 1983. These regulations were issued by the U.S. Environmental Protection Agency (EPA) after a wide range of regulatory options were considered. Briefly, the regulations for land disposal facilities contain a groundwater protection standard and certain design and operating requirements for each type of land disposal unit (e.g., landfill, land treatment, waste pile, etc.).

Part 264, Subpart M of the regulations specifically deals with HWLT units and applies to both new and existing land treatment units. Of key importance to HWLT is the treatment program established by the owner or operator to degrade, transform or immobilize the hazardous constituents (Appendix B) in the waste placed in the unit. The regulations define the three principal elements of the treatment program as the wastes to be disposed, the design and operating measures necessary to maximize degradation, transformation and immobilization of hazardous waste constituents, and the

TABLE 1.2   SOURCES OF INFORMATION ON LAND TREATMENT OF WASTE

Title	Author/Editor	Publisher (Date)	Area
Proceedings of the International Conference on Land for Waste Management	J. Tomlinson	Agricultural Institute of Canada (1974)	Overview of waste disposal and its interaction with soils with particular emphasis on northern areas.
Land Treatment and Disposal of Municipal and Industrial Wastewater	R. L. Sanks and T. Asano	Ann Arbor Science Publications, Inc. (1976)	Summary of land treatment technology as of March 1975.
Soils for Management of Organic Wastes and Waste Waters	T. F. Elliott and F. J. Stevenson	ASA, SSSA, and CSSA (1977)	A collection of papers dealing mainly with municipal and agricultural waste.
Land as a Wastewater Management Alternative	R. C. Loehr	Ann Arbor Science Publications, Inc. (1976)	Proceedings of a symposium dealing mainly with municipal and animal waste disposal.
Managing the Heavy Metals on the Land	G. W. Leeper	Marcel Dekker, Inc. (1978)	Summary of the movement and accumulation of soil applied metals.
Sludge Disposal by Land-Spreading Techniques	S. Torrey	Noyes Data Corp. (1979)	A collection of a group of government sponsored research projects dealing with sewage sludge disposal.
Design of Land Treatment Systems for Industrial Wastes—Theory and Practice	M. R. Overcash and D. Pal	Ann Arbor Science Publications, Inc. (1979)	Provides information on land disposal techniques for both hazardous and nonhazardous industrial wastewaters.
Decomposition of Toxic and Non-Toxic Organic Compounds in Soils	M. R. Overcash	Ann Arbor Science Publications, Inc. (1981)	Provides information on the terrestrial effect of various organic compounds.

unsaturated zone monitoring program. HWLT units are also required to have a groundwater monitoring program.

A treatment demonstration is required to establish that the combination of operating practices at the unit (given the natural constraints at the site, such as soil and climate) can be used to completely degrade, transform or immobilize the hazardous constituents of the wastes managed at the unit. The treatment demonstration will be used to determine unit-specific permit requirements for wastes to be disposed and operating practices to be used.

HWLT units must be designed, constructed, and operated to maximize degradation, transformation and immobilization of hazardous constituents. In addition, HWLT units must have effective run-on and runoff controls and the treatment zone must be designed to minimize runoff. Runoff collection facilities must be managed to control the water volume generated by a 25 year, 24 hour storm. Wind dispersal of particulate matter must be controlled. If food chain crops are grown, the owner or operator must demonstrate that the crops meet certain criteria.

HWLT units must follow a groundwater monitoring program similar to that followed by all disposal facilities. The goals of the groundwater monitoring program are to detect and correct any groundwater contamination. HWLT units must also have an unsaturated zone monitoring program, including both soil core and soil-pore liquid monitoring, to provide feedback on the success of treatment in the treatment zone.

The regulations also set forth requirements for closure and post-closure care. The owner or operator must continue managing the HWLT unit to maximize degradation, transformation, and immobilization during the closure period. A vegetative cover capable of maintaining growth without excessive maintenance is generally required. During the closure and post-closure care period the owner or operator must continue many of the activities required during the active life of the unit including: control of wind dispersal, maintenance of run-on and runoff controls, continuance of food chain crop restrictions, and soil core monitoring. Soil-pore liquid monitoring may be suspended 90 days after the date of the last waste application. The post-closure care regulations also contain a variance which allows the owner or operator to be relieved from complying with the vegetative cover requirements and certain post-closure regulations if it is demonstrated that hazardous constituents within the treatment zone do not significantly exceed background values.

The regulations also contain requirements for recordkeeping, reactive and ignitable wastes, and incompatible wastes. In addition to the general recordkeeping requirements for all hazardous waste disposal units (Part 264, Subpart E (EPA, 1981)), records must be kept of waste application date and rate to properly manage the HWLT unit. Special recordkeeping requirements for wastes disposed by land treatment are necessary to ensure that the treatment processes are not inhibited.

CHAPTER 1 REFERENCES

Elliott, T. F. and F. J. Stevenson. 1977. Soils for management of organic wastes and waste waters. Am. Soc. Agron., Soil Sci. Soc. Am., and Crop Sci. Soc. Am. Madison, WI. 650 p.

EPA. 1980. Damages and threats caused by hazardous material sites. Oil and Special Materials Control Division, EPA. Washington, D.C. EPA 430/9-80-004.

EPA. 1981. Standards for owners and operators of hazardous waste treatment, storage and disposal facilities: Subpart E - Manifest system, recordkeeping, and reporting. 40 CFR 264.70-264.77.

K. W. Brown and Associates, Inc. 1981. A survey of existing hazardous waste land treatment facilities in the United States. Submitted to the U.S. EPA under contract no. 68-03-2943.

Leeper, G. W. 1978. Managing the heavy metals on the land. Marcel Dekker Inc., New York. 121 p.

Loehr, R. C. (ed.) 1976. Land as a waste management alternative. Ann Arbor Science Publ. Inc. Ann Arbor, Michigan. 811 p.

Overcash, M. R. and D. Pal. 1979. Design of land treatment systems for industrial wastes-theory and practice. Ann Arbor Science Publ. Inc. Ann Arbor, Michigan. p. 481-592.

Overcash, M. R. (ed.) 1981. Decomposition of toxic and non-toxic organic compounds in soils. Ann Arbor Science Publ. Inc. Ann Arbor, Michigan.

Rosenberg, D. G., R. J. Lofy, H. Cruse, E. Weisberg, and B. Beutler. 1976. Assessment of hazardous waste practices in the petroleum refining industry. Jacobs Engineering Co. Prepared for the U.S. EPA. PB-259-097.

Sanks, R. L. and T. Asano (eds.) 1976. Land treatment and disposal of municipal and industrial wastewater. Ann Arbor Science Publ. Inc. Ann Arbor, Michigan. 300 p.

Tomlinson, J. (ed.) 1974. Proceedings of the international conference of land for waste management. Ottawa, Canada. October 1973. Agricultural Institute of Canada. 388 p.

Torrey, S. 1979. Sludge disposal by landspreading techniques. Noyes Data Corp., New Jersey. 372 p.

THE DYNAMIC DESIGN APPROACH 2

Gordon B. Evans, Jr.
K. W. Brown
Beth D. Frentrup

This chapter outlines a comprehensive land treatment design strategy based on sound environmental protection principles. Basic elements of the design are described as they fit into a total system approach. An understanding of this dynamic design approach is essential and is the key to using this document. The remaining chapters more thoroughly describe the specific components of the strategy and show how each component is important to an effective hazardous waste land treatment (HWLT) unit design.

Anyone involved with some aspect of land treatment of hazardous waste, whether treatment unit design, permit writing, or site management, should understand the basic concepts behind land treatment. The primary mechanisms involved in land treatment are degradation, transformation and immobilization of hazardous constituents in the waste so that the waste is made less hazardous. Land treatment is considered a final treatment and disposal process rather than a method for long-term storage of hazardous materials. Thus, facilities are designed to prevent acute or prolonged harm to human health and the environment. Land treatment of wastes is a dynamic process. Waste, site, soil, climate and biological activity interact as a system to degrade or immobilize waste constituents, and the properties of each of these system components varies widely, both initially and temporally. Furthermore, land treatment is an open system which, if mismanaged or incorrectly designed, can potentially lead to both on-site and off-site problems with groundwater, surface water, air, or food chain contamination. Therefore, design, permitting and operation of HWLT units should take a total system approach including adequate monitoring and environmental safeguards, rather than an approach which appraises the facility only as a group of unrelated components.

The dynamic design approach discussed in this Chapter is based on a logical flow of events from the initial choice of waste stream to be land treated and potential site through operation and closure. This design approach is used throughout the document and is presented as an appropriate method for approaching the design of HWLT units. This approach assures that all critical aspects of hazardous waste land treatment are addressed and provides the designer with a comprehensive understanding of the individual HWLT unit. This document has been written with consideration given to current Federal regulations, but it is important to note that the approach presented here fully addresses all land treatment systems regardless of regulatory constraints because this approach is firmly based on scientific and engineering principles.

This strategy for designing and evaluating HWLT units is patterned after a computer flow diagram (Fig. 2.1) and suggests the essential design elements and choices to be made. Several others have dealt with comprehensive planning, and their basic considerations are comparable to this suggested strategy, although the format and emphasis of each vary (Phung et al., 1978a & b; Overcash and Pal, 1979; Loehr et al., 1979a & b). For a given application, the particular approach may likewise vary somewhat from Fig. 2.1 depending on the background of the facility planner or conditions unique to the specific waste or site. However, all of the elements introduced in the figure and discussed below should be considered, and in all cases, conclusions must be supported by appropriate evidence.

2.1   PRELIMINARY SITE ASSESSMENT

The first fundamental decision to be made is locating the facility. The preliminary assessment of a site involves a two faceted approach to evaluating technical site characteristics (i.e., hydrogeology, topography, climatology, soils, etc.) and socio-geographic factors (i.e., land use and availability, proximity to the waste generator, public relations, local statutes, etc.). In designing and permitting HWLT units, evaluation of the technical site characteristics is emphasized since these factors directly affect the environmental acceptability of a proposed site. The owner or operator considers the socio-geographic factors to determine the feasibility of land treatment among the available waste management options. In situations where an HWLT unit will be located near a large population center or where waste will be hauled long distances over public roads, socio-geographic factors are also important to environmental protection. Chapter 3 deals with the factors considered in the preliminary site assessment in greater detail. However, the final choice of site often cannot be made without considering the specific waste to be treated, the results of waste-site interaction studies, and the preliminary management design; these topics are discussed in Chapters 4 through 8.

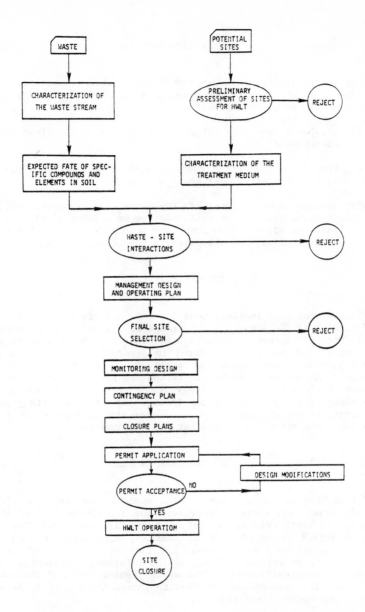

Figure 2.1.   Essential design elements and potential areas of rejection to be considered when planning and evaluating HWLT systems.

THE TREATMENT MEDIUM

Soil is the treatment medium for HWLT. Although soils are considered during the preliminary site assessment, a more thorough analysis of the treatment medium is necessary to:

(1)   develop a data base for pilot laboratory and/or field experiments; and

(2)   identify any limiting conditions which may restrict the use of the site as an HWLT unit.

The major components of interest are the variations in biological, physical and chemical properties of the soil. Native or cultivated plants, if used, and the climate modify the treatment medium. Methods for evaluating soil, as the treatment medium, are discussed in Chapter 4.

2.3   THE WASTE STREAM

Since wastes vary in their constituents, hazards and treatability, one must determine if the waste is (1) hazardous and (2) land treatable. The determination of whether a waste is hazardous is based on general knowledge of the industrial processes involved in generating the waste and on the chemical, physical and biological analyses of the waste as required by regulation. Knowledge of waste generating and pretreatment processes helps determine which compounds are likely to be present. In some cases, the treatability of a waste stream can be improved by controlled pretreatment or in-plant process changes. Chapter 5 presents information to be used in evaluating waste streams proposed for land treatment.

2.4   EXPECTED FATE IN SOIL

Information on the expected fate of specific compounds and elements in the soil, drawn from current literature and experience in land treatment, is presented. This information helps to identify waste constituents which may be resistant to degradation or that may accumulate in soils. Since waste streams are complex mixtures, the fate of the waste mixture in the environment can be estimated based on the information presented in Chapter 6. However, to specifically define waste treatability and the suitability of the land treatment option, waste-site interactions need to be evaluated by laboratory and/or field studies.

The key to the successful design of land treatment units for hazardous waste is the interpretation of the data emanating from preliminary waste-site interaction pilot studies. To justify using land treatment, the owner or operator must demonstrate that degradation, transformation, or immobilization will make the waste less hazardous. In addition, preliminary testing establishes the following:

(1)　the identity of waste constituents that limit short-term loading rates and the total allowable amount of waste over the life of the HWLT unit;

(2)　the assimilative capacity of soils for specific waste constituents;

(3)　criteria for management;

(4)　monitoring parameters to indicate possible contaminant migration into groundwater, surface water, air and cover crops;

(5)　the land area required to treat a given quantity of waste; and

(6)　the ultimate fate of hazardous constituents.

The laboratory, greenhouse and field tests are set up to determine degradability, mobility and toxicity of the waste in the land treatment system (Chapter 7). The amount of testing required depends on the amount of available information on the specific waste disposed at similar sites. Waste-site interaction studies are the major focus of HWLT design, since the independent inputs of waste and site converge here and the results form the foundation for subsequent planning and engineering.

2.6　　　　　　　　　　DESIGN AND OPERATING PLAN

The design and operation of an HWLT unit are based largely on the results obtained from the waste-site interaction studies. Management decisions include design of both the structure of the physical plant and the strategy for its operation. The various components considered in the management plan, include:

(1)　water control, including run-on control and runoff retention and treatment;

(2)　waste application, including technique, scheduling, storage, and monitoring for uniform distribution;

(3)　air emissions control which is closely related to waste application considerations, including control of odor, particulates, and and volatile constituents;

(4) erosion control, involving largely agricultural practices which are employed to limit wind and water erosion;

(5) vegetative cover and cropping practices; and

(6) records, reporting and inspections.

The management plan must adequately control waste loading and to provide effective waste treatment under varied environmental conditions; these topics are discussed in Chapter 8.

2.7   FINAL SITE SELECTION

Where more than one potential site is being considered for an HWLT unit, adequate knowledge of site limitations and facility economics, developed at this point in the design process (Fig. 2.1), provides the basis for deciding the location. Detailed management plans need not be prepared to determine the final site; however, consideration should be given to the topography, method of waste application, and required controls to manage water. These considerations affect the management, environmental protection, and the operating costs of the proposed facility and so should be considered during site selection. Where severe environmental or treatment constraints have not already limited the choice of sites, the decision will be based partly on economics and partly on the preferences of the owner or operator. Since it is likely that no site will be ideally suited, final site selection is often based on the best judgment of the owner or operator after careful review of all the data.

2.8   MONITORING

Monitoring is intended to achieve the threefold purpose of (1) determining whether the land treatment process is indeed decreasing the hazard of a waste, (2) identifying contaminant migration, and (3) providing feedback data for site management. Comprehensive monitoring includes following hazardous constituents along all of the possible routes of contaminant migration. Soils are generally sampled in the treatment zone to characterize waste treatment processes. Analysis of soil cores and soil- pore liquid in the unsaturated zone below the treatment zone aids the soil monitoring program in detecting the occurrence of contaminant leaching. Surface runoff may be analyzed. Air sampling may be advisable where volatile wastes are being land treated. Finally, since vegetation can translocate some hazardous compounds into the food chain, crops should be monitored when they are raised for human or animal consumption. Methods and requirements for monitoring the possible routes of contamination are discussed in Chapter 9.

After final site selection and before the owner or operator of a proposed HWLT unit applies for a permit to begin construction, the final design must be completed and several additional considerations must be addressed (Chapter 10). Routine health and safety procedures must be developed as well as preparedness for environmental emergencies. Contingency plans must also be developed to determine the remedial actions that will be taken in the event of:

(1) waste spill;

(2) soil overload;

(3) breach of surface water control structures;

(4) breakthrough to groundwater; or

(5) fire or explosion.

In addition, since the land treatability of a particular waste stream is approved on the basis of the results from preliminary testing, the decision to dispose of an alternate waste or to drastically change the composition of the waste stream may need to be accompanied by further data demonstrating that the new treatment combination also meets the land treatment objectives. Designs and the associated permits must then be amended as appropriate. The amount of additional testing required will depend on the waste stream, but the requirements may range from simple loading rate adjustments to a complete preapplication experimental program.

2.10   PLANNING FOR SITE CLOSURE

Plans for closure significantly affect facility design and must therefore be decided before a design can be completed. Site closure relies on the philosophy of nondeterioration of the native resource and emphasizes the eventual return of the land to an acceptable range of potential uses (Chapter 11). Plans must include the method of closure and procedures for site assessment and monitoring following closure. In addition, costs of closure and post-closure activities should be estimated.

2.11   PERMIT APPLICATION/ACCEPTANCE

In Fig. 2.1, an application-modification-acceptance feedback loop illustrates the permit application process. Because of the need for treatability data and the complexity of the design of any HWLT unit, the permit writer and the owner or operator are encouraged to cooperate in interpreting results from preliminary studies, evaluating data and modifying the HWLT unit design. The permitting process may vary depending on whether the U.S. Environmental Protection Agency or a State agency has the authority for permit issuance. Administrative procedures of the permitting process are not discussed in this document.

2.12   HWLT OPERATION

After receiving the appropriate permit, the owner or operator of an HWLT unit may begin operations following the design and monitoring plans outlined in the permit. Wastes delivered to the unit should be tested to determine if they contain the chemicals that are expected and for which the unit was designed. Monitoring and inspections must be carried out during the operation of the HWLT unit.

2.13   SITE CLOSURE

When the site capacity for which the HWLT unit has been designed is reached, the unit must be properly closed. HWLT units may also be closed for other reasons before this time. The closure plans included with the permit must be followed. The owner or operator is responsible for implementing these plans and is financially liable for closure costs, including any costs resulting from ensuing off-site groundwater pollution. Site closure requirements are discussed in detail in Chapter 11.

CHAPTER 2 REFERENCES

Loehr, R. C., W. J. Jewell, J. D. Novak, W. W. Clarkson, and G. S. Fried-man. 1979a. Land application of wastes. Vol. 1. Van Nostrand Reinhold Co., New York. 308 p.

Loehr. R. C., W. J. Jewell, J. D. Novak, W. W. Clarkson, and G. S. Fried-man. 1979b. Land application of wastes. Vol. 2. Van Nostrand Reinhold Co., New York. 431 p.

Overcash, M. R., and D. Pal. 1979. Design of land treatment systems for industrial wastes - theory and practice. Ann Arbor Sci. Publ. Inc. Ann Arbor, Michigan. 684 p.

Phung, T., L. Barker, D. Ross, and D. Bauer. 1978a. Land cultivation of industrial wastes and municipal solid wastes: state-of-the-art-study. Vol. 1. EPA-600/2-78-140a. PB 287-080/AS.

Phung, T., L. Barker, D. Ross, and D. Bauer. 1978b. Land cultivation of industrial wastes and municipal solid wastes: state-of-the-art-study. Vol. 2. EPA-600/2-78-140b. PB 287-081/AS.

PRELIMINARY ASSESSMENT OF SITES

3

K. W. Brown
Beth D. Frentrup
James C. Thomas

The assessment of sites proposed as locations for hazardous waste land treatment units involves a technical evaluation of the characteristics of each site and an evaluation of socio-geographic factors including area land use. The following objectives are fundamental to decision-making:

(1) Site characteristics should minimize the probability of off-site contamination via groundwater, surface water, or atmospheric emissions.

(2) Site characteristics should minimize the associated risk to the public and the environment in case of accidental fire, explosion, or release of hazardous substances.

Chapter 2 presented a model showing the flow of events from site assessment through site closure (Fig. 2.1). Figure 3.1 expands that model to indicate the aspects of site assessment and selection discussed in this Chapter.

Careful selection of sites is critical because, once the HWLT unit is in operation, the owner or operator has little control over natural processes (e.g., water table fluctuations, floods, winds) or over external societal influences (e.g., urban or industrial development). The operator of an existing HWLT unit can only adjust management practices to respond to these influences since the unit cannot be relocated without great cost.

Site analysis is essentially the same for both existing and proposed facilities. In permitting existing HWLT units, the permit evaluator must determine the appropriateness of continued operation. For existing units, the site assessment will indicate the aspects of the design or management that need to be modified to assure protection of human health and the environment. For example, a unit where excessive water during the wet season

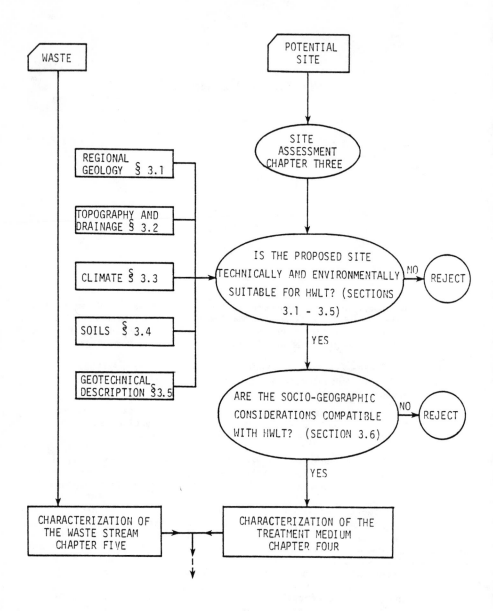

Figure 3.1. Factors considered during site selection.

has historically caused odor problems due to system anaerobicity might be allowed to continue operation if water control devices and water management were modified. In this case, reduction of wet season waste applications and modification of water management techniques might be required before permit approval.

In addition to determining the suitability of a given site for land treatment, predesign site analysis provides input for the design of demonstration studies and for subsequent management design. Site data also establish background conditions and furnish knowledge of the likely routes of contaminant migration for damage assessment in the event of accidental discharges. Table 3.1 shows how the information gained from the site assessment can be used throughout the design and management of the unit.

Evaluating the technical acceptability of a site involves establishing threshold conditions beyond which land treatment is not feasible, and the failure of a site to meet any one of these criteria may eliminate land treatment as an option. Threshold values are determined on the basis of a point or level beyond which the site constraints cannot be reasonably overcome by management. In formulating criteria, some threshold values appear rather arbitrary, even though an attempt has been made to remain flexible to account for the diversity of needs and circumstances. However, many limitations are ultimately a question of management extremes versus economics. For example, where alternate treatment or disposal techniques are not reasonably available, an industry may, for economic reasons, choose land treatment and use extreme management procedures to overcome site restrictions. The factors which determine the technical suitability of a site are discussed in Sections 3.1 to 3.5. These sections present general guidelines based on a moderate level of management, and the designer must recognize that exceptions to these could be acceptable. Section 3.6 discusses socio-geographic factors associated with the site selection process.

3.1   REGIONAL GEOLOGY

An understanding of the regional geology of the area in which the HWLT unit is located is an essential part of the site assessment. Knowledge of the geology of the site also helps determine the proper design and monitoring needs of the unit. Geologic information, published by federal and state geological surveys, describes the location, physical make-up, thickness and boundaries of geologic units which may be aquifers (EPA, 1977). A map of the proposed site(s) should be prepared to show the significant geologic features of the area, including:

(1) depth to bedrock;

(2) characteristics of the unconsolidated materials above the bedrock;

(3) characteristics of the bedrock;

TABLE 3.1 USE OF PRELIMINARY SITE ASSESSMENT INFORMATION

FACTORS CONSIDERED IN THE SITE ASSESSMENT PHASE	INFORMATION GATHERED IN THE SITE ASSESSMENT PHASE USED IN DECISION-MAKING OF LATER PHASES:				
	Waste-Soil Interaction Studies	Management Design	Monitoring Design	Final Site Selection	Closure Planning
Regional Geology	o determine effect on the ability of the soil to remain aerobic		o determine the placement of monitoring wells	o determine if the unit lies in a floodplain or aquifer recharge zone, over a fault zone, etc. o determine the local availability of suitable materials for pond and levee construction	o consider long-term stability of the site
Topography and Drainage	o determine effect on the ability of the soil to remain aerobic o determine the risk of mobile constituen's being leached to groundwater	o determine facility layout--plots roads, retention basins, etc. o consider modifications to natural topography	o determine the placement of unsaturated zone monitoring devices	o choose site to minimize amount of soil to be moved o avoid unstable areas	o consider drainage patterns needed at time of closure
Climate	o determine effect of temperature and moisture regimes on waste degradation	o determine waste application methods o determine waste storage capacity required due to wet or cold conditions o determine need to control wind dispersal of contaminants o determine (optimal) timing of operations	o determine the placement of air monitoring devices (optional)	o choose location downwind of major population centers	o consider the potential for acid rain and possible effects on waste constituent mobility

—continued—

TABLE 3.1 (continued)

FACTORS CONSIDERED IN THE SITE ASSESSMENT PHASE	INFORMATION GATHERED IN THE SITE ASSESSMENT PHASE USED IN DECISION-MAKING OF LATER PHASES:				
	Waste-Soil Interaction Studies	Management Design	Monitoring Design	Final Site Selection	Closure Planning
Soils	o determine effect of physical and chemical soil properties on waste degradation, transformation, and immobilization	o determine erosion hazards, calculate terrace spacings o consider horizonation	o consider how the leaching potential of soil will affect the choice and placement of monitoring devices	o determine overall suitability of soils as a treatment medium for HWLT	o consider erosion potential of soils following waste application
Geotechnical Description	o determine if groundwater will adversely affect treatment zone		o determine the placement of upgradient and downgradient monitoring wells o consider existing quality of water in underlying aquifers	o consider depth to water table o consider other potential sources of groundwater pollution in the area	
Sociogeographic		o consider how to minimize public risk from operations o determine need for buffer zones		o consider public opinion, zoning, current and future land use, etc. o avoid special use areas o choose a site close to waste generator	o consider public opinion and future land use when determining closure method

(4) outcrops;

(5) aquifer recharge zones; and

(6) discontinuities such as faults, fissures, joints, fractures, sinkholes, etc.

The depth to bedrock and the characteristics of the unconsolidated materials above the bedrock affect the conditions of the soil where treatment of wastes will take place, such as the ability of the soil to remain aerobic. Shallow water tables often occur in fine-grained geologic materials with low hydraulic conductivities. This does not necessarily make the site unacceptable for HWLT because these fine-grained materials may not provide a groundwater resource. Fine-grained materials are more effective than coarse-grained materials in slowing the movement of leachate and removing contaminants and are, therefore, more effective in protecting aquifers (Cartwright et al., 1981). The characteristics of the bedrock underlying the HWLT site also help to determine the potential for wastes to reach the groundwater unchanged. For example, a site underlain by limestone bedrock may be unacceptable because it may contain solution channels or develop sinkholes through which wastes could be rapidly transmitted to groundwater.

Outcrops of rock on or near the proposed site may indicate aquifer recharge zones. If water in a shallow aquifer is of high quality, or is being used as a drinking water source, this may be an unacceptable location for an HWLT unit. In addition, if any discontinuities exist, they should be carefully investigated to determine if they will allow contaminated leachate to reach groundwater (EPA, 1975). Hazardous waste facilities are required to be located at least 61 m (200 ft) away from a fault which has had displacement in Holocene time (EPA, 1981). How the groundwater directly beneath the site is connected to regional groundwater systems and drinking water aquifers is also an important consideration for choosing a site and designing effective monitoring systems.

3.2   TOPOGRAPHY AND DRAINAGE

Sites selected for HWLT units should not be so flat as to prevent adequate surface drainage, nor so steep as to cause excessive erosion and runoff problems; however, in selecting a site, it is important to remember that topography can be modified to some extent by facility design. The advantages of a relatively flat location include the ability to make waste applications by surface flooding in a slurry, minimization of erosion potential, and easy access by equipment. A 1% grade is usually sufficient to avoid standing water and prevent anaerobic conditions. One advantage of rolling terrain is that with careful design, less earth needs to be moved to construct retention basins and roads can be placed along ridges, providing all-weather site access. Slopes steeper than 4% may require special management practices to reduce erosion hazards. Management designs for different terrains are discussed in Chapter 8.

Generally the most desirable areas for HWLT units are upland flat and terrace landforms where the probability for washouts is low. Washouts are more likely in areas that are adjacent to stream beds or gullies or are in a floodplain. Site assessment and/or selection can be done by analyzing a topographic map for the area surrounding the HWLT site. The map should include the location of all springs, rivers and surface water bodies near the proposed site. Drainage patterns for the area should be determined. If the site lies within the 100-year floodplain, the level of the flood should be indicated on the map. Management of HWLT units located in the 100-year floodplain must include provisions to prevent washout of hazardous wastes (EPA, 1982).

The characteristics of the soil also affect the ability of the soil to remain aerobic and to support traffic. Aerobic conditions are necessary for the degradation of many wastes, so well drained or moderately well drained soils are needed. Poorly drained soils may become anaerobic and may limit the use of heavy equipment, and very well drained soils in humid regions may encourage rapid leaching of contaminants. Soil characteristics are discussed in Section 3.4

3.3   CLIMATE

Although climate greatly influences waste treatment, climatic conditions are not necessarily a major consideration in site selection. The principal reason for this is that the owner or operator of a proposed or existing unit has little choice about site location with respect to climate since conditions do not usually vary greatly within a given region and long distance waste shipment could be risky as well as uneconomical. An additional reason is that few regions within the United States exhibit such restrictive climatic conditions that land treatment is economically or technically infeasible. Careful design and a moderate level of management can safely overcome most climatic restrictions. An exception to this reasoning would be where inadequate land is available to treat the given waste stream based on climatic constraints (i.e., extended periods of low temperatures or excessive wetness).

The atmosphere directly affects the land treatment system by providing transport mechanisms for waste constituents, and acts indirectly as a modifier of soil-waste interactions. Table 3.2 lists these effects and the controlling atmospheric parameters which are important considerations for site selection. HWLT design and management plans should receive particular scrutiny if a temperature or moisture regime is present which would greatly influence treatment effectiveness. As a general rule, less land is required to treat a given quantity of waste if the unit is located in a warm, humid climate than in a cold, arid climate.

Since few if any HWLT sites have a sufficient historical data base to make reliable design decisions, climatic data must be extrapolated from a reporting station exhibiting conditions similar to those of the proposed site. For reliable climatological data it is best to choose an official

TABLE 3.2 THE INFLUENCE OF ATMOSPHERIC VARIABLES ON LAND TREATMENT OPERATIONS AND PROCESSES

Operation or Process	Atmospheric Variable	Effect
Biodegradation	Temperature	Indirect - controls soil temperature which controls microbial populations and activity
	Precipitation-Evapotranspiration	Indirect - controls soil moisture which controls (1) soil aeration, the supply of oxygen for microbes, and (2) adequacy of water supply
Waste application	Temperature	Direct - cold temperatures increase waste viscosity, thus decreasing ease of handling and hot temperatures may restrict application due to waste volatility hazard
		Indirect - cold temperatures keep soil temperature low, which can limit soil workability and waste degradation, and may increase the amount of runoff
	Precipitation-Evapotranspiration	Indirect - soil wetness can inhibit field accessability and enhance the waste leaching hazard
	Winds	Direct - hazard of off-site pollution due to transport of particulates and volatile constituents
	Atmospheric stability	Direct - surface inversions can lead to fumigation of the surface layer by volatile waste constituents
Site selection	Winds	Direct - potential hazard to public from advected particulates and volatile constituents

National Weather Service reporting station. These stations have standard-ized instrumentation, scrupulous instrument placement, and trained observational personnel. It is not always easy to choose a Weather Service reporting station that has a similar climate. Simply extrapolating from the nearest station is not necessarily acceptable. Due to orographic effects and major climatic modifiers, such as large bodies of water, a weather station 50 km from the proposed HWLT site may better match local conditions than observations made at a station only 5 km away from the site. Based on these considerations, the owner or operator of an HWLT unit or the permit writer should consult the services of a professional meteorologist.

3.3.1   Winds

Winds directly control site selection because of the need to minimize public risk from treatment operations. Although management strives to reduce air emissions to a minimum, atmospheric transport of contaminants may unavoidably occur when:

(1) hot weather or recent waste applications cause volatiliza-tion of waste constituents;

(2) aerosols from spray irrigation or suspended particulates from surface erosion are carried by high winds; or

(3) noxious vapors are released due to an accident such as fire or explosion.

Therefore, HWLT units should be placed downwind of major population centers whenever possible. Methods to control wind dispersal of contaminants are discussed in Section 8.4 and are particularly important during parts of the year when winds may blow toward a population center.

Siting with regard to winds is based on an analysis of prevailing winds during the waste application season. The application season is of particular importance since fresh wastes have the greatest potential for atmospheric emissions and applications often coincide with warm weather, which increases volatility and ignitability. Atmospheric stability at the time of waste application is also important. Accidents are more probable during waste handling operations and in case of fire or other emergency that release air contaminants, a knowledge of wind direction and speed helps the operator to assess the hazard and plan the response. Wind is a vector quantity, described by both magnitude and direction. Consequently, a frequency analysis to determine prevailing winds uses a two-way frequency distribution (Table 3.3) to construct a standard wind rose, (Fig. 3.2) which simultaneously considers wind speed and direction.

TABLE 3.3   TWO-WAY FREQUENCY DISTRIBUTION OF WIND SPEED AND DIRECTION*

Rating	SPEED, m/sec	S	SW	W	NW	N	NE	E	SE	
Weak	1.8 – 3.1			2		1		1		4
	3.2 – 4.4	6	8	2		16	13	17	2	64
Moderate	4.5 – 5.8	11	12	5	4	16	8	15	7	78
	5.9 – 7.1	11	16	10	14	21	7	6	2	87
	7.2 – 8.5	5	8	9	22	8	1	5	5	63
Strong	8.6 – 9.8	1	5	6	37	8		1		58
	9.9 – 11.2		1	5	26	2		2	1	37
	11.3 – 12.5	1		4	11	2				17
	12.6 – 13.9	1	1	4	14				2	22
	14.0 – 15.2			2	4					6
	15.3 – 16.6			1		2				3
	16.7 – 17.9				5					5
	18.0 – 19.3			1	1					1
		35	41	50	138	76	29	47	19	445

* Modified from Panofsky and Brier (1958).

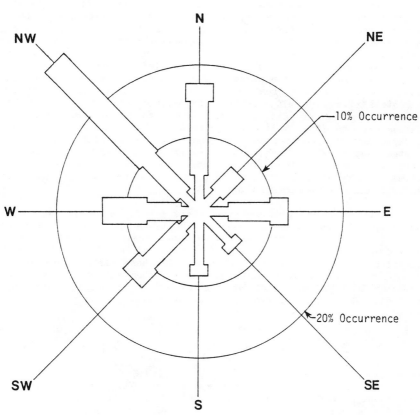

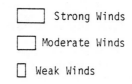

Figure 3.2. Standard wind rose using data presented in Table 3.3
(Modified from Panofsky and Brier, 1958). Reprinted
by permission of Pennsylvania State University.

Temperature and Moisture Regimes

Although climatic variables other than wind have a very limited effect on site suitability, two additional factors should be considered during the site assessment since management of HWLT units is greatly influenced by climate.   An appreciation of two broad climatic relationships can illuminate regions where particular scrutiny is required to determine if the design properly accounts for climatic effects.   First, the degradation of organic wastes effectively ceases when soil temperatures remain below 5°C (Dibble and Bartha, 1979).   Therefore, units located in cold northern or mountainous regions (Fig. 3.3) may have seasonal treatment restrictions and will need to have storage capacities, pretreatment methods and/or land areas that are adequate to handle the projected quantity of waste.   Second, when soil moisture content exceeds field capacity, aerobic decomposition, which is the primary treatment mechanism active in land treatment, is inhibited (Brown et al., 1980).   Seasonally wet climates promote soil anaerobicity and may also restrict access to the field.   Regions with excess moisture (Fig. 3.3) may require special designs or operational procedures such as increased waste storage capacity, field drainage systems to control water table depth, major runoff and run-on control structures, careful waste application timing, and/or vehicles equipped with flotation tires.   A more detailed discussion of how management must respond to climatic influences appears in Chapter 8.

As noted above, in some areas there may be seasonal restrictions on waste application based on climate.   The waste application season may be restricted in the northern and mountainous regions because of prolonged periods of low temperatures.   The Southeast and Pacific Northwest may have restrictions due to seasonal wetness.   If these restrictions are severe enough to halt the application of wastes, then sufficient waste storage capacity must be provided for the wastes being produced during these periods.   Section 8.8.1 discusses how to determine the waste application season.

3.4   SOILS

Since soil is the treatment medium for HWLT, careful consideration must be given to selecting a site with soil properties suitable for retention and degradation of the wastes to be applied.   The potential for erosion and leaching of hazardous constituents must be evaluated.

3.4.1   Soil Survey

A detailed soil survey conducted according to standard U.S. Soil Conservation Service (SCS) procedures should be completed to identify and map the soil series on sites proposed for HWLT units.   For each soil series, a general description of soil properties is needed to select potential areas

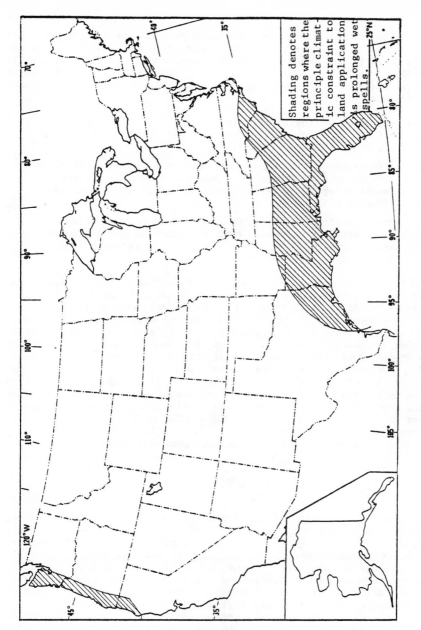

Shading denotes regions where the principle climatic constraint to land application is prolonged wet spells.

Figure 3.3.   Areas where waste application may be limited by excess moisture.

Soils   67

for waste application and to determine uniform areas for monitoring. Soil samples should be taken to adequately characterize the site and to determine the physical and chemical properties required for design (Chapter 5). Information, usually included in soil survey descriptions, that is useful during various phases of the design and management of HWLT units includes the following:

(1) estimates of the erodibility of the soil (Section 3.4.2), used to calculate terrace spacings and other erosion control structures (Section 8.5);

(2) information on the depth and texture of subsoils (Section 3.4.5), used to determine if suitable soil is available for constructing clay berms and clay lined retention ponds (Section 8.3); and

(3) measurements of surface texture, used to estimate acceptable waste application rates, water retention capacity, and types and amounts of constituents that will be retained (Section 3.4.3).

An SCS soil survey may also contain information on the average and/or seasonal water table height. Additional information on the historical water table height can be gained from a visual inspection of the soil horizons. Differences in soil color and patterns of soil color such as mottling and the gray colors that accompany gleying (a process that occurs in soils that are water saturated for long periods) are good indicators of poorly drained soils (USDA, 1951). Poor drainage can result from a seasonally high water table, a perched water table, or the internal drainage characteristics of the soil. In this inspection it is important to realize that the soil color may indicate past conditions of poor drainage and that drainage may be improving. In this case, soils will gradually become more oxidized as indicated by red, yellow and reddish brown colors. Geotechnical investigations described in Section 3.5 should be designed to verify water table fluctuations if soil color indicates poor drainage.

3.4.2   Erosion

Erosion is a function of the climate, topography, vegetative cover, soil properties and the activities of animals and man. The Universal Soil Loss Equation (USLE) is commonly used to estimate soil lost due to erosion; it is an empirical formula based on years of research and actual field work. The equation includes factors that affect soil loss and considers management alternatives to control soil loss. The USLE calculates loss from sheet and rill erosion. This is not the same as sediment yield at some downstream point; it equals sediment yield plus the amount of soil deposited along the way to the place of measure (Wischmeier and Smith, 1978). The USLE equation and tables for each factor use English units rather than metric for two reasons, 1) the USLE has traditionally used English units and direct conversion to metric units produces numbers that are awkward to use, and 2) data to be used in the USLE is more readily available in English units. The value of soil lost per acre per year can be multiplied by 2.24 to convert the value to metric tons per hectare per year. Wischmeier and Smith (1978) provide additional guidance on using the USLE with metric units for all factors. Although the soil losses calculated are estimates rather than absolute data, they are useful for selecting sites. Choosing management practices that minimize the factors in the equation will minimize erosion. The USLE is written as:

$$A = RKLSCP \qquad\qquad (3.1)$$

where

A = Soil-loss in tons/acre/year;

R = Rainfall factor;

K = Soil-erodibility factor;

L = Slope-length factor;

S = Slope-gradient factor;

C = Cropping management factor; and

P = Erosion control practice factor.

Rainfall (R). The amount, intensity and distribution of precipitation determine the dispersive action of rain on soil, the amount and velocity of runoff, and the losses due to erosion. Maps of the United States with iso-erodent lines, indicating equally erosive annual rainfall have been prepared; the R factor can be read off these maps. Wischmeier and Smith (1978) developed a map for the continental U.S. (Fig. 3.4).

Soil-erodibility (K). Some soils erode more readily than others even when all other factors are equal. This difference, due to the properties of the soil itself, is called soil erodibility. K values have been determined experimentally and can be obtained from nomographs (Fig. 3.5).

Slope-length and Slope-gradient (LS). These factors are closely interrelated and are considered as one value. Slope length is the distance from the point of origin of overland flow to the point where the slope gradient decreases to the extent that deposition begins or to the point where runoff enters a well-defined channel. The soil loss per unit area increases as the slope length increases. As slope gradient becomes steeper, the velocity of the runoff water increases, increasing the power of the runoff to detach particles from the soil and transport them from the field. Figure 3.6 shows how to determine the LS factor for a given site.

Cropping Management (C). This factor shows the combined effect of all the interrelated cover and management variables. The C factor is the ratio of soil loss from land managed under specified conditions to the corresponding loss from continuously fallow land. Values vary widely as shown in Table 3.4. Vegetation to be selected for levees and land treated areas between applications, or at closure, should have a minimum C value.

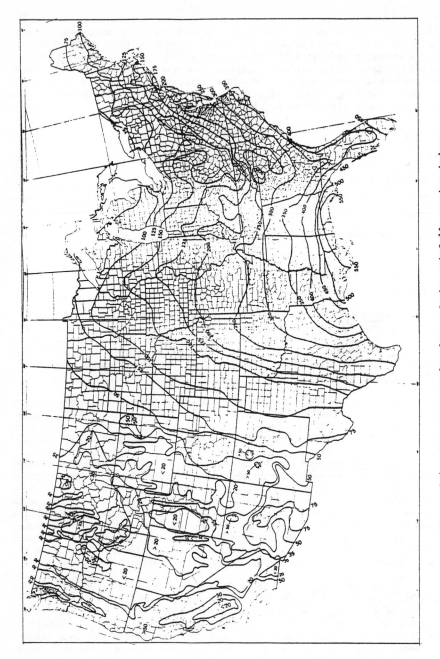

Figure 3.4. Average annual values of the rainfall erosion index (Wischmeier and Smith, 1978).

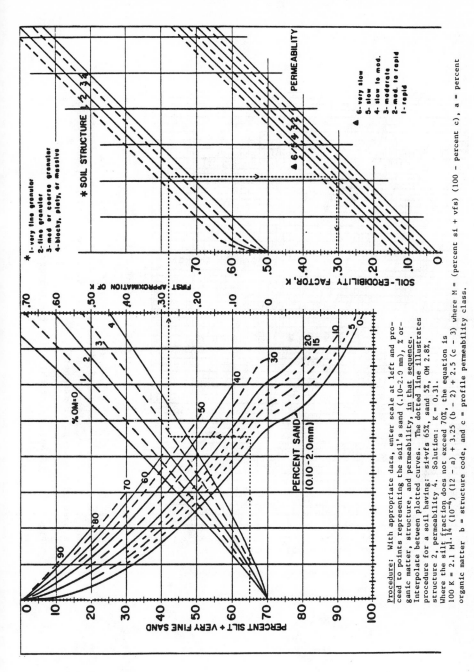

Procedure: With appropriate data, enter scale at left and proceed to points representing the soil's sand (.10-2.0 mm), % organic matter, structure, and permeability, in that sequence.
Interpolate between plotted curves. The dotted line illustrates
procedure for a soil having: si+vfs 65%, sand 5%, OM 2.8%,
structure 2, permeability 4. Solution: K= 0.31.
Where the silt fraction does not exceed 70%, the equation is
100 K = 2.1 M.14 (10⁻⁴) (12 - a) + 3.25 (b - 2) + 2.5 (c - 3) where M = (percent si + vfs) (100 - percent c), a = percent
organic matter b = structure code, and c = profile permeability class.

Figure 3.5. The soil erodibility nomograph (Wischmeier and Smith, 1978).

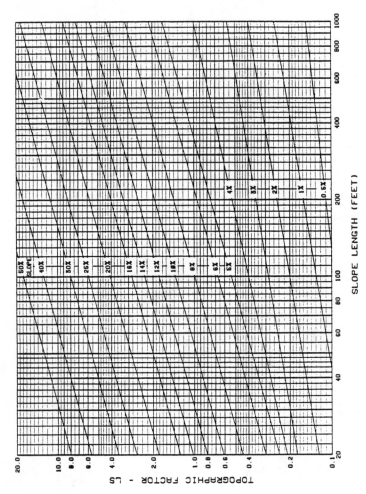

Note: $LS = (\lambda/72.6)^m (65.41 \sin^2\theta + 4.56 \sin\theta + 0.065)$ where λ = slope length in feet; θ = angle of slope; and $m = 0.2$ for gradients < 1 percent, 0.3 for 1 to 3 percent slopes, 0.4 for 3.5 to 4.5 percent slopes, and 0.5 for slopes of 5 percent or steeper.

Figure 3.6. Slope-effect chart for the topographic factor, LS (Wischmeier and Smith, 1978).

72   Site Assessment

A dense stand of permanent vegetation will give a C value of 0.01 after establishment.

TABLE 3.4 TYPICAL VALUES FOR THE C FACTOR

	Cover	C Factor
1.	Bare soil conditions freshly disced to 15-20 cm	1.00
	After one rain	0.89
	Undisturbed except scraped	0.66-1.30
	Sawdust 5 cm deep, disced in	0.61
2.	Seedings	
	Temporary, 0 to 60 days	0.40
	Temporary, after 60 days	0.05
	Permanent, 0 to 60 days	0.40
	Permanent, 2 to 12 months	0.05
	Permanent, after 12 months	0.01
3.	Weeds and brush	
	No appreciable canopy, 100% ground cover	0.003
	No appreciable canopy, 29% ground cover	0.24
	75% canopy cover* of tall weeds or short brush, 100% ground cover	0.007
	75% canopy cover of brush or bushes, 100% ground cover	0.007
4.	Undisturbed wood land	
	100% canopy cover with forest litter on 100% of area	0.0001
	20% canopy cover with forest litter on 40% of area	0.009

* Portion of total area that would be hidden from view by canopy projection.

Erosion Control Practice (P). This factor is the ratio of soil loss with the supporting practice to the soil loss with straight uphill and downhill plowing. Support practices that slow the runoff water and reduce the amount of soil it can carry include contour tillage, contour strip cropping, and terrace systems (Wischmeier and Smith, 1978). Tables 3.5 through 3.7 show the P values that have been prepared for various conservation practices.

TABLE 3.5 P VALUES AND SLOPE-LENGTH LIMITS FOR CONTOURING*

Land Slope (%)	P Value	Maximum Length[†] (feet)
1 to 2	0.60	400
3 to 5	0.50	300
6 to 8	0.50	200
9 to 12	0.60	120
13 to 16	0.70	80
17 to 20	0.80	60
21 to 25	0.90	50

* Wischmeier and Smith (1978).

[†] Limit may be increased by 25% if residue cover after crop seedlings will regularly exceed 50%.

TABLE 3.6 P VALUES, MAXIMUM STRIP WIDTHS, AND SLOPE LENGTH LIMITS FOR CONTOUR STRIPCROPPING*

Land Slope (%)	P Values[†]			Strip Width[#] (feet)	Maximum Length (feet)
	A	B	C		
1 to 2	0.30	0.45	0.60	130	800
3 to 5	0.25	0.38	0.50	100	600
6 to 8	0.25	0.38	0.50	100	400
9 to 12	0.30	0.45	0.60	80	240
13 to 16	0.35	0.52	0.70	80	160
17 to 20	0.40	0.60	0.80	60	120
21 to 25	0.45	0.68	0.90	50	100

* Wischmeier and Smith (1978).

[†] P values:

A For 4-year rotation of row crop, small grain with meadow seeding, and 2-years of meadow. A second row crop can replace the small grain if meadow is established in it.

B For 4-year rotation of 2-years row crop, winter grain with meadow seeding, and 1-year meadow.

C For alternate strips of row crop and small grain.

[#] Adjust strip-width limit, generally downward, to accomodate widths of farm equipment.

TABLE 3.7 P VALUES FOR CONTOUR-FARMED, TERRACED FIELDS*[†]

LAND SLOPE Percent	Farm Planning		Computing Sediment Yield[#]	
	Contour Factor[+]	Stripcrop Factor	Graded Channels Sod Outlets	Steep Backslope Underground Outlets
1 to 2	0.60	0.30	0.12	0.05
3 to 8	0.50	0.25	0.10	0.05
9 to 12	0.60	0.30	0.12	0.50
13 to 16	0.70	0.35	0.14	0.05
17 to 20	0.80	0.40	0.16	0.06
21 to 25	0.90	0.45	0.18	0.06

* Wischmeier and Smith (1978).

[†] Slope length is the horizontal terrace interval. The listed values are for contour farming. No additional contouring factor is used in the computation.

[#] These values include entrapment efficiency and are used for control of off-site sediment within limits and for estimating the field's contribution to watershed sediment yield.

[+] Use these values for control of interterrace erosion within specified soil loss tolerances.

3.4.3   General Soil Properties

The description of each soil series should include information on soil texture, permeability, available water holding capacity and the shrink-swell potential. Soil texture is an important consideration in the site selection process because texture influences many other soil properties, including the infiltration and subsoil percolation rates and aeration. Table 3.8 presents advantages and disadvantages of various soil textures for use in land treatment units. In general, HWLT units should not be established on extremely deep, sandy soils because of the potential for waste migration to groundwater. Similarly, silty soils with crusting problems should not be selected since they have the potential for excessive runoff. Generally, the soils best suited to land treatment of hazardous waste fall into one of the following categories: loam, silt loam, clay loam, sandy clay loam, silty clay loam, silty clay, or sandy clay. The leaching potential of soils, discussed in Section 3.4.4, depends greatly on soil texture.

Permeability of each horizon or zone should be determined by the methods discussed in Section 4.1.1.5, from available soil surveys of the area, or by the methods listed in other sources (Bouwer, 1978; Bouwer and Jackson, 1974; Linsley et al., 1975). Permeability is an indication of the

TABLE 3.8 SUITABILITY OF VARIOUS TEXTURED SOILS FOR LAND TREATMENT OF
HAZARDOUS INDUSTRIAL WASTES

Texture	Advantages	Disadvantages
sand	very rapid infiltration usually oxidized & dry low runoff potential	very low CEC very high hydraulic conductivity low available water poor soil structure
loamy sand	high infiltration low to medium runoff	low CEC moderate to high hydraulic con- ductivity rate low to medium available water
loam	moderate infiltration fair oxidation moderate runoff potential generally accessible good CEC	fair structure
silt loam	moderate infiltration fair oxidation moderate runoff potential generally accessible good CEC	some crusting fair to poor structure
silt	low infiltration fair to poor oxidation good CEC good available water	high crusting potential poor structure high runoff
silty clay loam	medium to low percolation fair structure high CEC	medium to low infiltration some crusting potential
silty clay	good to high available water	moderate runoff often wet fair oxidation
clay loam	medium to low percolation good structure medium to poor aeration high CEC high available water	medium to low filtration moderate to high runoff often wet
clay	low percolation high CEC high available water	low infiltration often massive structure high runoff sometimes low aeration
sandy clay	medium to low percolation medium to high CEC	fair structure moderate to high runoff
sandy clay loam	medium to high available water good aeration	medium infiltration

length of time the mobile constituents of the waste will remain in the soil (Sommers et al., 1978), and thus, is an indicator of the potential for groundwater contamination. High permeabilities of 2.5 cm/hr indicate rapid transmission of water associated with wastes and thus a high potential for groundwater contamination. The permeability of lower horizons influences the amount of water that will remain in the surface horizon following rainfall or irrigation. A textural discontinuity from coarse texture to fine texture or vice versa will result in greater amounts of water being retained above the discontinuity than would be retained in a deep uniform profile, thus resulting in wetter conditions than would otherwise be expected. Permeabilities of less than 0.05 cm/hr for the most restrictive layer in the top 1 m of soil may require artificial drainage.

Available water holding capacity (AWC) is a measure of the amount of water held against the pull of gravity. High AWC reduces the chance of runoff under high antecedent moisture conditions by permitting more moisture to be held. Water holding capacity also affects the amount of leaching. The higher the AWC the lower the chances for rapid contamination of groundwater. For example, a medium textured soil, when dry enough so that plants begin to wilt, with an AWC of 15-20% can adsorb 20-30 cm of water from sludge, wastewater or rainfall in the upper 1.5 m of the soil profile before transmitting the water to an underlying aquifer (Hall et al., 1976). Acceptable values for the AWC of the top 1.5 m of the profile would be 7.5 to 20 cm for humid regions and no less than 7.5 cm for arid regions (Sommers et al., 1978).

Shrink-swell potential, especially in montmorillonitic clay soils, can increase groundwater contamination hazard due to formation of cracks deep in the soil during extended periods of dry weather. Soils with a low to moderate shrink-swell potential are preferred for HWLT.

3.4.4   Leaching Potential

Based on the minimum infiltration rate of bare soil after prolonged wetting the SCS has developed a classification system which divides the soils into four hydrologic groups, A through D (USDA, 1971). These groups indicate the potential for water to flow through the entire soil profile. They may also be used as an indicator for the transmission of contaminants through the soil. Hydrologic Group A consists mainly of sands and gravels that are well drained, have high infiltration rates and high rates of water transmission. The greatest leaching potential is with Group A soils. The danger from leaching is highest with deep sandy soils which may connect with shallow aquifers. These soils have low cation exchange capacity (CEC) and high infiltration and hydraulic conductivity and will not be as effective in filtering water as will a finer soil with a higher CEC, lower infiltration and lower hydraulic conductivity (Groups B and C).

Group B soils are moderately deep to deep, moderately well to well drained, and moderately fine to moderately coarse in texture. They have moderate infiltration rates and water transmission rates. Group C soils are moderately fine to fine textured soils with a layer that impedes downward water movement. Both infiltration rates and water transmission rates are slow in this group.

Group D soils have the lowest leaching potential and one will need to be very cautious in applying liquids to avoid excessive runoff because these soils have very slow rates of infiltration and transmission. Group D soils are generally clays with high swelling potential, soils with a permanent high water table, soils with a claypan near the surface, or shallow soils over nearly impervious materials.

Leaching of applied wastes can be minimized by good design and management. High volume applications of liquid effluent to sandy soil may be permissible only if there is no evidence of leaching or groundwater contamination by mobile constituents such as nitrates or mobile organic compounds. In most cases, soils in hydrologic Group C, or possibly D, are best suited for the land treatment of hazardous wastes.

Soil structure as well as texture influences the leaching of waste constituents. If an organic waste is applied to a soil via irrigation or if the waste contains a high percentage of liquids, soils with very porous structure (such as crumb) or a high percentage of pore space to soil particles (low bulk density) have a high leaching potential. Leaching is increased in these soils because the detention time of the organic waste in the soil is decreased and the surface area of soil particles available to react with the waste is also decreased. Leaching of this nature can be expected when the moisture holding capacity of the soil is exceeded.

3.4.5   Horizonation

Surface soil characteristics alone are not sufficient to assess the suitability of a site for land treatment of hazardous waste. Many soil profiles have properties which make them a poor choice for use as a disposal facility. The specific properties that need to be examined include the depth to bedrock, a slowly impermeable layer and/or the groundwater table, and the presence of an inadequate textural sequence within the soil.

The profile depth to bedrock should be approximately three times the depth of the waste incorporation or 1.2 m (6 ft), whichever is greater. Soils having a slowly permeable layer or a deep groundwater table may be well suited to HWLT. If a slowly permeable layer is present, it should be at a depth of 1.5 m or greater to allow sufficient soil profile to treat the waste. Although data is available on which to base estimates of needed profile depth to the groundwater table for nontoxic sludges (Parizek, 1970), none is available for hazardous waste. Certainly, further work is needed to clarify these needs. The presence of a sand or loam layer in the profile, within 3 m of the surface, overlying a fine textured clay pan also creates a potential for horizontal flow and contamination of adjacent areas. Such a profile is thus unsuited for use as a hazardous waste disposal medium without special precautions.

While deep soils of relatively uniform physical and chemical characteristics are occasionally found, more often soils are characterized by distinct horizons which differ in texture, water retention, permeability, CEC and chemical characteristics. Appendix C lists the major horizons that may be present in a soil. Most of the biological activity and the waste decomposition is accomplished in the treatment zone which may range from several inches to one foot. Therefore, the characteristics of this horizon will be an important design consideration. Lower horizons will influence the rate of downward water movement and may serve to filter and remove other waste constituents or their degradation products which would otherwise move below these depths.

There are advantages to selecting soils which have coarser textural surface horizons over those with fine textured slowly permeable surface materials. Such soils will generally have greater infiltration rates and may be easier to work and incorporate large amounts of waste than those with clay surfaces. A clay subsoil will, however, slow the movement of leachate and protect groundwater. When such soils are selected, it is essential that water retaining levees are keyed into the less permeable subsurface materials.

3.5   GEOTECHNICAL DESCRIPTION

A geotechnical description which characterizes the subsurface conditions at the site should be prepared during the site assessment. The factors that need to be evaluated are the groundwater depths and flow directions, existing wells, springs, and other water supplies, and other activities located near the facility boundaries that might affect or come into contact with the groundwater. Any nearby sources of potential groundwater pollution other than the HWLT unit should also be considered. All data should be compiled on a map to assess the subsurface conditions at the site.

Some estimate of the groundwater recharge zone needs to be made during the site assessment. Whenever possible, it is desirable to locate HWLT units over areas with an isolated body of groundwater. If this is not possible, estimates of mixing between aquifers which may be impacted need to be made.

3.5.1   Subsurface Hydrology

Hydrologic characteristics of the soil and subsoil govern the speed and direction of fluid movement through the soil. Surface and subsurface hydrology are interrelated processes which are very important in evaluating the feasibility of using a given site for HWLT. The depth of soil to the seasonal water table is an important factor for judging potential groundwater contamination. The soils at the site should be deep enough so that the desired degree of treatment is attained within the treatment zone so that hazardous constituents do not percolate through the soil and reach groundwater. Shallow soils especially over karst formations and those with a sand classification have a high potential for transmitting hazardous wastes to groundwater. The maximum depth of the treatment zone should be 1.5 m and at least 1 m (3 ft) above the seasonal high water table to prevent contamination of the water table with untreated waste, and to provide sufficient soil aeration to allow microbial treatment and degradation of hazardous wastes, and to provide room to install an unsaturated zone monitoring system.

3.5.2   Groundwater Hydrology

Water table data are needed to position upgradient and downgradient monitoring wells and to determine if the water table is so close to the surface that it will interfere with land treatment. The depth of the water table tends to vary with surface topography and is usually shallower in relatively impermeable soils than in permeable soils. Since local water table depths and gradients cannot be accurately estimated from available regional data, it may be necessary to install observation wells at various locations within and surrounding the land treatment area. Sampling frequency of these observation wells should be chosen to account for seasonal changes. If care is taken in locating and properly installing these initial observation wells, future groundwater monitoring can use these same wells, minimizing the requirement and cost of additional well placement. Torrey (1979) recommends collection and analysis of three monthly samples from each well prior to waste application at new sites. For existing sites, only the upgradient well is useful for establishing background values. More information on groundwater monitoring can be found in Chapter 9.

3.5.3   Groundwater Quality

Current uses of groundwater in the area should also be noted. Where state regulations vary based on the current or potential uses of groundwater, groundwater quality may be an important concern during site selection. Information on groundwater quality, available from the U.S. Geological Survey and state agencies, can be used for preliminary site investigations, but site specific background quality data are needed for each HWLT unit.

3.6   SOCIO-GEOGRAPHIC FACTORS

Land use considerations generally have little impact on the technical grounds for site selection. Instead, land use encompasses the restraints imposed by the public and local or regional governmental authorities on the use of a parcel of land for HWLT. Occasionally past land use diminishes the ability to manage the area as an HWLT unit. For example, areas formerly used for landfills or areas contaminated with persistent residues from past chemical spills are likely to be unsuitable for HWLT units.

Evaluation of land use at and near a proposed or existing HWLT unit is primarily the responsibility of the owner or operator. There are a number of legal constraints that affect facility siting. Factors to consider include zoning restrictions, special ecological areas, historic or archaeological sites, and endangered species habitats. Local, state and federal laws concerning these factors will affect the siting of an HWLT unit. The proximity of the unit to the waste generator and the accessibility of the site both affect the transportation requirements. Ideally, a land treatment operation would be located on-site or immediately adjacent to the waste generator. If wastes must be transported to an off-site HWLT unit via public roads, rail systems or other means, the transporter must comply with 40 CFR Part 263, under the jurisdiction of the EPA, and 49 CFR Subchapter C, enforceable by the Department of Transportation. The operator may also want to route the waste through industrial areas rather than through residential neighborhoods.

In addition to the legal constraints to be considered, there are a number of social factors which must often be dealt with during the evaluation of proposed sites. How the owner or operator handles these issues may determine whether the public accepts or rejects the location of the unit. Social factors may include wooded areas and bodies of water that may be important visually or for recreational purposes, prime agricultural lands, existing neighborhoods, etc. Although facility design should strive to prevent deterioration of local resources while maximizing public and environmental protection, the possibility for conflict exists since most sites are less than ideal and are often situated near populated areas or in zones of high growth potential. Some potential areas of conflict include:

(1) proximity of the site to existing or planned community or industrial developments;

(2) zoning restrictions;

(3) effects on the local economy; and

(4) relocation of residents.

Socio-geographic considerations and interactions with the public are beyond the scope of this manual, except for the above discussion which points out the importance of including the public in the permitting process. It is the responsibility of the owner or operator to maintain an open and credible dialogue with local public officials and with individuals who will be directly affected by the HWLT unit. The role of the regulatory agencies in this respect is simply to assess whether the plans, as proposed, are technically and environmentally sound.

CHAPTER 3 REFERENCES

Bouwer, H. 1978. Groundwater hydrology. McGraw - Hill Book Company, New York. 480 p.

Bouwer, H., and R. D. Jackson. 1974. Determining soil properties. p. 611-673. In Tom Van Schilfgaarde (ed.) Drainage for agriculture. Number 17, Agron. Soc. Amer. Madison, Wisconsin.

Brown, K. W., K. C. Donnelly, J. C. Thomas, and L. E. Deuel, Jr. 1980. Factors influencing the biodegradation of API separator sludges applied to soils. Final report to EPA. Grant No. R 805474-10.

Cartwright, K., R. H. Gilkeson, and T. M. Johnson. 1981. Geological considerations in hazardous waste disposal. Journal of Hydrology, 54:357-369.

Dibble, J. T., and R. Bartha. 1979. Effect of environmental parameters on the biodegradation of oil sludge. Appl. and Environ. Micro. 37:729-739.

EPA. 1975. Evaluation of land application systems. EPA 430/9-75-001.

EPA. 1977. Process design manual for land treatment of municipal wastewater. EPA 625/1-77-008. PB 299-665/1BE.

EPA. 1981. Standards for owners and operators of hazardous waste treatment, storage, and disposal facilities. Federal Register Vol. 46, No. 7, p. 2848. January 12, 1981.

EPA. 1982. Standards for owners and operators of hazardous waste treatment, storage, and disposal facilities. Federal Register Vol. 47, No. 143, p. 32350. July 26, 1982.

Hall, G. F., L. P. Wilding, and A. E. Erickson. 1976. Site selection considerations for sludge and wastewater application on agricultural land. In Application of sludges and wastewaters on agricultural lands: A planning and educational guide. (Research Bulletin 1090) B. D. Knezek and R. H. Miller (eds.) Ohio Agricultural Research and Development Center, Wooster, Ohio.

Linsley, R. K. Jr., M. A. Kohler, and J. L. H. Paulhus. 1975. Hydrology for engineers. McGraw - Hill Inc., New York. 482 p.

Loehr, R. C., W. J. Jewell, J. D. Novak, W. W. Clarkson, and G. S. Friedman. 1979. Land application of wastes, Vol. 1. Van Nostrand Reinhold Environmental Engineering Series, New York. 308 p.

Panofsky, H. A., and G. W. Brier. 1958. Some applications of statistics to meteorology. The Pennsylvania State Univ. Press. University Park, Pennsylvania. 224 p.

Parizek, R. R. and B. E. Lane. 1970. Soil-water sampling using pan and depp pressure-vacuum lysimeters. J. of Hydrology 11:1-21.

Sommers, L. E., R. C. Fehrmann, H. L. Selznick, and C. E. Pound. 1978. Principles and design criteria for sewage sludge application on land. Prepared for U.S. EPA, Environmental Research Information Center Seminar entitled Sludge Treatment and Disposal.

Torrey, S. 1979. Sludge disposal by landspreading techniques. Noyes Data Corp., New Jersey. 372 p.

USDA. 1951. Soil Survey Manual. Handbook No. 18. Agricultural Research Administration. U.S. Government Printing Office, Washington, D.C.

USDA, Soil Convervation Service. 1971. SCS national engineering handbook. Section 4, hydrology. U.S. Government Printing Office, Washington, D.C.

Whiting, D. M. 1976. Use of climatic data in estimating storage days for soils treatment systems. U.S. EPA, Ada, Oklahoma. EPA 600/2-76-250. PB 263-597/7BE.

Wischmeier, W. H., and D. D. Smith. 1978. Predicting rainfall erosion losses - a guide to conservation planning. U.S. Dept. of Agriculture. Agr. Handbook No. 537. 58 p.

THE TREATMENT MEDIUM 4

Christy Smith
D. Craig Kissock
James C. Thomas
K. C. Donnelly

Soil characterization is essential to the design of hazardous waste land treatment units since soil is the waste treatment medium. When generally acceptable values for the various system properties are known, analyses may reveal conditions that make land treatment unsuitable, and consequently, may eliminate a proposed site (Chapter 3). In addition, analysis of the treatment medium will aid in efficiently designing laboratory or field waste treatability experiments. Preliminary soil characterization can be used for the following:

(1) to choose the soil parameters to be studied that will be most important in waste treatment;

(2) to determine the practical range of these parameters and the specific levels at which tests will be made;

(3) to choose the extremes to be measured; and

(4) to provide background data for comparison against later sampling results.

Many of the processes that occur in soils that treat the waste and render it less hazardous are the same processes that are used in industrial waste treatment plants. Table 4.1 lists soil treatment processes that are similar to the categories of treatment to be used by industries in describing their processes (from Appendix I of 40 CFR Part 264).

TABLE 4.1   TREATMENT PROCESSES OF SOIL IN A LAND TREATMENT UNIT

Absorption	Flocculation
Chemical fixation	Thickening
Chemical oxidation	Blending
Chemical precipitation	Distillation
Chemical reduction	Evaporation
Degradation	Leaching
Detoxification	Liquid ion exchange
Ion exchange	Liquid–liquid extraction
Neutralization	Aerobic treatment
Photolysis	Anaerobic treatment
Filtration	

The treatment medium is a part of the larger system including soil, plants and atmosphere.   Plants and atmospheric conditions can modify the processes occurring in the treatment medium.   Plants can protect the treatment zone from the adverse effects of wind and water.   Plants may also take up water and waste constituents and, if not harvested, supply the soil with additional organic matter.   Atmospheric conditions control the water content and temperature of the soil and consequently affect waste degradation rates and constituent mobility.   The modifying effects of plants and atmosphere are briefly discussed.   Figure 4.1 illustrates how the information presented in this chapter fits into the overall design process for HWLT units (Fig. 2.1).

4.1   SOIL PROPERTIES

Soil characterization is commonly done by conducting a soil survey, either in conjunction with the Soil Conservation Service (SCS) or by a certified professional soil scientist (Section 3.4.1).   In such an endeavor, the soil series present at a given site are identified and sampled.   Soil series are generally named for locations and are based on both physical and chemical characteristics.   These characteristics vary widely from place to place, and classification distinguishes one soil from another based on recognized limits in soil properties.

4.1.1   Physical Properties

Physical properties of a soil are defined as those characteristics, processes or reactions of a soil that are caused by physical forces and are described by physical terms or equations.   Physically, a mineral soil is a porous mixture of inorganic particles, decaying organic matter, air, and water.   The percentage of each of these components as well as the type of inorganic and organic particles determine the behavior of the soil.

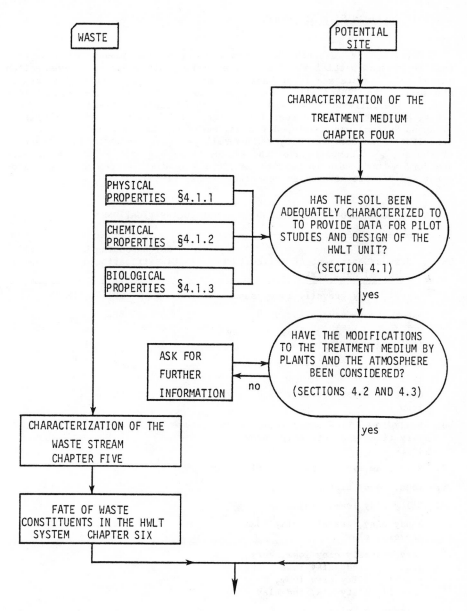

Figure 4.1. Characterization of the treatment medium for HWLT.

4.1.1.1 Particle Size Distribution

Particle size distribution is a measure of the amounts of inorganic soil separates (particles < 2 mm) in a soil. This property is most often called soil texture and is probably the most important physical property of the soil. The USDA (United States Department of Agriculture) classification is generally accepted and used by agricultural workers, soil scientists, and most of the current literature. The USCS (Unified Soil Classification System) was developed for engineers and is based on particle size distribution as influenced by the overall physical and chemical properties of the soil. A comparison of the two systems is given in Table 4.2. The standard methods used to measure particle size distribution are the hydrometer and pipette methods as described by Day (1965).

TABLE 4.2 CORRESPONDING USDA AND USCS SOIL CLASSIFICATIONS*

United States Department of Agriculture (USDA) Soil Textures	Corresponding Unified Soil Classification System (USCS) Soil Types
1. Gravel, very gravelly loamy sand	GP, GW, GM
2. Sand, coarse sand, fine sand	SP, SW
3. Loamy gravel, very gravelly sandy loam, very gravelly loam	GM
4. Loamy sand, gravelly loamy sand, very fine sand	SM
5. Gravelly loam, gravelly sandy clay loam	GM, GC
6. Sandy loam, fine sandy loam, loamy very fine sand, gravelly sandy loam	SM
7. Silt loam, very fine sandy clay loam	ML
8. Loam, sandy clay loam	ML, SC
9. Silty clay loam, clay loam	CL
10. Sandy clay, gravelly clay loam, gravelly clay	SC, GC
11. Very gravelly clay loam, very gravelly sandy clay loam, very gravelly silty clay loam, very gravelly silty clay and clay	GC
12. Silty clay, clay	CH
13. Muck and peat	PT

* Fuller (1978).

88   The Treatment Medium

The three dominant soil particles are sand, silt and clay. Sand and gravel particles are the coarse separates. Coarse textured soils usually have low water holding capacity, good drainage, high permeability and aeration, and generally have a loose and friable structure. Sand grains may be rounded or irregular depending on the amount of abrasion they have received. They do not have the capacity to be molded (plasticity) as does clay.

The silt and clay particles are the fine separates. Silt particles are irregularly fragmental, have some plasticity, and are predominantly composed of quartz. A high percentage of silt is undesirable and leads to physical problems such as soil crusting. Clay particles are very small, less than 0.002 mm in diameter, and therefore have a very high surface area. Clays are plate-like, highly plastic, cohesive, and have a very high adsorptive capacity for water, ions and gases. This high adsorptive capacity may be very useful to hold ions, such as heavy metals, in an immobile form and prevent their movement.

The USDA has devised a method for naming soils based on particle size analysis. The relationship between textural analysis and class names is shown in Fig. 4.2 and is often referred to as a textural triangle. When the percentages of at least two size separates are known, the name of the compartment where the two lines intersect is the textural class name of the soil being evaluated.

4.1.1.2 Soil Structure

Soil structure is the grouping of soil particles of a general size and shape into aggregates, called peds. Structure generally varies in different soil horizons and is greatly influenced by soil texture and organic matter content. The arrangement of the primary soil separates greatly influences water movement, aeration, porosity and bulk density (Pritchett, 1979). Addition of organic matter and the use of sod crops helps build and maintain good soil structure. Other factors which promote aggregation include 1) wetting and drying, 2) freezing and thawing, 3) soil tillage, 4) physical activity of plant roots and soil organisms, 5) influence of decaying organic matter, and 6) the modifying effects of adsorbed cations (Brady, 1974). Sandy soils need to be held together, into granules, by the cementing action of organic matter to stabilize the soil surface and increase water retention. Fine textured soils also need adequate structure to aid in water and air movement in the soil. Some types of organic waste additions may help soil structure by increasing aggregation.

Four primary types of soil structure are recognized: platy, prism-like, block-like and spheroidal. All structural types except platy have two subtypes each. Subgroups for the prism-like structure are, prismatic and columnar; for block-like, cube-like blocky and subangular blocky; and for spheroidal, granular and crumb. The names of the categories imply the form or shape of the aggregates, with crumb being the smallest structural aggregate. Two or more of the structural conditions may exist in the same

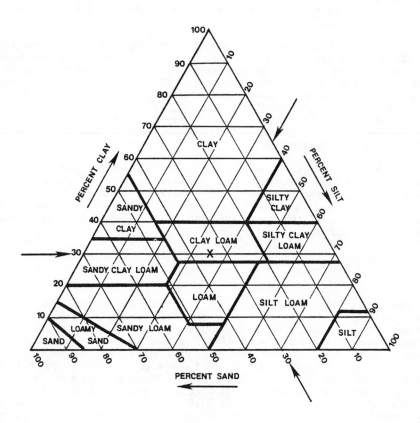

Figure 4.2. Textural triangle of soil particle size separates. Shown is an example of a soil with 35% silt, 30% clay and 30% sand, which is classified as a clay loam.

soil, for example, a soil may have a granular surface horizon with a sub-surface horizon that is subangular blocky.

Porosity and pore size distribution are related to soil structure as well as soil texture. Nonaggregated (poor structured) fine-textured soils have small pores with a narrow range of pore sizes. Nonaggregated coarse textured soils have large pores also with a narrow range of pore sizes. An intermediate situation is desirable in soils chosen for land treatment, such as a soil with texture to give several pore sizes as well as good structure for a wide distribution of sizes.

4.1.1.3 Bulk Density

Bulk density is a weight measurement in which the entire soil volume is taken into consideration. It is defined as the mass of a unit volume of soil and is generally expressed as gm/cm^3 (lb/ft^3). This measurement takes into account both the volume of the soil particles and the pore space between them. Techniques for measuring bulk density are outlined by Blake (1965).

Soils that are loose and porous will have low weights per unit volume, and thus, low bulk densities. Soils that are more compact will have high bulk density values. Soil bulk density generally increases with depth because there is less organic matter and less aggregation with depth and greater soil compression due to the weight of overlying soil. Bulk density is also influenced by soil texture and structure. Sandy soils which have particles that are close together, that is, have poor structure, have high bulk densities usually in the range of 1.20 to 1.80 g/cm^3. Fine textured soils generally have a higher organic content, better structure, more pore space and thus, lower bulk densities. Bulk densities for fine textured soils generally range from 1.0 to 1.6 g/cm^3 (Brady, 1974).

Good soil management procedures will decrease surface bulk density because the factors that build and maintain good soil structure will gener-ally increase with management. Conversely, intensive cultivation and excessive traffic by equipment generally increases bulk density values. Land treatment management should minimize unnecessary tillage and traffic, and maximize structural formation through organic matter additions and vegetative covers. Good structure and relatively low soil bulk densities promote good aeration and drainage, which are desirable conditions for waste treatment.

4.1.1.4 Moisture Retention

Moisture retention or moisture holding capacity is a measure of the amount of water a given soil is capable of retaining and is generally expressed as a weight percentage. The most common method of expressing soil moisture percentage is grams of water associated with 100 grams of dry soil. Soil tensions from the strong chemical attraction of polar water molecules are responsible for the adsorption of pure water in a soil. Water commonly considered to be available for plant and microbial use is held at tensions between 1/3 and 15 atm. This water is retained in capillary or extremely small soil pores. Moisture retained at tensions greater than 1/3 atm is termed gravitational or superfluous water (Fig. 4.3). Gravitational water moves freely in the soil and generally drains to lower portions of the profile carrying with it a fraction of plant nutrients and/or waste constituents. After all water has drained from the large soil pores and the water is held in the soil at 1/3 atm the soil is at field capacity. Moisture retained at tensions greater than 15 atm is termed unavailable or hygroscopic water because it is held too tightly to be used by plants. A soil is said to be at the permanent wilting point when the water is held at >15 atm. Generally, finer textured high organic content soils will retain the most water while sandy, low organic content soils will retain only very small amounts of available water.

For management of a land treatment unit, knowledge of the moisture retention of the soil is needed to help determine water loading rates that will not cause flooding or standing water, to predict possible irrigation needs, and to estimate leaching losses and downward migration of waste constituents. At a minimum, the values for 1/3 and 5 atm of suction should be measured to give an estimation of how much water will be available for plant and soil chemical reactions. Moisture retention can be measured by the pressure plate technique as outlined by Richards (1965).

4.1.1.5 Infiltration, Hydraulic Conductivity and Drainage

Infiltration is the entry of water into the soil surface, normally measured in cm/hr. Knowledge of this parameter is critical for a land treatment unit since application of a liquid at rates exceeding the infiltration rate will result in runoff and erosion, both of which are undesirable in such a system. Infiltration rates are also needed when calculating the water balance of an area.

Permeability, also called hydraulic conductivity, is the ease with which a fluid or gas can pass through the soil and is measured in cm/hr. Once a substance enters a soil, its movement is governed, in part, by soil permeability. Permeability is closely associated with particle size, pore space, and bulk density. Table 4.3 lists the classes of hydraulic conductivity for soils. Fine textured clays with poor structure and high bulk densities usually have very low permeabilities. Knowledge of the permeability is necessary to estimate the rate of movement of water or potential pollutants through the soil of the land treatment unit. The potential for a given chemical to alter the permeability of the soils on-site needs to be determined as a safeguard to prevent deep leaching and reduce the potential for groundwater contamination.

Hydraulic conductivity (K) is conventionally measured in the laboratory by either the constant head or falling head techniques as outlined by

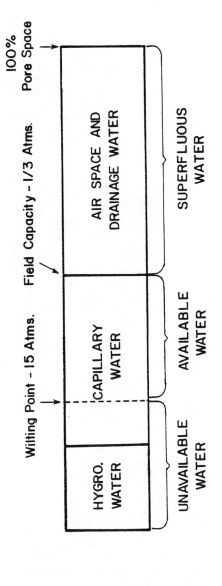

Figure 4.3. Schematic representation of the relationship of the various forms of soil moisture to plants (Buckman and Brady, 1960). Reprinted by permission of the Macmillan Publishing Co., Inc.

TABLE 4.3 SATURATED HYDRAULIC CONDUCTIVITY CLASSES FOR NATIVE SOILS

Class*	Saturated Hydraulic Conductivity* cm/hr	Description
Very high	>36	Soils transmit water downward so rapidly that they remain wet for extremely short periods. Soils are coarse textured and dominated by coarse rock fragments without enough fines to fill the voids or have large permanent cracks or worm holes.
High	3.6 - 36	Soils transmit water downward rapidly so that they remain saturated for only a few hours. Soils are typically coarse textured with enough fines to fill the voids in the coarse material. Soil pores are numerous and continuous.
Moderate	0.36 - 3.6	Soils transmit water downward very readily so that they remain wet for a few days after thorough wetting. Soil layers may be massive, granular, blocky, prismatic or weak platy and contain some continuous pores.
Moderately low	.036 - 0.36	Soils transmit water downward readily so they remain wet for several days after thorough wetting. Soils may be massive, blocky, prismatic, or weakly platy with a few continuous pores.
Low	0.0036 - 0.036	Soils transmit water downward slowly so they remain wet for a week or more after a thorough wetting. Soils are structureless with fine and discontinuous pores.
Very low	<.0036	Soils transmit water downward so slowly that they remain wet for weeks after thorough wetting. Soils are massive, blocky, or platy with structural plates or blocks overlapping. Soil pores are few, fine, and discontinuous.

* USDA (1981).

Klute (1965). For more exact, on-site determinations, field techniques are available. If the soil is above the water table, the double tube or "permeameter" method (Boersma, 1965a) is used; if below the water table, the auger hole or the piezometer method is used (Boersma, 1965b). More extensive reviews of field and laboratory methods for measuring hydraulic conductivity are given by the American Society of Agricultural Engineers (1961) and Bouma et al. (1982). These reviews cover most methods currently used to measure permeability.

Drainage refers to the speed and extent of the removal of water from the soil by gravitational forces in relation to additions by surface run-on or by internal flow. Soil drainage, as a condition of a soil, refers to the frequency and duration of periods of saturation or partial saturation of the soil profile. Drainage is a broad concept that encompasses surface runoff, internal soil drainage, and soil hydraulic conductivity. Seven classes of natural soil drainage are recognized in Table 4.4. Drainage may be controlled to maintain an aerobic environment and to minimize leaching hazards. Surface drainage can be managed by diversion structures, surface contouring, and ditches or grassed waterways to remove excess water before it totally saturates the soil. An understanding of these principles is necessary since rainfall and runoff must be managed and directed to appropriate locations. Subsurface drainage systems use underground drains to remove water from the upper portion of the soil profile and can also be successfully used to lower the water table and drain the treatment zone. Section 8.3 provides additional information on managing water at HWLT units.

4.1.1.6 Temperature

Soil temperature regulates the rate of many soil chemical and biological reactions. Most biological activity is greatly reduced at 10°C and practically ceases at 5°C, as illustrated in Fig. 4.4. Waste degradation during the cool spring and fall months is lower than in summer when the soil biological activity is at its peak. Thus, loading rates in some areas of the country need to be varied according to the soil temperature on a site-specific basis. In general, locations where soil temperatures are at or near freezing for much of the year will need seasonal adjustments in the amount of waste applied per application. Moreover, soil temperatures should be considered when estimating application rates and the land area required to treat the waste.

Freezing of the soil also changes many physical and chemical properties. Infiltration and percolation are nearly stopped when soil water becomes frozen so that surface waste applications need to be curtailed (Wooding and Shipp, 1979). Subsurface injection of wastes may be successful in some cases if the soil is not frozen below a 10-15 cm depth. Figure 4.5 illustrates the area of the country where frost penetration is a consideration.

TABLE 4.4 SEVEN CLASSES OF NATURAL SOIL DRAINAGE

Class*	Physical Description	Use
Excessively drained	Water is very rapidly removed from the soil as a result of very high hydraulic conductivity and low water holding capacity. Soils are commonly very coarse textured, rocky or shallow. All soils are free of mottling related to wetness.	Soils are not suited to crop production without supplemental irrigation. Soils not suited for land treatment due to possible high leaching of constituents.
Somewhat excessively drained	Water is removed from the soil rapidly as a result of high hydraulic conductivity and low water holding capacity. Soils are commonly sandy shallow and steep. All are free of mottling related to wetness.	Soils are suited for crop production only with irrigation but yield will be low. Soils are poorly suited for land treatment due to leaching and low water holding capacity.
Well drained	Water is removed from the soil readily, not rapidly, and the soils have an intermediate water holding capacity. Soils are commonly medium textured and mainly free of mottling.	Soils are well suited for crop production since water is available through most of the year and wetness does not inhibit growth of roots for significant periods of the year. Soils are well suited for land treatment.
Moderately well drained	Water is removed from the soil somewhat slowly. Soils commonly have a layer with low hydraulic conductivity, a wet state relatively high in the profile, receive large volumes of water, or a combination of these.	Soils are poorly suited for crop production without artificial drainage since free water remains close enough to surface to limit growth and management during short periods of the year. Soils are not well suited for land treatment as a result of free water being at or near the surface for short periods of time.

--continued--

TABLE 4.4 (Continued)

Class	Physical Description	Use
Somewhat poorly drained	Water is removed slowly enough that the soil is wet for significant periods during the year. Soils commonly have a slowly pervious layer, a high water table, an addition of water from seepage, nearly continuous rainfall, or a combination of these.	Soils are not suited for crop production without artificial drainage since free water remains at or near the surface for extended periods. Soils are poorly suited for land treatment since they remain saturated for extended periods.
Poorly drained	Water is removed so slowly that the soil is saturated for long periods. Free water is commonly at or near the surface but the soil is not continuously wet directly below plow depth (6"). Poor drainage is a result of a high water table, slowly pervious layer within the profile, seepage, continuous rainfall or a combination of these.	Soils are not suited for production under natural conditions since they remain saturated during much of the year. Land treatment operations are greatly limited due to free water remaining at or near the surface for long periods.
Very poorly drained	Water is removed so slowly from the soil that free water remains at or below the surface during much of the year. Soils are commonly level or depressed and frequently ponded yet in areas with high rainfall they can have moderate to high slope gradients.	Soils are suitable for only rice crops since they remain saturated during most of the year. Soils are not acceptable for land treatment unless artificially drained due to excessive wetness.

* USDA (1981).

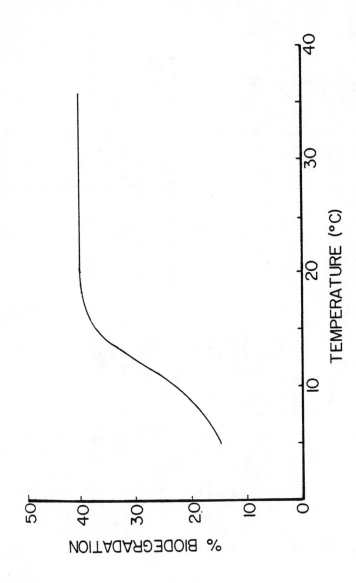

Figure 4.4. Effect of temperature on hydrocarbon biodegradation in oil sludge-treated soil (Dibble and Bartha, 1979). Reprinted by permission of the American Society of Microbiology.

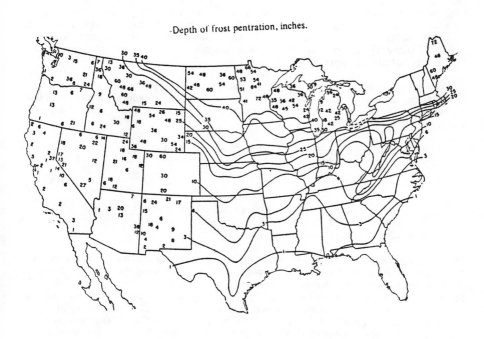

Figure 4.5. Average depth of frost penetration across
the United States (Stewart et al., 1975).

Reliable predictions of soil temperature are needed for a sound HWLT management plan, but there are few sources of soil temperature information. Only recently have soil temperature measurements been taken routinely. The owner or operator should check with the state climatologist to see if soil temperature data are available for the area of the proposed HWLT unit. The lack of extensive historical records is further complicated by the fact that most observations have been only seasonal as they related to agricultural needs. Therefore, a stochastic approach to soil temperatures in facility design is not possible for most locations. No attempt has been made to directly correlate soil temperatures with atmospheric parameters for which better records exist.

Work by Fluker (1958) is the only published study of an attempt to predict the annual soil temperature cycle. Fluker presented a mathematical expression to calculate soil temperature at a given depth from the mean annual soil temperature, as follows:

$$\theta_{zt} = \text{avg. annual soil temp.} + 12.0e^{-0.1386z}\sin\left(\frac{2\ t}{364}-1.840-0.132z\right) \quad (4.1)$$

where

$\quad \theta_{zt}$ = the average soil temperature in °C at depth z;
$\quad z$ = the depth in the soil in feet; and
$\quad t$ = time in days after Dec. 31.

The average annual soil temperature can be approximated as equal to, or slightly higher than, the average annual air temperature. The term used to represent the change in temperature with depth is $12e^{-0.1386z}$. The factor of 12 is defined as one-half the difference between the maximum and minimum average soil temperatures. Short of measuring these values, an estimate can be obtained by using the difference between the maximum and minimum air temperatures and adding 20%. Although the equation was developed empirically for a particular locale, the coefficients may be similar for other sites. The equation, however, should be used with caution, particularly in extremely cold climates.

Based on the lack of better predictive tools for soil temperatures, one approach is to collect data from one year at an on-site recording station and use it as a reasonable approximation of future conditions. Since a demonstration of waste treatability is required before an HWLT unit may be permitted, there would generally be time to take soil temperature measurements at the 10 cm depth. Climatic records can be consulted for guidance as to how the recorded year compares with other years; however, site topography and other factors cause local soil temperature variations.

4.1.2   Chemical Properties

Chemical reactions that occur between the soil and waste constituents must be considered for proper HWLT management. There are large numbers of complex chemical reactions and transformations which occur in the soil including exchange reactions, sorption and precipitation, and complexation. By understanding the fundamentals of soil chemistry and the soil components that control the reactions, predictions can be made about the fate of a particular waste in the soil. Fate of specific waste constituents is discussed in more detail in Chapter 6.

4.1.2.1 Cation Exchange

Cation exchange capacity (CEC) is the total amount of exchangeable cations that a soil can sorb and is measured in meq/100 g of soil. These cations are bound on negatively charged sites on soil solids through electrostatic bonding and are subject to interchange with cations in the soil solution. Among the exchangeable cations are some of the essential plant nutrients including calcium, magnesium, sodium, potassium, ammonium, aluminum, iron and hydrogen. In addition to these, the soil can also sorb nonessential cations and effectively remove and retain heavy metals (Brown et al., 1975). The CEC depends on the amount of specific types of clay, the amount and chemical nature of the organic matter fraction, and the soil pH (Overcash & Pal, 1979). The cation exchange reactions take place very rapidly and are usually reversible (Bohn et al., 1979).

Cation exchange capacity is associated with the negatively charged surface of the soil colloids which arises from isomorphic substitutions (e.g., Al^{3+} for Si^{4+}) in many layer silicate minerals. The total charge of soil colloids consists of a permanent charge as well as a pH dependent charge. All cations, however, are not retained on the soil colloid to the same degree. Usually, trivalent and divalent cations are more tightly held than monovalent cations with the exception of hydrogen (H^+) ions. Also, ions are less tightly held as the degree of hydration increases (Bohn et al., 1979). Generally, clays have large surface areas and a high CEC. Sands, being relatively low in surface area, are usually low in CEC.

Ions may also be bound to soil solids by covalent, rather than electrostatic bonding. When this type of bonding predominates, specific sorption is observed for many cations as well as anions. This phenomenon has been observed with clays, aluminum and iron oxides, and organic matter. Specific sorption is a more permanent type of sorption than cation exchange and is not always related to CEC.

Measurement of the CEC is necessary to give an estimation of the ability to the soil to sorb and retain potential pollutants. Methods used to measure CEC are ammonium or sodium saturation (Chapman, 1965a), however, laboratories in each region of the country may have developed other appropriate techniques for their area. If the ammonium displacement technique is used to determine CEC, exchangeable bases can also be measured in the extract (Chapman, 1965b).

4.1.2.2 Organic Carbon

Residual organic carbon found in soil is a result of the decay of former plant and animal life. The organic fraction is in a constant state of flux with more organic matter being added by roots, crop residues, and dying plants, animals and microorganisms and organic matter being removed by further decay. In the soil, microbial activity is constantly working to decompose organic residues, resulting in the evolution of carbon dioxide (CO_2). Figure 4.6 illustrates the carbon cycle.

The effect of organic matter on the physical properties of soils has already been discussed. It improves soil structure by increasing aggregation, reduces plasticity and cohesion, increases the infiltration rate and water holding capacity, and imparts a dark color to the soil. The organic fraction of the soil has a very high CEC, and consequently, increasing the organic matter content of a soil also increases the CEC. However, increases in organic carbon from large waste applications cannot be relied upon to provide long-term increases in soil sorption capacity since the organic matter decomposes over time and ultimately, the organic content of the soil will return to near the original concentration. Measurement of the amount of soil organic matter is normally done by using the Walkley-Black method as outlined by Allison (1965).

Native soil organic matter is comprised of humic substances which have a large influence on the soil chemistry. Soil organic matter exhibits a high degree of pH-dependent affinity for cations in solution by a variety of complexation reactions. Humic substances with high molecular weights complex with metals to form very insoluble precipitates, however, low molecular weight organic acids have high solubility in association with metals. A discussion of the reaction of organic matter with metals is found in Chapter 6.

4.1.2.3 Nutrients

There are sixteen elements essential for plant growth. Of these, carbon (C), hydrogen (H_2), and oxygen (O_2) are supplied from air and water, leaving the soil to supply the other thirteen. Six of the essential elements, nigrogen (N), phosphorus (P), potassium (K), calcium (Ca), magnesium (Mg), and sulfur (S), are required in relatively large amounts. Nitrogen, P and K are considered primary plant nutrients while Ca, Mg and S are referred to as secondary plant nutrients.

All three of the primary plant nutrients (N, P and K) are normally included in inorganic fertilizers. Nitrogen is of prime importance since, if deficient, it causes plants to yellow and exhibit stunted growth. Nitrogen deficiencies also greatly inhibit the degradation of hazardous organic wastes because N is also essential for microorganisms. If N is in excess, it is readily converted to nitrate (NO_3) which is a mobile anion that can leach and contaminate groundwater. Phosphorus is normally present

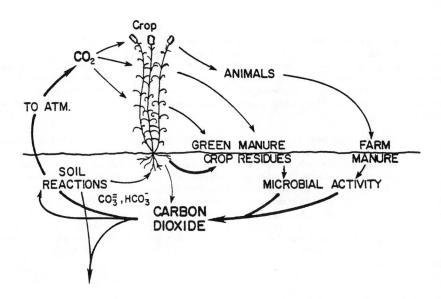

Figure 4.6. Diagramatic representation of the transformations of carbon,
commonly spoken of as the carbon cycle. Note the stress
placed on carbon dioxide both within and without the soil
(Buckman and Brady, 1960). Reprinted by permission of the
Macmillan Publishing Co., Inc.

in low concentrations and is specifically sorbed by soil colloids. The amount of K in the soil is sometimes adequate but often it is present in a form that is unavailable for plant use.

Each state generally has an extension soil testing laboratory that will analyze soil samples for primary and secondary plant nutrients. Nitrogen analysis is usually done by the Kjeldahl method (Bremner, 1965) and P and K are usually analyzed in an ammonium acetate extract as outlined by Chapman (1965a, 1965b).

Calcium and Mg are also required in relatively large amounts when plants are grown. Deficiencies in Ca usually occur in acid soils and can be corrected by liming. Most lime contains some Mg, but if the soil is deficient in Mg, the use of dolomitic lime is suggested. Sulfur, although required by plants in large amounts, is usually found in sufficient quantities in soils. Small amounts of S are normally in fertilizers as a constituent of one of the other components. Sulfur compounds can be used to lower soil pH.

Elements required by plants in relatively small amounts include iron, manganese, boron, molybdenum, copper and zinc, and chloride ions. Most of these micronutrients occur in adequate amounts in native soils. Excess concentrations of certain elements often cause nutrient imbalances that will adversely affect plant survival. Therefore, careful control of waste loading rates and routine monitoring of soil samples for these elements is essential to prevent buildup of phytotoxic concentrations when plants are to be grown during the active life or at closure. The single most important management consideration is pH since the solubility of each of these elements is pH dependent. Chapter 6 discusses this issue in greater detail for each element.

4.1.2.4 Exchangeable Bases

The exchangeable bases in a soil are those positively charged cations, excluding hydrogen, held on the surface exchange sites that are in equilibrium with the soil solution. These cations are available for plant use as well as for reaction with other ions in the soil solution. As they are absorbed by plants, more cations are released into solution from the exchange sites. This is a type of cation exchange reaction (discussed in Section 4.1.2.1). The major cations include calcium (Ca), magnesium (Mg), sodium (Na), and potassium (K). Plants can tolerate a fairly wide ratio of cations but the optimum ratio, as calculated by Homes (1955) is 33 K:36 Ca: 32 Mg. This ratio can be varied on a field scale as necessary by additions of lime, $Ca(CO_3)$; dolomite, $CaMg(CO_3)_2$; or potash fertilizer.

Laboratory analysis for exchangeable bases can be done by the ammonium acetate extraction procedure as outlined by Chapman (1965b) followed by measurement of Ca, Mg, Na and K in the extract using atomic absorption spectroscopy. The sum of the exchangeable bases expressed in meq/100 g is multiplied by 100 and divided by the CEC to give the percent base saturation. In essence, this tells what percentage of the CEC is occupied by bases. The percentage of the CEC that is not occupied by bases is predominantly filled by hydrogen ions which form what is called the reserve acidity. Percent base saturation depends on the climatic conditions, the materials from which the soil was formed, and the vegetation growing on the site (Pritchett, 1979). Generally, the percent base saturation increases as the pH and fertility of the soil increases.

4.1.2.5 Metals

Analysis of soil samples for metals content is normally done using an air dried sample ground with a porcelain mortar and pestle to pass a 2 mm sieve and digested using concentrated HNO_3 (EPA, 1979) or hydrofluoric acid in an acid digestion bomb (Bernas, 1968). Extracts can be analyzed for arsenic, cadmium, copper, chromium, iron, manganese, molybdenum, lead and zinc using atomic absorption spectrophotometry. Boron is normally measured in a hot water extract as described by Wear (1965). Selenium determinations can be done according to a procedure outlined by Fine (1965). The EPA has also established methods for analyzing arsenic, barium, beryllium, boron, cadmium, chromium, copper, cyanide, iron, lead, magnesium, manganese, mercury, molybdenum, nitrogen, nickel, potassium, selenium, sodium, vanadium, and zinc (EPA, 1979). The normal ranges for metals in soil and plants are presented in Chapter 6 (Tables 6.52 and 6.49). Prior to waste disposal by land treatment, the concentrations of various metals in the soil and waste should be measured. From these data, loading rates for waste can be calculated and background concentrations established.

4.1.2.6 Electrical Conductivity

Electrical conductivity (EC) is used to measure the concentration of salts in a solution. Since electrical currents are carried by charged ions in solution, conductance increases as electrolyte concentration increases. The standard method for assessing the salinity status of a soil is to prepare a saturated paste extract and measure the EC using standard electrodes (USDA, 1954). This can be related to the actual salt concentration in the soil solution that might be taken up by plants. The EC measurement of the saturated paste extract is considered to be one-half the salt concentration at field capacity and one-fourth of that at the permanent wilting point (-15 bars). As a general rule, where saturated paste extract EC values are less than 4 mmhos/cm salts have little effect on plant growth. In soils with EC values between 4 to 8 mmhos/cm salts will restrict yields of many crops. Only a small number of tolerant species can be grown on soils with EC values above 8 mmhos/cm.

When selecting a site and evaluating it for land treatment, careful attention should be given to the soluble salt content of both the soil and the proposed waste stream. Applications of large amounts of salty wastes to an already alkaline soil may decrease microbial degradation and result in barren conditions. These problems are most common to low rainfall, hot areas and to areas near large bodies of salt water. Remedial actions to be taken in the event of accidental salt buildup include stopping the addition of all salt containing materials, growing salt tolerant crops, and if practical, leaching the area with water. In some cases leaching salts may not be acceptable because hazardous constituents would also leach.

4.1.2.7 pH

Soil pH is probably the most informative and valuable parameter used to characterize the chemical property of a soil. Standard measurement procedures are given by Peech (1965). There are three possible basic soil conditions: acidic (pH<7.0), neutral (pH=7.0), and alkaline (pH>7.0). Acidic soils are formed in areas where rainfall leaches the soluble bases deep into the soil profile. Alkaline soils form in areas where rainfall is small and evaporation is high, allowing the accumulation of salts and bases in the soil profile.

Large amounts of lime or other neutralizing agents are needed to raise the pH of acidic soils. In general the pH should be maintained between 6 and 7 to have adequate nutrient availability for plants and microbes without danger of toxicity or deficiency. The addition of large quantities of organic wastes may require liming over and above that required by the native soil since many organic and inorganic acids are formed and released from the decomposing of organic wastes. The decision to add large quantities of fertilizer should be based on the potential for soil acidification, for example, ammonium sulfate may lower the soil pH.

Geographic areas of low rainfall and high evaporation tend to have alkaline soils where cations (Ca, Mg and K) predominate. When base saturation is above 90%, the formation of hydroxide is favored resulting in high pH. These conditions alter the nutrient availability since boron, copper, iron, manganese, phosphorus and zinc are only slightly available at a pH of 8.5 and above.

Measures commonly used for altering soil pH include liming and sulfur applications. Liming is the most common procedure used to raise soil pH. Normal agricultural lime, $CaCO_3$ is most often used, but dolomite $CaMg(CO_3)_2$ is also available for soils of limited Mg content. Lowering soil pH is much less commonplace, but can be accomplished by addition of ferrous sulfate or flowers of sulfur. Both of these compounds result in the formation of H_2SO_4, a strong acid. Sulfur flowers have a much higher potential acidity; however, in special situations, sulfuric acid may be used directly. Management of soil pH at HWLT units is discussed in Section 8.6.

4.1.2.7.1 Acid Soils. As exchangeable bases are leached from the soil in areas of high rainfall, surface soils gradually become more acidic. Local acid conditions can also result from oxidation of iron pyrite and other sulfides exposed by mining. Many conifers grow best at low soil pH and simultaneously take up and hold basic cations from the soil while dropping fairly acidic pine needles, thus, pine forests tend to increase soil acidity. Continued use of ammonia (NH_3) or ammonium (NH_4^+) fertilizers may also lead to a gradual increase in acidity as this reaction takes place in the soil:

$$NH_4^+ + 2O_2 \text{ -----> } 2H^+ + NO_3^- + H_2O \qquad \text{(Brady, 1974)}$$

Many plants grow poorly in acid soils due to high concentrations of soluble aluminum (Al) or manganese (Mn). Aluminum at a solution concentration of 1 ppm slows or stops root growth in some plants. Solution concentrations of 1-4 ppm Mn produce symptoms of toxicity in many plants (Black, 1968). Although most plants can tolerate slightly higher levels of Mn than Al, Mn levels in flooded or poorly drained acid soils can reach 10 ppm (Bohn et al., 1979).

4.1.2.7.2 <u>Buffering Capacity of Soils</u>. The ability of the soil solution to resist abrupt pH changes (buffering capacity) is due to presence of hydrolyzable cations, specifically Al^{3+}, on the surface of the clay colloid. Thus, the buffering capacity is proportional to the cation exchange capacity if other factors are equal (Brady, 1974).

In the soil environment Al^{3+} ions sorbed on the clay surface maintain equilibrium with Al^{3+} ions in the soil solution. As solution Al^{3+} ions are hydrolyzed and precipitated as $Al(OH)_3$, surface-bound Al^{3+} ions migrate into solution to maintain equilibrium. As the Al^{3+} ions hydrolyze and remove OH^- from solution, the solution pH tends to remain stable. Simultaneously as the sorbed Al^{3+} ions migrate into solution, other cations replace the Al^{3+} ions on the soil colloid. Cations such as Na^+, Ca^{2+} and and Mg^{2+} are defined as basic cations because of their difficulty in hydrolyzing in basic solution as compared to Al^{3+}. As the pH of the soil solution is increased, the percentage of the cation exchange complex occupied by basic cations (base saturation) increases. There is a gradual rise in pH and the percent base saturation increases.

At the high and low extremes of base saturation in soils, the degree of buffering is lowest. Buffering capacity is greatest at about 50% base saturation (Peech, 1941). Titration curves vary somewhat for individual soils. The pH of soils dominated by montmorillontic clay is 4.5-5.0 at 50% base saturation. At 50% base saturation soils dominated by kaolinite or halloyite are at a pH 6.0-6.5 (Mehlich, 1941).

Soils resist a sharp decrease in pH. When acid is added to a neutral soil, $Al(OH)_3$ dissolves, enters the soil solution, and the available Al^{3+} ions replace the basic cations on the exchange complex. The decrease in pH is gradual (Tisdale and Nelson, 1975) because of the stoichiometry of the neutralization reaction.

Plants and microorganisms depend upon a relatively stable environment. If the soil pH were to fluctuate widely, they would suffer numerous ill effects. The buffering capacity of the soil stabilizes the pH and protects against such problems (Brady, 1974).

4.1.3   Biological Properties

The soil provides a suitable habitat for a diverse range of organisms which help to render a waste less hazardous. Hamaker (1971) reports that biological action accounts for approximately 80% of waste degradation in soil. The types and numbers of decomposer organisms present in a waste amended soil are dependent on the soil moisture content, available oxygen and nutrient composition.

The population establishment of decomposer organisms following the land application of a waste material begins with bacteria, actinomycetes, fungi and algae (Dindal, 1978). These organisms have diverse enzymatic capabilities and can withstand extremes in environmental conditions. Following establishment of microbial decomposers, the second and third level consumers establish themselves and feed on the initial decomposers and each other (Fig. 4.7). Secondary and tertiary consumers include worms, nematodes, mites and flies. As these organisms use waste components, energy and nutrients from organic materials are released and distributed throughout the immediate environment.

4.1.3.1 Primary Decomposers

4.1.3.1.1 Bacteria. Soils contain a diverse range of bacteria which can be used to degrade a wide range of waste constituents. Bacteria are the most abundant of soil microorganisms, yet they account for less than half of the total microbiological cell mass (Alexander, 1977). Bacteria found in soil may be indigenous to the soil or invaders which enter via precipitation, diseased tissue, or land applied waste. The genera of bacteria most frequently isolated from soil include Arthrobacter, Bacillus, Pseudomonas, Agrobacterium, Alcaligenes, and Flavobacterium (Alexander, 1977).

Bacterial growth or inhibition is influenced by moisture, available oxygen, temperature, pH, organic matter content, and inorganic nutrient supply. In temperate areas, bacterial populations are generally greatest in the upper layers of soil, although in cultivated soils the population is less dense at the surface due to the lack of moisture and the bactericidal action of sunlight (Alexander, 1977). Bacterial activity is usually greatest in the spring and autumn months but decreases during the hot, dry summer and during cold weather.

Soil bacteria may require organic nutrients as a source of carbon and energy, or they may obtain carbon from carbon dioxide (CO_2) and energy

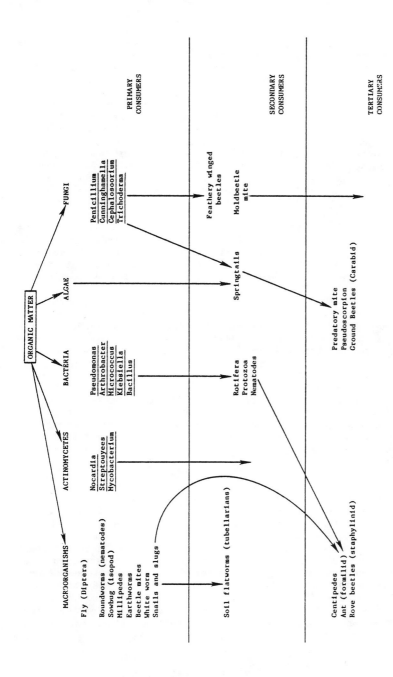

Figure 4.7.   Cycle of organisms that degrade land applied waste. (Jensen and Holm, 1972; Perry and Gerniglia, 1973; Dindal, 1978; Austin et al., 1977)

from the sun. Fungi, protozoa, animals, and most bacteria use organic carbon as a source of energy. Autotrophs, which obtain carbon from CO_2, obtain energy from sunlight or the oxidation of inorganic materials.

4.1.3.1.2 Actinomycetes. Under conditions of limited nutrient supply, actinomycetes become the predominate microorganism and use compounds which are less susceptible to bacterial attack. They are heterotrophic organisms that utilize organic acids, lipids, proteins, and aliphatic hydrocarbons. These organisms are a transitional group between bacteria and fungi, and appear to dominate other microbes in dry or cultivated areas (Alexander, 1977). Primary ecological influences on actinomycetes include moisture, pH, temperature, and amount of organic matter present. Addition of organic matter to the soil greatly increases the density of these organisms. Following the addition of organic matter, they undergo a lag phase of growth after which they show increased activity indicating that they are effective competitors only when the more resistant compounds remain. In addition, actinomycetes seem to influence the composition of the microbial community due to their ability to excrete antibiotics and their capacity to produce enzymes capable of inhibiting bacterial and fungal populations (Alexander, 1977).

4.1.3.1.3 Fungi. This group of heterotrophic organisms is affected by the availability of oxidizable organic substrates. Other environmental influences affecting the density of fungal populations include moisture content, pH, organic and inorganic nutrients, temperature, available oxygen, and vegetative composition. Fungi can withstand a wide range of pH and temperatures. They also have the ability to survive in a quiescent state when environmental conditions are no longer favorable for active metabolism. These organisms, because of their extensive mycelial or thread-like network, usually compose a significant portion of the soil biomass. One of the major activities of fungi in the mycelial state is the degradation of complex molecules. In addition, fungi are active in the formation of ammonium and simple nitrogen compounds.

4.1.3.1.4 Algae. This group of organisms uses light as a source of energy and CO_2 as a source of carbon. Thus, algae are abundant in habitats where light is plentiful and moisture is available. The population of algae is normally smaller than bacteria, actinomycetes or fungi. Because of the inability of algal populations to multiply beneath the zone of soil receiving sunlight, the most dense populations are found between 5 to 10 cm deep. Algae can generate organic matter from inorganic substances. Normally, they are the first to colonize barren surfaces, and the organic matter produced by the death of algae provides a source of carbon for future fungal and bacterial populations. Surface blooms produced by algae bind together soil particles contributing to soil structure and erosion control.

4.1.3.2 Secondary Decomposers

4.1.3.2.1 <u>Worms</u>. The major importance of small worms in decomposing organic material is their abundance and relatively high metabolic activity. When sewage sludge is land applied, the total number of earthworms in the biomass is enhanced with increasing treatment. Increased earthworm populations also enhance soil porosity and formation of water stable soil aggregates, thus improving the structure and water holding capacity of the soil.

Mitchell et al. (1977) found sludge decomposition was increased two to five times by the manure worm. Specific physical and biological characteristics improved by the manure worm include: 1) removal of senescent bacteria, which results in new bacterial growth; 2) enrichment of the sludge by nitrogenous excretions; 3) enhancement of aeration; 4) addition of mineral nutrients; and 5) influence on the carbon and nutrient flux produced by interactions between the microflora, nematodes and protozoa. In a later study they found that fresh anaerobic sludges killed earthworms, although aging the anaerobic sludge for two months removed this toxicity (Mitchell et al., 1978).

4.1.3.2.2 <u>Nematodes, Mites and Flies</u>. As these organisms use waste components, energy and nutrients are released and made available to other decomposers. Nematodes harvest bacterial populations while processing solid waste material. Both nematode and bacterial populations in sewage sludge are increased by the feeding of the isopod <u>Oniscus sellus</u> (Brown et al., 1978). Mold mites will feed on yeast and fungi. Beetle mites and springtails will also feed on molds, but usually under drier and more aerobic conditions. Flies are vital in the colonization of new organic deposits. These insects are used to transport the immobile organisms from one site to another.

4.1.3.3 Factors Influencing Waste Degradation

Following the land application of a hazardous waste, macrobiological activity is suppressed until the microorganisms stabilize the environment. The full range of soil organisms are important to waste degradation, however, habitation by macroorganisms depends on microbial utilization and detoxification of waste constituents. The rate at which microbes attack and detoxify waste constituents depends on many factors including the effect of environmental conditions on microbial life and the presence of certain compounds which are resistant to microbial attack (Alexander, 1977).

The adverse effects of land treatment on the soil fauna may be reduced by a carefully planned program which may involve modifications of certain waste characteristics or environmental parameters. Through the use of pretreatment methods of in-plant process controls (Section 5.2) certain waste characteristics may be modified to improve the rate of waste degradation. The factors affecting degradation which may be adjusted in the design and operation of a land treatment unit include soil parameters (moisture content, temperature, pH, available nutrients, available oxygen, and soil texture or structure) and design parameters (application rate and frequency).

In most cases, it is not feasible to adjust the soil moisture content in the field to enhance degradation. However, when soil moisture is low, it may be advantageous to add moisture through irrigation and when the moisture content is high, to delay waste application until the soil moisture content is more favorable for waste degradation. Water, although essential for microbial growth and transport, has a limited effect on the rate of waste degradation over a broad range of soil moisture contents. Only under excessively wet or dry conditions does soil moisture content have a significant effect on waste degradation (Brown et al., 1982). Dibble and Bartha (1979) found a negligible difference in the microbial activity of oil-amended soil at moisture contents between 30 and 90% of the water holding capacity of the soil.

Both moisture content and temperature will exert a significant effect on the population size and species composition of microorganisms in waste amended soil. The influence of temperature on the metabolic capabilities of soil bacteria was observed in a study by Westlake et al. (1974) in which enrichment cultures of soil bacteria grown on oil at 4°C were able to utilize the same oil at 30°C, while enrichment cultures obtained at 30°C exhibited little capacity for growing on the same oil at 4°C. At 4°C, the isoprenoid compounds phytane and pristane were not biodegraded, while at 30°C the bacteria metabolized these compounds (Westlake et al., 1974). In a six month laboratory study evaluating the rate of biodegradation of two API-separator sludges in soil, the rate of biodegradation of both wastes doubled between 10° and 30°C, but decreased slightly at 40°C (Brown et al., 1982). Similarly, a 50 day laboratory study by Dibble and Bartha (1979) showed little or no increase in the rate of hydrocarbon biodegradation above 20°C. The influence of temperature on the biodegradation of oil sludge in these laboratory studies is presented in Fig. 4.8. These results indicate that the optimum temperature for degradation of these oily wastes is between 20° and 35°C; and, that biodegradation increases with decreasing application rates. While temperature adjustments in the field are impractical, enhanced biodegradation rates may be achieved by delaying or reducing waste applications according to the soil temperature. Measurement of soil temperature is discussed in Section 4.1.1.6.

Through management activities such as the addition of lime, the soil at a land treatment unit is generally maintained at or above 6.5 to enhance the immobilization of certain waste constituents. This pH is also within the optimum range for soil microbes. Verstraete et al. (1975) found the optimum pH for microbial activity to be 7.4 with inhibition occurring at a pH of 8.5. In addition, Dibble and Bartha (1979) found that lime applications favored oil-sludge biodegradation.

Another soil parameter which may be readily adjusted at a land treatment unit is nutrient content. The land application of sludges with a high

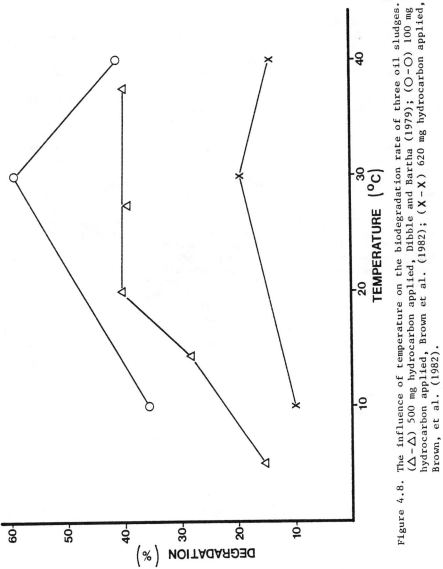

Figure 4.8. The influence of temperature on the biodegradation rate of three oil sludges. (△-△) 500 mg hydrocarbon applied, Dibble and Bartha (1979); (O-O) 100 mg hydrocarbon applied, Brown et al. (1982); (X-X) 620 mg hydrocarbon applied, Brown, et al. (1982).

hydrocarbon content stimulates microbial activity and results in the depletion of available nitrogen which eventually slows degradation. Through the addition of nitrogen containing fertilizers the C:N ratio can be reduced, thus stimulating microbial activity and maintaining the rate of biodegradation. It appears that optimum use is made of fertilizer when the application is delayed until after the less resistant compounds have been degraded. In a field study by Raymond et al. (1976), the rate of biodegradation in fertilized plots was not increased until a year after waste application. The rate of fertilizer needed depends on the characteristics of the waste. While the addition of proper amounts of nutrients can increase biodegradation, excessive amounts, particularly of nitrogen, provide no benefit and may contribute to leaching of nitrates. Dibble and Bartha (1979) determined that the optimum C:N ratio for the oily waste they studied was 60:1; while, in a study by Brown et al. (1982) a refinery waste exhibited optimum degradation at a C:N ratio of 9:1, and a petrochemical waste at 124:1. Thus, it appears that optimum degradation rates can be achieved when the fertilizer application rate is determined on a case-by-case basis.

The texture and structure of the soil exerts a significant influence on the rate of waste biodegradation. Although the choice of soil will in many cases be restricted, a careful evaluation of the rate of biodegradation using the specific soil and waste of the land treatment unit will result in the most efficient use of the land and minimize environmental contamination. In a laboratory study evaluating the biodegradation rates of two wastes in four soils, the most rapid degradation occurred in the silt loam soil and the least rapid in the clay (Table 4.5) (Brown et al., 1982). In fine textured soils where the availability of oxygen may limit degradation, frequent tilling may increase aeration and enhance degradation; although, excessive tilling can promote erosion.

TABLE 4.5   THE EFFECT OF SOIL TEXTURE ON THE BIODEGRADATION OF REFINERY AND PETROCHEMICAL SLUDGE*

Soil	Total Carbon Applied[†] (mg)	% Carbon Degraded as Determined by	
		CO_2-C Evolved	Residual C
Refinery Waste			
Norwood sandy clay	350	60	63
Nacogdoches clay	350	44	54
Lakeland sandy loam	350	37	45
Bastrop clay	350	37	47
Petrochemical Waste			
Norwood sandy clay	2,100	15	34
Nacogdoches clay	2,100	9	32
Lakeland sandy loam	2,100	13	30
Bastrop clay	2,100	0.3	19

* Brown et al. (1982)

† Sludge was applied at a rate of 5% (wt/wt) to soils at field capacity and incubated for 180 days at 30°C.

The frequency and rate of application are design parameters that can be used to enhance waste biodegradation. The amount of residual sludge in the soil influences both the availability of oxygen and the toxic effects of waste constituents on soil microbes. When small amounts of waste are applied frequently, the toxic effects of the waste on the microbes are minimized and microbial activity is maintained at an optimum level. Brown et al. (1982) observed that repeated applications of small amounts of waste resulted in greater degradation over the same time than occurred if all of the waste was applied at one time (Fig. 4.9). These results agree with those of Dibble and Bartha (1979) and Jensen (1975) who found maximum degradation at application rates of oily waste of less than 5% (wt/wt). Thus, it appears that the best results will be obtained when a balance is reached between the most efficient use of the land treatment area and the optimum application rate and frequency. Calculations are described in Sections 7.2.1.5 and 7.5.3.1.4 which can be used to assist in determining these parameters.

Land treatment of hazardous waste is a dynamic process requiring careful design and management to maintian optimum degradation and prevent environmental contamination. The laboratory studies described in Sections 7.2-7.4 can be used to evaluate the value of each parameter that will allow optimum biodegradation. In situations where an equivalent waste has been handled at an equivalent land treatment unit such testing may not be necessary. However, due to the variability of waste streams, soils, and climatic conditions, a careful evaluation of environmental parameters is

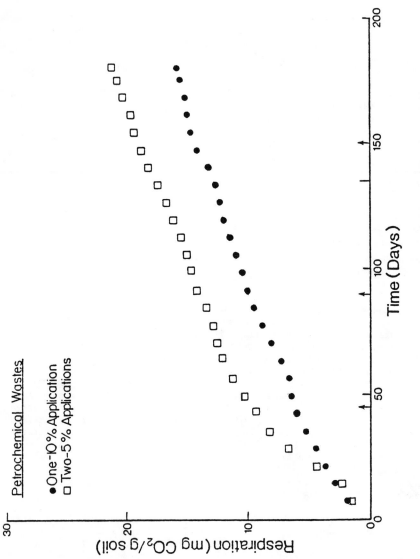

Figure 4.9. Effect of treatment frequency on the evolution of CO_2 from Norwood soil amended with petrochemical sludge and incubated for 180 days at 30 C and 18% moisture (Brown et al., 1982).

required in order to obtain maximum degradation rates using the minimum land area.

Environmental modifications to enhance biodegradation may take the form of amendments applied to the soil, as discussed above, or they may take the form of a microbial spike added to act on a specific class of compounds. Soil particles in sludges may hold bacteria or fungi in a resistant state. Once these organisms become acclimated to waste constituents, they may flourish whenever environmental conditions are improved. In most cases, the addition of limited amounts of organic matter to the soil results in increased microbial activity. Excessive additions of organic matter, however, can result in microbial inhibition because of the nature of the organic matter.

Pretreatment of recalcitrant waste constituents by chemical, physical, or biological degradation may render a waste more amenable to degradation in the soil. For example, pretreatment of PCB containing wastes by photo-decomposition can remove one or two chlorine atoms per molecule (Hutzinger et al., 1972). Since the most significant factor in the relative degradability of PCB wastes the degree of chlorination (Tucker, 1975), pretreatment of PCBs could render the waste more susceptible to microbial attack. Methods of pretreatment that may be useful for HWLT are discussed in Section 5.2.

4.1.3.4 Waste Degradation by Microorganisms

It is difficult to predict the effect of a hazardous waste on the microbial population of the soil. Most hazardous wastes are complex mixtures which contain a variety of toxic compounds, resistant compounds, and compounds susceptible to microbial attack. The application of a readily available substrate to the soil stimulates the microbial population and should provide a more diverse range of organisms to deal with the resistant compounds once the preferred substrate has been degraded. Davies and Westlake (1979) found that the inability of an asphalt based crude oil to support growth was due to the lack of n-alkanes rather than the presence of toxic compounds. Therefore, it appears that the effect of toxic inorganic and organic compounds on microorganisms will be reduced if there is a readily available substrate which can be used by these organisms.

Many hazardous wastes contain substantial quantities of toxic inorganic compounds, such as heavy metals. Kloke (1974) suggests that concentrations of lead in soil above 2000 mg/kg inhibit microbial activity. In addition, the recommended limit for total lead plus four times total zinc plus forty times total cadmium is 2000 mg/kg (Kloke, 1974); however, this calculation fails to account for both the synergistic effects between these cations and the effect of soil characteristics. Doelman and Haanstra (1979) found that a lead concentration of 7500 mg/kg had no effect on microbial activity in a peat soil with a high cation exchange capacity. These results were verified by Babich and Stotzky (1979) who found that lead toxicity was reduced by a high pH (greater than 6.5), the addition of phosphate or carbonate anions, a high cation exchange capacity, and the presence of soluble organic matter. Thus, it is evident that no fixed limit on heavy metal concentration can be generally applied to all waste-soil mixtures. Inorganic toxicity can be better determined empirically on a case by case basis. Similarly, the toxicity of organic compounds in a hazardous waste is dependent on the concentration of organic and inorganic constituents and the properties of the receiving soil. Under certain circumstances, the application of toxic organic compounds to soil may stimulate fungal or actinomycete populations while depressing bacterial populations. Applications of 5000 mg/kg 2,4-D reduced the number of bacteria and actinomycetes, but had little effect on the fungal population (Ou et al., 1978). Since many hazardous wastes can have an adverse effect on biological forms in the soil, land treatment should be carefully planned and monitored to ensure that the biological forms responsible for degradation have not been adversely affected.

There are indications that after long-term exposure to toxic compounds, microbes can adapt and utilize some of these compounds. Results of numerous experiments indicate that microbes have the capacity to adapt and use introduced substrates. The majority of these studies, however, have dealt with microbial utilization of a relatively pure substrate and even those dealing with the use of crude oil are examining a substrate which is predominantly composed of saturated hydrocarbons.

Poglazova et al. (1967) isolated a soil bacterium capable of destroying the ubiquitous carcinogen benzo(a)pyrene. This study indicated that the ability of soil bacteria to degrade benzo(a)pyrene may be enhanced by prolonged cultivation in media containing hydrocarbons. This indicates that the land treatment of hazardous wastes may stimulate the growth of microorganisms with the increased enzymatic capabilities to deal with toxic waste constituents. Jensen (1975) states that the most common genera of bacteria showing an increase in activity due to the presence of hydrocarbons in the soil include Corynebacterium, Brevibacterium, Arthrobacter, Mycobacteria, Pseudomonas and Nocardia. Of all groups of bacteria, Pseudomonas appear to have the most diverse enzymatic capabilities, perhaps due to the presence of plasmids which increase their ability to use complex substrates (Dart and Stretton, 1977). Friello et al. (1976) have transferred hydrocarbon degradive plasmids to a strain of Pseudomonas which gives the bacterium a broader range of available substrates. Enrichment cultures of such organisms may be useful for rapidly degrading certain classes of compounds. It may be useful to apply this type of an enrichment culture to enhance the degradation of a particular recalcitrant compound or group of compounds, although in the case of many complex wastes, a mixed microbial population is required to co-metabolize the various waste constituents.

Large additions of chlorinated hydrocarbons into the environment exert selective pressure on microorganisms to detoxify or utilize these compounds (Chakrabarty, 1978). As a result, bacteria are frequently isolated which have the capacity to use compounds previously thought to be resistant to microbial attack. For example, mixed or enrichment cultures of bacteria have been shown to degrade PCBs (Clark et al., 1979), DDT (Patil et al.,

1970), polyethylene glycol (Cox and Conway, 1976), and all classes of oil hydrocarbons (Raymond et al., 1976). However, some compounds, such as hexachlorobenzene, appear to be resistant to microbial attack (Ausmus et al., 1979).

Various strains of actinomycetes are capable of degrading hazardous compounds. Walker et al. (1976) isolated petroleum degrading actinomycetes from polluted creek sediments which composed over 30% of all the organisms isolated. In addition, Chacko et al. (1966) isolated several strains of actinomycetes that could use DDT.

Fungi capable of degrading the persistent pesticide dieldrin were isolated in a study by Bixby et al. (1971). Perry and Cerniglia (1973) found fungi able to degrade greater quantities of oil during growth than bacteria. This capability was probably due to the ability of fungi to grow as a mat on the surface of the oil. The most efficient hydrocarbon using fungi isolated by Perry and Cerniglia (1973) utilized 30–65% of an asphalt based crude oil. Davies and Westlake (1979) also isolated fungi that could use crude oil. The genera most frequently isolated in their study were Penicillium and Verticillium.

4.2   PLANTS

Plants modify the treatment functions that occur in soil. Primarily, a crop cover on the active treatment site, protects the soil-waste matrix from adverse impacts of wind and water, namely erosion and soil crusting. Plants also function to enhance removal of excess water through transpiration. Some of the more mobile, plant-available waste constituents may be absorbed along with the water and then altered within the plant. Absorbed wastes ultimately are returned to the soil as the decaying plants supply organic matter. The organic matter, in turn, enhances soil structure and cation exchange capacity. The plant canopy may range from spotty to complete coverage and may vary with the season or waste application schedule. Also, cover crops are not required during the operation of an HWLT unit so management decisions about the selection of species, time of planting, desired periods of cover, or whether or not plants are even desirable are all left to the discretion of the owner or operator. A cover crop is advantageous in many cases but it is not essential. The functions plants serve can be divided into two classes, protective functions and cycling and treatment functions.

Plants protect the soil by intercepting and dampening the effects of rainfall and wind. In climates where wetness is a problem for land treatment, a plant canopy can intercept precipitation and prevent significant amounts of water from ever reaching the soil; however, this depends on plant species, completeness of cover, rainfall intensity, and atmospheric conditions. Plants also decrease the erosive effects of raindrop impact on the soil, preventing detachment of particles from the soil and decreasing the splash transport of soil and waste particles. Plants enhance infiltration and lessen runoff transport of waste constituents by decreasing surface flow velocities and by filtering particulates from runoff water. Wind erosion is reduced since the plant canopy dampens wind speed and turbulent mixing at ground level.

Cycling and treatment functions include translocation of substances from soil to plant, transformations within plants, and loss from plants to the atmosphere or back to the soil. Land treatment in a wet climate can benefit from an established crop cover to enhance water loss through uptake and transpiration. Certain soluble, plant available waste constituents and plant nutrients can also be absorbed through plant roots. If testing of plant tissues indicates no food chain hazard from these absorbed constituents, crop harvest can be a removal pathway. However, crops may not be harvested either because tissue analyses have indicated unacceptable concentrations of hazardous constituents or because the expense of plant monitoring exceeds any potential benefit from harvesting. In such cases, the crop residues can be returned to the soil organic matter pool.

Where it has been determined that cover crop is desirable, proper selection of plant species or mixture of species can maximize the desired function. The choice of plant species will vary depending on the season and the region of the country. It is a good idea to consult with area agronomists from the State Agricultural Extension Service, U.S. Department of Agriculture, or the agronomy department at a nearby university to obtain information on varieties and cultural practices which are suited to a given region. Section 8.7 provides additional information on species selection.

4.3   ATMOSPHERE

The atmosphere primarily operates as a modifier of treatment processes in the soil. Atmospheric conditions control the water content and temperature of the soil which in turn control biological waste degradation rates and waste constituent mobility. Winds act along with the heat balance and moisture content to provide for gas exchange, such as the movement of oxygen, carbon dioxide, water vapor, and waste volatiles between soil and atmosphere. In addition to soil-atmosphere interactions, the atmosphere exchanges gases with plants and transmits photosynthetically active radiation to plants. Finally, shortwave radiation may be responsible for some degree of photodegradation of some waste organics exposed at the soil surface. Comprehension of soil, plants and atmosphere interactions and of the various active treatment functions directs attention to those system properties which influence treatment effectiveness and which should be examined more thoroughly.

The important climatic parameters affecting land treatment should be understood from the perspective of site history for design purposes. On-site observations are essential as an input to management decisions (Chapter 8). An off-site weather reporting station will ordinarily be the source of climatic records. Section 3.3 discusses the selection of reliable sources of information that will be representative of site conditions. During the operational life of the HWLT unit it may be useful to install an instrument package and make regular observations of important climatic parameters, such as temperature, rainfall, pan evaporation and wind velocity. Measurement of soil temperature and moisture and particulate emissions may also be useful.

Climate affects the management of hazardous waste facilities. Air temperature influences many treatment processes but has an especially profound effect on the length of the waste application season, the rate of biodegradation, and the volatilization of waste constituents. On an operational basis, temperature observations can aid in application timing for volatile wastes and surface irrigated liquid wastes. Wind, atmospheric stability and temperature determine application timing for volatile wastes. The moisture budget at an HWLT unit is critical to timing waste applications and determining loading rates and storage requirements. Climatic data can be used in the hydrologic simulation to predict maximum water application rates, and to design water retention and diversion structures. A discussion of how the management of the unit can be developed to respond to climatic influences is included in Chapter 8.

Alexander, M. 1977. Introduction to soil microbiology. John Wiley and Sons Inc., New York. p. 1-467.

Allison, L. E. 1965. Organic carbon. pp. 1367-1378. In C. A. Black (ed.) Methods of soil analysis. Part 2. Chemical and microbiological properties. Am. Soc. Agron. Monogr. No. 9. Madison, Wisconsin.

American Society of Agricultural Engineers. 1961. Measuring saturated hydraulic conductivity of soils. Am. Soc. Agr. Engr. St. Joseph, Michigan. p. 1-18.

Ausmus, B., S. Kimbrough, D. R. Jackson, and S. Lindberg. 1979. The behaviour of hexachlorobenzene in pine forest microcosms: transport and effects on soil processes. Applied Science Publ. Ltd., England. Environ. Pollut. 0013-9327/79/0020-0103.

Austin, B., J. J. Calomiris, J. D. Walker, and R. R. Colwell 1977. Numerical taxonomy and ecology of petroleum degrading bacteria. Appl. and Environ. Micro. 34(1):60-68.

Babich, H., and G. Stotzky. 1979. Abiotic factors affecting the toxicity of lead to fungi. Appl. and Environ. Micro. 38(3):506-513.

Bernas, B. 1968. A new method for decomposition and comprehensive analysis of silicates by atomic absorption spectrometry. Anal. Chem. 40:1682-1686.

Bixby, M. W., G. M. Boush, and F. Matsumura. 1971. Degradation of dielorin to carbon dioxide by a soil fungus Trichoderma koningi. Bull. Environ. Contam. Toxicol. 6(6):491-495.

Black, C. A. (ed.). 1968. Methods of soil analysis. Part 2. Chemical and microbiological properties. Am. Soc. Agron. Monogr. No. 9. Madison, Wisconsin.

Blake, G. R. 1965. Bulk Density. pp. 374-390. In C. A. Black (ed.) Methods of soil analysis. Part 1. Physical and mineralogical properties including statistics of measurement and sampling. Am. Soc. Agron. Monogr. No. 9. Madison, Wisconsin.

Boersma, L. 1965a. Field measurement of hydraulic conductivity above a water table. pp. 234-252. In C. A. Black (ed.) Methods of soil analysis. Part 1. Physical and mineralogical properties including statistics of measurements and sampling. Am. Soc. Agron. Monogr. No. 9. Madison, Wisconsin.

Boersma, L. 1965b. Field measurement of hydraulic conductivity below a water table. pp. 222-233. In C. A. Black (ed.) Methods of soil analysis. Part 1. Physical and mineralogical properties including statistics of measurement and sampling. Am. Soc. Agron. Monogr. No. 9. Madison, Wisconsin.

Bohn, H. L., B. L. McNeal, and G. A. O'Connor. 1979. Soil chemistry. John Wiley and Sons, New York.

Bouma, J., R. F. Paetzold, and R. B. Grossman. 1982. Measuring hydraulic conductivity for use in soil survey. U.S. Department of Agriculture, Soil Conservation Service. Report No. 38. 14 p.

Brady, N. C. 1974. The nature and properties of soils. 8th ed. MacMillan Co., New York. 639 p.

Bremner, J. M. 1965. Inorganic forms of nitrogen pp. 149-176. In C. A. Black (ed.) Methods of soil analysis. Part 2. Chemical and microbiological properties. Am. Soc. Agron. Monogr. No. 9. Madison, Wisconsin.

Brown, B., B. Swift, and M. J. Mitchell. 1978. Effect of Oniscus aesellus feeding on bacteria and nematode populations in sewage sludge. Oikos 30:90-94.

Brown, K. W., L. E. Deuel, Jr., and J. C. Thomas. 1982. Soil disposal of API pit wastes. Final Report of a Study for the Environmental Protection Agency (Grant No. R805474013). 209 p.

Brown, K. W., C. Woods, and J. F. Slowey. 1975. Fate of metals applied in sewage at land wastewater disposal sites. Final report AD 43363 submitted to the U.S. Army Medical Research and Development Command, Washington, D.C.

Buckman, H. O., and N. C. Brady. 1960. The nature and properties of soils. MacMillan Co., New York. 239 p.

Chacko, C. I., J. L. Lockwood, and M. Zabick. 1966. Chlorinated hydrocarbon pesticides - degradation by microbes. Science 154:893-894.

Chakrabarty, A. M. 1978. Molecular mechanisms in the bio-degradation of enviromental pollutants. Am. Soc. Micro. News 44(12):687-690.

Chapman, H. D. 1965a. Cation exchange capacity. pp. 891-900. In C. A. Black (ed.) Methods of soil analysis. Part 2. Chemical and microbiological properties. Am. Soc. Agron. Monogr. No. 9. Madison, Wisconsin.

Chapman, H. D. 1965b. Total exchangeable bases. pp. 902-904. In C. A. Black (ed.) Methods of soil analysis. Part 2. Am. Soc. Agron. Monogr. No. 9. Madison, Wisconsin.

Clark, R. R., E. S. K. Chian, and R. A. Griffin. 1979. Degradation of poly-cholorinated biphenyls by mixed microbial cultures. App. Environ. Microbiol. 37(4):680-685.

Cox, D. P. and R. A. Conway. 1976. Microbial degradation of some polyethylene glycols. pp. 835-841. In J. M. Sharpley and A. M. Kaplan (ed.) Proceedings of the Third International Bio-degradation Symposium. Appl. Sci. Publ., London.

Dart, R. K. and R. J. Streton. 1977. Microbial aspects of pollution control. Elsevier Sci. Publ. Co., Amsterdam. pp. 180-215.

Davies, J. S. and D. W. S. Westlake. 1979. Crude oil utilization by fungi. Can. J. Microbiol. 25:146-156.

Day, P. R. 1965. Particle fractionation and particle size analysis pp. 545-566. In C. A. Black (ed.) Methods of soil analysis. Part 1. Physical and mineralogical properties including statistics of measurement and sampling. Am. Soc. Agron. Monogr. No. 9. Madison, Wisconsin.

Dibble, J. T. and R. Bartha. 1979. Effect of environmental parameters on biodegradation of oil sludge. Appl. Environ. Micro. 37:729-738.

Dindal, D. L. 1978. Soil organisms and stabilizing wastes. Compost Sci./ Land Utilization. 19(4):8-11.

Doelman, P. and L. Haanstra. 1979. Effects of lead on the decomposition of organic matter. Soil Biol. Biochem. 11:481-485.

EPA. 1979. Methods for chemical analysis of water and wastes. Environmental Monitoring and Support Laboratory. Office of Research and Development, EPA. Cincinnati, Ohio. EPA 600/4-79-020. PB 297-686/8BE.

Fine, L. O. 1965. Selenium. pp. 1117-1123. In C. A. Black (ed.) Methods of soil analysis. Part 2. Chemical and microbiological properties. Am. Soc. Agron. Monogr. No. 9. Madison, Wisconsin.

Fluker, B. J. 1958. Soil temperatures. Soil Sci. 86:35-46.

Friello, D. A., J. R. Mylroie, and A. M. Chakrabarty. 1976. Use of genetically engineered multi-plasmid microorganisms for rapid degradation of fuel hydrocarbons. pp. 205-214. In CRC Critical Reviews in Micro.

Fuller, W. H. 1978. Investigations of landfill leachate pollutant attenuation by soils. Municipal Environmental Research Laboratory. Office of Research and Development, EPA. Cincinnati, Ohio. EPA 600/2-78-158.

Hamaker, J. W. 1971. Decomposition: quantitative aspects. pp. 253-4340. In C. A. I. Goring and J. W. Hamaker (ed.) Organic chemicals in the soil environment. Vol. 2. Marcel Dekker, Inc., New York.

Homes, M. V. 1955. A new approach to the problem of plant nutrition and fertilizer requirements. Soils Fertilizers 18:1.

Hutzinger, O., S. Safe, and V. Zitko. 1972. Photochemical degradation of chlorobiphenyls (PCB's). Environ. Health Perspect. 1:15-20.

Jensen, V. 1975. Bacterial flora of soil after application of oily waste. Oikos 26:152-158.

Jensen, V. and E. Holm. 1972. Aerobic chemoorganotrophic bacteria of a Danish beech forest. Oikos 23:248-260.

Kloke, A. 1974. Blei-zink-cadmium: anreicher in boden and pflanzen. Staub 34:18-21.

Klute, A. 1965. Laboratory measurement of hydraulic conductivity of saturated soil. pp. 210-221. In C. A. Black (ed.) Methods of soil analysis. Part 1. Physical and mineralogical properties including statistics of measurement and sampling. Am. Soc. Agron. Monogr. No. 9. Madison, Wisconsin.

Mehlich, A. 1941. Base saturation and pH in relation to soil type. Soil Sci. Soc. Am. Proc. 6:150-156.

Mitchell, M., R. M. Mulligan, R. Hartenstein, and E. F. Neuhauser. 1977. Conversion of sludges into "topsoils" by earthworms. Compost Sci. 18(4):28-32.

Mitchel, M. J., R. Hartenstein, B. L. Swift, E. F. Neuhauser, B. I. Abrams, R. M. Mulligan, B. A. Brown, D. Craig, and D. Kaplan. 1978. Effects of different sewage sludges on some chemical and biological characteristics of soil. J. Environ. Qual. 14:301-311.

Ou, Li-tse, D. F. Rothwell, W. B. Wheeler, and J. M. Davidson. 1978. The effect of high 2,4-D concentrations on degradation and carbon dioxide evolution in soils. J. Environ. Qual. 7(2):241-246.

Overcash, M. R. and D. Pal. 1979. Design of land treatment systems for industrial wastes--theory and practice. Ann Arbor Science Publ. Ann Arbor, Michigan. 684 p.

Patil, K. C., F. M. Matsumura, and G. M. Boush. 1970. Degradation of Endrin, Aldrin, and DDT by soil microorganisms. Appl. Micro. 19(5):879-886.

Peech, M. 1941. Availability of ions in light sandy soils as affected by soil reaction. Soil Sci. 51:473-486.

Peech, M. 1965. Lime requirement. Agron. 9:927-932.

Perry, J. J. and R. Cerniglia. 1973. Studies on the degradation of petroleum by filamentous fungi. In D. C. Ahearn and S. P. Meyers (ed.) The microbial degradation of oil pollutants. Louisiana State Univ. Center for Wetland Resources. Baton Rouge, Louisiana. LSU-SG-7301.

Poglazova, M. N., G. E. Fedoseeva, and A. J. Klesina, M. N. Meissel, and L. M. Shabad. 1967. Destruction of benzo (a) pyrene by soil bacteria. Life Sci. 6:1053-1067.

Pritchett, W. L. 1979. Properties and management of Forest Soils. John Wiley and Sons, New York.

Raymond, R. L., J. O. Hudson, and V. W. Jamison. 1976. Oil degradation in soil. Appl. Environ. Microbiol. 31(4):522-535.

Richards, L. A. 1965. Physical condition of water in soil. pp. 128-152. In C. A. Black (ed.) Methods of soil analysis. Part 1. Physical and mineralogical properties, including statistics of measurement and sampling. Am. Soc. Agron. Monogr. No. 9. Madison, Wisconsin.

Stewart, B. A., D. A. Woolhiser, W. H. Wischmeier, J. H. Carow, and M. H. Frere. 1975. Control of water pollution from cropland: Vol I. A manual for guideline development. U.S. Department of Agriculture. Report ARS-H-5-1. Hyattsville, Maryland. 111 p.

Tisdale, S. L. and W. L. Nelson. 1975. Soil fertility and fertilizers. 3rd ed. MacMillan Publishing Co., New York.

Tucker, E. S., V. W. Saeger, and O. Hicks. 1975. Activated sludge primary biodegradation of polychlorinated biphenyls. Bull. Environ. Contam. Toxic. 14(6):705-713.

USDA. 1954. Diagnosis and improvement of saline and alkali soils. L. A. Richards (ed.) Agriculture Handbook. No. 60. 160 p.

USDA. 1981. Examination and description of soils in the field. Revised Chapter 4, pp. 4-31-4-37. In Soil survey manual. U.S. Soil Conservation Service, Washington, D.C.

Verstraete, W., R. Vanlooke, R. de Borger, and A. Verlinde. 1975. Modeling of the breakdown and mobilization of hydrocarbons in unsaturated soil layers. pp. 98-112. In J. M. Sharpley and A. M. Kaplan (eds.) Proceedings of the Third International Biodegradation Symposium. Applied Science Publ., London.

Walker, J. D., R. R. Colwell, and L. Petraskis. 1976. Biodegradation of petroleum by Chesapeake Bay sediment bacteria. Can. J. Microbiol. 22:423-428.

Wear, J. I. 1965. Boron. pp. 1059-1063. In C. A. Black (ed.) Methods of soil analysis. Part 2. Chemical and microbiological properties. Am. Soc. Agron. Monogr. No. 9. Madison, Wisconsin.

Westlake, D. W. S., A. Jobson, R. Phillippee, and F. D. Cook. 1974. Biodegradability and crude oil composition. Can. Jour. Microbial. 20:915-928.

Wooding, H. N. and R. F. Shipp. 1979. Agricultural use and disposal of septic tank sludge in Pennsylvania - information and recommendations for farmers, septage haulers, municipal officials, and regulatory agencies. Penn. State Univ. Cooperative Ext. Ser. Special Circular 257.

HAZARDOUS WASTE STREAMS 5

Jeanette Adams
K. C. Donnelly
David C. Anderson

This chapter presents information to be used in evaluating waste streams proposed for land treatment. There are three main factors that need to be considered when evaluating the information on waste streams submitted with a permit application for an HWLT unit. These three factors are the characterization of the wastes, the pretreatment options available and the techniques used for sampling and analysis. Figure 5.1 shows how each of these topics fits into the decision-making framework for evaluating HWLT units, first presented in Chapter 2 (Fig. 2.1).

Each section in this chapter focuses on one of the topics shown in Fig. 5.1. Section 5.1 briefly discusses sources of hazardous waste. A number of pretreatment options are available that can reduce the hazards associated with certain waste streams; Section 5.2 discusses these options. Finally, in order to accurately predict the fate of a given waste in an HWLT unit, the permit evaluator must know what analytical techniques were used by the applicant in performing the waste analysis. Section 5.3 discusses procedures that are appropriate for analyzing hazardous wastes.

5.1 SOURCES OF HAZARDOUS WASTE

The first step in evaluating a waste stream is to determine what the expected waste constituents are based on what is known about the sources of the waste. Hazardous waste sources fall into two broad categories as follows:

(1) Specific industrial sources that generate waste streams peculiar to the feedstocks and processes used by that industry, such as leather, rubber or textiles; and

127

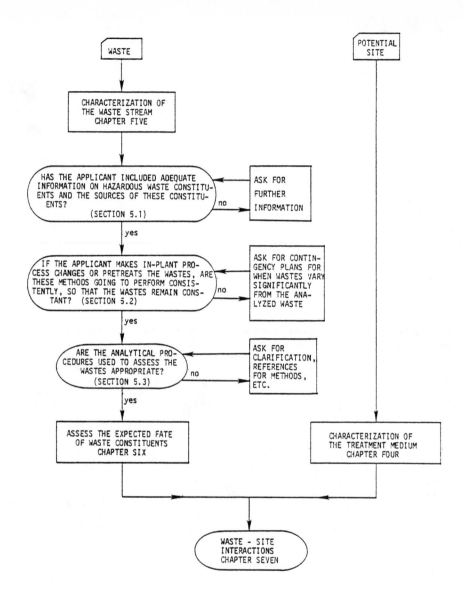

Figure 5.1. Characterization of the waste stream to be land treated.

(2) Nonspecific sources of waste that cut across industrial categories, but may still be characterized according to the raw materials and processes used, such as solvent cleaning or product painting.

5.1.1   Specific Sources

Industries that produce a waste unique to that industry are "specific sources" of that waste. Examples of "specific" industrial sources are textiles, lumber, paper, inorganic chemicals, organic chemicals, petroleum products, rubber products, leather products, stone products, primary metals and others. Table 5.1 ranks most of the specific sources according to the volume of hazardous waste each is projected to generate in 1985.

5.1.2   Nonspecific Sources of Hazardous Waste

There are several hazardous waste generating activities that are not specific to a particular industry. For instance, many manufactured products are cleaned and painted before they are marketed. Product cleaning is usually done with solvents and, consequently, many industries generate spent solvent wastes. Similarly, industrial painting generates paint residues. Eighteen nonspecific wastes are listed in Table 5.2. There are three main categories of hazardous constituents generated by these nonspecific sources which are solvents, heavy metals and cyanide, and paint (Fig. 5.2).

5.1.3   Sources of Information on Waste Streams

The designers can use published information on the chemical analysis of similar hazardous wastes to help them determine the constituents expected in the wastes to be land treated. In some cases, this information may indicate the presence of constituents which may need to be pretreated before they are disposed in an HWLT unit.

There is little information on the waste streams from the organic chemicals industry because each plant uses a unique collection of feedstocks and unit chemical processes to produce its line of products. However, some information about the nature of the waste can be gained if information is known about the chemical feedstocks and unit process used (Herrick et al., 1979).

An unpublished document has been prepared for EPA by K. W. Brown and Associates, Inc. that pulls together information on waste streams generated by the industries that produce hazardous wastes. The document presents chemical analyses (where available) and information on the hazardous constituents contained in the waste streams of these industries according to

TABLE 5.1   PROJECTED 1985 HAZARDOUS WASTE GENERATION BY INDUSTRY*

SIC Code	Industry	1980 Estimate	Low#	High+
		Annual Volume of Waste Generated†		
			1985 Projection	
28	Chemicals & Allied Products	25,509	24,564	30,705
33	Primary Metals	4,061	3,699	4,624
--	Nonmanufacturing Industries	1,971	1,882	2,352
34	Fabricated Metal Products	1,997	1,807	2,259
29	Petroleum & Coal Products	2,119	1,789	2,236
37	Transportation Equipment	1,240	1,309	1,636
26	Paper & Allied Products	1,295	1,201	1,501
36	Electric & Electronic Equipment	1,093	1,145	1,431
31	Leather & Leather Tanning	474	342	428
35	Machinery, Except Electrical	322	330	413
39	Miscellaneous Manufacturing	318	299	374
30	Rubber & Miscellaneous Plastic Products	249	226	282
22	Textile Mill Products	203	162	203
27	Printing & Publishing	154	145	182
38	Instruments & Related Products	90	99	124
24	Lumber & Wood Products	87	75	94
25	Furniture & Fixtures	36	29	36
32	Stone, Clay & Glass Products	17	15	19
	TOTAL	41,235	39,118	48,899

* Booz-Allen and Hamilton, Inc. and Putnam, Hayes and Bartlett, Inc. (1980).

† In thousands of wet metric tons.

Based on a reasonable estimate of the potential reduction (20%) in waste generation.

+ Based on the industrial growth rate used to calculate 1980 and 1981 estimates.

TABLE 5.2 POTENTIALLY HAZARDOUS WASTE STREAMS GENERATED BY NONSPECIFIC INDUSTRIAL SOURCES

Modified SIC Code	Hazardous Waste Number	Activity	Waste Stream	LAND TREATMENT POTENTIAL* Rate (R) or Capacity (C) Limiting Components
	F001	Degreasing operations (halogenated solvent)	Spent halogenated solvents & sludge	Tetrachloroethylene (C); carbon tetrachloride (C); Trichloroethylene (C); 1,1,1-trichloroethane (C); Methylene chloride (C); chlorinated fluorocarbons (C)
	F002	Halogenated solvent recovery	Spent halogenated solvents & still bottoms	Tetrachloroethylene (C); methylene chloride (C); Trichloroethylene (C); 1,1,1-trichloroethane (C); 1,1,2-trichloro-1,2,2-fluoroethane (C) Chlorobenzene (C) o-dichlorobenzene (C); trichlorofluoroethane (C)
	F003	Nonhalogenated solvent recovery	Spent nonhalogenated solvents & still bottoms	Flammable solvents (R)
	F004	Nonhalogenated solvent recovery	Spent nonhalogenated solvents & still bottoms	Cresols (R) and cresylic acid (R); nitrobenzene (C)
	F005	Nonhalogenated solvent recovery	Spent nonhaolgenated solvents & still bottoms	Methanol (R); toluene (R); methyl ethyl ketone (R); Methyl isobutyl ketone (R); carbon disulfide (R); Isobutanol (R); pyridine (R)
3471.1	F006	Electroplating	Wastewater treatment sludge	Cadmium (C); chromium (C); nickel (C); Cyanide (complexed) (C)
3471.2	F007	Electroplating	Spent plating bath	Cyanide salts (C)
3471.3	F008	Electroplating	Plating bath bottom sludge	Cyanide salts (C)
3471.4	F009	Electroplating	Spent stripping & cleaning bath solutions	Cyanide salts (C)
3398.1	F010	Metal heat treating	Quenching oil bath sludge	Cyanide salts (C)
3398.2	F011	Metal heat treating	Spent salt bath solutions	Cyanide salts (C)
3398.3	F012	Metal heat treating	Wastewater treatment sludge	Cyanide (complexed) (C)
	F013	Metal recovery	Flotation trailings	Cyanide (complexed) (C) and metals from the ore
	F014	Metal recovery	Cyanidation wastewater treatment tailing pond bottom sediments	Cyanide (complexed) (C)
	F015	Metal recovery	Spent cyanide bath solutions	Cyanide salts (C)
3312.1	F016	Operations involving coke ovens & blast furnaces	Air pollution control scrubber sludge	Cyanide (complexed) (C)
3479.1	F017	Industrial painting	Paint residues	Cadmium (C); chromium (C); lead (C); cyanides (C); toluene (R); tetrachloroethylene (C)
3479.2	F018	Industrial painting	Wastewater treatment sludge	Cadmium (C); chromium (C); lead (C); cyanide (C); toluene (R); tetrachloroethylene (C)

* Values for waste constituents may vary; hence, loading rates and capacities should be based on the analysis of the specific waste to be land treated and on the results of the pilot studies performed. Organic compounds are labeled (C) when it is believed that there may be some soil conditions under which the compound may not degrade rapidly enough to prevent toxicity hazards, either due to accumulation in soil or migration via water or air.

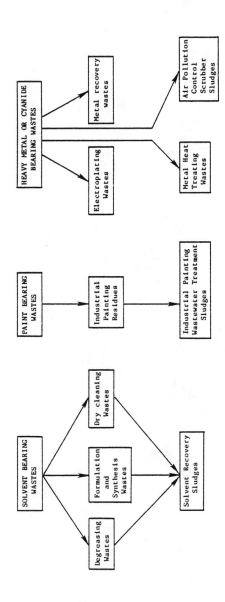

Figure 5.2.   Categories of hazardous constituents generated by nonspecific sources.

the standard industrial classification. This document together with waste analyses supplied to EPA should form a basis for a better understanding of hazardous waste streams.

5.2   WASTE PRETREATMENT

Pretreatment processes may be used to render a waste more amenable to land treatment. This can be accomplished by altering the waste in a way that either changes its physical properties or reduces its content of the waste constituents that limit the land treatment operation. Physical alterations include premixing the waste with soil and reducing the unit size of waste materials. Specific waste constituents can limit the ultimate capacity, yearly loading rate, or the single application dosage of a waste disposed in an HWLT unit (Section 7.5.1). Pretreatment processes are available that will reduce the concentration of a limiting constituent. Pretreatment may improve both the economic and environmental aspects of the HWLT unit. When waste form or waste constituents warrant examining pretreatment options, in-plant process changes should also be explored.

It is beyond the scope of this document to review all the available pretreatment techniques and their treatment efficiencies for the thousands of pollutant species. However, EPA (1980a) has recently published a five volume manual that exhaustively covers the following topics that can be used to evaluate pretreatment.

(1) Volume one is a compendium of treatability data, industrial occurrence data, and pure species descriptions of metals, cyanides, ethers, phthalates, nitrogen containing compounds, phenols, mono and polynuclear aromatics, PCBs, halogenated hydrocarbons, pesticides, oxygenated compounds, and a number of miscellaneous organic compounds. This volume focuses on the 129 priority pollutants and other compounds that are prevalant in industrial wastewaters and that do not readily degrade or disappear from aqueous environments, which are the ultimate receivers of leachates generated by land treatment units.

(2) Volume two is a collection of industrial wastewater discharge information and includes data for both raw and treated wastewaters.

(3) Volume three is a compilation of available performance data for existing wastewater treatment technologies.

(4) Volume four is a collection of capital and operating cost data for the treatment technologies described in volume three.

(5) Volume five is an executive summary and describes the use of information contained in volumes one through four.

To determine the most desirable mix of pretreatments for a land treatment system, total costs should be weighed against the degree of treatment required. Possible pretreatment steps for enhancing the land treatability of waste as presented by Loehr et al. (1979), are discussed below.

(1) Preliminary treatment (coarse screening or grinding) is used to remove large objects such as wood, rags and rocks to protect piping and spray systems.

(2) Primary treatment usually involves the removal of readily settleable and floatable solids. The primary treatment effluent can then be land treated by spray irrigation or overland flow. Since the removed solids can clog both spray nozzles and the soil surface, these solids are usually land treated by soil incorporation.

(3) Secondary treatment includes several biological treatments (such as aerated lagoons, anaerobic digestion, composting and activated sludge) and any subsequent solids settling. Secondary pretreatment systems may be necessary where it is desirable to remove soluble organics or suspended solids that may clog the soil. Secondary treatment effluents are usually suitable for spray irrigation while the secondary treatment sludges can be incorporated into the soil. Land treatment of a waste often results in the breakdown of organics as rapidly as secondary treatment but the additional treatment may be necessary for some refractory organics.

(4) Disinfection is the treatment of effluents to kill disease causing organisms such as pathogenic bacteria, viruses and amoebic cysts. Chlorination effectively kills pathogens but may also generate chlorinated organics and have undesirable effects on cover crops and leachate quality. Ozonation is more expensive than chlorination, but effectively disinfects a waste stream without the undesirable effects of chlorination. Coupling ozonation with irradiation by ultraviolet light may improve its economic feasibility and enhance overall waste treatment. Compounds normally refractory to ozone alone are rapidly converted to carbon dioxide and water when subjected to the combination (Rice and Browning, 1981).

(5) Advanced (tertiary) wastewater treatment refers to processes designed to remove dissolved solids and soluble organics that are not adequately treated by secondary treatment. Land treatment usually exceeds the results obtainable through tertiary treatment for removal of nitrogen, phosphorous and soluble organics. In these cases a tertiary treatment may not be useful; however, tertiary treatment for the removal of dissolved salts (such as reverse osmosis or distillation) may produce an effluent of drinking water quality and circumvent the need for land treatment.

Table 5.3 lists the different pretreatment methods and their applicability to hazardous waste treatment. Although, in many cases, pretreatment of the waste is not necessary prior to land treatment, pretreatments with the most potential for enhancing the land treatability of wastes are examined in the following sections (5.2.1 through 5.2.6). Neutralization, dewatering, degradation processes, premixing with soil, and size reduction may greatly increase the effectiveness of land treatment for a given waste; however, in-plant process changes may also be effective in reducing troublesome waste constituents. In all cases, care must be taken when pretreatment processes are being considered to evaluate the cost effectiveness of the process and to determine if the process (which may have originally been developed to render a waste compatible with another disposal option) is appropriate for land treatment operations.

5.2.1   Neutralization

Neutralization (pH adjustment) may be a desirable pretreatment for strongly acidic or alkaline wastes being land treated. Biological treatment systems, such as land treatment, rely on microbial degradation as the major treatment mechanism for organic constituents in the waste. Microbial growth and, hence, treatment efficiency are optimized by maintaining the pH near neutral.

Neutralization involves the reaction of a solution with excess hydronium or hydroxide ions to form water and neutral salts (Adams et al., 1981). Care should be taken to select a neutralizing agent that will not produce a neutral salt that is detrimental to the land treatment process. For instance, lime ($CaCO_3$) is vastly preferable to caustic soda (NaOH) as an agent to neutralize an acidic waste. Lime adds calcium to the waste which will improve the workability of the treatment soil. Calcium is also an essential nutrient for cover crops and microbes. Conversely, caustic soda adds sodium which can decrease the workability of the soil and, at high concentrations, sodium is toxic to cover crops and microbes.

It should be noted that the biological treatment process that occurs in land treated soils may itself change the pH of a waste-soil mixture. The pH of treated soil is reduced by the following (Adams et al., 1981):

(1) Hydroxide alkalinity is destroyed by the biochemical production of CO_2;

$$\text{Carbohydrate} + (n)O_2 \xrightarrow[\text{oxidation}]{\text{Biochemical}} (n)\ CO_2 + (n)\ H_2O$$

$$CO_2 + OH^- \longrightarrow HCO_3^-$$

TABLE 5.3 PRETREATMENT METHODS FOR HAZARDOUS WASTES*

Pretreatment Method	Heavy Metal Removal	Organic Removal	Organic Destruction	Waste Volume Reduction	Comments	Physical Form Treated
Activated sludge	Yes	Yes	Yes	Yes	Waste must have heavy metal content less than 1%	Liquid, slurry, sludge
Aerated lagoons	No	Yes	Yes	Yes	Used in temperate climates	Liquid, slurry, sludge
Anaerobic digestion	No	Yes	Yes	Yes	Very sensitive to toxic compounds	Slurry, sludge, liquid
Composting	No	Yes	Yes	Yes	Least sensitive method of biological treatment	Slurry, sludge, liquid
Enzymatic biological treatment	No	Yes	Yes	No	Only works for specific chemicals	Liquid, slurry, sludge
Trickling filters	No	Yes	Yes	Yes	Low efficiency for organic removal	Liquid, slurry, sludge
Waste stabilization ponds	No	Yes	Yes	Yes	Waste must have dilute concentrations of organic and inorganics	Liquid, slurry, sludge
Carbon adsorption	Yes	Yes	No	No	Efficient for wastes with less than 1% organics	Liquid
Resin adsorption	Possible	Yes	No	No	Extracts and recovers mainly organics solutes from aqueous waste	Liquid
Calcination	Possible	No	Yes	Yes	Will require volume of nonorganics and convert them into a form of low leachability	Liquid, slurry, sludge
Catalysis	No	No	Yes	No		Liquid
Centrifugation	Yes	No	No	Yes	Primarily used for dewatering sludge	Slurry
Chlorinolysis	No	No	Yes	Yes	Conversion of chlorinated hydrocarbons to carbon tetrachloride	Liquid
Dialysis	Yes	No	No	No	Separation of salts from aqueous	Liquid
Dissolution	Yes	No	No	No	Removal of heavy metals from fly ashes	Liquid, slurry, sludge
Distillation	No	Yes	No	Yes	Recovery of organic solvents	Liquid, slurry, sludge
Electrolysis	Yes	No	No	No	Removal of heavy metals from concentrated aqueous solution	Liquid

--continued--

TABLE 5.3 (continued)

Pretreatment Method	Heavy Metal Removal	Organic Removal	Organic Destruction	Waste Volume Reduction	Comments	Physical Form Treated
Electrodialysis	Possible	No	No	No	Recovery of inorganic salts	Liquid
Evaporation	No	Possible	No	Yes	Recovery of inorganic salts	Liquid
Filtration	Yes	No	No	Yes	Removal of metal precipitates	Slurry
Precipitation, flocculation, sedimentation	Yes	Yes	No	Yes	Removal or recovery of solids from aqueous solution	Liquid, slurry
Flotation-biological	Yes	No	No	No	Separation of solid particles suspended in a liquid medium	Slurry
Freeze crystallization	Yes	Yes	No	Yes	Desalination of water	Liquid, slurry, sludge
Freeze drying	No	No	No	Yes	Separation of pure water from solids	Liquid, slurry
Suspension freezing	Yes	No	No	Yes	Separation of suspended particles magnetic particles from liquids	Liquid
Hydrolysis	No	No	Yes	No	May increase toxicity of waste	Liquid, slurry, sludge
Ion exchange	Yes	Yes	No	No	Selective removal of heavy metals and hazardous anions	Liquids
Liquid ion exchange	Yes	No	No	No	Selective removal and/or separation of free and complexed metal ions in high concentrations	Liquid, slurry, sludge
Liquid-liquid extraction of organics	No	Yes	No	No	Solvent recovery	Liquid
Microwave discharge	No	Possible	Yes	Yes	Developmental stages; primarily for small quantities of toxic compounds	Liquid
Neutralization	No	No	No	No	Renders waste treatable by other	Liquid, slurry, sludge
Chemical oxidation	Possible	No	Yes	No	Detoxification of hazardous materials	Liquid
Ozonolysis	No	No	Yes	No	May be used to make toxic wastes more susceptible to biological action, especially chlorinated hydrocarbons	Liquid

--continued--

TABLE 5.3 (continued)

Pretreatment Method	Heavy Metal Removal	Organic Removal	Organic Destruction	Waste Volume Reduction	Comments	Physical Form Treated
Photolysis	No	No	Yes	No	Degradation of aromatic and chlorinated hydrocarbons	Liquid
Chemical reduction	Possible	No	No	No	Detoxification of hazardous materials	Liquid
Reverse osmosis	Yes	Yes	No	Yes	Purification dilute wastewaters	Liquid
Size reduction	No	No	No	No	For spill debris such as contaminated pallets and lumber	Solid
Soil mixing	No	No	No	No	Volume of waste will increase, this technique applies to stick or tarry waste	
Steam distillation	No	Yes	No	Yes	Solvent recovery	Liquid, slurry, sludge
Air stripping	No	Possible	No	No	Recovery of volatile compounds from aqueous solutions	Liquid, slurry
Steam stripping	No	Yes	No	No	Recovery of volatile compounds from aqueous solutions	Liquid, slurry
Ultra filtration	Yes	Yes	No	No	Separation of dissolved or suspended particles from a liquid stream	Liquid
Zone refining	Yes	Yes	No	No	Purification technique for obtaining high-purity organic and inorganic materials	Liquid

* De Renzo (1978).

(2) Reduced forms of sulfur can be biochemically oxidized to sulfuric acid; and

$$H_2S + 2O_2 \xrightarrow{\text{Biochemical oxidation}} H_2SO_4$$

(3) Oxidation of ammonium releases hydrogen ions.

$$NH_4^+ + 2O_2 \longrightarrow NO_3^- + 2H^+ + H_2O$$

The pH of treated soil is increased by the biochemical oxidation of organic acids as follows (Adams et al., 1981).

$$R - COOH + (n)O_2 \xrightarrow{\text{Biochemical oxidation}} (n)\ CO_2 + (n)H_2O$$

5.2.2   Dewatering

Dewatering is a broad term referring to any process that reduces the water content and, hence, the volume of a waste which increases the solids content of the remaining waste. The oldest, simplest and most economical method of dewatering a waste uses shallow evaporation ponds. However, for such a system to be feasible, adequate land area must be available and evaporation rates must exceed precipitation rates (Adams et al., 1981).

Evaporative rates can be increased by placing spray aerators on the surface of the pond. Spray aeration has the added advantages of increasing waste decomposition by exposing the wastewater to ultraviolet rays present in sunlight and encouraging aerobic decomposition using oxygen adsorbed during spraying.

A wastewater can be signficantly dewatered through freeze crystaliza- tion. This process is used to segregate a liquid waste stream into fresh- water ice cyrstals and a concentrated solution of the remaining heavy metals, cyanides and organics. The ice crystals can then be removed by mechanical means (Metry, 1980). Freeze crystalization is an especially attractive dewatering technique in northern sections of the U.S. where evaporative rates are low and the cold climate provides cost-free freezing.

Drying beds are shallow impoundments usually equipped with sand bottoms and tile drains. Typically, sludge is poured over the sand to a depth of 20 to 30 cm. Free drainage out of the tile drains occurs for several days and drying time ranges from weeks to months, depending on the weather and sludge properties (Ettlich et al., 1978).

Filtration is the mechanism used in several dewatering processes. It involves the separation of liquids and solids by forcing liquids through porous membranes (screen or cloth) or media as in the drying beds discussed above. Liquids are forced through by pressure, vacuum, gravity or centri- fugal force and the dewatered solids can then be land treated.

Various processes are used to increase the ease or extent to which sludge dewaters. The most widely used of these processes involves two steps. First, a chemical conditioner (such as lime, ferric chloride, aluminum chloride or a variety of organic polymers) is added to the wastewater that causes dissolved or suspended solids to clump together into suspended particles. Then these suspended particles clump together into larger particles which either settle out of solution or can be more easily removed by filtration.

5.2.3   Aerobic Degradation

Several aerobic degradation processes are used to pretreat land treated wastes. These processes can effectively reduce the quantity of volatile and highly mobile organic species in a waste stream. Aerobic processes discussed below are composting, activated sludge and aerated lagooning.

Composting involves the aerobic degradation of a waste material placed in small piles or windrows so that the heat produced by microbial action is contained. Maintenance of an abundant supply of oxygen in the compost pile, coupled with elevated temperature and sufficient moisture, results in a degradation process which is much more rapid than that which would otherwise occur. Pretreatment by composting can result in a product that can be easily stored until land treated. This is a particularly useful approach where a continuous stream of waste cannot be continuously land treated due to frozen or wet soil conditions.

The Beltesville method of composting uses forced aeration through windrows and has been used for composting oily wastes (Epstein and Taffel, 1979; Texaco Inc., 1979). In these studies, the oily waste is first mixed with a bulking agent, such as rice hulls or wood chips, to reduce the moisture content to 40-60%. Aeration of the mixed waste is maintained by drawing air through a perforated pipe located under the waste pile using an exhaust fan. The waste pile is covered with previously composted material which acts as an insulator and helps to maintain an elevated temperature. Air which has passed through the pile is filtered through another smaller pile of previously composted waste to reduce odors. Epstein and Taffel (1979) noted that composting of sewage sludge almost completely degraded the polycyclic aromatic hydrocarbons.

Activated sludge uses an aerobic microbial population that is acclimated to the particular waste stream to increase the rate of degradation. The acclimated population is recycled and kept in constant contact with incoming wastewater. Activated sludge has been extensively applied to industrial wastewaters for the degradation of organic wastewaters that have low heavy metal content. Tucker et al. (1975) demonstrated that PCBs can be degraded in the activated sludge process, but others have found heavily chlorinated molecules to be resistant to microbial degradation by this method. Use of microorganisms acclimated to these chlorinated waste constituents may improve efficiency of the activated sludge process for pre-treatment of wastes containing these types of resistant compounds.

As with activated sludge, aerated lagoons are used for the treatment of aqueous solutions with a low metals content. Aerobic lagooning is currently used by industry in temperate climates where sufficient land is available. This method of aerobic degradation is land intensive and slow compared to composting and activated sludge processes; however, it may be less expensive and it serves as a convenient method for storing wastes until weather or other limiting conditions are suitable for the waste to be land treated. A major drawback of aerated lagooning is that it presents a considerable risk of groundwater contamination. This risk has prompted regulatory requirements (discussed in Section 5.2.4) for lagoons.

5.2.4   Anaerobic Degradation

Anaerobic degradation involves microbes that degrade organics in the absence of oxygen. These microbes use metabolic pathways that differ from the pathways used by aerobic microbes and can, therefore, more effectively degrade some organics that are resistant to degradation in the aerobic soils of a land treatment unit. Two widely used methods for this type of degradation are anaerobic lagooning and anaerobic digestion.

Anaerobic and aerobic lagooning of wastes has been widely used for pretreatment and storage of wastes to be land treated. While the technique has been inexpensive, recent regulatory requirements for lining, monitoring and closing these facilities will increase the cost of lagooning hazardous waste. Other disadvantages associated with both types of lagooning include the following:

(1) wastes often require retention times of several months for effective treatment;

(2) due to the long retention times, large amounts of land may be required to handle all the waste; and

(3) there may be significant long-term liability associated with lagoons due to their potential for groundwater contamination.

Anaerobic digestion of waste uses enclosed tanks to anaerobically degrade waste under controlled conditions. Initially, the technique is capital intensive; however, there are several advantages compared to anaerobic lagooning, as follows:

(1) since the treatment process is completely enclosed, there would be few, if any, long-term liabilities;

(2) retention time for waste, although dependent on waste composition, may be less than 10 days (Kugelman and Jeris, 1981);

(3) short retention times mean less waste volume on hand at any time and consequently less land is required for treatment facilities; and

(4) useful by-products, such as methane and carbon dioxide, can be obtained from the process.

5.2.5   Soil Mixing

Several industries produce tarry wastes that may be too sticky or viscous to be easily applied to land. Examples of this physical state are coal tar sludge and adhesives waste. Mixing of these wastes with soil is difficult because the sticky wastes tend to ball-up or stick to the surface of discing implements. A treatment that eliminates most of these difficulties is the premixing of soil with the waste in a pug mill. Pug mills cut up the sticky mass as it combines with the soil, producing a soil-waste mixture that can be easily applied to land.

5.2.6   Size Reduction

Often bulky materials are contaminated with hazardous waste during production processes or accidental spills. Examples of contaminated bulk materials are pallets, lumber and other debris saturated or coated with hazardous materials. A common approach to making these wastes suitable for land treatment is to grind or pulverize the debris.

5.3   WASTE CHARACTERIZATION PROTOCOL

A waste characterization protocol serves an important function to prevent adverse health, safety, or environmental effects from land treatment of hazardous waste. It is required for the following reasons:

(1) to evaluate the feasibility of using land treatment for a particular waste;

(2) to define waste characteristics indicative of changes in composition;

(3) to evaluate results generated in pilot studies;

(4) to define management and design criteria;

(5) to determine application, rate, and capacity limiting constituents (These design parameters are further discussed in Chapter 7.);

(6) to determine if the treatment medium is effectively rendering the applied waste less nonhazardous; and

(7) to effectively monitor any environmental impact resulting
     from the HWLT unit.

To satisfy these requirements, the designer needs to perform an
acceptable characterization of the waste. Additionally, he needs to be
able to evaluate the results of the analyses to determine if the appropriate parameters have been addressed or if additional analyses are required.
This section provides the information needed to evaluate the waste characterization phase of the design process for HWLT.

Because of the complexity involved in both the characterization of
hazardous waste and the evaluation of the results, a set of guidelines or
analytical requirements is appropriate. The following step-by-step
approach to waste characterization will provide guidance to the facility
designer as he approaches the design of a new facility or the redesign of
an old one. The following sections are intended to reduce and simplify the
characterization and evaluation processes.

5.3.1   Preliminary Waste Evaluation

There are a tremendous number of industrial process wastes which contain a wide variety of complex chemical mixtures. Initial indicators of
the probable composition of a particular waste include the following:

(1) previous analytical data on waste constituents;

(2) feedstocks used in the particular industrial process; and

(3) products and by-products resulting from production processes.

By examining data presented on waste streams, the analytical requirements
for a particular waste may be sufficiently evaluated by the designer to
preclude any extensive, unwarranted analyses. One must realize, however,
that there may be toxic or recalcitrant constituents present in a given
hazardous waste that are either new or previously unnoted. Therefore, all
possible means need to be used to thoroughly characterize the constituents
found in waste samples.

5.3.2   Waste Analysis

The analytical chemistry associated with HWLT should include appropriate analyses of the waste in conjunction with preliminary soil studies,
compound degradation determinations, and monitoring needs (Chapters 4, 7,
and 9). Most of the following discussion refers primarily to a general
approach to be used for analyzing the waste itself. Physical, chemical and
biological waste analyses are discussed.

5.3.2.1 Sampling and Preparation

In sampling hazardous waste and other media relevant to HWLT, one must continually strive to ensure personal safety while correctly collecting representative samples that will provide an accurate assessment of the sample constituents. After obtaining some background information about the probable nature of the waste and the associated dangers, the analysis may then proceed using the appropriate safety measures, as outlined by de Vera et al. (1980). The person sampling a hazardous material must be aware that it may be corrosive, flammable, explosive, toxic or capable of releasing toxic fumes.

Since hazardous waste may be composed of a diverse mixture of organic and inorganic components present in a variety of waste matrices (i.e., liquids, sludges and solids), it is necessary to use specialized sampling equipment to ensure that the sample is representative of the waste in question. For instance, the Coliwasa sampler, which consists of a tube, shaft and rubber stopper, may be used for sampling layered liquids: after insertion of the tube into the liquid waste, the shaft is used to pull the stopper into place and retain the sample. Other examples of appropriate samplers that may be used for sampling various types of wastes are listed in Table 5.4. Additional information on sampling equipment, methods, and limitations can be found in EPA (1982a).

TABLE 5.4 SAMPLERS RECOMMENDED FOR VARIOUS TYPES OF WASTE*

Waste type	Waste Location or Container	Sampling Apparatus
Free flowing liquids and slurries	Drums, trucks, tanks Tanks, bins Pits, ponds, lagoons	Coliwasa Weighted Bottle Dipper
Dry solids or wastes	Drums, sacks, waste piles, trucks, tanks pits, ponds, lagoons	Thief, scoops, shovels
Sticky or moist solids and sludges	Drums, trucks, tanks, sacks, waste piles, pits, ponds, lagoons	Trier
Hard or packed wastes	Drums, sacks, trucks	Auger

* EPA (1982a).

It is very important that all sampling equipment be thoroughly cleaned and free of contamination both prior to use and between samples. Storage containers should be similarly free of contamination. Plastic or teflon may be used for samples to be analyzed for inorganic constituents. Glass, teflon or stainless steel may be used for samples intended for organic analysis. Caution should be observed that both the sampler and storage container materials are nonreactive with the waste. Ample room in the sample container must be left to allow for expansion of water if the sample is to be frozen in storage.

To ensure that the analytical methods employed in the waste characterization do not under or over-estimate either the potential impact or treatment effectiveness, representative samples must be obtained. A representative sample is proportionate with respect to all constituents in the bulk matrix. The probability of obtaining a representative sample is enhanced by compositing multiple samples. These composites can be homogenized prior to subsampling for subsequent analysis. Table 5.5 may be used to determine the number of samples to be taken when a waste is sampled from multiple containers. These numbers should be considered a minimum requirement. If large variability is encountered in the sample analysis, additional samples may be required. Similar precautions must be taken to ensure that the total waste substrate has been sampled. Table 5.6 suggests appropriate sampling points to be selected for sampling various waste containments. Descriptions of detailed statistical analyses for use in sampling can be found in EPA (1982a).

TABLE 5.5   MINIMUM NUMBER OF SAMPLES TO BE SELECTED FROM MULTIPLE CONTAINERS*[†]

Number of Containers	Number of Samples to be Composited	Number of Containers	Number of Samples to be Composited
1 to 3	all	1332 to 1728	12
4 to 64	4	1729 to 2197	13
65 to 125	5	2198 to 2744	14
126 to 216	6	2745 to 3375	15
217 to 343	7	3376 to 4096	16
344 to 512	8	4097 to 4913	17
513 to 729	9	4914 to 5832	18
730 to 1000	10	5833 to 6859	19
1001 to 1331	11	6860 or over	20

* ASTM D-270

[†] Numbering the containers and using a table of random numbers would give an unbiased method for determining which should be sampled.

Following sampling operations, all samples should be tightly sealed and stored at 4°C (except, in some cases, soils). Freezing may be required when organic constituents are expected to be lost through volatilization. This may be easily accomplished by packaging all samples in dry ice immediately after collection if other refrigeration methods are unavailable. Prior arrangements should be made with the receiving laboratory to ensure sample integrity until the time of analysis.

TABLE 5.6 SAMPLING POINTS RECOMMENDED FOR MOST WASTE CONTAINMENTS

Containment type	Sampling point
Drum, bung on one end	Withdraw sample from all depths through bung opening.
Drum, bung on side	Lay drum on side with bung up. Withdraw sample from all depths through bung opening.
Barrel, fiberdrum, buckets, sacks, bags	Withdraw samples through the top of barrels, fiberdrums, buckets, and similar containers. Withdraw samples through fill openings of bags and sacks. Withdraw samples through the center of the containers and different points diagonally opposite the point of entry.
Vacuum truck and similar containers	Withdraw sample through open hatch. Sample all other hatches.
Pond, pit, lagoons	Visually inspect the area. If there is evidence of differential settling of material as it enter the pond, this area needs to be estimated as a percentage of the pond and sampled separately. If the remaining area is free of differential settling, divide surface area into an imaginary surface, one sample at mid-depth or at center, and one sample at the bottom should be taken per grid. Repeat the sampling at each grid over the entire pond or site. A minimum of 5 grids should be sampled.
Waste pile	Withdraw samples through at least three different points near the top of pile and points diagonally opposite the point of entry.
Storage tank	Sample all depths from the top through the sampling hole.

5.3.2.2 Physical Analysis

The physical characteristic of hazardous waste that is most relevant to land treatment is density. Density determinations are required to convert the volumes of waste which will be treated into their corresponding masses. The mass measurements will then be used to determine loading rates and other application requirements (Section 7.5).

The density of a liquid waste may be determined by weighing a known volume of the waste. A water insoluble viscous waste may be weighed in a calibrated flask containing a known volume and mass of water. The water displaced is equivalent to the volume of waste material added. A similar technique may be used for the analysis of water soluble wastes by replacing water with a nonsolubilizing liquid for the volumetric displacement measurement. In this case, a correction must be made for the density of the solvent used.

5.3.2.3 Chemical Analysis

The chemical characterization of complex mixtures such as hazardous waste consists of chemically specific analytical procedures which need to be performed under a strict quality control program by well-trained personnel. Procedural blanks defining background contamination should be determined for all analytical techniques. Maximum background contamination should not exceed 5% of the detector response for any compound or element being analyzed. (For instance, if the concentration of a constituent results in 95% full-scale deflection on a recorder, the background level found in the analytical blank should not exceed 4.5% full-scale deflection.) The procedural blank should be taken through the complete analytical characterization, including all steps in collection and storage, extraction, evaporative concentration, fractionation, and other procedures that are applied to the sample. A general reference for the control of blanks in trace organic analysis is Giam and Wong (1972).

The accuracy and precision of all detailed analytical methodology need to be evaluated by no less than three reproducible, full procedural analyses of reference standards. All data on procedural recovery levels (accuracy) and reproducibility (precision) need to be reported as a mean plus or minus the standard deviation. Analytical data should be reliable to at least two significant figures or as defined by the measuring devices used. Other quality control and assurance guidelines may be found in EPA (1982a).

If a waste contains other hazardous constituents, not covered in either the following general chemical characterization protocol or EPA (1982a), it is the responsibility of the designer to determine an appropriate and reliable analytical technique for their determination. This may be accomplished through a literature search or consultation with regulatory officials or an analytical service. All techniques need to meet the quality control requirements of EPA (1982a).

The following sections are designed primarily to provide relevant information and explanations of chemical analytical techniques applicable to hazardous waste and land treatment. For the designer it is intended to provide some guidance and understanding of analytical chemistry and the role it plays in HWLT. Additionally, these sections should provide aid in understanding and evaluating the analytical data obtained as to its adequacy in addressing the relevant design questions.

In providing a general overview of the analytical chemistry, references are provided which describe specific methods which may be used for analyzing waste and other media relevant to HWLT. The U.S. EPA in Test Methods for Evaluating Solid Waste (EPA, 1982a) has developed detailed methodologies which may be acceptable by the EPA as methods for analyzing hazardous waste and used by the EPA in conducting regulatory investigations. However, many of the analytical methods described have not yet been tested on actual waste samples. Therefore, it is the responsibility of the individual laboratories to test all specific analytical methodologies under strict quality control and assurance programs to ensure that the analysis is providing an acceptable chacterization of the specific waste in question.

5.3.2.3.1 Inorganic Analysis. The inorganic chemical characterization of hazardous waste and other samples will cover a diverse range of elements and other inorganic parameters. Standard techniques that may be used for inorganic analyses are presented in the following sections and are discussed in more detail by the EPA (1982a).

5.3.2.3.1.1 Elements, present in the waste, may include a large variety of heavy metals and nutrients. Elemental analysis is necessary to determine the numerical values needed to calculate the constituents that limit the land treatment process (Section 7.5). The general method for determining metals, nutrients and salts consists of appropriate sample digestion followed by atomic absorption (AA) spectrophotometry or inductively coupled plasma are spectrometry (ICP). Specific techniques may be found in EPA (1982a), EPA (1979c) and Black (1965). Halides may be determined by various techniques (EPA, 1979c and 1982a; Stout and Johnson, 1965; Brewer, 1965). Boron may be determined by colorimetric techniques (EPA, 1979c; Wear, 1965). Total nitrogen may be analyzed by a Kjeldahl technique (EPA, 1979a; Bremner, 1965).

5.3.2.3.1.2 Electrical conductivity (EC) determination is necessary because it provides a numerical estimation of soluble salts which may limit the treatment process. EC may be directly determined on a highly aqueous waste. For organic wastes an aqueous extract may be analyzed, and with highly viscous or solid wastes, a water-saturated paste may be prepared and the aqueous filtrate analyzed for EC. Specific methods applicable to waste and other samples may be found in EPA (1979a) and Bower and Wilcox (1965).

5.3.2.3.1.3 pH and titratable acids and bases may be determined by various methods. The determination of hydrogen ion activity and the concentration of inorganic acids and bases is important to the treatment processes of HWLT due to possible adverse effects on soil structure, soil microbes, and constituent mobility. The measurements of pH may be made on aqueous waste suspensions and other samples according to procedures outlined in EPA (1979a) and Peech (1965). Titratable acids and bases may be determined on aqueous waste suspensions according to EPA (1979c). The use of indicators to determine equivalence points may result in erroneous values unless caution is taken to ensure that the titration is performed in a way which would be sensitive to all acid and base strengths (Skoog and West, 1979). This measurement may also determine titratable strong organic acids and bases.

5.3.2.3.1.4 Water may be a limiting constituent in the land treatment of certain wastes and so it is necessary to estimate the percent water (wet weight) of highly aqueous wastes. Determinations by such techniques as Karl Fischer titrations (Bassett et al., 1978) are unnecessary because water content is important only when it is present as an appreciable component of the waste. In an organic waste, water may be present as a discreet layer and thus may be easily quantitated. If water is present in an emulsion, salts may be added to disrupt the emulsion to determine the quantity of water. If water is the carrier solvent for a dissolved inorganic waste, water concentration may be estimated as 100%. For viscous inorganic wastes, in which water is present at a level comparable to the other inorganic constituents, heavy metals or sludge-like materials may be filtered from the aqueous phase following precipitation with a known amount of KOH.

5.3.2.3.2 Organic Analysis. The determination of organic constituents present in waste and other samples may be reported with respect to the following sample classes and constituents:

(1) Total organic matter (TOM);

    (a) Volatile organic compounds;

    (b) Extractable organic compounds (acids, bases, neutrals and water solubles); and

(2) Residual solids (RS).

The numerical concentrations should be reported on a wet weight basis for both gravimetric determination of each individual class and specific determination of each compound contained in each class.

5.3.2.3.2.1 <u>Total organic matter</u> derived from this determination will indicate the amount of organic matter available for microbial degradation in HWLT. The percent TOM (wet weight) may be used for estimating organic carbon necessary to calculate the C:N ratio. The percent TOM will be numerically equal to the sum of the gravimetric determinations of percentage of volatiles and extractables (acids, bases, neutrals, and water solubles).

5.3.2.3.2.1.1 <u>Volatile organic compounds</u> are sample constituents that are amenable to either purge and trap or head space determinations and generally have boiling points ranging from less than 0°C to about 200°C. This upper limit is not an exact cut-off point, but techniques that rely on evaporative-concentration steps may result in appreciable losses. Examples of typical organic compounds which may be found as volatile constituents in hazardous wastes are given in Table 5.7.

A gravimetric estimation of the concentration of these compounds should be reported as percent wet weight for calculating total organic matter (TOM). This may be accomplished by bubbling air through a vigorously stirred aqueous sample. The percentage loss in sample weight may be used to estimate percent volatiles. A highly viscous or solid waste may be suspended in a known weight of previously boiled water and similarly analyzed. If a 10 g sample is used (and suspended in perhaps 100 g of water), an accuracy to the nearest 0.1 g may be acceptable.

The two methods recommended for the specific determination of individual volatile sample constituents are head space analysis and purge-and-trap techniques (EPA, 1982a). In head space analysis, the sample is allowed to equilibrate at 90°C, and a sample of the head space gas is withdrawn with a gas-tight syringe (EPA, 1982a). The gaseous sample is then analyzed by gas-chromatography (GC) and/or GC-mass spectrometry (GC-MS). The major limitations to the method appear to be variability in detection limits, accuracy, and precision caused by the equilibrium requirement. For instance, detection limits may be reduced with both increasing boiling point and affinity of the compound for the sample matrix (EPA, 1982a).

The alternate technique using purge-and-trap methods appears to be the most reliable of the two. It requires more sophistication, but can be applied to a greater number of sample types and a larger range of compound volatility (EPA, 1982a). The major limitation is that only one analysis may be performed per sample preparation. Thus, if analysis by several GC detectors is required, several samples may need to be prepared.

A simplified example of the purge-and-trap technique follows. An aliquot of a liquid waste may be placed into an airtight chamber which is connected to a supply of inert gas and an adsorbent trap. The carrier gas is bubbled through the waste of room temperature and passes out of the chamber through an adsorbent specific for volatile organics. Following this purge step, the adsorbent trap may be flushed for a few minutes with clean carrier gas to remove any residual water and oxygen, attached to the injection port of a GC or a GC-MS, and heated to desorb the organics. As

TABLE 5.7   PURGABLE ORGANIC COMPOUNDS.*[†]

I.   Hydrocarbons

   A.   Alkanes (R_n)[#]--C_1-C_{10}

   B.   Alkenes (R=R')--C_1-C_{10}

   C.   Alkynes (R=R")--C_1-C_{10}

   D.   Aromatics (Ar)[#]--benzene, ethylbenzene, toluene, styrene

II.   Compounds containing simple functional groups

   A.   Organic halides (R-X, Ar-X)*--chloroform, 2-dichlorobenzene,
        trichlorofluoromethane, tetrachloroethylene, trichloroethylene,
        vinyl chloride, vinylindene chloride

   B.   Alcohols (R-OH; OH-R-R-OH)--methanol, benzyl alcohol, ethylene
        glycol, dichloropropanol

   C.   Phenols (Ar-OH)--phenol, cresols, o-chlorophenol

   D.   Ethers (R-O-R', Ar-O-R', C_4H_8O)--ethyl ether, anisole,
        ethylene oxide, dioxan, tetrahydrofuran, vinyl ether, allyl
        ether, bis(2-chloroethyl)ether

   E.   Sulfur-containing compounds

       1.   Mercaptans (R-SH)--methylmercaptan

       2.   Sulfides (R-S-R', C_4H_4S)--thiophene, dimethyl sulfide

       3.   Disulfides (R-SS-R')--diethyldisulfide, dipentyldash
            disulfide

       4.   Sulfoxides (R-SO-R')--Dimethyl sulfoxide

       5. Alkyl hydrogen sulfates (R-O-SO_3H)--methyl sulfate

   F.   Amines

       1.   Alkyl (R-NH_2, RR'-NH, RR'R"-N)--methylamine, triethylamine,
            benzylamine, ethylenediamine, N-nitrosoamine

       2.   Aromatic (Ar-NH_2, etc.)--aniline, acetanilide, benzidine

       3.   Heterocyclic (C_5H_5N)--pyridine, picolines

                          --continued--

TABLE 5.7 (Continued)

III. Compounds containing unsaturated functional groups

    A. Aldehydes (R-CHO, AR-CHO)--formaldehyde, phenylacetaldehyde, benzaldehyde, acrolein, furfural, chloroacetaldehyde, paraldehyde

    B. Ketones (R-CO-R')--acetone, methyl ethyl ketone, 2-hexanone

    C. Carboxylic acids (R-COOH)--C_1-C_5 carboxylic acids

    D. Esters (R-COO-R', AR-COO-R)--methylacetate, ethyl formate, phenylacetate

    E. Amides (R-CO-NHR')--acrylamide

    F. Nitriles (R-CN, Ar-CN)--acetonitrile, acrylonitrile, benzonitrile

* Hendrickson et al. (1970); Morrison and Boyd (1975).

[†] The following compound classes are not expected due to their instabilities either in air and/or water:
    acid halides and anhydrides
    imines
    oximes

[#] R= alkyl groups, eg., CH_3, CH_3CH_2-, etc.
Ar= aromatic groups, eg., C_6H_5-
X= halogen, eg., Cl, Br, etc.

the carrier gas passes through the heated trap, the volatiles are transferred onto the cooled head of the analytical GC column. Following heat desorption, the GC is temperature-programmed to facilitate resolution of all volatile compounds collected from the sample.

A variety of adsorbents may be used in this analysis (EPA, 1982a; Namiesnik et al., 1981; Russell, 1975), but Tenax-GC (registered trademark, Enka N.V., the Netherlands) appears to be the most widely used (Bellar and Lichtenberg, 1979; Dowty et al., 1979). It is a hydrophobic porous polymer which has a high affinity for organic compounds. Because of its high thermal stability (maximum 375°C), it can be easily cleaned before use and regenerated after use by heating and flushing with an inert gas. However, there are some problems with Tenax-GC due to its instability under certain conditions (Vick et al., 1977). Other general information concerning Tenax-GC may be found in "Applied Science Laboratories Technical Bulletin No. 24."

Tenax-GC has been shown to be an effective adsorbent for collection and analysis of volatile hazardous hydrocarbons, halogenated hydrocarbons, aldehydes, ketones, sulfur compounds, ethers, esters and nitrogen compounds (Pellizzari et al., 1976). Technical descriptions of usable techniques may be found in Pellizzari (1982), Reunanen and Kroneld (1982), Pellizzarri and Little (1980), EPA (1982a and 1979b), Pellizzari et al. (1978), Bellar and Lichtenberg (1979), and Dowty et al. (1979).

These methods may be used for a variety of hazardous wastes. Soils may be analyzed by the procedure for solid wastes. Air samples for monitoring activities may be taken directly by pulling a known volume of air through a similar adsorbent trap and analyzing it following heat desorption (Brown and Purnell, 1979; Pellizzari et al., 1976).

To accurately analyze the different classes of volatile organics present in samples, different GC detectors may be required. A flame ionization detector (FID) may be used for hydrocarbons, a flame photometric detector (FPD) for sulfur and/or phosphorus-containing compounds, an electron capture detector (ECD) for halogenated hydrocarbons and phthalates, and a nitrogen-phosphorus detector (NPD) for nitrogen and/or phosphorus-containing compounds. There are several other GC detectors on the market available for analyzing different classes of organics. The final confirmation, or even the complete analysis, of volatiles present in samples may be determined by GC-MS computer techniques. Some general references dealing with organic mass spectrometry are Safe and Hutzinger (1973), Middleditch et al. (1981) and McLafferty (1973).

5.3.2.3.2.1.2 Extractable organic compounds are organic constituents that are amenable to evaporative-concentration techniques and may be analyzed by methods based on the classical method of isolation according to functional group acid-base reactions. Other methods have been developed for the chromatographic fractionation of complex organic mixtures into individual compound classes (Miller, 1982; Boduszynski et al. 1982a and b; Later et al. 1981; Crowley et al., 1980; Brocco et al., 1973), but the liquid-liquid acid/base extraction method appears to be the easiest and least instrumentally intensive. This technique has been used in the analysis of a variety of complex organic mixtures (Colgrove and Svec, 1981), including fossil fuels (Buchanan, 1982; Matsushita, 1979; Novotny et al., 1981 and 1982) and environmental samples (Adams et al., 1982; Stuermer et al., 1982; Hoffman and Wynder, 1977; Grabow et al., 1981; Lundi et al., 1977). This method is also the basic technique recommended by the U.S. EPA (EPA, 1982a; Lin et al., 1979). Fractions derived from this analysis may be used in biological assays and other pilot studies (Grabow et al., 1981).

The liquid-liquid acid/base extraction method is based on the acidity constants (pK_as) of organic compounds. Compounds characterized by low pK_as are acidic; compounds with high pK_as are basic. If a complex mixture is equilibrated with an aqueous inorganic acid at low pH (<2), the organic bases should protonate to become water soluble positively-charged cations, while the organic acids remain unaffected and water insoluble (and thus extractable by an organic solvent). The neutral organics, which are not affected by either aqueous acids or bases, will remain in the organic solvent phase at all times. Similarly, if an aqueous inorganic base at high pH (>12) is added to a complex organic mixture, the organic acids should deprotonate to become water soluble negatively-charged anions, while the organic bases remain unaffected and water insoluble. Thus by selectively adjusting the pH of the aqueous phase, a complex mixture may be separated into its acidic, basic and neutral organic constituents. Table 5.8 lists some common organic chemicals and their pK_as.

TABLE 5.8   SCALE OF ACIDITIES*

Conjugate Acid	pK_a	Conjugate Base
$R-NH_3+$		$R-NH_2$
$RR'-NH_2+$	10	$RR'-NH$
$RR'R"-NH+$		$RR'R"-N$
$Ar-OH$	10	$Ar-O^-$
HCN	9.1	CN^-
C_5H_5N-H+	5.2	C_5H_5N
$Ar-NH_3+$	4.6	$Ar-NH_2$
RCOOH	4.5	$RCOO^-$
HCOOH +	3.7	$HCOO^-$
Ar_2-NH_2+	1.0	Ar_2-NH
2,4,6-Trinitrophenol	0.4	$(NO_2)_3-Ar-O^-$

* Hendrickson et al. (1970). Note: the most acidic compound is the conjugate acid with the lowest pK_a (i.e., 2,4,6-trinitro-phenol). Conversely, the most basic compound is the conjugate base with the highest pK_a (i.e., alkyl amines). Thus, at neutral pH, compounds with pK_as $\geq$ 9 9 should predominantly exist as their conjugate acids, and compounds with pK_as $\leq$ 5 should predominantly exist as their conjugate bases.

Figure 5.3 outlines the steps which may be taken in this initial class separation scheme. Table 5.9 lists typical organic compounds that may be present in hazardous waste and other samples which are amenable to this type of separation. Air samples collected on Florisil (registered trademark, Floridin Co.), glass fiber filters, or polyurethane foam may be first extracted with appropriate solvents and then the extract may be similarly analyzed by the above procedures (EPA, 1980b; Adams et al., 1982; Cautreels and van Cauwenbergh, 1976). Either diethylether or dichloromethane may be used as the organic solvent in the extraction procedures. Dichloromethane has been recommended (EPA, 1982a) and has the advantage that it is denser than water. Thus, it can be removed from the separatory funnel in the extraction procedure without having to remove the aqueous phase. However, it may be prone to bumping in evaporative concentration procedures (Adams, 1982). Ether, however, is more water soluble, and extra time is required in the extraction procedure to allow the phases to completely separate. Either solvent must be dried with an hydrous Na_2SO_4 prior to evaporative concentration. For either solvent, a few grains of Na_2SO_4 in the evaporation-concentration flask should facilitate boiling and reduce bumping (Adams et al., 1982). The EPA (1982a) has recommended the use of Kuderna-Danish evaporative concentrators equipped with three-ball Snyder columns for concentrating solvents. For the higher molecular weight compounds, this method should provide an easy, efficient and reproducible method for concentrating solvents. However, some researchers (Adams et al., 1982) have found that for microgram quantities of some lower molecular weight extractables (i.e., 2- and 3-ringed aza-aromatics), optimum recoveries in the concentration step were achieved by using a vacuum rotary evaporator at 30°C; the solvent receiving flask was immersed in an ice bath, and the condenser was insulated with glass wool and aluminum foil. In any case, samples for specific compound determination should not be evaporated to dryness as this may cause significant losses of even high molecular weight compounds such as benzo(a)pyrene (Bowers et al., 1981).

For each of the following classes isolated by this method, a separate aliquot of the sample extract may be analyzed gravimetrically for use in determining total organic matter. In this case, the solvent may be evaporated to dryness at room temperature. To minimize losses, the vaporation should be allowed to occur naturally without externally applied methods to increase solvent vaporization (e.g., N_2 blow-down, heat, etc.) as in Bowers et al. (1981).

The following sections describe specific methods which may be used in the analyses of the various classes obtained from the acid-base fractionation. Some general references which may be useful are McNair and Bonelli (1968), Johnson and Stevenson (1978), Packer (1975), Holstein and Severin (1981), Hertz et al. (1980), and Bartle et al. (1979).

Organic Acids. This class of compounds may include a variety of carboxylic acids, guaiacols, and phenols (Claeys, 1979). They frequently are determined following derivitization (Francis et al., 1978; Shackelford and Webb, 1979; EPA, 1982a; Cautreels et al., 1977). With diazomethane, the relatively non-volatile carboxylic acids are converted into esters which

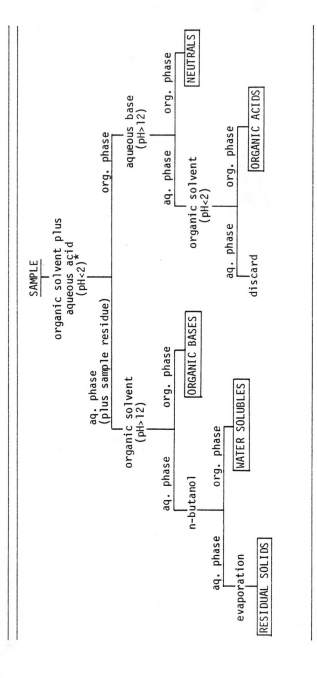

*Initial acidic extraction may lessen severity of emulsions (Mousa and Whitlock, 1979).

Figure 5.3. Typical acid-base extraction scheme for isolating organic chemical classes.

TABLE 5.9   TYPICAL HAZARDOUS ORGANIC CONSTITUENTS AMENABLE TO ACID-BASE
EXTRACTION TECHNIQUES

Extractable Neutral Organic Compounds

1,2-Dichlorobenzene	Benzo(g,h,i)perylene
1,3-Dichlorobenzene	4-Chlorophenyl phenyl ether
1,4-Dichlorobenzene	bis(2-Chloroethyl)ether
Hexachloroethane	Hexachlorocyclopentadiene
Hexachloropentadiene	bis(2-Chloroisopropyl)ether
Hexachlorobenzene	
1,2,4-Trichlorobenzene	Pesticides/PCB's
bis(2-Chloroethoxy)methane	
Naphthalene	α-Endosulfan
2-Chloronaphthalene	β-Endodsulfan
Isophorone	Endosulfan sulfate
Nitrobenzene	α-BHC
2,4-Dinitrotoluene	β-BHC
2,6-Dinitrotoluene	δ-BHC
4-Bromophenyl phenyl ether	γ-BHC
bis (2-Ethylhexyl)phthalate	Aldrin
Di-n-octyl phthalate	Dieldrin
Dimethyl phthalate	4,4'-DDE
Diethyl phthalate	4,4'DDD
Di-n-butyl phthalate	4,4'DDT
Acenaphthylene	Endrin
Acenaphthene	Endrin aldehyde
Butyl benzyl phthalate	Heptachlor
Fluorene	Heptachlor epoxide
Fluoranthene	Chlordane
Chrysene	Toxaphene
Pyrene	Aroclor 1016
Phenanthrene	Aroclor 1221
Anthracene	Aroclor 1232
Benzo(a)anthracene	Aroclor 1242
Benzo(b)fluoranthene	Aroclor 1248
Benzo(k)fluoranthene	Aroclor 1254
Benzo(a)pyrene	Aroclor 1260
Indeno(1,2,3-c,d)pyrene	2,3,7,8-Tetrachlorodibenzo-p-
Dibenzo(a,h)anthracene	dioxin (TCDD)

Extractable Basic Organic Compounds

3,3'-Dichlorobenzidine	Quinoline
Benzidine	Isoquinoline
1,2-Diphenylhydrazine	Acridine
N-Nitrosodiphenylamine	Phenanthridine
N-Nitrosodimethylamine	Benz[c]acridine
N-Nitrosodi-n-propylamine	

--continued--

TABLE 5.9 (continued)

Extractable Acidic Organic Compounds

Phenol	Abietic acid
2-Nitrophenol	Dehydroabietic acid
4-Nitrophenol	Isopimaric acid
2,4-Dinitrophenol	Pimaric acid
4,6-Dinitro-o-cresol	Oleic acid
Pentachlorophenol	Linoleic acid
p-Chloro-m-cresol	9,10-Epoxystearic acid
2-Chlorophenol	9,10-Dichlorostearic acid
2,4-Dichlorophenol	Monochlorodehydroabietic acid
2,4,6-Trichlorophenol	3,4,5-Trichloroguaiacol
2,4-Dimethylphenol	Tetrachloroguaiacol

may be determined by gas chromatography. Diazomethane similarly converts phenols into their corresponding anisoles (ethers). Pentaflourobenzylbromide converts phenols into their pentafluorobenzyl (PFB) derivatives.

Whereas carboxylic acids require derivitization prior to GC analysis, phenols may be determined directly by GC (EPA, 1982a; Shackelford and Webb, 1979; Mousa and Whitlock, 1979). The direct determination of phenols appears to be preferable because of problems encountered with both diazomethane and pentafluorobenzylbromide derivitization techniques (Shackelford and Webb, 1979). Guaiacols may be determined as in Knuutinen (1982).

These compounds may be characterized by GC with either capillary or packed columns. For packed-column GC, the polarity of these compounds requires the use of specially deactivated supports and liquid phases. SP-1240A (manufactured by Supelco, Inc., Supelco Park, Bellefonte, PA 16823) has been recommended for use (EPA, 1982a; Shackelford and Webb, 1979). Detection may be accomplished by either flame ionization or electron capture, depending on the compounds being determined. GC-MS may be used for further identification and/or confirmation.

Organic Bases. This fraction may contain a variety of nitrogen containing compounds including alkyl, aromatic, and aza-heterocyclic amines. These compounds may be directly characterized by GC with either FID or nitrogen-specific detection. As with the organic acids, either capillary or packed-column gas chromatography with specially deactivated packing materials may be used. For organic bases, Supelco, Inc. also manufactures a packing material, SP-2250 DB, which provides good packed-column resolution with a minimum of peak tailing. The analysis of this class of compounds should be performed soon after isolation because they tend to decompose and polymerize with time (Tomkins and Ho, 1982; Worstell and Daniel, 1981; Worstell et al. 1981). Additional GC-MS confirmation and identification may be performed.

Neutrals. This fraction may be composed of a variety of organic compounds including aliphatic and aromatic hydrocarbons, oxygenated and chlorinated hydrocarbons. This class may require further fractionation depending on whether the sample is to be analyzed for either hydrocarbons and more polar compounds by flame ionization, flame photometric, or nitrogen-phosphorus detection GC, or for chlorinated hydrocarbons and phthalic acid esters by electron-capture detection GC.

For FID, FPD or NPD-GC analysis, an aliquot of the neutral fraction may be separated into aliphatics, aromatics, and other semi-polar compounds and polar compounds by column chromatography. Lin et al. (1979) used 5% deactivated silica gel to separate neutral compounds isolated from drinking and waste treatment water: hexane eluted aliphatics; hexane/benzene eluted aromatics; dichloromethane eluted phthalic and fatty acid esters; methanol eluted aldehydes, alcohols, and hetones. Anders et al. (1975), using washed alumina, eluted hydrocarbons with pentane, moderately polar compounds with benzene, and more polar compounds with methanol. The polar fraction was then further characterized by chromatography on silica gel using increasing ratios of ethyl ether in pentane. Other researchers have used similar chromatographic methods for separating this class of compounds into its constituents (Giam et al., 1976; Gritz and Shaw, 1977). A good general review of methods applicable for this type of separation is (Altgelt and Gouw, 1979).

Since esters and other hydrolyzable compounds may be present in the aromatic and later fractions, the sample fractions may be analyzed prior to and following alkaline hydrolysis. (Hydrolyzable compounds may not withstand the original acid-base extraction and perhaps may be determined by other procedures). Alkaline hydrolysis may easily be accomplished by placing a small sample aliquot into a tightly capped vial containing 2% methanolic KOH and heating on a steam bath. After cooling, water is added to solubilize the resulting carboxylic acids and alcohols, and the organic phase is brought to original volume with solvent. The organic phase is then reanalyzed. The hydrolyzable compounds are thus confirmed through their disappearance, and interference in the analysis of the aromatics is removed.

For ECD-sensitive compounds, it may be possible to reduce analytical requirements if the previously described alumina/silica chromatographic separations can be co-adapted for use with halogenated hydrocarbons and phthalates (Holden and Marsden, 1969; Snyder and Reinert, 1971). Additionally, with appropriate technology, it may be possible to simultaneously detect both FID- and ECD-sensitive compounds in the GC analysis (Sodergren, 1978).

However, a separate aliquot of the neutral fraction may be analyzed for halogenated hydrocarbons and phthalates. (Some of these compounds may not withstand the original acid-base extraction and perhaps may be determined by other methods.) This procedure typically requires the use of Florisil to separate different polarities of halogenated compounds and phthalates (EPA, 1980b, 1979b and 1982a). If needed, clean mercury metal may be shaken with the various fractions to eliminate sulfur interference.

For compound confirmation these samples also may be analyzed by ECD-GC prior to and following alkaline hydrolysis. In this case, alkaline hydrolysis saponifies the phthalic acid esters and dehydrochlorinates many of the chlorinated organics. Table 5.10 lists compounds which can be confirmed by alkaline hydrolysis. The experimental conditions must be carefully controlled for obtaining reproducible results. Additional GC-MS confirmation, using selective ion monitoring (SIM) if necessary, may be performed.

Water Solubles. This class of compounds may consist of constituents which were not solvent extractable in any of the previously isolated organic fractions. The use of n-butanol as extracting solvent may serve to isolate this class of compounds (Stubley et al., 1979). Since further characterization of this class may be difficult, results of pilot studies may be used to determine further analytical requirements.

5.3.2.3.2.2 Residual solids may be determined by evaporating the water (110°C) from the original aqueous fraction isolated in the acid-base

TABLE 5.10 REACTONS OF VARIOUS COMPOUNDS TO ALKALINE HYDROLYSIS*

Compound	Chromatographic Appearance After Hydrolysis
Esters (phthalic and fatty acid)	Disappear
PCBs	Unchanged
Heptachlor	Unchanged (under mild conditions)
Aldrin	Unchanged
Lindane, other BHC isomers	Disappear
Heptachlor epoxide	Unchanged (under mild conditions)
Dieldrin	Unchanged
Endrin	Unchanged
DDE	Unchanged
DDT	Disappears as DDE appears
DDD	Disappears as DDE appears
Chlordane	Unchanged
HCB	Unchanged
Mirex	Unchanged (under mild conditions)
Endosulfan I and II	Disappear
Dicofol	Disappears
Toxaphene	Changed (other peaks appear)
Alkylhalides	Disappear[†]
Nitriles	Disappear[†]
Amides	Disappear[†]

* EPA (1980c).

[†] Predicted according to reactions typical of these compound types.

extraction procedure (Fig. 5.3). Residual solids (RS) may consist of both inorganics and relatively non-degradable forms of carbon such as coke, charcoal, and graphite. This value may be used in waste loading calculations and for determining the rate of waste solids buildup. A buildup of solids may increase the depth of the treatment zone.

5.3.2.4 Biological Analysis

A primary concern when disposing any waste material is the potential for adverse health effects. Toxic effects resulting from improper waste disposal either may be acute, becoming evident within a short period of time, or they may be chronic, becoming evident only after several months or years. Before a hazardous waste is disposed in an HWLT unit, biological analyses should be performed to determine the potential for adverse health effects. The complex interactions of the components of a hazardous waste make it impossible to predict the acute or chronic toxicity of any waste by chemical analysis alone. A solution to this problem is to use a series of biological test systems that can efficiently predict the reduction of the acute and chronic toxic characteristics of the waste. Biological systems can be used to determine the toxicity and treatability of the waste and to monitor the environmental impact of land treating the waste.

5.3.2.4.1 Acute Toxicity. The acute toxicity of a hazardous waste should be evaluated with respect to plants and microbes endemic to the land treatment site. This evaluation will indicate the effects on the immediate environment of the land treatment unit. Obviously, a waste which is toxic to microbes will not be degraded unless it is applied at a rate that will diminish these acute toxic effects. The acute toxicity of a waste with respect to soil bacteria and plants can be evaluated in treatability studies as described in Chapter 7. Specific methods for measuring acute toxicity are presented in Section 7.2.4.1.

5.3.2.4.2 Genetic toxicity. Hazardous wastes should be managed so that the public is protected from the effects of genotoxic agents in a waste. Genotoxic compounds in a hazardous waste should be monitored to minimize the accidental exposure of workers or the general public to mutagenic, carcinogenic, or teratogenic agents, and to prevent transmission of related genetic defects to future generations. Genetic toxicity may be determined using a series of biological systems which predict the potential of waste constituents to cause gene mutations and other types of genetic damage. A list of some of the prospective test systems and the genetic events which they can detect is given in Table 5.11. These are test systems for which a standardized protocol has been developed, and the genetic events detected are clearly understood.

The test systems used to detect gene mutations should be capable of detecting frameshift mutations, base-pair substitutions, and deletions. The systems that are used to detect other types of genetic damage should

TABLE 5.11 BIOLOGICAL SYSTEMS WHICH MAY BE USED TO DETECT GENETIC TOXICITY OF A HAZARDOUS WASTE

| Organism | Genetic Event Detected | | | References |
	Gene Mutation	Other Types of Genetic Damage	Metabolic Activation	
PROKARYOTES				
Bacillus subtilis	Forward, reverse	DNA repair	Mammalian	Felkner et al., 1979; Kada et al., 1974; Tanooka, 1977; Tanooka et al., 1978.
Escherichia coli	Forward, reverse	DNA repair	Mammalian plant	Green et al., 1976; Mohn et al., 1974; Slater et al., 1971; Speck et al., 1978; Scott et al., 1978.
Salmonella typhimurium	Forward, reverse	DNA repair	Mammalian plant	Ames et al., 1975; Plewa and Gentile, 1976; Skopek et al., 1978.
Streptomyces coelicolor	Forward	DNA repair	Not Developed	Carere et al., 1975.
EUKARYOTES				
Aspergillus nidulans	Forward, reverse	DNA repair, chromosome aberrations	Mammalian plant	Bignami et al., 1974; Roper, 1971; Scott et al., 1978; Scott et al., 1980.
Neurospora crassa	Forward	Not developed	Mammalian	DeSerres and Mailing, 1971; Ong, 1978; Tomlinson, 1980.

-- continued --

TABLE 5.11 (continued)

	Genetic Event Detected			
Organism	Gene Mutation	Other Types of Genetic Damage	Metabolic Activation	References
Saccharomyces cervisiae	Forward	Mitotic gene conversion	Mammalian	Brusick, 1972; Loprieno et al., 1974; Mortimer and Manney, 1971; Parry, 1977.
Schizosaccharomyces pombe	Forward	Mitotic gene conversion	Mammalian	Brusick, 1972; Loprieno et al., 1974; Mortimer and Manney, 1971; Parry, 1977.
PLANTS Tradescantia sp.	Forward	Chromosome aberrations	Plant	Nauman et al., 1976; Underbrink et al., 1973.
Arabidopsis thaliana	Chlorophyll mutation	Chromosome aberrations	Plant	Redei, 1975.
Hordeum vulgare	Chlorophyll mutation	Chromosome aberrations	Plant	Kumar and Chauham, 1979; Nicoloff et al., 1979.
Pisum sativua	Chlorophyll mutation	Chromosome aberrations	Plant	Ehrenburg, 1971.
Triticum sp.	Morphological mutation	Chromosome aberrations	Plant	Ehrenberg, 1971.
Glycine max	Chlorophyll mutation	Chromosome aberrations	Plant	Vig, 1975.

-- continued --

TABLE 5.11 (continued)

	Genetic Event Detected				
Organism	Gene Mutation	Other Types of Genetic Damage	Metabolic Activation	References	
Vicia faba	Morphological mutation	Chromosome aberrations	Plant	Kihlman, 1977.	
Allium cepa	Morphological mutation	Chromosome aberrations	Plant	Marimuthu, et al., 1970.	
INSECTS					
Drosophila melanogaster	Recessive lethels	Non-disjunction, deletions	Insect	Wurgler and Vogel, 1977.	
Habrobracon sp.	None developed	Dominant lethels	Insect	Von Borstel and Smith, 1977.	
MAMMALIAN CELLS IN CULTURE					
Chinese hamster ovaries	Forward, reverse	Chromosome aberrations	Mammalian	Neill et al., 1977; Beek et al., 1980.	
V79 Chinese hamster cells	Forward, reverse	Chromosome aberrations	Mammalian	Artlett, 1977; Soderberg et al., 1979.	
Chinese hamster lung cells	Forward	Chromosome aberrations	Mammalian	Dean and Senner, 1977.	
Human fibroblasts	Forward	DNA repair	Mammalian	Jacobs and DeMars, 1977.	
Human lymphoblasts	Forward	DNA repair	Mammalian	Thilly et al., 1976.	

-- continued --

TABLE 5.11 (continued)

| Organism | Genetic Event Detected | | | References |
	Gene Mutation	Other Types of Genetic Damage	Metabolic Activation	
L5178Y mouse lymphoma cells	Forward	Chromosome aberrations	Mammalian	Clive and Spector, 1975; Clive et al., 1972; Clive, 1973.
P388 mouse lymphoma cells	Forward	Chromosome aberrations	Mammalian	Anderson, 1975.
Human peripheral blood lymphocytes	Forward	Chromosome aberrations	Mammalian	Evans and O'Riordan, 1975.
Various organisms	None developed	Sister chromatid exchange	Mammalian	Perry and Evans, 1975; Stretka and Wolff, 1976.

exhibit a response to compounds that inhibit DNA repair and to those that cause various types of chromosome damage. A minimum of two systems should be selected that will respond to the types of genetic damage described above and which can incorporate metabolic activation into the testing protocol. All systems should include provisions for solvent control and positive controls to demonstrate the sensitivity of the test systems and the functioning of the metabolic activation system, and to act as an internal control for the biological system. Samples should be tested at a minimum of four equally spaced exposure levels, all of which will yield between 10 and 100% survival. Cell survival should be estimated by plating exposed cells on a supplemented minimal medium. The data from waste analysis should be in the form of mutation induction per survivor or per surviving fraction if the waste is overly toxic.

Typical results from mutagenicity testing using the Salmonella/microsome assay (Ames et al., 1975) on the subfractions of a wood-preserving bottom sediment and the liquid stream from the acetonitrile purification column are presented in Figs. 5.4 and 5.5 (Donnelly et al., 1982). These results demonstrate that constituents of these wastes have the ability to induce point mutations in bacteria; such constituents may be mutagenic, carcinogenic, or teratogenic (Kada et al., 1974).

The presence of genotoxic compounds in a waste indicates the need for monitoring land treatment units using biological analysis when genotoxic compounds are present in a waste stream. Bioassays can also be performed at various stages of the waste-site interaction studies to determine the reduction of genotoxic effects along with the other treatability data collected. The data obtained from biological analyses of waste-soil mixtures can be compared with the toxicity of the waste alone to determine the degree of treatment (see Section 7.2.4).

5.3.3   Summary of Waste Characterization Evaluation

To adequately address the needs of both the facility designer and the regulatory agency governing the facility, a standardized waste evaluation data processing procedure should be devised. For instance, Table 5.12 gives an example summary of the type of information (and appropriate section references to this manual) needed to fulfill the initial analytical requirements for an HWLT design. Numerous government agencies can provide assistance in the way of guidance documents which list analytical requirements. Ideally, all facility designers and regulatory officials would have access to a computerized data bank containing a compilation of data describing standard waste streams and analytical results derived from incoming permit applications. Thus, as analytical needs are evaluated and fulfilled, future facility designers and regulatory agencies would have a continuous up-date on toxic or recalcitrant compounds determined in the wastes and analytical procedures acceptable for their determination. This should reduce the necessity for extensive analytical requirements in the future, as monitoring could be limited to those compounds either found to restrict rate, application or capacity of the HWLT unit, or to adversely affect environmental quality.

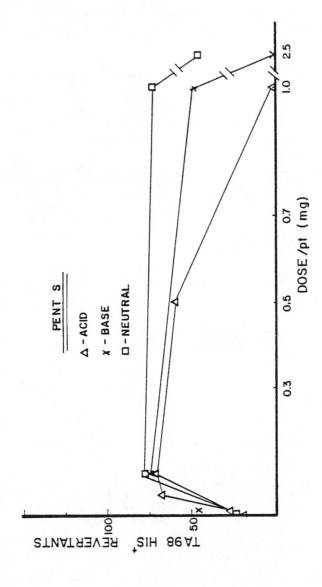

Figure 5.4. Mutagenic activity of acid, base, and neutral fraction of wood-preserving bottom sediment as measured with S. typhimurium TA 98 with metabolic activation (Donnelly et al., 1982).

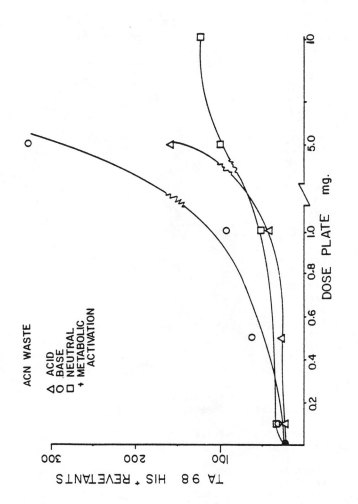

Figure 5.5. Mutagenic activity of liquid stream from the acetonitrile purification column as measured with S. typhimurium TA 98 with metabolic activation (Donnelly et al., 1982).

TABLE 5.12   HAZARDOUS WASTE EVALUATION

```
  I.  Applicant's Name
 II.  Waste SIC Code or Description of Source Process
III.  Analytical Laboratory
      A.  Person Responsible for Analyses
      B.  Quality Control Certification
 IV.  Analytical Results
      A.  Method of Collection and Storage (5.3.2.1)
      B.  Density and Method of Measurement (5.3.2.2)
      C.  Chemical Analyses
          1.  Brief Description of Analytical Methods
          2.  Recoveries & Reproducibilities of Methods
          3.  Inorganics (6.1 and 5.3.2.3.1)
              a.  Elements (5.3.2.3.1.1)
                  (1)  Metals (6.1.6)
                  (2)  Nutrients (6.1.2)
                       (a)  Nitrogen (N)
                       (b)  Phosphorus (P)
                       (c)  Sulphur (S)
                       (d)  Boron (B)
                  (3)  Salts (6.1.4)
                       (a)  Calcium (Ca)
                       (b)  Magnesium (Mg)
                       (c)  Potassium (K)
                       (d)  Sodium (Na)
                       (e)  Sulfate ($SO_4^{-2}$)
                       (f)  Bicarbonate ($CO_3^{-2}$)
                  (4)  Halides (6.1.5)
                       (a)  Flouride ($F^-$)
                       (b)  Chloride ($Cl^-$)
                       (c)  Bromide ($Br^-$)
                       (d)  Iodide ($I^-$)
              b.  EC (5.3.2.3.1.2)
              c.  pH and Titratable Acids & Bases (5.3.2.3.1.3)
              d.  Water (6.1.1 and 5.3.2.3.1.4)
          4.  Organics (6.2, Table 6.53 and 5.3.2.3.2)
              a.  Total Organic Matter (TOM) (5.3.2.3.2.1)
              b.  Volatiles (5.3.2.3.2.1.1)
              c.  Extractables (5.3.2.3.2.1.2)
                  (1)  Organic Acids
                  (2)  Organic Bases
                  (3)  Neutrals
                  (4)  Water solubles
              d.  Residual Solids (RS) (5.3.2.3.2.2)
      D.  Biological Analysis
          1.  Acute Toxicity (5.3.2.4.1 and 7.2.4)
          2.  Genetic Toxicity (5.3.2.4.2)
```

A critical question within the broad scope of waste stream character-istics is whether all wastes are land treatable, given the proper design and operation, or if there are any waste streams which should be unequivoc-ably prohibited from land treatment. In view of this, one must be cogni-zant of the acceptable treatment processes for HWLT units: degradation, transformation and immobilization (EPA, 1982b).

Few compounds remain unchanged when incorporated into the active sur-face horizons of soils. As previously established (Section 4.1.3), the primary pathway of organic waste degradation in soils is biological, sup-plemented by chemical alteration and photodecomposition. In contrast, many inorganic waste constituents are adsorbed, complexed or precipitated to innocuous forms within reasonable limits. Any given waste can, however, be unacceptable for land treatment if proposed soils or sites lack the ability to render the constituents less hazardous. For example, a highly volatile waste may not be adequately treated in a coarse textured soil, or the application of an acidic waste to an already acidic soil may present a high mobility hazard for toxic constituents. In addition, some compounds, such as hexachlorobenzene, may not be altered within a reasonable time by soil processes or may be mobile and subject to volatilization or leaching.

Dilution is not an acceptable primary treatment process for land treatment. Dilution may in some cases serve as a secondary mechanism associated with degradation, transformation or immobilization. Volume reduction (i.e., evaporation of water) is also not acceptable as the pri-mary treatment process in a land treatment system. Although evaporation may be an important mechanism, application of hazardous waste to land purely for dewatering should, in general, be restricted to lined surface impoundments which are designed with ground and surface water protection in mind. In an acceptable HWLT design, evaporative losses should, therefore, be of secondary importance and only one among several mechanisms operating.

In any case, one must be hesitant to set arbitrary prohibitions on particular waste streams until their unacceptability has been adequately demonstrated. Where dilution is functioning, supportive to treatment, the question of what constitutes adequate dilution also requires restraint to avoid setting arbitrary standards.

Due to the myriad of components and the complexities associated with possible interactions, chemical analytical data may not adequately predict acceptability of land treatment for a waste liquid, slurry or sludge. Acceptability is perhaps best derived empirically. Thus, the final deci-sion as to the acceptability of a waste needs to be based on evaluations derived from the integrated results of waste analysis, preliminary experi-ments such as waste degradability, sorption and mobility in soils, toxic-ity, mutagenicity, and field pilot studies, and the ultimate design and monitoring criteria relevant to HWLT. The following chapters are designed to aid the evaluation and decision processes by addressing the integration of these parameters.

CHAPTER 5 REFERENCES

Adams, C. E., Jr., D. L. Ford, and W. W. Eckenfelder, Jr. 1981. Development of design and operational criteria for wastewater treatment. Enviro Press, Inc. Nashville, Tennesee.

Adams, J. 1982. Unpublished results.

Adams, J., E. L. Atlas, and C. S. Giam. 1982. Ultratrace determination of vapor-phase nitrogen heterocyclic bases in ambient air. Anal. Chem. 54: 1515-1518.

Altgelt, K. H. and T. H. Gouw (eds.) 1979. Chromatography in petroleum analysis. Marcel Dekker, New York, New York. 500 p.

Ames, B. N., J. McCann, and E. Yamasaki. 1975. Methods for detecting carcinogens and mutagens with the Salmonella/mammalian microsome mutagenicty test. Mutat.Res. 31:347.

Anders, D. E., F. G. Doolittle, and W. E. Robinson. 1975. Polar constituents isolated from Green River oil shale. Geochim. Cosmochim. Acta 39: 1423-1430.

Anderson, D. 1975. The selection and induction of 5-iodo-2-deoxyuridine and thymidine variants of P388 mouse lymphoma cells with agents which are used for selection. Mutat. Res. 33:399.

Atlas, E. and C. S. Giam. 1981. Global transport of organic pollutants: ambient concentrations in the remote marine atmosphere. Science 211:163-165.

Arlett, C. F. 1977. Mutagenicity testing with V79 Chinese hamster cells. p. 175-191. In B. J. Kilbey, M. Legator, W. Nichols, and C. Ramel (ed.) Handbook of mutagenicity test procedures. Elsevier Biomedical Press, Amsterdam.

Bartle, K. D., G. Collin, J. W. Stadelhofer, and M. Zander. 1979. Recent advances in the analysis of coal-derived products. J. Chem. Tech. Biotechnol. 29:531-551.

Bassett, J., R. C. Denney, G. H. Jeffery, and J. Mendham. 1978. Vogel's textbook of quantitative inorganic analysis. Longman, London. pp. 687-690.

Beek, B., G. Klein, and G. Obe. 1980. The fate of chromosomal aberrations in a proliferating cell system. Biol. Zentralbl. 99(1):73-84.

Bellar, T. A. and J. J. Lichtenberg. 1979. Semiautomated headspace analysis of drinking waters and industrial waters for purgable volatile organic compounds. pp. 108-129. In C. E. Van Hall (ed.) Measurement of organic pollutants in water and wastewater. American Society for Testing and Materials ASTM STP 686.

172   **Waste Streams**

Bigani, M., G. Morpurgo, R. Pagliani, A. Carare, G. Conte, and G. DiGuideppe. 1974. Non-disjunction and crossing over induced by pharmaceutical drugs in Aspergillus nidulans. Mutat. Res. 26:159.

Black, C. A. (ed.). 1965. Methods of soil analysis, part 2. Chemical and microbiological properties. Am. Soc. Agron. Monogr. No. 9. Madison, Wisconsin. 802 p.

Boduszynski, M. M, R. J. Hurtubise, and H. F. Silver. 1982a. Separation of solvent-refined coal into solvent derived fractions. Anal. Chem. 54:372-375.

Boduszynski, M. M., R. J. Hurtubise, and H. F. Silver. 1982b. Separation of solvent-refined coal into compaound-class fractions. Anal. Chem. 54:375-381.

Booz-Allen and Hamilton, Inc. and Putnam, Hayes and Bartlett, Inc. 1980. Hazardous waste generation and commercial hazardous waste management capacity, an assessment. Prepared for the U.S. EPA. SW-894. U.S. Government Printing Office, Washington, D.C.

Bower, C. A. and L. V. Wilcox. 1965. Soluble Salts. pp. 936-940. In C. A. Black (ed.) Methods of soil analysis, part 2. Chemical and microbiological properties. Am. Soc. Agron. Monogr. No. 9. Madison, Wisconsin.

Bowers, W. D., M. L. Parsons, R. E. Clement, and F. W. Karasek. 1981. Component loss during evaporation-reconstitution of organic environmental samples for gas-chromatographic analysis. J. Chromatogr. 207:203-211.

Bremner, J. M. 1965. Total nitrogen, pp. 1149-1178, In C. A. Black (ed.) Methods of soil analysis. Part 2. Chemical and microbiological properties. Am. Soc. Agron. Monogr. No. 9. Madison, Wisconsin.

Brewer, R. F. 1965. Fluorine. pp. 1135-1148. In C. A. Black Methods of soil analysis, part 2. Chemical and microbiological properties. Am. Soc. Agron. Monogr. No. 9. Madison, Wisconsin.

Brocco, D., A. Cimmino, and M. Possanzini. 1973. Determination of aza-heterocyclic compounds in atmospheric dust by a combination of thin-layer and gas chromatography. J. Chromatogr. 84:371-377.

Brown, R. H. and C. J. Purnell. 1979. Collection and analysis of trace organic vapour pollutants in ambient atmospheres. The performance of a Tenax-GC adsorbent tube. J. Chromatogr. 178:79-90.

Brusick, D. J. 1972. Induction of cyclohexamide resistant mutants in Saccharomyces cerevisiae with N-methyl-N-nitro-N-nitrosoguanine and ICR-170. J. Bacteriol. 109:1134.

Buchanan, M. V. 1982. Mass spectral characterization of nitrogen-containing compounds with ammonia chemical ionization. Anal. Chem. 54:570-574.

Carere, A., G. Morpurgo, G. Cardamone, M. Bignami, F. Aulicino, G. DiGuisepie, and C. Conti. 1975. Point mutations induced by pharmaceutical drugs. Mutat. Res. 29:235.

Cautreels, W. and K. van Cauwenberghe. 1976. Extraction of organic compounds from airborne particulate matter. Water, Air, Soil Pollut. 6:103-110.

Cautreels, W., K. van Cauwenberghe, and L. A. Guzman. 1977. Comparison between the organic fraction of suspended matter at a background and an urban station. Sci. Tot. Environ. 8:79-88.

Claeys, R. C. 1979. Colloquium--the impact of the consent decree on analytical chemistry in industry. pp. 20-21. In C. E. Van Hall (ed.) Measurement of organic pollutants in water and wastewater. American Society for Testing and Materials. ASTM STP 686.

Clive, D. W. 1973. Recent developments with the L5178Y TK heterozygote mutagen assay system. Environ. Hlth. Perspec. 6:119.

Clive, D. W., W. G. Flamm, M. R. Machesko, and N. H. Bernheim. 1972. Mutational assay system using the thymidine kinase locus in mouse lymphoma cells. Mutat. Res. 16:77.

Clive, D. W. and J. F. S. Spector. 1975. Laboratory procedure for assessing specific locus mutations at the TK locus in cultured L5178Y mouse lymphoma cells. Mutat. Res. 31:17.

Colgrove, S. G. and H. G. Svec. 1981. Liquid-liquid fractionation of complex mixtures of organic components. Anal. Chem. 53:1737-1742.

Crowley, R. J., S. Siggia, and P. C. Uden. 1980. Class separation and characterization of shale oil by liquid chromatography and capillary gas chromatography. Anal. Chem. 52:1224.

Dean, B. J. and K. R. Senner. 1977. Detection of chemically induced mutation in Chinese hamsters. Mutat. Res. 46:403.

DeRenzo, D. J. 1978. Unit operations for treatment of hazardous industrial wastes. pp. 1-920. Noyes Data Corporation, Park Ridge, New Jersey.

DeSerres, F. J. and H. V. Malling. 1971. Measurement of recessive lethal damage over the entire genome and at two specific loci in the ad-3 region of a two component heterokaryon of Neurosporacrassa. 2:311-341. In A. Hollaender (ed.) Chemical mutagens, principles and methods for their detection. Plenum Press, New York.

de Vera, E. R., B. P. Summons, R. D. Stephens, and D. L. Storm. 1980. Samplers and sampling procedures for hazardous waste streams. Municipal Environmental Research Lab. Office of Research and Development, EPA. Cincinnati, Ohio. EPA 600/2-80-018. PB 80-1B5353.

Donnelly, K. C., K. W. Brown, and B. R. Scott. 1982. The use of bioassays to monitor land treatment of hazardous waste. In M. D. Waters, S. S. Sandhu, J. Lewtas, and L. Clayton (eds.) The application of short-term bioassays in the analysis of complex environmental mixtures. Plenum Press (in press).

Dowty, B. J., S. R. Antoine, and J. L. Laseter. 1979. Quantitative and qualitative analysis of purgable organics by high resolution gas chromatography and flame ionization detection. pp. 24-35. In C. E. Van Hall (ed.) Measurement of organic pollutants in water and wastewater. American Society for Testing and Materials. ASTM STP 686.

Ehrenburg, L. 1971. Higher plants. 2:365-386. In A. Hollaender (ed.) Chemical mutagens, principles and methods for their detection. Phenum Press, New York.

EPA. 1979a. Development document for effluent limitations, guidelines, and standards: leather, tanning and finishing point source category. EPA 440/1-79-016.

EPA. 1979b. Guidelines establishing test procedures for the analysis of pollutants, proposed regulations. Fed. Reg. Vol. 44. No. 233. Dec. 3, 1979.

EPA. 1979c. Methods for chemical analysis of water and wastes. EPA 600/4-79-020. PB 297-686/8BE.

EPA. 1980a. Treatability manual (5 volumes). Office of Research and Development. U.S. Environmental Protection Agency, EPA-600-8-80-042 a-e.

EPA. 1980b. Analysis of pesticide residues in human and environmental samples. EPA, Research. Triangle Park, North Carolina. EPA 600/8-80-038.

EPA. 1980c. Interim status standards for owners and operators of hazardous waste treatment, storage, and disposal facilities. Fed. Reg. Vol. 45, No. 98. Book 2 of 3. May 19, 1980.

EPA. 1980d. Subtitle C, Resource conservation and recovery act of 1976. Listing of hazardous waste, background document. U.S. EPA Office of Solid Waste. Washington, D.C.

EPA. 1982a. Test methods for evaluating solid waste: physical/chemical methods. 2nd. ed. U.S. EPA. SW-846. July 1982.

EPA. 1982b. Hazardous waste management system; permitting requirements for land disposal facilities. Federal Register Vol. 47, No. 143, pp. 32274-32388. July 26, 1982.

Epstein, E. and W. Taffel. 1979. A study of rapid biodegradation of oily wastes through composting. U.S. Dept. of Transportation. U.S. Coast Guard Report No. CG-D-83-79. 57 p.

Ettlich, W. F., D. J. Hinrichs, and T. S. Lineck. 1978. Sludge handling and conditioning. U.S. Environmental Protection Agency, EPA 430/9-78-002.

Evans, H. J. and M. L. O'Riodan. 1975. Human peripheral blood lymphocytes for analysis of chromosome aberrations in mutagen tests. Mutat. Res. 31:135.

Felkner, I. C., K. M. Hoffman, and B. C. Wells. 1979. DNA – damaging and mutagenic effect of 1,2 – Dimethylhydrazine on Bacillus subtilis repair-deficient mutants. Mutat. Res. 28:31-40.

Florisil Properties Applications Bibliography. Floridin Co., 3 Penn Center, Pittsburgh, Pennsylvania 15235.

Francis, A. J., E. D. Morgan, and C. F. Poole. 1978. Flophemesy derivatives of alcohols, phenols, amines and carboxylic acids and their use in gas chromatography with electron-capture detection. J. Chromatogr. 161:111-117.

Giam, C. S., H. S. Chan, and G. F. Neff. 1976. Distribution of n-parrafins in selected marine benthic organisms. Bull. Environ. Contam. Toxicol. 16:37-43.

Giam. C. S. and M. K. Wong. 1972. Problems of background contamination in the analysis of open ocean biota for chlorinated hydrocarbons. J. Chromatogr. 72:283-292.

Grabow, W. O. K., J. S. Burger, and C. A. Hilner. 1981. Comparison of liquid-liquid extraction and resin adsorption for concentrating mutagens in Ames Salmonella/microsome assays on water. Bull. Environ. Contam. Toxicol. 27:442-449.

Green, M. H. L., W. J. Muriel, and B. A. Bridges. 1976. Use of a simplified fluctuation test to detect low levels of mutagens. Mutat. Res. 38:33.

Gritz, R. L. and D. G. Shaw. 1977. A comparison of methods for hydrocarbon analysis of marine biota. Bull. Environ. Contam. Toxicol. 17(4):408-415.

Hendrickson, J. B., D. J. Cram, and G. S. Hammond. 1970. Organic chemistry. 3rd. ed. McGraw-Hill Book Co., New York. 1279 p.

Herrick, E., C. J. A. King, R. P. Quellette, and P. N. Cheremisinoff. 1979. Unit process guide to organic chemical industries, Vol. 1 of unit process series. Ann Arbor Sci. Publ. Inc., Ann Arbor, Michigan.

Hertz, H. S., J. M. Brown, S. N. Chesler, F. R. Guenther, L. R. Hilpert, W. E. May, R. M. Parris, and S. A. Wise. 1980. Determination of individual organic compounds in shale oil. Anal. Chem. 52:1650-1657.

Hoffman, D. and E. L. Wynder. 1977. Organic particulate pollutants-chemical analysis and bioassays for carcinogenecity. pp. 361-455. In A. C. Stern (ed.) Air Pollution. Vol. II. 3rd ed. Academic Press, New York, New York.

Holden, A. V. and K. Marsden. 1969. Single-stage clean-up of animal tisue extracts for organochlorine residue analysis. J. Chromatogr. 44:481-492.

Holstein, W. and D. Severin. 1981. Liquid chromatography of coal oil fractions. Anal. Chem. 53:2356-2358.

Jacobs, L. and R. DeMars. 1977. Chemical mutagenesis with diploid human fibroblasts. pp. 193-220. In B. J. Kilbey, M. Legator, W. Nichols and C. Ramel (ed.) Handbook of mutagenicity test procedures. Elsevier Biomedical Press, Amsterdam.

Johnson, E. L. and R. Stevenson. 1978. Basic liquid chromatography. Varian Associates, Inc. Palo Alto, California.

Kada, T., M. Morija, and Y. Shirasu. 1974. Screening of pesticides for DNA interactions by REC-assay and mutagenic testing and frameshift mutagens detected. Mutat. Res. 26:243.

Kihlman, B. A. 1977. Root tips of Vicia fava for the study of the induction of chromosomal aberrations. pp. 389-410. In B. J. Kilbey, M. Legator, W. Nichols, and C. Ramel (ed.) Handbook of mutagenicity test procedures. Elsevier Biomedical Press, Amsterdam.

Knuutinen, J. 1982. Analysis of chlorinated guaiacols in spent bleach liquor from a pulp mill. J. Chromatogr. 248:289-295.

Kugelman, I. J. and J. S. Jeris. 1981. Anaerobic digestion. pp. 211-278. In W. W. Eckenfelder, Jr. and C. J. Santhanam (eds.) Sludge treatment. Marcel Dekker, Inc., New York, New York.

Kumar, R. and S. Chauhan. 1979. Frequency and spectrum of chlorophyll mutations in a 6 rowed barley Hordeum vulgare. Indian J. Agri. Sci. 49(11):831-834.

Later, D. W., M. L. Lee, K. D. Bartle, R. C. Kong, and D. L. Vassilaros. 1981. Chemical class separation and characterization of organic compounds in synthetic fuels. Anal. Chem. 53:1612-1620.

Lin, D. C. K., R. L. Foltz, S. V. Lucas, B. A. Petersen, L. E. Slivon, and R. G. Melton. 1979. Glass capillary gas chromatographic-mass spectrometric analysis of organics in drinking water concentrates and advanced waste treatment water concentrates--II. pp. 68-84. In C. E. Van Hall (ed.) American Society for Testing and Materials. ASTM STP 686.

Loehr, R. C., W. J. Jewell, J. D. Novak, W. W. Clarkson, and G. S. Friedman. 1979. Land application of wastes. Vol. I. Van Hostrand Reinhold Co., New York, New York. 308 p.

Loprieno, N., R. Barale, C. Bauer, S. Baroncelli, G. Bronzetti, A. Cammellini, A. Cinci, G. Corsi, C. Leporini, R. Nieri, M. Mozzolini, and C. Serra. 1974. The use of different test systems with yeasts for the evaluation of chemically induced gene conversions and gene mutations. Mutat. Res. 25:197.

Lundi, G., J. Gether, N. Gjos, and M. B. S. Lande. 1977. Organic micropollutants in precipitation in Norway. Atmos. Environ. 11:1007-1014.

Marimuthu, K. M., A. H. Sparrow, and L. A. Schairer. 1970. The cytological effects of space flight factors, vibration, clinostate, and radiation on root tip cells of Tradescantia and Allium cepa. Radiation Res. 42:105.

Matsushita, H. 1979. Micro-analysis of polynuclear aromatic hydrocarbons in petroleum. Am. Chem. Soc., Div. Fuel Chem. 24:292-298.

Metry, A. A. 1980. The handbook of hazardous waste management. Technomic Publishing Company. Westport, Connecticut.

McLafferty, F. W. 1973. Interpretation of mass spectra. W. A. Benjamin, Inc., London.

McNair, H. M. and E. J. Bonelli. 1968. Basic gas chromatography. Varian Associate, Inc., Palo Alto, California.

Middleditch, B. S., S. R. Missler, and H. B. Hines. 1981. Mass spectrometry of priority pollutants. Plenum Press. New York, New York.

Miller, R. 1982. Hydrocarbon class fractionation with bonded-phase liquid chromatography. Anal. Chem. 54:1742-1746.

Mohn, G., J. Eltenberger, and D. B. McGregor. 1974. Development of mutagenicity tests using Escherichia coli K-12 as an indicator organism. Mutat. Res. 23:187.

Morrison, R. T. and R. N. Boyd. 1975. Organic chemistry. Allyn and Bacon, Inc., Boston, Massachusetts.

Mortimer, R. K. and T. R. Manney. 1971. Mutation induction in yeast. 1:289-310. In A. Hollaender (ed.) Chemical mutagens, principles and methods for their detection. Plenum Press, New York.

Mousa, J. J. and S. A. Whitlock. 1979. Analysis of phenols in some industrial wastewaters. pp. 206-220. In C. E. Van Hall (ed.) Measurement of organic pollutants in water and wastewater. American Society for Testing and Materials. ASTM STP 686.

Namiesnik, J., L. Torres, E. Kozlowski, and J. Mathieu. 1981. Evaluation of the suitability of selected porous polymers for preconcentration of volatile organic compounds. J. Chromatogr. 208:239-252.

Nauman, C. H., A. H. Sparrow, and L. A. Schairer. 1976. Comparative effects of ionizing radiation and two gaseous chemical mutagens on somatic mutation induction in one mutable and two non-mutable clones of Tradescantia. Mutat. Res. 38:53-70.

Neill, J. P., P. A. Brimer, R. Machanoff, G. P. Hirseh, and A. W. Hsie. 1977. A quantitative assay of mutation induction at the HGPRT locus in Chinese hamster ovary cells: development and definition of the system. Mutat. Res. 45:91.

Nicoloff, H., K. I. Gecheff, and L. Stoilov. 1979. Effects of caffeine on the frequencies and location of chemically induced chromatid aberrations in barley. Mutat. Res. 70:193-201.

Novotny, M. J. W. Strand, S. L. Smith, D. Wiesler, and F. J. Schwende. 1981. Compositional studies of coal tar by capillary gas chromatography/ mass spectrometry. Fuel 60:213-220.

Novotny, M., D. Wiesler, and F. Merli. 1982. Capillary gas chromatography/ mass spectrometry of azaarenes isolated from crude coal tar. Chromatographia 15:374-377.

Ong, T. N. 1978. Use of the spot, plate and suspension test systems for the detection of the mutagenicity of environmental agents and chemical carcinogens in Neurospora crassa. Mutat. Res. 54:121-129.

Packer, K. (ed.) 1975. Nanogen index, a dictionary of pesticides and chemical pollutants. Nanogens International. Freedom, California.

Parry, J. M. 1977. The use of yeast cultures for the detection of environmental mutagens using a fluctuation test. Mutat. Res. 46:165.

Peech, M. 1965. Hydrogen-ion activity. Agron. 9:914-926.

Pellizzari, E. D. 1982. Analysis for organic vapor emissions near industrial and chemical waste disposal sites. Environ. Sci. Technol. 16:781-785.

Pellizzari, E. D., J. E. Bunch, R. E. Berkley, and J. McRae. 1976. Determination of trace hazardous organic vapor pollutants in ambient atmospheres by gas chromatography/mass spectrometry/computer. Anal. Chem. 48:802-806.

Pellizzari, E. D., N. P. Castillo, S. Willis, D. Smith, and J. T. Bursey. 1978. Identification of organic constituents in aqueous effluents from energy-related processes. Am. Chem. Soc., Div. Fuel Chem. 23:144-155.

Pellizzari, E. D. and L. Little. 1980. Collection and analysis of purgable organics emitted from wastewater treatment plants. EPA. Cincinnati, Ohio. EPA 600/2-80-017.

Perry, P. and H. J. Evans. 1975. Cytological detection of mutagen-carcinogen exposure by sister chromatid exchange. Nature (London). 258:121.

Plewa, M. J. and J. M. Gentile. 1976. The mutagenicity of atrazine: a maize-microbe bioassay. Mutat. Res. 38:287-292.

Redei, G. P. 1975. Arabadopsis as a genetic tool. Ann. Rev. Genet. 9:111-125.

Reunanen, M. and R. Kroneld. 1982. Determination of volatile halocarbons in raw and drinking water, human serum, and urine by election capture GC. J. Chromatogr. Sci. 20:449-454.

Rice, R. G. and M. E. Browning. 1981. Ozone treatment of industrial wastewater. Pollution Technology Review No. 84. pp. 31-35. Noyes Data Corporation. Park Ridge, New Jersey.

Russell, J. W. 1975. Analysis of air pollutants using sampling tubes and gas chromatography. Environ. Sci. Technol. 9:1175-1178.

Safe, S. and O. Hutzinger. 1973. Mass spectrometry of pesticides and pollutants. CRC Press. Cleveland, Ohio.

Scott, B. R., E. Kafer, G. L. Dorn, and R. Stafford. 1980. Aspergillus nidulans: systems and results of test for induction of mutation and mitotic segregation. Mutat. Res. (in press)

Scott, B. R., A. H. Sparrow, S. S. Lamm, and L. Schairer. 1978. Plant metabolic activation of EDB to a mutagen of greater potency. Mutat. Res. 49:203-212.

Shackelford, W. M. and R. G. Webb. 1979. Survey analysis of phenolic compounds in industrial effluents by gas chromatography-mass spectrometry. pp. 191-205. In C. E. Van Hall (ed.) Measurement of organic pollutants in water and wastewater. American Society for Testing and Materials. ASTM STP 686.

Skoog, D. A. and D. M. West. 1979. Analytical chemisty, 3rd ed. pp. 186-246. Holt, Rinehart and Winston, New York.

Skopek, T. R., J. L. Liber, J. J. Krowleski, and W. G. Thilly. 1978. Quantitative forward mutation assay in Salmonella typhimurium using 8-azaguanine resistance as a genetic marker. Proc. Nat'l. Acad. Sci. 75:410.

Slater, E., M. D. Anderson, and H. S. Rosenkranz. 1971. Rapid detection of mutagens and carcinogens. Cancer Res. 31:970.

Snyder, D. and R. Reinert. 1971. Rapid separation of polychlorinated biphenyls from DDT and its analogues on silica gel. Bull. Environ. Contam. Toxicol. 6:385-387.

Soderberg, K., J. T. Mascarello, G. Breen, and I. E. Scheffler. 1979. Respiration deficient Chinese hamster cell mutants genetic characterization. Somatic Cell Genet. 5(2):225-240.

Sodergren, A. 1978. Simultaneous detection of halogenated and other compounds by electron-caputre and flame-conization detectors combined in series. J. Chromatogr. 160:271-276.

Speck, W. T., R. M. Santella, and H. S. Rosenkranz. 1978. An evaluation of the prophage induction (inductest) for the detection of potential carcinogens. Mutat. Res. 54:101.

Stetka, D. G. and S. Wolff. 1976. Sister chromatid exchange as an assay for genetic damage induced by mutagen-carcinogens. Mutat. Res. 41:333.

Stout, P. R. and C. M. Johnson. 1965. Chlorine and bromine. pp. 1124-1134. In C. A. Black (ed.) Methods of soil analysis, part 2. Chemical and microbial properties. Am. Soc. Agron. Madison, Wisconsin.

Stubley, C., J. G. P. Stell, and D. W. Mathieson. 1979. Analysis of azanaphthalenes and their enzyme oxidation products by high-performance liquid chromatography, infrared spectroscopy, and mass spectrometry. J. Chromatogr. 177:313-322.

Stuermer, D. H., D. J. Ng, and C. J. Morris. 1982. Organic contaminants in groundwater near an underground coal gasification site in northeastern Wyoming. Environ. Sci. Technol. 16:582-587.

Tanooka, J. 1977. Development and applications of Bacillus subtillus test system for mutagens involving DNA repair deficiency and suppressible auxotrophic mutations. Mutat. Res. 48:367.

Tanooka, H., N. Munakata, S. Kitahara. 1978. Mutation induction with UV- and X-radiation in spores and vegetative cells of Bacillus Subtillis. Mutat. Res. 49:179-186.

Texaco, Inc. 1979. Composting of oily sludges. Report to the Solid Waste Committee, Environ. Affairs Dept. of the Am. Petroleum Institute. 22 p.

Thilly, W. G., J. G. DeLuca, I. V. H. Hoppe, and B. W. Penmann. 1976. Mutation of human lymphoblasts by methylnitrosourea. Chem. Biol. Interact. 15:33.

Tucker, E. S., V. W. Saeger, and O. Hicks. 1975. Activated sludge primary biodegradation of polychlorinated biphenlys. Bull. Environ. Contam. Toxicol. 14(6):705-713.

Underbrink, A. G., L. A. Schairer, and A. H. Sparrow. 1973. Tradescantia stamen hairs. A radiobiological test system applicable to chemical mutagenesis. 3:171-207. In A. Hollaender (ed.) Chemical mutagens, principles and methods for their detection. Plenum Press, New York.

Vick, R. D., J. J. Richard, H. J. Svec, and G. A. Junk. 1977. Problems with Tenax-GC for environmental sampling. Chemosphere 6:303-308.

Vig, B. K. 1975. Soybean (Glycine max): a new test system for study of genetic parameters as affected by environmental mutagens. Mutat. Res. 31:49-56.

Von Borstel, R. C. and R. H. Smith. 1977. Measuring dominant lethality in Habrobracon. pp. 375-387. In B. J. Kilbey, M. Legator, W. Nichols and C. Ramel (ed.) Handbook of mutagenicity testing procedures. Elsevier Biomed. Press, Amsterdam.

Wear, J. I. 1965. Boron. pp. 1059-1063. In C. A. Black (ed.) Methods of soil analysis, part 2. Chemical and microbial properties. Am. Soc. Agron. Madison, Wisconsin.

Worstell, J. H. and S. R. Daniel. 1981. Deposit formation in liquid fuels. 2. The effect of selected compounds on the storage stability of Jet A turbine fuel. Fuel 60:481-484.

Worstell, J. H., S. R. Daniel, and G. Frauenhoff. 1981. Deposit formation in liquid fuels. 3. The effect of selected nitrogen compounds on diesel fuel. Fuel 60:485-487.

Wurgler, F. E. and E. Vogel. 1977. Drosophila as an assay system for detecting genetic changes. pp. 335-373. In B. J. Kilbey, M. Legator, W. Nichols, and C. Ramel (ed.) Handbook of mutagenicity test procedures. Elsevier Biomed. Press, Amsterdam.

FATE OF CONSTITUENTS IN THE SOIL ENVIRONMENT

David C. Anderson
Christy Smith
Stephen G. Jones
K. W. Brown

An understanding of the behavior of the various waste streams in the soil environment at an HWLT unit may be derived from a knowledge of the specific constituents that compose the waste. Chapter 5 provided general information on the characterization of waste streams. After determining the constituents present in the waste, this chapter can be used to gain a better understanding of the fate of the wastes disposed by HWLT.

Knowledge about the specific components expected to be found in a given waste stream can be gained from information on the sources of the waste, any pretreatment or in-plant process changes, and waste analyses. Although only hazardous constituents are regulated by EPA, there may be other waste constituents, not listed as hazardous, that are nevertheless significant. Once waste characterization (Section 5.3) has confirmed the presence of a specific compound or element, this chapter will serve as a source of information on the environmental fate, toxicity and land treatability of individual components of the waste. Figure 6.1 indicates the topics discussed and the organization of the material presented in this chapter. Additional literature references are cited which can be used when more detailed information is desired.

6.1   INORGANIC CONSTITUENTS

Although inorganic chemical soil reactions have been more thoroughly studied than organic, comprehensive information is still limited on the behavior of some inorganic chemicals in the heterogeneous chemical, physical and biological matrix of the soil. Agriculturally important compounds have received greater scrutiny than others. For instance, metals have only recently begun to attract widespread interest as the use of land treatment

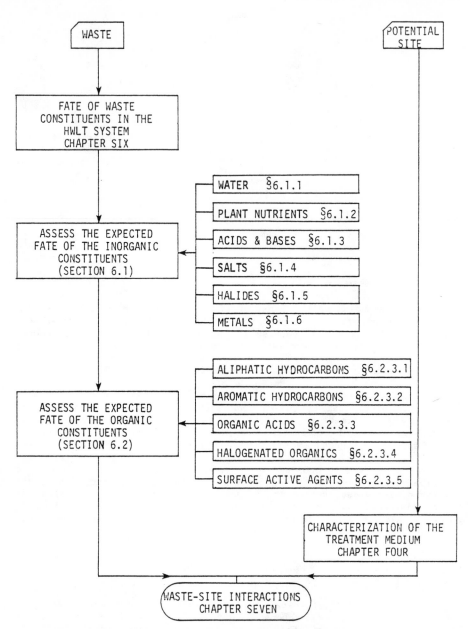

Figure 6.1. Constituent groups to be considered when assissing the fate of wastes in the land treatment system.

for municipal wastes has increased. The information developed from treating municipal wastes does not, however, address the entire range of constituents that may be present in hazardous industrial wastes.

6.1.1   Water

Water is practically ubiquitous in hazardous waste streams and often constitutes the largest waste fraction. In a land treatment system, water has several major functions. As a carrier, water transports both dissolved and particulate matter through both surface runoff and deep percolation. Water also controls gas exchange between the soil and the atmosphere. Thus, water may be beneficial by controlling the release rate of volatile waste constituents. For example, where aeration is poor due to high soil water content, biological decomposition of waste constituents is inhibited and may be accompanied by acute odor problems. A lack of soil water can also inhibit waste degradation.

Since the application of waste may contribute significant amounts of water in addition to precipitation inputs, a complete hydrologic balance including the water content of the waste must be developed. Techniques for calculating the hydrologic balance are presented in Section 8.3.1; these calculations are used to estimate waste storage requirements, waste application rates, and runoff retention and treatment needs.

6.1.2   Plant Nutrients

Many of the elements essential to plant growth may have detrimental effects when excessive concentrations are present in soil. Some may be directly toxic to plants, while others may induce toxic responses in animals. Further problems may involve damage to the soil physical properties or to surface water ecosystems. Consequently, plant nutrients, present in significant concentrations in the waste, that may adversely affect the environment should be considered in determining the feasibility of land treatment and appropriate waste loading rates. This section deals with the plant essential elements not classified and discussed as metals or halides, which may cause problems in an HWLT unit.

6.1.2.1 Nitrogen (N)

Land application of a waste high in nitrogen requires an understanding of the various forms of N contained in the waste, the transformations that occur in soils, and the rates associated with these transformations. A knowledge of N additions to and losses from the disposal site can then be used to calculate a mass balance equation which is used to estimate the amount and rate of waste loading.

Wastes high in N have typically included sewage sludges, wastewaters, and animal wastes. Table 6.1 lists the N content of several sewage types and Table 6.2 gives the N analysis of manure samples. Pharmaceutical and medicinal chemicals manufacturing generate wastes high in ammonia, organonitrogen and soluble inorganic salts. In sewage and animal manure, N is usually found as ammonium or nitrate. Industrial wastes often contain N in small quantities incorporated in aromatic compounds, such as pyridines.

TABLE 6.1   CHEMICAL COMPOSITION OF SEWAGE SLUDGES*[†]

| Component | Number of Samples | Concentration[#] | | | Coefficient of Variability (%)[+] |
		Range (%)	Median (%)	Mean (%)	
Total N	191	0.1 - 17.6	3.3	3.9	85
NH_4-N	103	0.1 - 6.8	0.1	0.7	171
NO_3-N	45	0.1 - 0.5	0.1	0.1	158

* Sommers (1977).

[†] Data are from numerous types of sludges (anaerobic, aerobic, activated, lagoon, etc.) in seven states: Wisconsin, Michigan, New Hampshire, New Jersey, Illinois, Minnesota, Ohio.

[#] Oven-dry solids basis.

[+] Standard deviation as a percentage of the mean. Number of samples on which this is based may not be the same as for other columns.

TABLE 6.2   CHEMICAL ANALYSES OF MANURE SAMPLES TAKEN FROM 23 FEEDLOTS IN TEXAS*[†]

Element	Range (%)	Average (%)
N	1.16 - 1.96	1.34
P	0.32 - 0.85	0.53
K	0.75 - 2.35	1.50
Na	0.29 - 1.43	0.74
Ca	0.81 - 1.75	1.30
Mg	0.32 - 0.66	0.50
Fe	0.09 - 0.55	0.21
Zn	0.005 - 0.012	0.009
H_2O	20.9 - 54.5	34.5

* Mathers et al. (1973).

[†] All values based on wet weight.

Precipitation adds to the N that reaches the surface of the earth and several attempts have been made to quantify this. Additions of N from precipitation are greater in the tropics than in humid temperate regions and larger in humid temperate regions than in semiarid climates. Table 6.3 lists N values in precipitation from various locations. A study by Gamble and Fisher (1964) revealed that most of the N reaching the earth is in the NO_3^- and NH_4^+ forms. Concentrations of N in the rain resulting from a thunderstorm are shown in Fig. 6.2. The initial concentrations of NO_3^- are 8 ppm and decrease sharply as the precipitation cleanses the air of N containing dust, eroded soil, and incomplete combustion products.

TABLE 6.3 AMOUNTS OF NITROGEN CONTRIBUTED BY PRECIPITATION*

Location	Years of Record	kg/ha/yr		
		Rainfall (cm)	Ammoniacal Nitrogen	Nitrate Nitrogen
Harpenden, England	28	73.2	2.96	1.49
Garford, England	3	68.3	7.20	2.16
Flahult, Sweden	1	82.6	3.72	1.46
Groningen, Holland	--	70.1	5.08	1.64
Bloemfontein and Durban, South Africa	2	--	4.50	1.56
Ottawa, Canada	10	59.4	4.95	2.42
Ithaca, N.Y.	11	74.9	4.09	0.77

* Lyon and Bizzell (1934).

Nitrogen exists in waste, soil and the atmosphere in several forms. Organic N, such as alkyl or aromatic amines, is bound in carbon-containing compounds and is not available for plant uptake or leaching until transformed to inorganic N by microbial decomposition. Humus and crop residues in the soil contain organic N.

Inorganic N is found in various forms such as ammonia, ammonium, nitrite, nitrate and molecular nitrogen. Ammonium (NH_4+) can be held in the soil on cation exchange sites because of its positive charge. Ammonium is used by both plants and microorganisms as a source of N. Ammonia (NH_3) exists as a gas, and NH_4^+ may be converted to NH_3 at high pH values. Nitrite (NO_2^-) is a highly mobile anion formed in soils as an intermediate in the nitrification process discussed in Section 6.1.2.1.3. Nitrite is toxic to plants in small quantities. Nitrate (NO_3^-) is a highly mobile anion readily used by plants and microorganisms. Nitrates may be readily leached from the soil and may present a health hazard. (The term NO_3-N is read nitrate-nitrogen and is not the same as NO_3 (10 mg/l NO_3-N = 44.3 mg/l NO_3). Molecular nitrogen (N_2) is a gas comprising nearly 80% of the normal atmosphere.

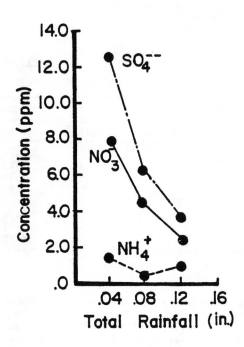

Figure 6.2. Chemical composition of thundershower
samples (Gamble and Fisher, 1964).
Reprinted by the permission of the
American Geophysical Union.

The nitrogen cycle (Fig. 6.3) is often used to illustrate the additions and removals of N from the soil system and the subsequent changes in form due to the prevailing soil environment. In addition to the N added to the soil by wastes and precipitation (discussed previously), the nitrogen cycle is affected by the processes of mineralization, nitrogen fixation, nitrification, plant uptake, denitrification, volatilization, storage in the soil, immobilization, runoff, and leaching. The amount of N added or removed by each of these mechanisms, the rate at which they occur, and the optimum soil conditions for each are discussed below.

6.1.2.1.1 <u>Mineralization</u>. The process of mineralization involves the conversion of the plant unavailable organic forms of N to the available inorganic state by microbial decomposition. Mineralization includes the ammonification process which oxidizes amines into NO_2^- or NO_3^-. Organic N contained in wastes is not available for plant uptake or subject to other losses until mineralization occurs. Only a portion of the organic N in the waste will be converted to the available inorganic form during the first year after application, and only smaller amounts will be mineralized in subsequent years.

Table 6.4 shows an estimated decay series, or fractional mineralization, for a given waste application. The table also shows a ratio of N inputs necessary to supply a constant mineralization rate. The table, developed by Pratt et al. (1973), is an estimate of decomposition based on the type of animal waste and amount of weathering the waste has undergone. For example, dry corral manure containing 2.5% N has an estimated decay series of 0.40, 0.25, and 0.06 which means that at any given application, 40% of the N applied will be mineralized the first year, 25% of the remaining N will become available the second year, and 6% of the remaining N will be mineralized in the third and all subsequent years. If 22.5 metric tons/ha of this manure (dry weight basis) were applied, of the 560 kg total N, 224 kg would be mineralized the first year, 63.75 kg the second, 12.4 kg the third, 11.6 kg the fourth, 10.9 the fifth, and 10.2 the sixth year (Pratt et al., 1973). The ratios shown in Table 6.4 are useful for estimating the amount of N that will be available given a decay series. In the example above, 2.5 kg of total N must be added to furnish 1 kg of available N the first year. If manure is added to the same field next year, only 1.82 kg must be added to provide 1 kg of available N, and so on.

Research by Hinesley et al. (1972) shows that considerable amounts of organic N in sludge and soil organic matter are mineralized during a growing season. This research indicates that about 25% of the organic N in sludge is mineralized in the first year of application, and 3–5% of the organic N is converted to inorganic N during the next three years.

Another decay series of mineralization is given in Table 6.5 where the values are calculated on the basis of having 3% of the remaining or residual organic N released as available inorganic N during the second, third, and fourth growing seasons. For example, if 5 metric tons/ha of sludge containing 3.5% (175 kg) of organic N were applied to a soil one year, during the following growing season, 0.9 kg/metric ton of sludge would become

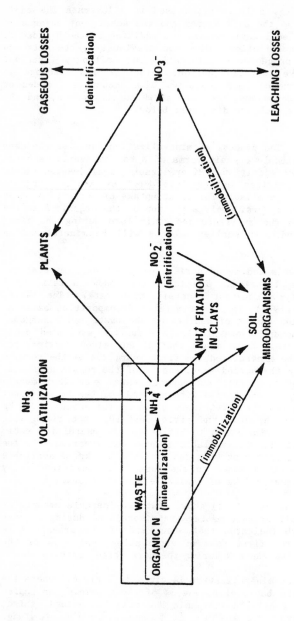

Figure 6.3. Nitrogen cycle illustrating the fate of sludge nitrogen (Beauchamp and Moyer, 1974).

TABLE 6.4 RATIO OF YEARLY NITROGEN INPUT TO ANNUAL NITROGEN MINERALIZATION RATE OF ORGANIC WASTES*†

Decay Series	Typical Material#	Time (years)							
		1	2	3	4	5	10	15	20
		—————————N input/mineralization ratio—————————							
0.90, 0.10, 0.05	Chicken manure	1.11	1.10	1.09	1.09	1.08	1.06	1.05	1.04
0.75, 0.15, 0.10, 0.05	Fresh bovine waste, 3.5% N	1.33	1.27	1.23	1.22	1.20	1.15	1.11	1.06
0.40, 0.25, 0.06	Dry corral manure, 2.5% N	2.50	1.82	1.74	1.58	1.54	1.29	1.16	1.09
0.35, 0.15, 0.10, 0.05	Dry corral manure, 1.5% N	2.86	2.06	1.83	1.82	1.72	1.40	1.23	1.13
0.20, 0.10, 0.05	Dry corral manure, 1.0% N	5.00	3.00	2.90	2.44	2.17	1.38	1.13	1.04
0.35, 0.10, 0.05	Liquid sludge, 2.5%	2.86	2.33	2.19	2.03	1.90	1.45	1.22	1.11

* Pratt et al. (1973).

† This ratio is for a constant yearly mineralization rate for six decay series for various times after initial application. The ratio equals kilograms of N input required to mineralize 1 kg of N annually.

The N content is on a dry weight basis.

available. Therefore, for a 5 metric ton/ha rate, 4.3 kg N/ha would be mineralized to the inorganic form (Sommers and Nelson, 1976).

TABLE 6.5   RELEASE OF PLANT-AVAILABLE NITROGEN DURING SLUDGE DECOMPOSITION IN SOIL*

Years After Sludge Application	Organic N Content of Sludge, %						
	2.0	2.5	3.0	3.5	4.0	4.5	5.0
	kg residual N release per metric ton sludge added						
1	0.5	0.6	0.7	0.85	0.95	1.1	1.2
2	0.45	0.6	0.7	0.8	0.9	1.05	1.15
3	0.45	0.55	0.65	0.75	0.85	1.0	1.1

* Sommers and Nelson (1976).

Microbial degradation of complex aromatic compounds containing N depends on the structure, nature, and position of functional groups. General results of many investigations are summarized as follows: short chain amines are more resistant to mineralization than those of higher molecular weight; unsaturated aliphatic amines tend to be more readily attacked than saturates; resistance to decomposition increases with the number of chlorines in the aromatic ring; and branched compounds are more resistant than unbranched compounds (Goring et al., 1975).

6.1.2.1.2 Fixation. The process by which atmospheric nitrogen (N_2) is converted to available inorganic N by bacteria is called nitrogen fixation; it may either be symbiotic or nonsymbiotic. Symbiotic N fixation is the conversion of N_2 to NH_4^+ by Rhizobium bacteria, which live in root nodules of leguminous plants. Nonsymbiotic fixation involves the conversion of N by free-living bacteria, Clostridium and Azotobacter. Fixation by leguminous bacteria accounts for the great majority of N fixation (Brady, 1974). Table 6.6 reports the N fixation of various legumes in kg/ha/yr.

TABLE 6.6   NITROGEN FIXED BY VARIOUS LEGUMES*

Crop	(kg/ha/yr)	Crop	(kg/ha/yr)
Alfalfa (Medicago sativa)	281	Soybeans (Glycine max)	118
Sweet clover (Melilotus sp.)	188	Hairy vetch (Vicia villosa)	76
Red clover (Trifolium (pratense)	169	Field beans (Phaseolus vulgaris)	65
Alsike clover (Trifolium hybridum)	158	Field peas (Pisum arvense)	53

* Lyon and Bizzell (1934).

The amount of N fixed by Rhizobium depends on many factors. Soil conditions favorable for microbial populations include good aeration, adequate moisture, and a near neutral pH. A high N containing waste or fertilizer may actually discourage nodulation and thereby reduce fixation (Fig. 6.4).   Therefore, N input from N-fixing bacteria is of minor significance on land receiving waste applications.

The exact amount of N fixed by nonsymbiotic bacteria in soils is very difficult to determine because other processes involving N are taking place simultaneously.   Experiments in several areas of the U.S. indicate that 20-60 kg N/ha/yr may be fixed by nonsymbiotic organisms (Moore, 1966). Table 6.7 lists amounts of N fixed nonsymbiotically.

TABLE 6.7   NITROGEN GAINS ATTRIBUTED TO NONSYMBIOTIC FIXATION IN FIELD EXPERIMENTS*

Location	Period (years)	Description	Nitrogen Gain (kg/ha/yr)
Utah	11	Irrigated soil and manure	49
Missouri	8	Bluegrass (Poa sp.) sod	114
California	10	Lysimeter experiment	54
California	60	Pinus ponderosa stand	63
United Kingdom	20	Monoculture tree stands	58
Australia	3	Solonized soil	25
Nigeria	3	Latosolic soil	90
Michigan	7	Straw mulch	56

* Moore (1966).

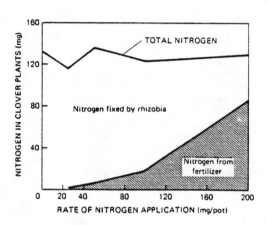

Figure 6.4. Influence of added inorganic nitrogen on the total
nitrogen in clover plants, the proportion supplied
by the fertilizer and that fixed by the rhizobium
organizations associated with the clover roots.
Increasing the rate of nitrogen application decreased the amount of nitrogen fixed by the organisms in this greenhouse experiment (Walker, 1956).
Reprinted by permission of the author.

6.1.2.1.3 Nitrification. The process of nitrification involves the conversion of NH_4^+ to NO_2^- by Nitrosomonas and the conversion of NO_2^- to NO_3^- by Nitrobacter via reactions that occur in rapid sequence and preclude any great accumulation of NO_3^-. These nitrifying organisms are auto-trophic (obtaining energy from oxidation or inorganic NH_4^+ or NO_2^-) in con-trast to the heterotrophic organisms involved in the mineralization proc-ess. These organisms are strictly aerobic and can not survive in saturated soils. The optimum temperature for nitrification is in the range of 30–36°C (Downing et al., 1964). Maximum oxidation rates for Nitrosomonas are found at pH 8.5–9.0 (Downing et al., 1964) and at pH 8.9 for Nitro-bacter (Lees, 1951). The activity of these bacteria may cease altogether where the pH is 4.0–4.5 or below. Nitrification occurs at a very rapid rate under conditions ideal for microbial growth. Daily rates of 7–12 kg N/ha have been found when 110 kg ammonium nitrate/ha were added (Broadbent et al., 1957).

The nitrification curves for most soils are sigmoid-like curves when NO_3^- production is plotted against time. A typical nitrification pattern is shown in Fig. 6.5. The NH_3-N concentration decreases sigmoidally until it disappears. The NO_2^- and NO_3^- concentrations start rising from the first day, but by the fourth day, the concentration of NO_2-N more than doubles that of the NO_3-N. A steady state is reached after the seventh day when the NO_2-N concentration approaches zero and the NO_3-N approaches total nitrogen.

6.1.2.1.4 Plant Uptake. Crop uptake of N by harvestable crops constitutes a significant removal of N. Table 6.8 lists the N uptake for various crops in kg/ha. Nitrogen is returned to the soil by crop residues (Table 6.9). The fraction of total NO_3^- in the soil that is assimilated by the roots of growing plants varies depending on the depth and distribution of root-ing, nitrogen loading rate, moisture movement through the root zone, and species of plant. In general, the efficiency of uptake is not high, and grasses tend to be more efficient than row crops. Excess available N in the soil does not cause phytotoxicity, yet corn silage and other grass forages that contain greater than 0.25% NO_3-N may cause animal health problems (Walsh et al., 1976).

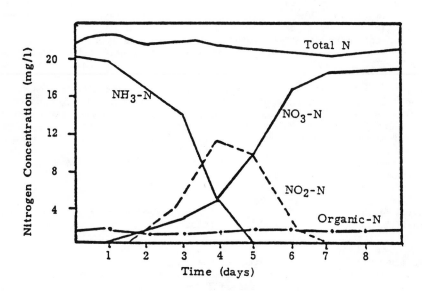

Figure 6.5.   Typical sigmoid pattern of nitrification
in soil (De Marco et al., 1967).
Reprinted by permission of the American
Water Works Association.

TABLE 6.8   REMOVAL OF NITROGEN FROM SOILS BY CROPS AND RESIDUES*[†]

Crop	Annual Crop Yield (metric/tons/ha)	Nitrogen Uptake (kg/ha/yr)
Corn (Zea mays)	9.4	207
Soybeans (Glycine max)	3.4	288[#]
Grain sorghum (Sorghum bicolor)	9	280
Peanuts (Archis hypogaea)	2.8	105
Cottonseed (Gossypium hirsutum)	2	69
Wheat (Triticum aestivum)	4.3	140
Rice (Oryza sativa)	6.7	87
Oats (Arena sativa)	3.6	168
Barley (Hordeum vulgare)	5.4	168
Corn silage (Zea mays)	71.7	224
Sugarbeets (Beta vulgaris)	56	24
Alfalfa (Medicago sativa)	17.9	504[#]
Alfalfa hay (Medicago sativa)	15.7	372
Coastal bermuda hay (Cynodon dactylon)	21.3	272
Orchard grass (Dactylis glomerata)	13.4	336
Bromegrass (Bromus sp.)	11.2	186
Tall fescue (Festuca arundinacea)	7.8	151
Reed canary grass (Phalaris arundinacea)	13.4	493
Reed canary grass hay (Phalaris arundinacea)	15.7	189
Bluegrass (Poa sp.)	6.7	224
Tomatoes (Lycopersicon esculentum)	44.8	80
Lettuce (Lactuca sativa)	28	38
Carrots (Daucus carota)	44.8	65
Loblolly pine (Pinus taeda)	annual growth	10

* Hart (1974).

[†] Where only grain is removed, a significant proportion of the nutrients is left in the residues.

[#] While legumes can get most of their N from the air, if mineral nitrogen is available in the soil, legumes will use it at the expense of fixing N from the air.

TABLE 6.9 THE NITROGEN RETURNED TO THE SOIL FROM UNHARVESTED OR UNGRAZED
PARTS OF STUBBLE ABOVE THE GROUND*

Crop	Nitrogen Returned to Soil (kg/metric ton)
Corn (Zea mays)	9
Wheat (Triticum aestivum)	7
Rye (Secale cereale)	7
Oats (Avena sativa)	6
Alfalfa (Medicago sativa)	24

* McCalla and Army (1961).

6.1.2.1.5 Denitrification. The microbial process whereby NO_3^- is
reduced to gaseous N compounds such as nitrous oxide and elemental nitrogen is termed denitrification. This reaction is facilitated by heterotrophic, facultative anaerobic bacteria living mainly in soil micropores where
oxygen is limited. As a waste is applied on land, the rate and extent of
denitrification is likely to be governed by the organic matter content,
water content, soil type, pH, and temperature of the soil. The degree of
water saturation has a profound influence on the rate of denitrification.
The critical moisture level is about 60% of the water holding capacity of
the soil, below which practically no denitrification occurs, and above this
level denitrification increases rapidly with increases in moisture content.
The amount of N lost through denitrification as a function of water content
(described as percentage of the water holding capacity) is illustrated in
Fig. 6.6 (Bremner and Shaw, 1958).

The rate of denitrification is also greatly affected by the pH and
temperature of the soil. It tends to be very slow at pH below 5.0. The
rate increases with increasing soil pH and is very rapid at pH 8-8.5. The
optimum temperature for denitrification is about 25°C. The rate of denitrification increases rapidly when the temperature is increaed from 2° to
25°C. Figure 6.7 illustrates the effect of temperature on N lost as gas
over time.

Organic matter content also affects the amount and rate of denitrification. Denitrification of NO_3^- by heterotrophic organisms cannot
occur unless the substrate contains an organic compound that can support
the growth of the organisms. The rate of denitrification for these materials varies with their resistance to decomposition by soil microorganisms
(Table 6.10). The rate is most rapid with cellulose and slowest with
lignin and sawdust.

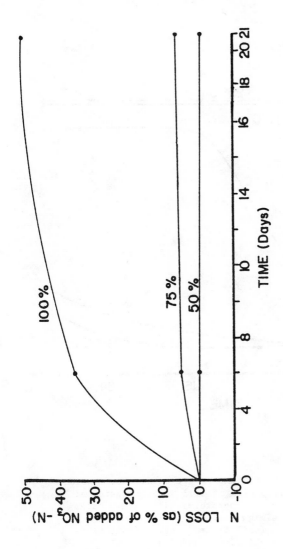

Figure 6.6. Effect of soil water content on denitrification. 5 g. samples of soil 4 in 300 ml. Kjeldahl flasks were incubated at 25° C. with 5 mg. NO₃-N (as KNO₃) and 15 mg. C (as glucose) dissolved in different volumes of water. Water content of soil is expressed on each graph as percentage of waterholding capacity of soil (Bremner and Shaw, 1958). Reprinted by permission of the Journal of Agricultural Science.

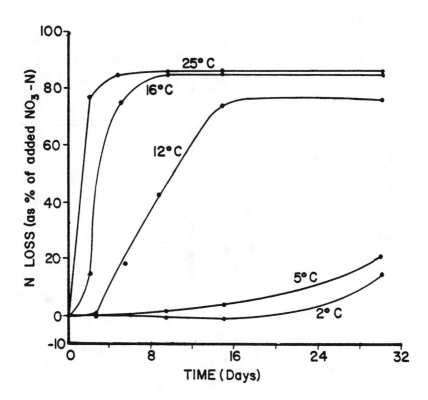

Figure 6.7. Effect of temperature on denitrification.
5 g. samples of soil were incubated at
various temperatures with 11 ml. water
containing 5 mg. $NO_3.N$ (as KNO_3) and 15
mg. C (soil 1) or 25 mg. C (soil 6) as
glucose (Bremner and Shaw, 1958).
Reprinted by permission of the Journal
of Agricultural Science.

TABLE 6.10   PERCENTAGE OF ADDED NITROGEN LOST DURING INCUBATION OF WATER-LOGGED SOIL WITH NITRATE AND DIFFERENT AMOUNTS OF ORGANIC MATERIALS AT 25°C*

Organic Materials Added	N Lost (% of added NO_3-N)											
	50 mg added				100 mg added				200 mg added			
	4[†]	12[†]	20[†]	30[†]	4[†]	12[†]	20[†]	30[†]	4[†]	12[†]	20[†]	30[†]
Lignin	2	3	6	8	5	6	8	11	7	7	9	15
Sawdust	5	7	8	9	6	9	10	12	9	11	16	18
Grass	6	8	11	13	14	27	30	36	27	37	49	60
Straw	7	10	12	14	16	28	33	37	20	44	56	84
Cellulose	5	29	83	90	5	37	87	91	5	39	88	90

* Bremner and Shaw (1958).

[†] Length of incubation period in days.

Denitrification can be a major source of N removal from an HWLT unit containing a high inorganic nitrogenous waste or an organic nitrogenous waste that has been mineralized.   Under the optimum conditions of neutral to alkaline pH, high soil water or small pores filled with water, warm temperatures, and the presence of easily decomposable organic matter, almost 90% of the NO_3-N in the waste can be converted to gaseous N and lost from the system (Bremner and Shaw, 1958).

6.1.2.1.6 Volatilization. Another mechanism for N loss is volatilization. Ammonium salts such as $(NH_4)_2CO_3$ can be converted to gaseous ammonia ($2HN_3 + H_2CO_3$) when sludge is surface applied to coarsely textured alkaline soils.   The magnitude of such losses is highly variable, depending on the rate of waste application, clay content of the soil, soil pH, temperature, and climatic conditions.   In a greenhouse study, Mills et al. (1974) reported that when pH values were above 7.2, at least half of the N applied to a fine sandy loam was volatilized as NH_3, generally within two days of the application.   In a laboratory study, Ryan and Keeney (1975) reported NH_3 volatilization from a surface applied wastewater sludge containing 950 mg/l of ammonium-nitrogen.   Volatilization values ranged from 11 to 60% of the applied NH_3-N.   The greatest losses occurred in low clay content soils with the highest application rate. Incorporating the sludge into the soil decreases volatilization losses.

6.1.2.1.1 Storage in Soil. Both the organic and inorganic soil fractions have the ability to fix NH_4^+ in forms unavailable to plants or even microorganisms.   Clay minerals with a 2:1 type structure have this capacity, with clays of the vermiculite group having the greatest capacity, followed by illite and montmorillonite. Ammonium ions fixed into the crystal lattice of the clay do not exchange readily with other cations and are not accessible to nitrifying bacteria (Nommik, 1965). The quantity of NH_4^+ fixed depends on the kind and amount of clay present. Figure 6.8 illustrates the amount of NH_4^+ fixed by three soils receiving five consecutive applications of a 100 mg/l solution of NH_4^+-N. The Aiken clay, primarily kaolinite, fixed no NH_4^+ and the Columbia and Sacramento soils containing vermiculite and montmorillonite were capable of NH_4^+ fixation (Broadbent et al., 1957).

Like other cations in the waste, NH_4^+ can be adsorbed onto the negatively charged clay and organic matter colloids in soil. Retention in this exchangeable form is temporary, and NH_4^+ may become nitrified when oxygen and nitrifying bacteria are available.

6.1.2.1.8 _Immobilization_. The process of immobilization is the opposite of mineralization; it is the process by which inorganic N is converted to an unavailable organic form. This requires an energy source for microorganisms such as decomposable organic matter with a carbon to N ratio greater than 30 to 1. This condition may exist with certain industrial wastes or cannery wastes and some crop residues, straws or pine needles. In a study of immobilization of fertilizer N, only 2.1 kg/ha was immobilized during the first 47 days after fertilization with 328 kg/ha. As soil temperature increased above 22°C, the rate increased to an additional 60 kg/ha immobilized by day 107 (Kissel et al., 1977).

6.1.2.1.9 _Runoff_. At an HWLT unit containing a nitrogenous waste, the runoff water may remove a significant amount of N, potentially polluting adjacent waterways. However, a well designed and managed disposal site should have minimum runoff since waste application rates would not exceed soil infiltration capacity. Though surface runoff from HWLT units is collected, it may be important to keep the runoff water of high quality if the facility has a discharge permit. Soil and cropping management practices, rate of waste application, and the time and method of application control the amount of runoff. Of these factors, a highly significant correlation between N loading rate and its average concentration in runoff water was shown in a linear regression analysis (Khaleel et al., 1980). Application of waste during winter and on the surface results in less rapid decomposition and high concentrations of N in runoff water. Reincorporation of plant material into the soil decreases N concentrations in runoff by one-third over areas where all plant residues are removed at harvest (Zwerman et al., 1974). Table 6.11 provides a summary of N concentrations in runoff from areas receiving animal waste.

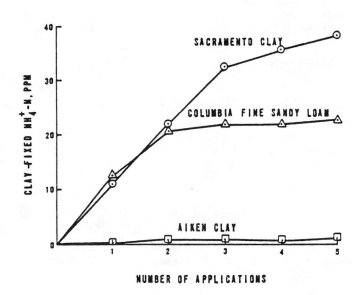

Figure 6.8. Clay-fixed NH$_4^+$ in three soils resulting from
five applications of a solution containing
100 mg/1 NH$_4^+$-N, without intervening drying
(Broadbent, 1976). Reprinted by permission of
the Division of Agricultural Sciences, University of California.

TABLE 6.11   TRANSPORT OF TOTAL NITROGEN IN RUNOFF WATER FROM PLOTS
RECEIVING ANIMAL WASTE*

Location	Type of Manure	Total N Applied	Total N Runoff	Remarks	Reference
Wisconsin	Fresh dairy liquid	120	12.7	8 Plots, 10- 17% slope,	Minshall et al. (1970)
		95	3.6	silt loam	
Alabama	Liquid dairy	5661	13.8	12 Plots, 3.3%	McCaskey et al. (1971)
		3774			
		1782			
	Dry dairy	7769	18.3		
		5179	17.7		
		2590	7.5		
N. Carolina	Swine lagoon	1344	23.4	9 Plots, 1-3% slope, sandy loam, coastal	Khaleel et al. (1980)
	effluent			bermuda	
New York	Dairy	478	18.4	24 Plots corn, continuous study	Klausner et al. (1976)

* Total N = organic N+NH$_4$-N + NO$_3$-N in ppm.

6.1.2.1.10 Leaching. Of all the losses of N from an HWLT unit, leaching
is the potentially most serious. Groundwater can become contaminated, and
drinking water containing greater than 10 mg/l nitrate-nitrogen may cause
human health problems. Not only should high concentrations of N in leachate be avoided, but also large amounts of leachate with a low concentration
of N. Methemoglobinemia, a reduction in the oxygen-carrying capacity of
the blood, can develop in infants when nitrate-nitrogen levels in drinking
water are greater than 10 ppm (or greater than 45 ppm nitrate).

Most studies of N leachate agree that the amount of N in percolating
water is site-dependent and difficult to extrapolate from one site to
another. Parameters that have the most direct effect on N content in
leachate are N application rate, cropping system, soil water content, soil
texture, and climate. A number of these parameters can be controlled or
modified by management practices.

A study by Bielby et al. (1973) investigates the quantity and concentration of NO$_3^-$ in percolates from lysimeters receiving liquid poultry
manure over three years. Nitrogen removal by corn (Zea mays), plus that in
the leachate, accounts for less than 25% of the amount applied to the soil.
The average concentration of NO$_3^-$ in percolates from all treatments
exceeded the drinking water standard (10 ppm).

6.1.2.2 Phosphorus (P)

Phosphorus is a key eutrophication element and may be transported in such forms as adsorbed phosphate and soluble phosphate by surface runoff and groundwater, respectively. Enrichment of lake waters and sediments with high P concentrations may create a potential for water quality impairment and eventual extinction of aquatic life in a lake or stream. The critical level above which eutrophication may occur has been set at 0.01 mg/l of P. This level may be exceeded when surface runoff levels are greater than 10 kg/ha/yr (Vollenweider, 1968). Runoff P concentrations from well-managed agricultural lands are typically less than 0.1 kg/ha/yr (Khaleel et al., 1980). Municipal wastewaters generally have total P concentrations ranging from 1.0 to 40 mg/l (Hunter and Kotalik, 1976; Bouwer and Chaney, 1974; Pound and Curtis, 1973), while concentrations of less than 20 mg/l are average (Ryden and Pratt, 1980).

Phosphorus concentrations in waste streams that range from 0.01 to 50 mg/l P pose little runoff or leachate hazard. However, P concentrations found in waste from rock phosphate quarries, fertilizers and pesticides are high enough to potentially contaminate runoff water or leach into the groundwater beneath a soil with low P retention capacity. Once the waste-soil parameters of P are adequately assessed, land treatment of P laden hazardous wastes may be managed to successfully reduce soluble P concentrations to the levels usually found in soil.

The soluble P concentration in the unsaturated zone of normal soil ranges between 3 and 0.03 mg/l (Russel, 1973), where the lower value is at the normal level of groundwater (Reddy et al., 1979). Barber et al. (1963) report that this value generally decreases with depth in the soil profile. Surface soil layers tend to have a greater P adsorption capacity than lower levels of the profile (Fig. 6.9).

Decomposition of organic wastes and dissolution of inorganic fertilizers provide a variety of organic and soluble forms of P in soil. Phosphorus may be present in such forms as soluble orthophosphate, condensed phosphate, tripolyphosphate, adsorbed phosphate or crystallized phosphate, thus, reflecting the chemical composition of the source and its phosphorus content. Hydrolysis and mineralization convert most of the condensed and polyphosphate forms to the soluble phosphate ion which is readily available to plants and soil microorganisms. Hence, soluble orthophosphate is released from organic wastes and soil humus through weathering and mineralization. On the other hand, it is expected that organic compounds resistant to decomposition will immobilize P, especially when the carbon:phosphorus ratio exceeds 300:1.

Given sufficient time, net mineralization will release P from organic substrates and this solubilized P generally may be used as a nutrient source by microbial populations degrading other carbonaceous substrates. Degradation of organic P compounds accounts for only 10–15% of the removal

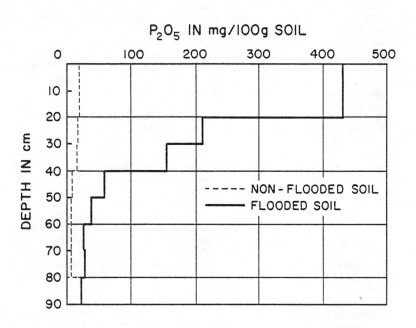

Figure 6.9. Phosphate distribution with depth in non-
flooded soil and soil flooded with sewage
water (Beek and de Haan, 1973). Reprinted
by permission of the Canadian Society of
Soil Science.

efficiency; however, microbes appear to be highly efficient in mobilizing the natural P reservoir in soil. Phosphorus concentrations in soil in quantities greater than the nutrient requirements for growth and substrate decomposition will be attenuated on the adsorption sites in the soil profile or reduced by dilution in the groundwater. Given sufficient retention time, P will precipitate as iron, aluminum or calcium phosphate (Ballard and Fiskell, 1974). The iron and aluminum oxides and hydrous oxides (e.g., hematite and gibbsite) are of primary importance since they have extremely high absorptive capacities (Ryden and Pratt, 1980).

Retention efficiency of the soil for P is related to the soil pH, cation exchange capacity, clay content and mineralogical composition. The equilibrium time for soil-phosphorus interactions is influenced by the retention time of the waste in soil, which is dependent on the soil infiltration capacity and permeability. The presence of organic anions and high pH will tend to decrease P sorption (Ryden and Syers 1975). Subbarao and Ellis (1977) and John (1974) report precipitation of calcium phosphates following liming usually control the solubility of P in acidic soils.

Phosphorus released from point sources will move radially by diffusion (Sawhney and Hill, 1975), thus increasing the P adsorption capacity through additional underground travel distance. Retention time may be positively influenced when waste leachate is slowed by the increased tortuosity or some relatively impermeable layer. If insufficient soil volume is available above the water table, the equilibration time in shallow soil can be drastically reduced and penetration to groundwater is likely to occur.

Phosphorus supplied in waste applications augmented over time may saturate the P adsorption capacity of the soil, thus creating the potential for extreme discharges to the groundwater. Adriano et al. (1975) showed evidence of perched water table contamination by P from daily application of food processing waste in quantities that exceed the adsorption maxima. Lund et al. (1976) observed that coarsely textured soil is enriched with P to a depth of 3 meters below sewage disposal ponds. Since soil has a finite capacity to fix P, attention should be directed to the long-term effect of waste applications containing P on the adsorption mechanisms.

The Langmuir isotherm has been used to estimate the P adsorption maximum of several soils (Table 6.12). To prepare a Langmuir isotherm test, standard amounts of soil are shaken with a known concentration of KH_2PO_4 over a dilution range of 0 to 100 mg/l of P. When the mass of the P adsorbed per gram of soil is linear with the equilibrium concentration of the P remaining in solution, the sorption maximum can be calculated from the slope. The Langmuir equation is:

$$C/b = C/b_{max} + (1/Kb_{max}) \tag{6.1}$$

where

C = equilibrium P concentration ($\mu g/ml$);
b = P adsorbed on soil surface ($\mu g/g$ soil);
b_{max} = adsorption maximum of the soil ($\mu g/g$ soil); and
K = constant related to the bonding energy.

The Langmuir adsorption maximum must be evaluated with the mineralogy, since P retention is known to improve when aluminum and iron are present in the soil. Successive P sorptions (Fig. 6.10) have been found to decrease the P sorption capacities of the soil (Sawhney and Hill, 1975). After wetting and drying treatments, the P sorption capacity may be reestablished in some soils such as the Merrimac sandy loam. In the Buxton silty clay loam the P sorption capacity was only partially reestablished. Thus, P in waste leachate in quantities that exceed the adsorption capacity can be expected to pass through the profile to groundwater.

TABLE 6.12 SUMMARY OF PHOSPHORUS ADSORPTION VALUES*

Compound Location	No. of Soil Samples	Notes	Sorption Capacity or b max. mgP/100 g soil
Michigan	29–100	Average for 1 m depth	1.81–49.0
Florida	6	Average for 50 cm depth	nil – 28.0
New Brunswick	24	Soils from upper B horizon	227–1760
New Jersey	17	A, B and C horizons	0.165–355
Maine	3	From column tests	26–71
	2	" " "	13.3–25.9
	5	" " "	3.8–51.0
	31	Average for 31 soils	12.0
New York	240	A, B and C horizons and deeper	0.3–278
Wisconsin	5	A, B and C horizons	2.5–20

* Tofflemire and Chen (1977).

Harvested forage crops may be used to remove as much as 50 to 60% of the P applied (Russel, 1973), however, annually harvested crops normally remove less than 10% of the annual P application (Ryden and Pratt, 1980). Furthermore, as the application of P increases, crop removal of the element decreases (Ryden and Pratt, 1980). Maximum crop removal is dependent on crops having a large rooting mass such as various grasses (Table 6.13). Moreover, studies have shown that P is the most limiting plant nutrient for production of legumes (Vallentine, 1971; Brady, 1974; Heath et al., 1978; Chessmore, 1979). A grass-legume mixture with legume species dominating may be a viable alternative to enhance P uptake in many land treatment units. Various herbaceous species may be clipped either two or three times a year, thus allowing significantly greater P removal.

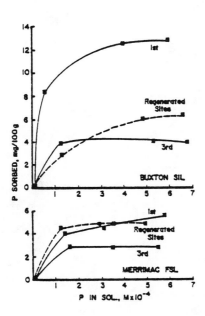

Figure 6.10. General Langmuir isotherms of Merrimac
sandy loam and Buxton silt loam after
successive P sorptions and following
wetting and drying treatments for regeneration of P sorption sites (Sawhney and
Hill, 1975). Reprinted by permission of
the American Society of Agronomy, Inc.

TABLE 6.13 REMOVAL OF PHOSPHORUS BY THE USUAL HARVESTED PORTION OF
SELECTED CROPS

Crop	Annual Crop Yield (Metric tons/ha)	Phosphorus Uptake (kg/ha/yr)
Corn (Zea mays)	11	35
Cotton (Gossypium hirsutum) Lint and seed	4.1	19
Wheat (Triticum aestivum)	5.2	22
Rice (Oryza sativa)	7.8	22
Soybeans (Glycine max)	3.0	25
Grapes (Vitus sp.)	27	11
Tomatoes (Lycopersicon esculentum)	90	34
Cabbage (Brassica oleracea)	78	18
Oranges (Citrus sp.)	60	11
Small grain, corn-hay rotation	---	32
Reed canary grass (Phalaris arundinacea)	---	45
Corn silage (Zea mays)	---	30–40
Poplar trees (Populus sp.)	---	26–69
Barley (Hordeum vulgare)- sudan grass (Sorghum sudanense) rotation for forages*	---	84–95
Johnson grass (Sorghum halepense)	27	94
Guinea grass (Panicum maximum)	26	50
Tall fescue (Festuca arundinacea)	7.8	32

* Unpublished data for barley in the winter followed by sudan grass in the
  summer. P.F. Pratt and S. Davis, University of California, and USDA-ARS,
  Riverside, California.

Application of P from wastewater may be described as either a low application rate system, usually less than 10 mg/1 or a high rate application system, consisting of greater than 10 mg/1 (Ryden and Pratt, 1980). Low rate systems use crop uptake as a sink for both the P and wastewaters applied. The P rates applied and the crop yields are comparable to those attained under good agronomic management of intensive cropland. Movement of P in this type of system is generally very slow since the P is retained near the zone of incorporation. The essential features of a low rate system are removal of a large amount of P by a forage crop, control of surface runoff to prevent erosion, and reduction of P concentrations to a desirable level by using a long pathway of highly sorptive materials between the soil surface and the discharge point of water into surface or groundwaters (Ryden and Pratt, 1980).

High-rate wastewater treatment systems usually have large quantities of water moving through the soil profile and the quantities of P applied are higher than those normally used on intensively farmed croplands. Thus, this system usually requires coarse gravelly soils which can maintain high infiltration rates (Ryden and Pratt, 1980). Generally, a cycle of flooding and drying is used to maintain the infiltration capacity of the system and increase the P sorption capacity by enhancing the oxidation-reduction cycle. Soils with a high sand or organic content that have low contents of iron and aluminum hydrous oxides associated with a low surface area are most likely to have the greatest leaching of P (Syers and Williams, 1977). Ryden and Pratt (1980) report P removal by harvested crops, in a high rate system, to be insignificant unless P concentrations are less than 1 mg/1.

6.1.2.3 Boron (B)

The B concentration in rocks varies from 10 ppm in igneous rocks to 100 ppm in sandstones. The average soil concentration of B is 10 ppm (Bowen, 1966). High levels of B are most likely to occur in soil derived from marine sediments and arid soils. In most humid region soils, B is bound in the form of tourmaline, a borosilicate that releases B quite slowly. Most of the available soil B is held by the organic fraction where it is tightly retained. Boron is released as the organics decompose and is quite subject to leaching losses. Some B is adsorbed by iron and aluminum hydroxy compounds and clay minerals. Finer textured soils retain added B longer than do coarse, sandy soils. Therefore, less B can be applied to sandy soil than to fine-textured soil (Tisdale and Nelson, 1975). Boron sorption by clay minerals and iron and aluminum oxides is pH dependent, with maximum sorption in the pH range 7-9. The amount of B adsorbed depends on the surface area of the clay or oxide and this sorption is only partially reversible, indicating the retention is by covalent bonding.

Boron is frequently deficient in acid soils, light-textured sandy soils, alkaline soils, and soils low in organic matter. Boron availability to plants is decreased by liming, but the increase of pH alone is not sufficient to decrease B absorption. Fox (1968) found that both high levels of calcium and high pH values reduced B uptake by cotton by nearly 50%, but that high calcium concentrations or high pH studied separately had little influence on reducing B uptake.

Boron in plants is involved in protein synthesis, nitrogen and carbohydrate metabolism, root system development, fruit and seed formation, and the regulation of plant water relations (Brady, 1974). The symptoms of B deficiency vary somewhat from one plant species to another. Symptoms often include dieback, chlorotic spotting of leaves and necrosis in fruits and roots (Bradford, 1966).

The difference between the amount of B which results in deficiencies and that which is toxic is very small. Boron-sensitive plants can tolerate between 0.5 and 1.0 ppm available B in soils while boron-tolerant plants usually show toxicity symptoms at 10 ppm B (Bingham, 1973). Table 6.14 shows the tolerance limits of several plant species to boron. The first symptoms of B injury are generally leaf-tip yellowing, followed by a progressive necrosis of the leaf. Leaching of B below the root zone is recommended in the case of moderate toxicity. Moderate liming of the soil or liberal application of nitrogen fertilizers may be beneficial (Bradford, 1966).

If B can be leached from the soil at concentrations acceptable for groundwater discharge, B may be applied continously in small amounts as long as it does not accumulate to toxic levels. No drinking water standard has been set for human consumption; however, water used for cattle should contain less than 5 ppm B.

6.1.2.4 Sulfur (S)

The earth's crust contains about 600 ppm S and soils have an average S content of 700 ppm (Tisdale and Nelson, 1975). Since S is a constituent of some amino acids, it is an important plant nutrient. The widespread occurrence of S in nature ensures that it will be a common industrial waste product. Wastes from kraft mills, sugar refining, petroleum refining, and copper and iron extraction all contain appreciable amounts of S (Overcash and Pal, 1979).

Because of its anionic nature and the solubility of most of its salts, leaching losses of S can be quite large. Leaching is greatest when monovalent cations such as potassium and sodium predominate and moderate leaching occurs where calcium and magnesium predominate. When the soil is acidic and appreciable levels of exchangeable iron and aluminum are present, S leaching losses are minimal (Tisdale and Nelson, 1975).

Land application sites where wastes containing large amounts of S are disposed must be well drained. The hydrogen sulfide formed in reducing conditions is toxic and has an unpleasant odor. Since acid is formed by oxidation of S compounds, the pH of the site must be monitored and regulated. In the soil under aerobic conditions, bacteria oxidize the more

TABLE 6.14 CROP TOLERANCE LIMITS FOR BORON IN SATURATION EXTRACTS OF SOIL*[†]

Tolerant	Semitolerant	Sensitive
4.0 ppm B	2.0 ppm B	1.0 ppm B
Athel (Tamarix aphylla)	Sunflower (Hellanthus annus)	Pecan (Carya illnoensis)
Asparagus officinalis	Potato (Solanum tuberosum)	Walnut, Black and Persian, or
Palm (Phoenix canariensis)	Cotton, Acala and Pima	English (Juglans spp.)
Date palm (P. dactylifera)	(Gossypium sp.)	Jerusalem artichoke
Sugarbeet (Beta vulgaris)	Tomato (Lycopersicon esculentum)	(Hellanthus tuberosus)
Mangel (Beta vulgaris)	Sweetpea (Lathyrus odoratus)	Navy bean (Phaseolus vulgaris)
Garden beet (Beta vulgaris)	Radish (Raphanus sativus)	American elm (Ulmus americana)
Alfalfa (Medicago sativa)	Field pea (Pisum sativum)	Plum (Prunus domestica)
Gladiolus (Gladiolus sp.)	Ragged-robin rose (Rosa sp.)	Pear (Pyrus communis)
Broadbean (Vicia faba)	Olive (Olea europaea)	Apple (Malus sylvestris)
Onion (Allium cepa)	Barley (Hordeum vulgare)	Grape, Sultanina and Malaga
Turnip (Brassica rapa)	Wheat (Triticum aestivum)	(Vitus sp.)
Cabbage (Brassica oleracea	Corn (Zea mays)	Kodata fig (Ficus carica)
var. capitata)	Milo (Sorghum bicolor)	Persimmon (Diospyros virginiana)
Lettuce (Lactuca sativa)	Oat (Avena sativa)	Cherry (Prunus sp.)
Carrot (Daucus carota)	Zinnia (Zinnia elegans)	Peach (Prunus persica)
	Pumpkin (Cucurbita spp.)	Apricot (Prunus armeniaca)
	Bell Pepper (Capsicum annuum)	Thornless blackberry (Rubus sp.)
	Sweet potato (Ipomoea batatas)	Orange (Citrus sinensis)
	Lima bean (Phaseolus lunatus)	Avocado (Persea americana)
		Grapefruit (Citrus paradisi)
		Lemon (Citris limon)
2.0 ppm B	1.0 ppm B	0.3 ppm B

* Bresler et al. (1982).

[†] For each group, tolerant, semitolerant, and sensitive, the range of tolerable boron is indicated; tolerance decreases in descending order in each column.

reduced forms of S to form sulfate which will decrease the pH. In water-logged soils, anaerobic bacteria reduce sulfides, generating hydrogen sulfide.

Some soils have the capacity of retain sulfates in an adsorbed form. At a given pH, adsorption is least when the cation adsorbed on the clay is potassium, moderate when the adsorbed cation is calcium, and greatest when the adsorbed cation is aluminum (Tisdale and Nelson, 1975). Adsorption by clay minerals is ranked as kaolinite<illite<bentonite (Chao et al., 1963).

When soils contain large amounts of carbon and nitrogen, but little S, immobilization of added S may occur when S is incorporated into proteins by soil microorganisms. Organic S may also be mineralized in which the organic form becomes the plant available SO_4 (Brady, 1974). Sulfur behaves much like nitrogen as it is absorbed by plants and microorganisms and moves through the S cycle.

Management techniques for land treatment systems receiving large amounts of S can improve the S assimilation capacity of soils. A slightly acidic pH will minimize leaching losses, but it must not be so much below neutral that mineralization and plant uptake are reduced. The amount of S which can be applied to a particular soil depends on the ability of that soil to neutralize the acidity resulting from the addition. If acid-tolerant plants are chosen, a larger addition is possible. Active pH monitoring and pH correction, when required, is essential.

6.1.3   Acids and Bases

Waste acids and/or bases can be disposed by land treatment. These wastes should, if at all possible, be neutralized before they are applied to the soil. According to the Lowry-Bronsted theory of acid-base reactions, an acid is any material which produces hydronium (H_3O^+) ions when dissolved in water. Conversely, a base is a material which produces hydroxyl (OH^-) ions in water. Thus, when an acid and base are combined, the net neutralization reaction can be expressed as:

$$H_3O + OH^- \rightleftharpoons 2H_2O$$

As the neutralization reaction occurs, the cations and anions from the original acidic and basic species combine to form a salt. With strong acids and bases, the aqueous reaction equilibrium strongly favors dissociation into hydronium and hydroxyl ions. With weaker species, however, the dissociation equilibrium will depend on the strengths of the ionization constants (Bohn et al., 1979).

The buffering capacity of the soil should be determined and used as a guide to loading rates. If the buffering capacity is exceeded, the soil pH must be adjusted by appropriate liming or addition of acid. When both acidic and basic wastes exist, the basic waste should be applied first and mixed with the soil, then the acidic waste can be applied. This method will prevent the solubilization and leaching of metals in the soil. Addition of acids and bases to the soil can increase the concentration of soluble salts in the system. For a discussion of salts, refer to Section 6.1.4. Management of soil pH is discussed in Section 8.6.

6.1.4   Salts

By definition, a salt is any substance that yields ions upon dissolution other than hydrogen ions or hydroxyl ions. For all practical purposes in agriculture and land treatment, this definition has been narrowed to include only the major dissolved solids in natural waters and soils. The principal ions involved are calcium, magnesium, sodium, potassium, chloride, sulfate, bicarbonate and occasionally nitrate. Salts occur naturally in many soils and are a common constituent of hazardous and nonhazardous wastes. Salt inputs to the soil may occur from fertilizer applications, precipitation, and irrigation. Typical irrigation practices may result in annual salt applications to soil which exceed 4000 kg/ha. Table 6.15 lists the salinity classes of water.

The behavior of salts in soil and their influence on plant growth has been studied by agricultural scientists for many years and is still the topic of extensive research. The U.S. Salinity Laboratory Staff (USDA, 1954) and Bresler et al. (1982) have reviewed various aspects of soil salinity, including diagnosis and management of salt affected soils. Salinity problems may result from the bulk osmotic effects of salts on the soil-plant system and the individual effects of specific ions, especially sodium.

6.1.4.1 Salinity

The concentration of salt in water can be expressed in terms of electrical conductivity (EC), total dissolved solids (TDS), osmotic pressure, percent salt by weight, and normality. Electrical conductivity in mmhos/cm is the preferred measurement for solutions of common salts or combinations of salts. The following factors are useful for obtaining an approximate conversion of units.

(0.35) x (EC mmhos/cm) = Osmotic pressure in bars
(651) x (EC mmhos/cm) = TDS mg/l
(10) x (EC mmhos/cm) = Normality meq/l
(0.065) x (EC mmhos/cm) = Percent salt by weight

Measuring the concentration of salts in soil first requires that an aqueous soil extract be obtained. Extracts taken from soils at field moisture content will seldom provide a sufficient quantity for analysis. On the other hand, exhaustive leaching or extraction at very high moisture contents will yield a sample that is not typical of the soil solution

TABLE 6.15 WATER CLASSES IN RELATION TO THEIR SALT CONCENTRATION*

Class of Water	Electrical Conductivity micromho per cm at 25°C	Milligrams per liter	Kilograms per hectare-30 cm	Comments
Low salinity water	0– 400	0– 250	0– 800	These waters can be used for irrigating most crops with a low probability that salt problems will develop. Some leach is required, but this generally occurs with normal irrigation practices.
Moderate salinity water	400–1,200	250– 750	800–2,200	These waters can be used if a moderate amount of leaching occurs. Plants with moderate salt tolerance can be grown in most instances without special practices for salinity control.
High salinity water	1,200–2,250	750–1,450	2,200–3,300	These waters should not be used on soils with restricted drain age. Special management is required even with adequate drainage. Plants tolerant to salinity should be grown. Excess water must be applied for leaching.
Very high salinity water	2,250–5,000	1,450–3,200	3,300–9,600	These waters are not suitable for irrigation except under very special circumstances. Adequate drainage is essential. Only very salt-tolerant crops should be grown. Considerable excess water must be applied for leaching.

* Bresler et al. (1982).

because of the effect of ion exchange and mineral dissolution. As a compromise, soil saturation has been selected for obtaining aqueous extracts (USDA, 1954). A sufficient amount of solution can usually be extracted with vacuum from 200-300 grams of soil. The concentration of salts in soil is, therefore, commonly expressed as the EC of a saturated soil paste extract. The relationship of salt concentration in the soil to the EC of a saturation extract is influenced by the moisture holding capacity of the soil as illustrated in Fig. 6.11. The EC of a saturation extract does not directly reflect the salinity of the soil solution, but the saturation extract is the best practical means to obtain such a measurement. Under a typical irrigated crop system, the average salinity of the soil solution is approximately twice the salinity of the saturation extract (Rhoades, 1974); however, use of the saturation extract is so widely practiced that it is the measure best correlated in the literature to plant growth responses, soil structure, and other observations of soil condition.

In the absence of adequate rainfall or irrigation and subsequent drainage, applied or naturally occurring salts can accumulate on the soil surface and in upper horizons of the soil. Salt concentrations in the soil that exceed 4 mmhos/cm can inhibit growth of sensitive plants and may retard microbial activity. Physical and chemical characteristics of the soil are also affected by salt accumulation. Severe salt accumulation can be disastrous to a land treatment system and may require costly remedial action. Furthermore, soluble salts are relatively mobile in the soil and can easily migrate to ground or surface waters, resulting in pollution. Management of salts applied in irrigation water or waste materials therefore requires that salt accumulation be controlled, while at the same time pollution of ground or surface waters is prevented.

Many schemes for managing salt accumulation and migration assume steady state conditions and that applied salts do not interact with the soil matrix. Salts do, however, interact with the soil matrix. They may be precipitated as insoluble compounds, sorbed by soil colloids, or dissolved in the soil solution. The extent of precipitation, sorption and dissolution depends upon the salt concentration in the soil, the ionic species present, soil physical and chemical properties, and the moisture content of the soil. Predicting the concentration of salts in the soil solution at any given time for a particular soil is therefore difficult. The assumptions of steady state and no interactions may be valid in an irrigated crop system, but is not applicable to many land treatment systems, especially those receiving relatively heavy and infrequent waste applications. Understanding soil and salt interactions may, and should, be quantified and included in the waste application rate design.

Where inadequate water or poor soil drainage prevent leaching of salts from the treatment zone or the plant root zone, salts will concentrate in the soil through evaporation. The soil surface behaves like a semi-permeable membrane allowing soil water to enter the atmosphere through evaporation while leaving dissolved salts at or near the soil surface. Once salts are deposited at the soil surface in this manner, additional soil water and its dissolved salts are driven to the surface by osmotic forces in addition to evaporative demand. For this reason, many saline soils will

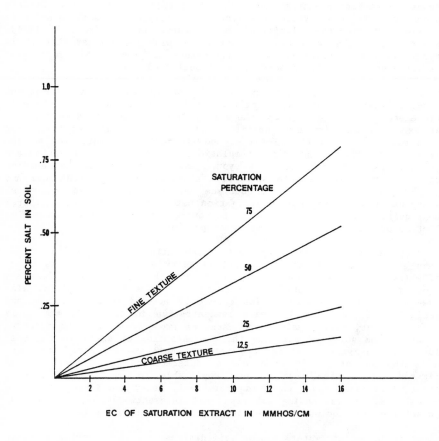

Figure 6.11. Correlation of salt concentration in the soil to the EC of saturation extracts for various soil types (USDA, 1954).

appear to be moist, when in reality there is little or no water available for plants or waste decomposing microbes.

Soil salinity inhibits plant growth by restricting plant uptake of water. As the osmotic gradient between the soil solution and plant roots increases, the plant uptake of water and nutrients decreases. This same mechanism may also adversely affect the growth of soil microbes. Crop sensitivity to salt damage varies between different species and varieties depending on the specific salts present. See Table 6.16 for general crop response to soil salinity and Table 6.17 for the salt tolerance of various crops. For specific choice of the proper plant species, other factors, such as drought tolerance and regional adaptation, must be considered. Additional guidance on species selection is provided in Section 8.7.

TABLE 6.16 GENERAL CROP RESPONSE AS A FUNCTION OF ELECTRICAL CONDUCTIVITY*

EC (mmhos/cm)	Degree of Problem
0-2	None
2-4	Slight to none
4-8	Many crops affected
8-16	Only tolerant crops yield well
greater than 16	Only very tolerant crops yield well

* USDA (1954).

Salts that accumulate in surface soils may be reduced by precipitation, irrigation, and to a small extent by crop uptake. In the presence of adequate precipitation or irrigation, the salts dissolve and are then carried away in runoff or are leached into the subsoil. Leached salts may be transported back to the soil surface as a result of evaporation if subsequent precipitation or irrigation does not occur. If a sufficient quantity of drainage water passes through the soil profile, leached salts may be carried farther into the subsurface and may intercept groundwater. The concentration and quantity of salts present in drainage water and that remaining in the surface soil may be approximated by a mass balance approach such as that proposed by Rhoades (1974).

In general, management of the soil-plant system to prevent damaging salt accumulation in surface soils includes the following:

(1) limiting the amount of salt applied to the soil in irrigation water or waste;

(2) using salt tolerant crops;

(3) maintaining a healthy vegetative cover or mulching;

(4) properly scheduling irrigation and waste applications; and

TABLE 6.17 THE RELATIVE PRODUCTIVITY OF PLANTS WITH INCREASING SALT CONCENTRATION IN THE ROOT ZONE*†

Plant	Relative Productivity, % at Selected EC mmho/cm									% Productivity decrease per mmho/cm increase	Salinity Threshold EC
	1	2	3	4	5	6	7	8	9		
SENSITIVE:											
Algerian ivy (Hedera canariensis)'	100	81								--	1.0
Almond (Prunus dulcis)	100	91	73	55	36	18	0			18	1.5
Apple (Malus sylvestris)'	100	91	91							--	1.0
Apricot (Prunus armeniaca)	100	91	68	45	27	0				23	1.6
Avocado (Persea americana)	100	90	70							--	1.0
Bean (Phaseolus vulgaris)	100	81	62	43	25	6	0			18.9	1.0
Blackberry (Rubus spp.)	100	89	67	44	22	0				22.2	1.5
Boysenberry (Rubus ursinus)	100	89	67	44	22	0				22.2	1.5
Burford holly (Ilex cornuta)	100	82	59	36	14	0				--	1.0
Carrot (Daucus carota)	100	86	72	58	44	30	15	1	0	14.1	1.0
Celery (Apium graveolens)'	100	90	75							--	1.0
Grapefruit (Citrus paradisi)	100	97	81	65	48	32	16	0		16.1	1.8
Heavenly bamboo (Nandina domestica)	100	88	75	61	47	34	20	7	0	--	1.0
Hibiscus (Hibiscus rosa-sinensis)'	100	86	72	58	42	28	15	0		--	1.0
Lemon (Citrus limon)	100	91	75							--	1.0
Okra (Abelmoschus esculentus)'	100	90	71							--	
Onion (Allium cepa)	100	87								16.1	1.2
Orange (Citrus sinensis)	100	95	79	63	48	32	16	0		15.9	1.7
Peach (Prunus persica)	100	94	73	52	31	10	0			18.8	3.2
Pear (Pyrus spp.)	100	91	75							--	1.0
Pineapple guava (Feijoa selloviana)	100	71	34	0						--	1.2
Plum (Prunus domestica)	100	91	73	55	36	18	0			18.2	1.5
Prune (Prunus domestica)'	100	91	75							--	1.0
Pittosporum (Pittosporum tobira)†	100	89	79	69	60	50	40	30	20	--	1.0
Raspberry (Rubus idaeus)'	100	60	62							--	1.0
Rose (Rosa spp.)	100	74	36	0						--	1.0
Strawberry (Fragaria sp.)	100	67	33	0						33.3	1.0

--Continued--

TABLE 6.17 (continued)

Plant	Relative Productivity, % at Selected EC mmho/cm																								% Productivity decrease per mmho/cm Increase	Salinity Threshold EC
	1	2	3	4	5	6	7	8	9	10	11	12	13	14	15	16	17	18	19	20	21	22	23	24		
Star Jasmine (Trachelospermum jasminoides)	100																								—	1.6
MODERATELY SENSITIVE:																										
Alfalfa (Medicago sativa)	100	83	61	40	18	0																			7.3	2.0
Arborvitae (Thuja orientalis)+	100	100	93	85	78	71	64	56	49	42	34	27	20	12											—	2.0
Bottlebrush (Callistemon viminalis)*	100	94	85	77	68	59	50	41	33																—	1.5
Boxwood (Buxus microphylla var. Japonica)	100	96	86	76	65	54	43	32	21	11	0														10.8	1.7
Broadbean (Vicia faba)	100	96	87	77	67	58	48	38	29	19	10	0													9.6	1.6
Cauliflower (Brassica oleracea)*	100	100	93	85																					—	2.5
Cabbage (Brassica oleracea)	100	98	88	79	69	59	50	40	30	20	11	1	0												9.7	1.8
Clover, alsike, ladino red, strawberry (Trifolium spp.)	100	94	82	70	58	40	34	22	10	0															12.0	1.5
Corn, forage (Zea mays)	100	99	91	84	76	69	61	54	47	39	32	24	17	10											7.4	1.8
Corn, grain, sweet (Zea mays)	100	96	84	72	60	48	36	24	12	0															12.0	1.7
Cowpea (Vigna unguiculata)	100	90	76	61	47	33	19	4	0																14.3	1.3
Cucumber (Cucumis sativus)	100	100	94	81	68	55	42	29	16	3	0														13.0	2.5
Dodonaea (Dodonia viscosa var. Atropurpurea)	100	94	86	77	68	59	51	42	33	25	17	9	0												7.8	1.0
Flax (Linum usitatissimum)	100	96	84	72	60	48	36	24	12	0															12.0	1.7
Grape (Vitis spp.)	100	95	86	76	66	57	47	38	28	18	9	0													9.5	1.5
Juniper (Juniperus chinensis)	100	91	81	72	63	54	45	36	27	18	9	0													9.5	1.5
Lantana (Lantana camera)	100	92	82	72	62	52	41	30	20	9	0														—	1.8
Lettuce (Lactuca sativa)	100	91	78	65	52	39	26	13	0																13.0	1.3
Lovegrass (Eragrostis spp.)	100	100	92	83	75	66	58	49	41	32	24	15	7	0											8.5	2.0
Meadow foxtail (Alopecurus pratensis)	100	95	85	76	66	56	47	37	27	17	8	0													9.7	1.5
Muskmelon (Cucumis melo)*	100	100	95	80																					—	2.5

--continued--

TABLE 6.17 (continued)

Plant	Relative Productivity, % at Selected EC mmho/cm																								% Productivity decrease per mmho/cm increase	Salinity Threshold EC
	1	2	3	4	5	6	7	8	9	10	11	12	13	14	15	16	17	18	19	20	21	22	23	24		
Oleander (*Nerium oleander*)+	100	100	93	86	79	72	65	58	51	44	37	30	24												---	2.0
Pea (*Pisum sativum*)#	100	100	90																						---	2.5
Peanut (*Arachis hypogaea*)	100	100	100	77	49	20	0																		28.6	3.2
Pepper (*Capsicum annuum*)	100	93	79	65	51	37	23	8	0																14.1	1.5
Potato (*Solanum tuberosum*)	100	96	84	72	60	48	36	24	12	0															12.0	1.7
Pyracantha (*Pyracantha braperi*)	100	99	90	81	72	62	53	43	34	24	14	6	0												9.1	2.0
Radish (*Raphanus sativus*)	100	90	77	64	51	38	25	12	0																13.0	1.2
Rice, Paddy (*Oryza sativa*)	100	100	100	88	76	63	51	39	27	15	2	0													12.2	3.0
Sesbania (*Sesbania exaltata*)	100	100	95	88	81	74	67	60	53	47	40	33	26	19											7.0	2.3
Spinach (*Spinacia oleracea*)	100	100	92	85	77	70	62	55	47	39	32	24	17	9											7.6	2.0
Squash (*Cucurbita maxima*)#	100	100	90	74																					---	2.5
Sugarcane (*Saccharum officinarum*)	100	98	92	86	81	75	69	63	57	51	45	39	34	28											5.9	1.7
Silverberry (*Elaeagnus pungens*)	100	95	87	78	69	59	50	40	29	18	7	0													---	1.6
Sweet potato (*Ipomoea batatas*)	100	95	84	73	62	51	40	29	18	7	0														11.0	1.5
Texas privet (*Ligustrum lucidum*)	100	94	85	75	66	56	46	36	26	16	7	0													9.1	2.0
Tomato (*Lycopersicon esculentum*)	100	100	95	85	75	65	55	46	36	26	16	6	0												9.9	2.5
Trefoil, Big (*Lotus uliginosus*)	100	100	87	68	49	30	11	0																	18.9	2.3
Vetch, Common (*Vicia sativa*)	100	100	100	89	78	67	56	44	33	22	11	0													11.1	3.0
Viburnum (*Viburnum* spp.)	100	90	73	58	44	32	20	10	0																13.2	1.4
Xylosma (*Xylosma senticosa*)	100	94	81	67	54	40	27	14	0																13.3	1.5
MODERATELY TOLERANT:																										
Alkali sacaton (*Sporobolus airoides*)#	100	100																							---	---
Barley, forage (*Hordeum vulgare*)	100	100	100	100	100	100	93	86	79	72	65	58	51	44	37	30	23	15	8						7.0	6.0
Beet, garden (*Beta vulgaris*)	100	100	100	100	91	82	71	64	55	46	38	29	20	11	2	0									9.0	4.0
Broccoli (*Brassica oleracea* var. Capitata)	100	100	98	89	80	71	61	52	43	34	25	16	6	0											9.1	2.8

---continued---

222   Environmental Fate

TABLE 6.17 (continued)

Plant	1	2	3	4	5	6	7	8	9	10	11	12	13	14	15	16	17	18	19	20	21	22	23	24	% Productivity decrease per mmho/cm increase	Salinity Threshold EC
Clover, berseem (Trifolium alexandrinum)	100	97	91	86	80	74	69	63	57	51	46	40	34	29	23	17	11	6	0						5.8	1.5
Dracaena endivisa (Dracaena endivisa)	100	100	100	94	85	76	67	58	49	40	31	22	13	4	0										9.1	4.0
Euonymus (Euonymus japonica var. grandiflora)							100																		--	7.0
Fescue (Festuca elatior)	100	100	100	99	94	89	84	78	73	68	62	57	52	47	41	36	31	25	20						5.3	3.9
Fig (Ficus carica)	100	100	100	100	90	85	82	74	67	59	52	44	36	29	21	14	6	0							--	4.2
Hardinggrass (Phalaris tuberosa)	100	100	100	100	100	89	82	74	67	59	52	44	36	29	21	14	6	0							7.6	4.6
Kale (Brassica campestris)	100	100	100	100	100	100	90																		--	6.5
Olive (Olea europaea)	100	100	100	100	100	85																			--	4.0
Orchard grass (Dactylis glomerata)	100	97	91	84	78	72	66	60	53	47	41	35	29	22	16	10	4	0							6.2	1.5
Pomegranate (Punica granatum)	100	100	100	100	90	85																			--	--
Ryegrass, perennial (Lolium perenne)	100	100	100	100	100	97	89	82	74	67	59	52	44	36	29	21	14	6	0						7.6	5.6
Safflower (Carthamus tinctorius)[a]	100	100	100	100	100	100	97	90	85	80	75	50													--	6.5
Sorghum (Sorghum bicolor)[a,b]	100	100	100	100	98	90	84	78	70	63	56	50	43	36	29	22	15	8	0						4.8	4.8
Soybean (Glycine max)	100	100	100	100	100	80	80	40	20	0															20.0	5.0
Sudangrass (Sorghum sudanense)	100	100	99	95	91	86	82	78	73	69	65	61	56	52	48	43	38	35	30						4.3	2.8
Trefoil, birdsfoot (Lotus corniculatus tenuifolium)	100	100	100	100	100	100	80	70	60	50	40	30	20	10	0										10.0	5.0
Wheat (Triticum aestivum)	100	100	100	100	100	100	93	86	79	71	64	57	50	43	36	29	21	14	7						7.1	6.0
Wildrye, beardless (Elymus triticoides)	100	100	100	92	86	80	74	68	62	56	50	44	38	32	26	20	14	8	2						6.0	2.7
TOLERANT:																										
Barley, grain (Hordeum vulgare)					100	100	100	100	95	90	85	80	75	70	65	60	55	50	45	40	35				5.0	8.0
Bermudagrass (Cynodon dactylon)					100	100	99	93	87	80	74	67	61	54	48	42	35	29	22	16	10	3	0		6.4	6.9
Bougainvillea (Bougainvillea spectabilis)							100																		--	8.5
Cotton (Gossypium hirsutum)				100	100	100	100	98	93	88	83	78	73	67	62	57	52	47	41	36	31	26	21	16	5.2	7.7
Date (Phoenix dactylifera)				100	96	93	89	86	82	78	75	71	68	64	60	57	53	49	46	42	39	35	31	28	3.6	4.0

--continued--

TABLE 6.17 (continued)

Plant	1	2	3	4	5	6	7	8	9	10	11	12	13	14	15	16	17	18	19	20	21	22	23	24	% Productivity decrease per mmho/cm increase	Salinity Threshold EC
Natal Plum (Carissa grandiflora)[#]								82		68															—	6.0
Rosemary (Rosmarinus lockwoodii)[#][**]					100	95	85	75																	—	4.5
Sugarbeet (Beta vulgaris)[+]				100	100	100	100	94	88	82	76	71	65	59	53	47	41	35	29	24	18	12	6	0	5.9	7.0
Wheatgrass, Crested (Agropyron desertorum)				98	94	90	86	82	78	74	70	66	62	58	54	50	46	42	38	34	30	26	22	18	4.0	3.5
Wheatgrass, Fairway (Agropyron cristatum)				100	100	100	100	97	90	83	76	69	62	55	48	41	34	28	21	14	7	0			6.9	7.5
Wheatgrass, tall (Agropyron elongatum)				100	100	100	100	98	94	89	85	81	77	73	68	64	60	56	52	47	43	39	35	31	4.2	7.5
Wildrye, Altai (Elymus angustus)				100	100	100																			—	—

* Bresler et al. (1982).

! Salt concentration is shown as the electrical conductivity of saturated soil extracts (EC).

Tabled values are estimates based on the EC for a relative yield of 90% and yield reductions for similar crops as EC increases.

+ The lower part of the yield curve approaches zero asymptotically to the abscissa; only linear data are shown.

§ Tabled values are based on three data points available in the literature.

** Tabled values are based on three data points, productivity drops sharply towards zero for the lower 50% productivity.

224   Environmental Fate

(5) prudent leaching of salts below the root zone through irrigation.

In addition, migration of unacceptable quantities of salts to ground or surface waters may be controlled by:

(1) using soil erosion and runoff control practices;

(2) avoiding locations with shallow unconfined aquifers;

(3) limiting the amount of applied salt through optimum waste application rates in conjunction with soil, soil water, and groundwater monitoring; and

(4) using effective irrigation practices.

Where salts are anticipated to be a problem in a given waste, choice of a site having at least moderately well drained soils is essential to maintain the usefulness of the land treatment unit. In soils where a high water table causes continued capillary rise of salts, subsurface drainage (e.g., drain tile or ditches) can be installed to lower the water table and the associated capillary fringe.

Aside from these general guidelines, there is no reliable and widely available means to quantify acceptable salt loading rates and management practices. The approach described by the Salinity Laboratory Staff (USDA, 1954) is inappropriate to the case of intentional salt applications, and, even if it were modified to better fit the given case, the method is too simplistic to reliably yield results that are accurate enough for design purposes. Therefore, it is recommended that this simplistic approach not be patently applied to all situations. Some, more complex, computer models which show promise are in developmental or modification stages (Dutt et al., 1972; Franklin, personal communication). These models, however, would require considerable alteration to apply generally and in a land treatment context. Based on the current lack of a definitive solution to the problem, salt management questions in a land treatment system should be referred to a soil scientist having specific experience regarding saline and sodic soils. Other useful information can be found in a book by Bresler and McNeal (1982).

6.1.4.2 Sodicity

Sodium, as a constituent of soluble salts contained in applied waste or irrigation water, deteriorates soil structure and exhibits direct toxic effects on sensitive crops. When soluble salts accumulate in the surface soil, sodium salts may be preferentially concentrated in the soil solution because of their higher solubility in comparison to the corresponding calcium, magnesium, or potassium salts. Sodium ions are, therefore, more available for plant uptake and to compete in cation exchange reactions with soil colloids. Sodic effects on soils and crops can be minimized by limiting the amount of applied sodium and by maintaining a favorable balance between sodium ions and other basic cations in the soil solution.

Sodium affects soil structure by dispersing flocculated organic and inorganic soil colloids. Dispersion occurs when sodium ions are adsorbed to clay surfaces and colloidal organic matter causing individual particles to repel one another. In addition, sodium ions can hydrolyze water molecules resulting in elevated soil pH and dissolution of soil organic matter that holds soil aggregates together (Taylor and Ashcroft, 1972). As soil aggregates are collapsed by raindrop impact and tillage, the infiltration capacity and hydraulic conductivity of the soil decrease significantly. Air and water entry into soil is then restricted so runoff increases, soil erosion increases, plants die, and oxidative waste degradation processes in the soil are slowed. Sodium affected soils can be reclaimed by adding various soil amendments and intensively managing the site. Reclamation efforts, however, can be costly and are often ineffective. The threshold sodium concentration of the soil solution that results in dispersion of soil colloids is influenced by several factors including the following:

(1) the relative concentration of sodium to calcium and magnesium is commonly expressed as the sodium adsorption ratio (SAR) where concentrations are expressed in normality (meq/1)

$$SAR = \frac{[Na]}{\left(\frac{[Ca] + [Mg]}{2}\right)^{1/2}} \qquad (6.2)$$

(2) the salinity of the soil solution;

(3) physical and chemical soil properties;

(4) cropping and tillage practices; and

(5) irrigation and waste application methods.

Prediction of a threshold value in terms of sodium application to the soil is therefore difficult. The USDA (1954) states that soil sodicity occurs when the percentage of exchangeable sodium exceeds 15 or the SAR of a saturated soil paste extract exceeds 12. Other researchers, however, have observed decreased infiltration rates when SAR values are as low as 5 (Miyamoto, 1979). Permeability is also decreased when the exchangeable sodium percentage (ESP) increases. Figure 6.12 illustrates that hydraulic conductivity is decreased by over 50% when the ESP is raised from 5 to 10%. As with soil salinity, management schemes for predicting and controlling sodicity have been developed for irrigated agriculture and assume steady state conditions. To the extent that these schemes apply to land treatment systems, the general approach assumes that the SAR should be maintained at or preferably below 12. Management to achieve this objective would logically fall into one of the following approaches:

(1) waste pretreatment or addition of calcium or magnesium salts to maintain the SAR of the waste below the critical level;

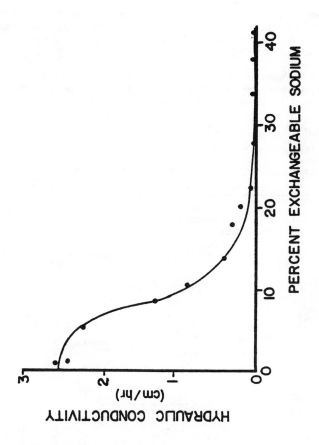

Figure 6.12.   Effect of increasing ESP upon hydraulic conductivity (Martin et al., 1964). Reprinted by permission of the Soil Science Society of America.

(2) calcium or magnesium salts (e.g., gypsum) amendements to soils;

(3) applications of waste to larger areas of land; and

(4) allow SAR to exceed critical levels, then take corrective action (the least attractive alternative).

Details of these approaches can be found in Overcash and Pal (1979). Sodium affected soils can be diagnosed by the occurrence of decreased infiltration rates, low aggregate stability, elevated levels of exchangeable sodium, and elevated soil pH.

The phytotoxicity of sodium to various crops is listed in Table 6.18. Sodium toxicity can occur through direct plant uptake of sodium and through nutrient imbalance caused by an unfavorable calcium to sodium ratio (USDA, 1954).

TABLE 6.18   SODIUM TOLERANCE OF VARIOUS CROPS*

Tolerance Range	Crop
Extremely Sensitive (Exchangeable Na = 2-10%)	Deciduous fruits Nuts Citrus Avocado (Persea americana)
Sensitive (Exchangeable Na = 10-20%)	Beans (Phaseolus spp.)
Moderatley Tolerant (Exchangeable Na = 20-40%	Clover (Trifolium spp.) Oats (Avena fatua) Tall fescue (Festuca arundinacea) Rice (Oryza sativa) Dallis grass (Paspalum dilatatum)
Tolerant (Exchangeable Na = 40-60%)	Wheat (Triticum aestivum) Cotton (Gossypium hirsutum) Alfalfa (Medicago sativa) Barley (Hordcum vulgare) Tomatoes (Lycopersicon esculentum) Beets (Beta vulgaris)
Most Tolerant (Exchangeable Na exceeds 60%)	Crested wheatgrass (Agropyron desertorum) Fairway wheatgrass (Agropyron cristatum) Tall wheatgrass (Agropyron elongatum) Rhodesgrass (Chloris gayana)

* Pearson (1960).

6.1.5   Halides

The halides are the stable anions of the highly reactive halogens, fluorine (F), chlorine (Cl), bromine (Br) and iodine (I). Although halides occur naturally in soils, overloading a land treatment facility with wastes high in halides poses a toxic threat to soil microbes, cover crops and grazing animals. Chloride, iodide, and probably fluoride are essential nutrients to animals, however, only chloride is essential to plants. Each of the halides is discussed below with respect to its sources in wastes, background levels, mobility in soils, and plant and animal toxicity. The fate of halogenated organic compounds is discussed in Section 6.2.3.4.

6.1.5.1 Fluoride

Fluoride is present in many industrial wastes including the process wastes from the production of phosphatic fertilizers, hydrogen fluoride, and fluorinated hydrocarbons and in certain petroleum refinery waste streams. Fluorides occur naturally in soils at levels ranging from 30-990 ppm (Table 6.19).

TABLE 6.19 TYPICAL TOTAL HALIDE LEVELS IN DRY SOIL

Halide	PPM (Dry Weight)		Reference
	(Mean)	(Range)	
Bromide	10	(2-100)	Bowen (1966)
		(10-40)	Martin (1966a)
Chloride	100		Bowen (1966)
Fluoride	200	(30-300)	Bowen (1966)
	240		Brewer (1966a)
	345	(70-990)	Gilpin and Johnson (1980)
Iodide	5		Bowen (1966)
	2.83	(2.5-3.9)*	Aston and Brazier (1979)
		(0.1-10)	Martin (1966b)

* Iodide deficient soils.

The mobility of fluoride in soil depends on the percentage of the total fluoride that is water soluble. Fluoride solubility is dependent on the kind and relative quantity of cations present in the soils that have formed salts with the fluoride ion (F^-). Sodium salts of fluoride (NaF) are quite soluble and result in high soluble fluoride levels in soils low in calcium. Calcium salts of fluoride (CaF_2) are relatively insoluble and serve to limit the amount of fluoride taken up by plants or leached from the soil.

Fluoride is not an essential nutrient to plants but may be essential for animals; however, soluble fluorides are readily taken up by plants at levels that may be toxic to grazing animals. The upper level of chronic lifetime dietary exposure of fluoride (dry weight concentration in the diet) that will not result in a loss of production for cattle is 40 ppm and for swine, 150 ppm (National Academy of Sciences, 1980). Chronic fluorosis, a disease in grazing animals caused by excess dietary fluoride, has reportedly resulted from industrial contamination of pastures and underground water sources. Fluorosis can occur in grazing animals from the consumption of water containing 15 ppm fluoride (Lee, 1975) or forage containing 50 ppm fluoride (Brewer, 1966).

Phytotoxic concentrations of fluoride based on plant tissue content and irrigation water fluoride content are given in Table 6.20. A tissue concentration of only 18 ppm (dry weight) was toxic to elm, a sensitive plant (Adams et al., 1957), yet, buckwheat survived tissue concentrations of 990–2450 ppm fluoride (Hurd–Karrer, 1950). Tissue concentrations toxic to various crops have been determined (Brewer, 1966a).

While liming a soil will temporarily decrease both plant uptake and leaching of fluoride, the loading capacity allowed for fluoride in a land treatment unit should take into account that liming will cease following closure. Soils with high cation exchange capacities (CEC) that are high in calcium and low in sodium have a higher long-term loading capacity for fluoride than soils with lower CECs or higher sodium content. Leachate concentrations of fluoride should not exceed the EPA drinking water standard. The EPA drinking water standard (Table 6.21) is dependent on climatic conditions because the amount of water (and consequently the amount of fluoride) ingested is primarily influenced by air temperature. The rationale behind limiting the leachate concentration of fluoride to the drinking water standard is that groundwater is a primary source of drinking water and since groundwater is likely to remain in the same climatic zone (with respect to where it may be used as drinking water) a graduated standard is a reasonable guide for leachate quality.

6.1.5.2 Chloride (Cl)

Chlorides occur to some extent in all waste streams either as a production by-product (i.e., chlorinated hydrocarbon production wastes, chlorine gas production, etc.) or as a contaminant in the water source used. A typical value for chloride in soil is 100 ppm (Table 6.19). Chloride is very soluble and will move with leachate water.

TABLE 6.20   PHYTOTOXICITY OF HALIDES FROM ACCUMULATION IN PLANT TISSUE AND
APPLICATIONS TO SOIL

	Tissue Content		
Halide	Plant	Toxic Level in Tissue (ppm dry wt.)*	Reference
Fluoride	Buckwheat (Fagopyrum esculentum)	2450–990	Hurd-Karrer (1950)
	Elm (Ulmus sp.)	18	Adams et al. (1957)
Chloride	Apple (Malus sp.)	0.24%	Dilley et al. (1958)
	Alfalfa (Medicago sativa)	0.27%	Eaton (1942)
Bromide	Cabbage (Brassica oleracea)	0.1%	Martin (1966a)
	Citrus seedling (Citrus sp.)	0.17%	Martin et al. (1956)
Iodide	Tomato (Lycopersicon esculentum)	8.05	Newton and Toth (1952)
	Buckwheat (Fagopyrum esculentum)	8.75%	Newton and Toth (1952)

	Soil Applied in Irrigation Water (IW) or Water Soluble (WS)		
Halide	Plant	Toxic Level (ppm)	Reference
Fluoride	Tomato (Lycopersicon esculentum)	100 (IW)	McKee and Wolf (1963)
	Red Maple seedlings (Acer rubrum)	380 (IW)	Maftoun and Sheilbany (1979)
Chloride	Pea (Pisium sativum)	9 (IW)	Eaton (1966)
	Oats (Avena sativa)	120 (IW)	Eaton (1966)
Bromide	Bean (Phaseolus vulgaris)	38 (WS)	Stelmach (1958)
	Cabbage (Brassica oleracea)	83 (WS)	Stelmach (1958)
Iodide	Tomato (Lycopersicon esculentum)	5 (WS)	Newton and Toth (1952)
	Buckwheat (Fagopyrum esculentum)	5 (WS)	Newton and Toth (1952)

* Unless otherwise noted.

† Possible Cl-salt effect on toxicity.

TABLE 6.21 EPA DRINKING WATER STANDARD FOR FLUORIDE*

Annual average of maximum daily air temperatures (Degrees C)[†]	Fluoride maximum (mg/l)
12 and below	2.4
12.1 to 14.6	2.2
14.7 to 17.6	2.0
17.7 to 21.4	1.8
21.5 to 26.2	1.6
26.3 to 32.5	1.4

* EPA (1976a).

[†] Based on temperature data obtained for a minimum of 5 years.

When soils are carefully managed to avoid leachate generation, chloride concentrations in the soil may increase rapidly. To avoid chloride buildup in soils, the amount applied in wastes and irrigation water should be balanced with the amount removed by cover crops and leached through the soil profile.

Chloride is an essential element to both plants and animals. Although, plants readily take up chloride, animals are generally unaffected by concentrations in forage. Phytotoxicity generally occurs before plant concentrations reach levels that would adversely affect grazing animals. Phytotoxic levels of chloride with respect to its concentration in plant tissue and irrigation water are given in Table 6.20.

Plant removal of chlorides can be increased by regularly harvesting the stalk and leafy portion of the cover crop. Corn plants remove only 3 kg/ha/yr of chloride when harvested as corn; however, when the same crop is harvested for silage over 35 kg/ha/yr of chloride is removed (Kardos et al., 1974). The concentration of chloride in soil solutions associated with yield reductions in various crops have been determined (Van Beekom et al., 1953; Van Dam, 1955; Embleton et al., 1978).

Loading rate considerations for chloride should include the amount removed by plant uptake and the amount lost in leachate while keeping the concentration in the soil below the phytotoxic level. Additionally, the leachate concentration should not exceed the EPA drinking water standard for chloride of 250 mg/l.

6.1.5.3 Bromide

Bromide is present in several industrial wastes including synthetic organic dyes, mixed petrochemical wastes, photographic supplies, production wastes, pharmaceuticals and inorganic chemicals. Hydrogen bromide is produced for use as a soil fumigant in agriculture. Naturally occurring

232   Environmental Fate

bromide concentrations in soil range from 2-100 ppm (Table 6.19). In addition to the bromide ion, other forms of this element that occur naturally in soils, though at smaller concentrations, are bromate (BrO_3^-) and bromic acid. Most bromide salts (CaBr, MgBr, NaBr and KBr) are sufficiently soluble to be readily leachable in water percolating through soils. Consequently, most of the bromide found in soils is organically combined.

Bromide is not an essential nutrient to plants or animals. Although bromide is strongly concentrated by plants, reports of toxicity to animals are scarce. Table 6.20 lists bromide concentrations that are phytotoxic with respect to plant tissue content and the water soluble content in soils. The upper level of chronic lifetime dietary exposure of bromide (dry weight concentration in the diet) that will not result in a loss of production for cattle and swine is 200 ppm (National Academy of Sciences, 1980). Loading rates for bromide should include consideration of plant uptake and leachate losses to maintain the concentration in the soil below phytotoxic levels.

6.1.5.4 Iodide

Iodide is present in several industrial wastes including those generated by the pharmaceutical industry and the analytical chemical industry. Iodides naturally occur in soils at levels ranging from 0.1-10 ppm (Table 6.19). It is only slightly water soluble (0.001 m) and is thought to be retained in soil by forming complexes with organic matter and possibly by being fixed with soil phosphates and sulfates (Whitehead, 1975).

Iodide is not essential for plant growth, but it is an essential nutrient for animals. Soluble iodide in wastes will be readily taken up by plants and animals consuming large quantities of iodide-rich forage may ingest toxic levels. Phytotoxic concentrations of iodide in plant tissues and of water soluble iodide in soils are given in Table 6.20. It should be noted that toxic responses may be partially a result of excess salts not iodide. The upper levels of chronic lifetime dietary exposure of iodide (dry weight concentration in the diet) that will not result in a loss of production for cattle is 50 ppm and swine, 400 ppm (National Academy of Sciences, 1980).

Loading rate calculations for the land treatment of wastes containing iodide should include iodide taken up by plants and leached, from the soil to maintain the concentration in the soil below phytotoxic levels.

6.1.6 Metals

The metallic components of waste are found in a variety of forms. Metals may be solid phase insoluble precipitates, sorbed or chelated by organic matter or oxides, sorbed on exchange sites of waste constituents or soil colloids, or in the soil solution. If an element is essentially

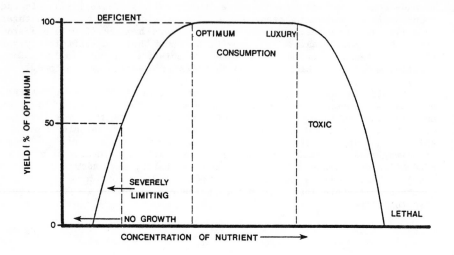

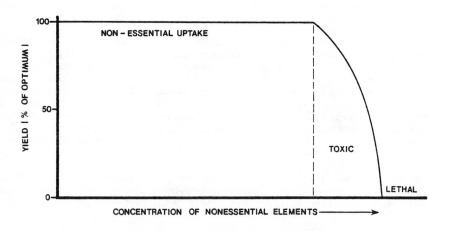

Figure 6.13. Schematic diagram of the yield response to an essential but toxic element (top diagram) and a nonessential toxic element (bottom diagram).

insoluble at usual soil pH ranges (5.5-8.0) then the metal has a low concentration in the soil solution and cannot be absorbed by plants or leached
at an appreciable rate. If the metal is strongly sorbed or chelated, even
though it is not precipitated, it will have low plant uptake and low leaching potential. If the metal is weakly sorbed and soluble, then it is
available for plant uptake or transport by leaching or runoff. When
present in this soluble form metals may accumulate in plants to excess.
Little specific information on metal immobilization is available so treatability tests should be designed to determine the mobility of a given metal
in a given waste-site environment (Chapter 7).

Although many HWLT units will not use plants as a part of the ongoing
management plan, plant uptake of metals is discussed extensively in this
section since closure of sites generally requires a vegetative cover (EPA,
1982). Metals may be applied in excess of the phytotoxic level if they
continue to be immobilized in the treatment zone. However, since a vegetative cover will be necessary at closure (unless hazardous constituents show
no increase over background), highly contaminated soils may need to be removed and disposed in another hazardous waste facility. This could
increase the cost associated with disposal and make consideration of more
land and lower loading rates a viable option.

Plants do not accumulate metals in a consistent proportional relationship to soil concentrations. Thus, prediction of the plant concentrations
of a metal resulting from growing on metal containing soil is extremely
difficult. Due to the variability of soil properties and conditions, and
plant species, lists are given for each metal, when available, to provide
the broadest range of operating conditions.

The reaction of plants to metals in the growth media depends on
whether or not the element is plant essential. The upper half of Fig. 6.13
shows the response of plants to an essential nutrient. At low concentrations the metal is deficient; at higher concentrations of the element the
plant reaches optimum growth and additional metal concentrations have
little effect; at very high concentrations the metal will become toxic.
The response of plants to nonessential metals, in which no deficiency
results, is shown in the lower half of Fig. 6.13.

Most positively charged metals remain in the treatment zone under
aerated conditions where they are immobilized, either temporarily or somewhat permanently, by the properties of the soil itself. The mechanisms of
metal retention by soil are described in Section 4.1.2.1 and include chemi-
sorption and electrostatic bonding. Chemical sorption is a more permanent
type of metal retention than electrostatic sorption and is primarily due to
the mineralogy of the soil. Electrostatic bonding, or ion exchange,
increases as the CEC of the soil increases and is reversible. A direct
comparison between CEC and the sorption capacity of the soil is not possible, however, since competition between ions in the waste or present in the
native soil will influence the quantity of metal ions sorbed by the soil.

A variety of mathematical relationships has been used to quantify
sorption of metals to soils. These models, generally called isotherms,

include the linear, Freundlich, Langmuir, two-surface Langmuir and various kinetic sorption isotherms. The models provide a reasonably good basis for interpolation of metal sorption and are extensively reviewed by Travis and Etnier (1981) who include numerous references for a variety of metals. Bohn et al. (1979) discuss isotherm theory in detail. Sorption isotherm experiments may be included as part of laboratory analysis for treatment demonstration of metal immobilization.

The partitioning of metals between various chemical forms is a dynamic process, regulated by equilibrium reactions. The initial behavior of the metal after addition to the soil largely depends on the form in which it was added, which in turn, depends on its source. A complex set of chemical reactions, physical and chemical characteristics of the soil, and a number of biological processes acting within the soil govern the ultimate fate of metallic elements.

This section discusses the sources of metal enrichment to the environment as well as background soil and plant concentrations. The soil chemistry of each metal including solubility, metal species and soil conditions governing the predominant form of the metal are presented. Following a review of metal chemistry, the fate of each metal in the soil, whether bio-accumulated, sorbed by soil or waste constituents, or transported, is discussed. Finally, recommendations for metal loading are given based on accumulation in the soil and plant and animal toxicity. These recommendations are generally based on the accumulation of the element within the upper 15 cm (6 in) of soil, or "plow layer," which is estimated to be 2×10^6 lb/acre or 2.2×10^6 kg/ha. In developing the recommendations, consideration was given to the 20-year irrigation standards developed by the National Academy of Sciences and National Academy of Engineering (1972) which are based on the tolerance of sensitive plants, to metal chemistry, and to other sources of information on plant and animal toxicity. There are more data available on plant and animal toxicity to metal concentrations in the soil than on the ability of the soil to immobilize a given element. Consequently, treatability studies are generally needed to determine if adequate immobilization of metals is occurring in a given soil since the factors affecting immobilization are very site-specific.

6.1.6.1 Aluminum (Al)

Hazardous wastes containing Al include paper coating pretreatment sludge and deinking sludge. It is one of the most abundant elements in soils, occuring at an average concentration of 71,000 ppm.

Aluminum exists in many forms in soil. There are several Al oxide and hydroxide minerals including $Al(OH)_3$ (amorphous, bayerite, and gibbsite) and $AlOOH$ (diaspore and boehmite) (Lindsay, 1979). In soils with pH less than 5.0, exchangeable Al is found as the trivalent ion (Bohn et al., 1979). In an alkaline medium, Al is present as $(Al)OH_4^-$. Aluminum in soil may be precipitated as Al phosphates; this reaction removes plant essential phosphate from the soil solution. Where the NaOH:Al ratio is greater than 3:0, polymerization of Al and hydroxide ions may lead to the formation of crystalline Al hydroxide minerals (Hsu, 1977).

The most soluble form of Al found in most soils is $Al(OH)_3$ (amorphous) and other Al oxides are somewhat less soluble. At pH 4.06, 96 ppm soluble Al may be found in a particular soil solution, yet when the pH is raised to 7.23, the Al concentration in the same soil solution is reduced to zero (Pratt, 1966a). Aluminum is highly unstable in the normal pH range of soils and readily oxidizes to Al^{3+} (Lindsay, 1979).

There is no evidence that Al is essential to plants. Sensitivity to Al varies widely and some plants may be harmed by low concentrations of the element in the growing media (Table 6.22). Very sensitive plants whose growth is depressed by soil concentrations of 2 ppm Al include barley (Hordeum vulgare), beet (Beta vulgaris), lettuce (Lactuca sativa) and timothy (Phleum pratense). Tolerant plants depressed by 14 ppm Al are corn (Zea mays), redtop (Agrostis gigantea) and turnip (Brassica rapa). An interesting Al indicator plant is the hydrangea which produces blue flowers if Al is available in the growth medium and pink flowers if Al is not available (Pratt, 1966a).

There are some accumulator plants that can tolerate large amounts of Al. Accumulator plants that transport Al to above-ground parts include club moss, sweetleaf (Symplocos tinctoria), Australian silk oak, and hickory (Juncus sp.). Aluminum concentrations of 3.0-30 ppm have been reported for ash (Fraxinus sp.) and hickory (Pratt, 1966a).

Loehr et al. (1979b) state that Al poses relatively little hazard to animals. Cattle and sheep can tolerate dietary levels of 1000 ppm Al. Poultry, considered sensitive to the element, can tolerate dietary levels of 200 ppm Al (National Academy of Sciences, 1980).

Aluminum levels in sludge seldom limit application rates, particularly if the pH is maintained above 5.5 and the soil is well aerated (Loehr et al., 1979b). With proper pH management, large amounts of Al may be land applied.

6.1.6.2 Antimony (Sb)

The major producers of hazardous wastes containing Sb are the paint formulation industry, textile mills, and organic chemical producers. Concentrations of Sb range from 0.5-5 ppm in coal and 30-107 ppm in petroleum, and urban air contains 0.05-0.06 ppm Sb (Overcash and Pal, 1979). The average concentration of Sb in plants is 0.06 ppm and the average range of Sb in dry soils is 2-10 ppm (Bowen, 1966).

Naturally occurring forms of Sb include Sb sulfides (stibinite) and Sb oxides (cervanite and valentinite). Antimony in soils usually occurs as Sb^{3+} or Sb^{5+} and is very strongly precipitated as Sb_2O_3 or Sb_2O_5 (Overcash and Pal, 1979).

TABLE 6.22 PLANT RESPONSE TO ALUMINUM IN SOIL AND SOLUTION CULTURE

Al Concentration (ppm)	Media	Species	Effect	Reference
1-2	Solution	Barley (Hordeum vulgare)	50% yield reduction	Pratt (1966a)
1-2	Solution	Sorghum (Sorghum bicolor)	50% yield reduction	Ibid.
2-5	Solution	Corn (Zea mays)	50% yield reduction	Ibid.
2-8	Solution	Kentucky bluegrass (Poa pratensis)	20% yield reduction	Ibid.
2-8	Solution	Yellow foxtail	20% yield reduction	Ibid.
4	Soil	Sugar beet (Beta vulgaris)	Significant root growth reduction	Keser et al. (1975)
6-8	Solution	Rye (Secale cereale)	31% yield reduction	Pratt (1966a)
6	Solution	Wheat (Triticum aestivum)	Tolerant	Kerridge et al. (1971)
7	Solution	Cabbage (Brassica oleracea)	No response	Pratt (1966a)
14	Solution	Turnip (Brassica rapa)	No response	Ibid.
12	Solution	Lovegrass (Eragrostis secundiflora) & tall fescue (Festuca arundinacea)	Serious injury	Fleming et al. (1974)
13	Solution	Pea (Pisum sativum)	Reduced growth	Klimashevsky et al. (1972)
20	Solution	Potato (Solanum tuberosum)	No response	Pratt (1966a)
20	Sand	Potato (S. tuberosum)	Depressed growth	Lee (1971a)
25	Acid soil	Cotton (Gossypium hirsutum)	Damage	Velly (1974)
32-80	Solution	Colonial bentgrass (Agrostis fenuis)	20% yield reduction	Pratt (1966a)

--continued--

TABLE 6.22 (continued)

Al Concentration (ppm)	Media	Species	Effect	Reference
32–80	Solution	Red top (Agrostis gigantea)	20% Yield reduction	Ibid.
60	Solution	Wheat (T. aestivum)	Chlorosis of leaves	Cruz et al. (1967)
100 kg/ha	Glacial till soil (pH 6.5)	Barley (H. vulgare)	Significant yield reduction	Hutchinson and Hunter (1979)
120–130	Acid soil	Maize (Zea mays)	Damage	Velly (1974)
2000	Solution	Peach seedlings (Prunus persica)	Severe toxicity	Edwards et al. (1976)

Very high concentrations of Sb may present a hazard to plants and animals, though little information is available. A concentration of 4 ppm Sb in culture solution has been shown to produce a toxic response in cabbage (_Brassica oleracea_) plants (Hara et al., 1977). Bowen (1966) points out that Sb in industrial smoke may cause lung disease.

6.1.6.3 Arsenic (As)

Arsenic is contained in wastes from the production of certain herbicides, fungicides, pesticides, veterinary pharmaceuticals and wood preservatives. Arsenic levels in municipal sewage are variable, ranging from 1-18 ppm (Loehr et al., 1979a). In addition, industries manufacturing glass, enamels, ceramics, oil cloth, linoleum, electrical semiconductors and photoconductors use As. The element is also used to manufacture pigments, fireworks and certain types of alloys (Page, 1974).

In soils, the total As concentration normally ranges from 1-50 ppm, though it does not generally exceed 10 ppm. Soils producing plants containing As at levels toxic to mammals are found in parts of Argentina and New Zealand (Bowen, 1966).

Research involving application of As compounds to agricultural soil-plant systems has dealt primarily with an anions arsenate (AsO_4^{-3}) and arsenite (AsO_3^{-3}). Arsenate is an oxidized degradation product from organoarsenic defoliants and pesticides. Arsenite may be formed both biologically and abiotically under moderately reduced conditions (Woolson, 1977). The reduced state of As (arsenite) is 4 to 10 times more soluble in soils than the oxidized arsenate and, consequently, more prone to leaching.

Cycling of As in the environment is dominated by sorption to soils, leaching and volatilization (Fig. 6.14). The most important mechanism for attenuation is sorption by soil colloids (Murrman and Koutz, 1972). Arsenic movement in soils may be reduced by sorption to, or precipitation by, iron (Fe) and aluminum (Al) oxides or calcium. The amount of As sorbed by the soil increases as pH and clay, Al, and Fe content increase (Jacobs et al., 1970). Movement of As in aquatic systems often results from As sorption to sediments containing Fe or Al (Woolson, 1977). Wind borne particles may also carry sorbed As. Reduction of Fe in flooded soils may resolubilize As from ferric arsenate or arsenite to arsine or methylarsines (Deuel and Swoboda, 1972).

Reduction of As compounds under saturated conditions can result in As volatilization. Some As may be reduced to As^{3-} and then lost as arsine, a toxic gas (Keaton and Kardos, 1940). In a study by Woolson (1977), however, only 1-2% of arsenate applied at a rate of 10 ppm was volatilized as dimethyl arsine $[(CH_3)_2AsH]$ after 160 days. High organic matter content, warm temperatures and adequate moisture are the conditions conducive to microbial and fungal growth. These conditions may cause the reduction of

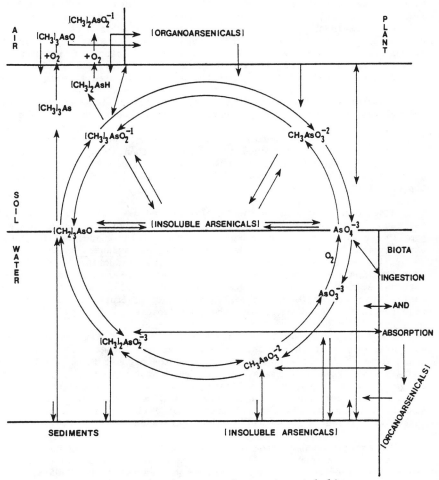

Figure 6.14. Cyclical nature of arsenic metabolism
in different environmental compartments
(Woolson, 1977). Reprinted by permission
of the National Institute of Environmental
Health Sciences.

As and can drive the reaction toward methylation and subsequent volatilization of As. Reducing conditions may also lead to an increase in As as
arsenite which increases the leaching potential of the element.

Biomagnification through the food chain does not occur with the arsenicals. Lower members of the aquatic food chain contain the highest As
residues (Woolson, 1977); typically brown algae contain about 30 ppm As and
mollusks contain about 0.005 ppm As (Bowen, 1966). In plants, the As concentration varies between 0.01-1.0 ppm. Even plants grown in soils
contaminated with As do not show higher concentrations of As than plants
grown on uncontaminated soil. The toxicity of As limits plant growth
before large amounts of As are absorbed and translocated (Liebig, 1966).

There is no evidence that As is essential for plant growth. Arsenic
accumulates in much larger amounts in plant roots than in the tops.
Arsenic in soils is most toxic to plants at the seedling stage where it
limits germination and reduces viability. The concentration of As that is
toxic to plants was determined to be greater than 10 ppm by the National
Academy of Sciences and National Academy of Engineering (1972). Initial
symptoms of As toxicity include wilting followed by reduction of root and
top growth (Liebig, 1966).

Arsenic at 1 ppm in nutrient solution reduces root and top growth of
cowpeas (Vigna unguiculata) and concentrations of soluble As as low as 0.5
ppm in nutrient solution produce an 80% yield reduction in tomatoes
(Lycopersicon esculentum). Sudan grass (Sorghum sudanense), considered to
be quite tolerant, does not show growth reduction until the As
concentration in the soil reaches 12 ppm (National Academy of Sciences and
National Academy of Engineering, 1972). Table 6.23 lists the response of
various crops to As levels in soil and solution culture, and it indicates a
wide response to As depending on the plant species.

The toxicity of As to animals results from its interaction with the
sulhydryl groups or SH radicals of some enzymes (Turner, 1965). The inorganic forms of As are much more toxic than the organic forms which are more
rapidly eliminated by animals. Frost (1967) states that a dietary level of
10 ppm As will be toxic to any animal. There is little evidence that As
compounds are carcinogenic in experimental animals (Milner, 1969) although
studies indicate that human subjects chronically exposed to As compounds
have a significantly increased incidence of cancer (Yeh, 1973).

The greatest danger from As to livestock is in drinking water where As
is present as inorganic oxides. An upper limit of 0.2 ppm As is recommended for livestock drinking water. A concentration of 0.05 ppm is the
upper allowable limit for As in water intended for human consumption
(National Academy of Sciences and National Academy of Engineering, 1972).

A review by Overcash and Pal (1979) indicates that As is toxic to
plants at soil application rates between 200 and 1000 kg/ha. However,
Table 6.23 indicates that some plant species may be affected by less than
100 ppm As in the soil. A soil accumulation of between 100 and 300 ppm
appears acceptable for most land treatment units.

TABLE 6.23 PLANT RESPONSE TO ARSENIC IN SOIL AND SOLUTION CULTURE

As Concentration (ppm)	Media	Species	Effect	Reference
2-26	Soil	Potatoes (Solanum tuberosum)	None	Steevens et al. (1972)
8	Sand	Rye (Secale cereale)	Translocated to shoots and leaves	Chrenekova E. (1973)
50	Clay loam	Horse bean (Vicia faba)	Decreased growth	Chrenekova C. (1977)
80	Silt loam	Maize (Zea mays)	Toxic	Jacobs and Keeney (1970)
85	Loamy sand	Blueberry	Plant injury	Anastasia and Kender (1973)
100	Soil	Reed canary grass (Phalaris arundinacea)	No effect	Hess and Blanchar (1977)
100	Soil	Apple (Malus sp.) trees	Decreased size	Benson et al. (1978)
450	Soil	Apple (Malus ap.) trees	Zero growth	Benson et al. (1978)

6.1.6.4 Barium (Ba)

Barium is found in waste streams from a large number of manufacturing plants in quantities that seldom exceed the normal levels found in soil. Normal background levels for soil range from 100-3000 ppm Ba (Bowen, 1966).

Although Ba is not essential to plant growth, soluble salts of Ba are found in the accumulator plant Aragalus lamberti. Barium accumulation in plants is unusual except when the Ba concentration exceeds calcium (Ca) and magnesium (Mg) concentrations in the soil, a condition which may occur when sulfate is depleted. Liming generally restores a favorable Ca:Ba balance in soil (Vanselow, 1966a). All the soluble salts of Ba, which exclude Ba sulfate, are highly toxic to man when taken by mouth. There is little information available on which to base a Ba loading rate for HWLT facilities.

6.1.6.5 Beryllium (Be)

Beryllium may be found in waste streams from smelting industries and atomic energy projects. The major source of Be in the environment is the combustion of fossil fuels (Tepper, 1972). Soil concentrations generally range from 0.1 to 40 ppm, with the average around 6 ppm.

Beryllium reacts similarly to aluminum. It undergoes isomorphic substitution as well as cation exchange reactions. It is strongly immobilized in soils by sorption. It is present in the soil solution as Be^{2+} and it may displace divalent cations already on sorption sites. It is readily precipitated by liming.

Beryllium becomes hazardous when found in soil solutions or groundwater supplies. It may be taken up by plants at levels that result in yield reduction; phytotoxicity of Be is caused by the inhibition of enzyme activity (Williams and LeRiche, 1968). The growth inhibiting effects usually recognized in higher plants are reduced as the pH is raised above 6.0, and it has been proposed that the decreased toxicity is caused by Be precipitation at high pH levels (Romney and Childress, 1965). The response of plants to Be applied to soil is given in Table 6.24 which indicated that 40 ppm Be in soil did not cause a yield decrease in neutral pH soils but substantially decreased plant yields in quartz soils. Table 6.25 illustrates that a very soluble Be salt will decrease plant yields substantially when present in soil concentrations of 20 ppm.

TABLE 6.24 YIELDS OF GRASS AND KALE WITH LEVELS OF BERYLLIUM IN QUARTZ AND SOIL*

Soil	pH	Soluble Be Added (ppm)	Mean Yield of Fresh Matter (G)	
			Grass	Kale
Lincolnshire	7.5	0	13.3	36.0
		0.4	17.2	46.0
		40.0	19.9	42.8
Hertfordshire	7.5	0	21.3	44.8
		0.4	31.0	55.6
		40.0	25.0	57.0
Quartz	†	0	6.4	2.8
		0.4	7.9	1.8
		40.0	0.1	0.1

* Williams and LaRiche (1968).

† Not available.

TABLE 6.25 YIELD OF BEANS GROWN ON VINA SOIL TREATED WITH BERYLLIUM SALTS DIFFERING IN SOLUBILITY*

Be Applied to Soil		Solubility of Be Salt g/100 ml Cold Water	Yield Dry Plant Tops (g)
Form	ppm		
BeO		2.3×10^{-5}	
	0		8.76
	10		8.72
	20		8.64
$(BeO_5) CO_2 5H_2O$		Insoluble	
	0		8.68
	10		8.36
	20		8.30
$BeSO_4 4H_2O$		42.5	
	0		8.81
	10		7.03
	20		5.92
$Be(NO_3)_2 3H_2O$		Very soluble	
	0		8.31
	10		6.09
	20		2.97

* Romney and Childress (1965).

Beryllium is a suspected carcinogen. Experimental data indicate Be causes cancer in animals and epidemiological studies report a significant increase in respiratory cancers among Be workers (Reeves and Vorwald, 1967; Mancuso, 1970).

Recommendations established in the National Academy of Science and National Academy of Engineering (1972) Water Quality Criteria limit irrigation over the short-term to water containing 0.50 ppm Be; water for long-term irrigation is limited to 0.20 ppm. The use of irrigation water containing the upper limit of the acceptable Be concentration recommended by the National Academy of Sciences and National Academy of Engineering (1972) is equivalent to an accumulation of 50 ppm Be in the soil. Table 6.24 shows that soil concentrations of 40 ppm do not cause a decrease in plant yields if applied to a neutral pH soil. Thus, a comparison of the irrigation water standard and the phytotoxic limit appears to provide a reasonable estimate of the acceptable cumulative soil Be level of 50 ppm.

6.1.6.6 Cadmium (Cd)

Cadmium is used in the production of Cd-nickel batteries, as pigments for plastics and enamels, as a fumicide, and in electroplating and metal coatings (EPA, 1980a). Wastes containing significant levels of Cd include paint formulating and textile wastes. The estimated mean Cd concentration of soil is 0.06 ppm, ranging from 0.01-0.7 ppm (Siegel, 1974).

The soil chemistry of Cd is, to a great extent, controlled by pH. Under acidic conditions Cd solubility increases and very little sorption of Cd by soil colloids, hydrous oxides, and organic matter takes place (Anderson and Nilsson, 1974). Street et al. (1977) found a 100-fold increase in Cd sorption for each unit increase in pH.

Solid phase control of Cd by precipitation has been reported under high pH conditions. Figure 6.15 illustrates that the formation of $Cd(OH)_4$ controls the equilibrium concentration of Cd at high pH values. Precipitation of Cd with carbonates ($CdCO_3$) and phosphates ($Cd_3(PO_4)_2$) may regulate Cd concentration in the soil solution at low pH values. Under reducing conditions, such as poorly drained soils, the precipitation of Cd sulfide may occur. Since this compound is relatively stable and slowly oxidized, a lag occurs between the formation of Cd sulfide and the release of Cd to the soil solution.

Cadmium may also be sorbed by organic matter in the soil as soluble or insoluble organometallic complexes or by sorption to hydrous oxides of iron and manganese (Peterson and Alloway, 1979). Evidence suggests that these sorption mechanisms may be the primary source of Cd removal from the soil solution except at very high Cd levels. Column studies by Emmerich et al. (1982) show that no leaching of Cd occurred from sewge sludge amended soils, all of which had CEC values between 5 and 15. Of the 25.5 ppm Cd applied to the Ramona soil, 24.7 ppm or 97% of the Cd was recovered from

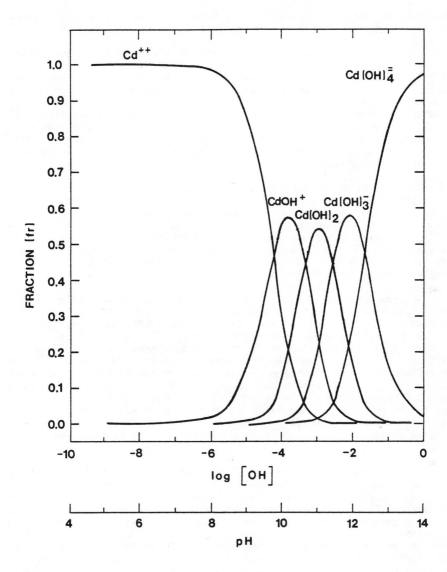

Figure 6.15. Distribution of molecular and ionic species of divalent cadmium at different pH values (Fuller, 1977).

the columns. Yet, as the equilibrium between sorbed Cd and soil solution Cd changes, some sorbed Cd may be released to the soil solution.

Land treatment of Cd containing waste can affect microbial populations as well as plant and animal life. Microorganisms exhibit varying degrees of tolerance or intolerance toward Cd. Williams and Wollum (1981) found that 5 ppm Cd in the growing media retards Actinomycete and soil bacteria growth, but at concentrations greater than 5 ppm, the microorganisms exhibited a tolerant response and the tolerant population attained dominance in the cultures. Borges and Wollum (1981) reported Rhizobium japonicum strains associated with soybean (Glycine max) plants showed tolerance to Cd and that after time, R. japonuim strains develop the ability to accomodate the element.

The long-term availability of Cd to plants is related to several soil properties, the presence of other ions in the soil solution, and the plant species. Soil organic matter, hydrous oxides, redox potential, and pH (the dominant factor) influence the concentration of Cd in the soil solution as well as its availability to plants. Liming reduces Cd uptake by plants and increases Cd sorption by soil (CAST, 1976), while acidification releases the Cd bound in hydrous oxides. High organic matter in soil reduces plant uptake of the element (White and Chaney, 1980).

Cadmium absorbed by plant roots is slowly translocated to the leaf and stem. The metabolic processes responsible for Cd absorption are influenced by temperature (Schaeffer et al., 1975; Haghiri, 1974) and other minerals in the nutritive solution (Cunningham et al., 1975; Miller et al., 1977). Chaney (1974) proposed that zinc-cadmium interactions reduce the amount of Cd taken up by plants when the concentration of Cd is less than 1% of the zinc (Zn) content in the sludge. This is due to the competition of Zn and Cd for -SH groups of proteins and enzymes in plants. Since the content of Zn and Cd taken up by plants is not always related to the concentration in waste, the principle of the Zn-Cd interrelationship should not be the sole basis for determining loading rates. Calcium has been shown to depress Cd content in plants because these divalent cations compete for adsorption by roots.

Crops differ markedly in their Cd accumulation, tolerance and translocation. The foliar Cd concentrations associated with phytotoxicity vary in different crops from 5 to 700 ppm, dry weight (Chaney et al., 1981) yet the phytotoxicity of Cd does not limit Cd in crops to acceptable limits for animal consumption. Soil additions of Cd at a rate of 4.5 kg/ha/yr for two consecutive years raised the Cd content of corn (Zea mays) leaves from 0.15 to 0.71 ppm, while the increase was less significant to grain (Overcash and Pal, 1979). Cadmium additions ranging from 11 to 7640 ppm in soil resulted in reduced yields of various forage crops (Table 6.26). Melsted (1973) suggested a tolerance limit of 3 ppm Cd in agronomic crops. The influence of Cd concentration on the growth of various plants is given in Table 6.27. The yield and Cd concentration in the leaves of bermudagrass grown in sewage sludge containing Cd are given in Table 6.28. Recently, Cd tolerance has been found in grasses in some populations from Germany and Belgium (Peterson and Alloway, 1979). Tomato (Lycopersicon esculentum) and cabbage

(Brassica oleracea) are considered Cd tolerant and soybean (Glycine max) is considered rather sensitive.

TABLE 6.26   CADMIUM ADDITION TO A CALCAREOUS SOIL ASSOCIATED WITH A 50%
YIELD REDUCTION OF FIELD AND VEGETABLE CROPS*

Crop	Cd Addition rate reducing yield 50% mg/kg
Soybean (Glycine max)	11
Sweet corn (Zea mays)	35
Upland rice (Oryza sativa)	36
Sudan grass (Sorghum sudanense)	58
Field bean (Phaseolus sp.)	65
Wheat (Triticum aestivum)	80
Turnip (Brassica rapa)	100
White clover (Trifolium sp.)	120
Alfalfa (Medicago sativa)	145
Swiss chard (Beta vulgaris var. Cicla)	320
Tall fescue (Festuca arundinacea)	320
Bermudagrass (Cynodon dactylon)	400
Paddy rice (Oryza sativa)	7,640

* Page et al. (1972).

Cadmium can be quite toxic to aquatic organisms, even in concentrations of less than 1 ppm Cd in water; therefore, runoff or movement of particles containing Cd into water must be avoided. Coombs (1979) reviewed the Cd content in fish, marine mammals, invertebrates, and plankton and determined the toxic levels of Cd for each species. Experimental data indicate that Cd causes cancer in animals (Lucis et al., 1972). However, there have not been any large scale epidemiological studies to show significant association between occupational exposure to Cd and cancer in workers (Sunderman, 1977). Acceptable Cd levels for crops used for animal feed or human consumption have not been established although adverse health effects from prolonged consumption of food grown on Cd enriched soils is well documented (Tsuchiya, 1978; Friberg et al., 1974).

The National Academy of Sciences and National Academy of Engineering (1972) and Dowdy et al. (1976) suggest maximum cumulative applications of Cd should not exceed 3 mg/kg or 10 ppm when added in sewage sludge. EPA cumulative criteria have adjusted application levels to 5 kg/ha Cd for soils with a pH less than 6.5 and for soils with a pH greater than 6.5,

TABLE 6.27 PLANT RESPONSE TO CADMIUM IN SOIL AND SOLUTION CULTURE

Cd Concentration (ppm)	Media	Species	Effect	Reference
1	Solution	Purple nutsedge	Growth reduction	Quimby et al. (1979)
1	Soil	Pin oak (Quercus palustris)	Chlorosis	Russo and Brennan (1979)
2	Rooting medium	Honeylocust (Gleditsia triacanthos)	Reduced root growth	Lamoreaux et al. (1978)
3–5	Soil	Soybean (Glycine max)	Depressed growth	Miller et al. (1976)
4	Sand	Soybean (G. max)	Severe growth reduction	Chaney et al. (1977)
5	Solution	Rice (Orzya sativa) seedlings	Growth redution	Saito and Takahashi (1978)
10	Soil	Wheat (Triticum aestivum)	Reduced growth	Keul et al. (1979)
25	Soil	Beans (Phaseolus aureus)	Growth inhibition	Jain (1978)
25	Soil	Maize (Zea mays)	Depressed growth	Hassett et al. (1976)
30	Soil	(Rudbecki hirta)	25% germination reduction	Miles and Parker (1979)
50	Soil	Oats (Avena sativa)	Chlorsis	Kloke and Schenke (1979)
50	Soil (pH 7.3)	Soybean (G. max)	Relatively resistant	Boggess et al. (1978)
65	Solution	Cotton (Gossypium hirsutum)	Yield reduction	Rehab and Wallace (1978d)
100	Sandy soil	Little bluestem (Schizachyrium scoparium)	Tolerant	Miles and Parker (1979)
100	Soil	White pine (Pinus strobus)	Reduced yield	Kelly et al. (1979)
600	Yolo silt loam	Cotton (G. hirsutum)	15% yield reduction	Rehab and Wallace (1978e)

TABLE 6.28 CADMIUM CONTENT OF BERMUDAGRASS ON THREE SOILS WITH DIFFERENT APPLICATIONS OF SEWAGE SLUDGE

Sludge applied per hectare, metric tons	Cd added per gram of soil, mg	Domino Soil		Harford Soil		Redding Soil	
		pH	Cd per gram of dry matter, mg	pH	Cd per gram of dry matter, mg	pH	Cd per gram of dry matter, mg
80	0.40	6.6	0.41	5.6	0.44	5.6	1.55
80	0.59	6.7	0.40	5.4	0.49	5.4	2.94
80	1.08	6.8	0.78	5.4	1.60	5.1	5.68
80	1.56	6.8	0.85	5.5	1.73	5.2	4.65
80	2.05	6.8	1.30	5.5	2.95	5.4	4.02
80	3.03	6.8	2.64	5.6	4.00	5.3	6.60
80	4.00	6.7	3.56	5.5	3.52	5.1	8.72

* Page (1974).

maximum cumulative amounts of Cd are allowed to increase with CEC (5 meq/100 g, 5 kg/ha; 5-15 meq/100 g, 10 kg/ha; and >15 meq/100 g, 20 kg/ha) (EPA, 1982). It is recommended that the level of Cd in wastes be reduced to below 15-20 mg Cd/kg waste by pretreatment if at all possible. This review indicates soil microbial populations can be affected by soil concentrations of 5 ppm, but plant populations exhibit a high tolerance for the element. Therefore, the basis for Cd loading should not be phytotoxic response but the ability of the soil to immobilize Cd. Liming the soil supplies carbonates and calcium ions which help immobilize Cd. Liming also serves to maintain an equilibrium between the soluble and precipitated forms of Cd in soil, thus reducing the hazard of Cd mobilization.

6.1.6.7 Cesium (Cs)

Cesium metals are used in research on thermoionic power conversion and ion propulsion. Cesium-137 contamination may occur by nuclear fallout. Cesium-137 is a beta emitter with a half life of 33 years. Soil concentrations range from 0.3-25 ppm Cs, with an average of 6 ppm (Bowen, 1966).

Although Cs is retained in field crops and grasses over long periods of time, phytotoxic levels have not been reported. One explanation of Cs tolerance may be that potassium (K) provides protection against plant contamination by Cs since the two monovalent cations may compete for plant absorption (Konstantinov et al., 1974). Cesium uptake in plants increases with nitrogen fertilization, possibly reflecting exchangeable Cs concentrations in soil. Fertilization with phosphorus and potassium decreases Cs concentrations in most plants. Weaver et al. (1981) found that kale (Brassica campestris) accumulated more Cs-137 in the early stages of growth than after four weeks of growth. The average concentration of Cs in plants is 0.2 ppm, and pytotoxicity would not be expected in Cs amended soils if adequate K is available.

6.1.6.8 Chromium (Cr)

The sources of Cr in waste streams are from its use as a corrosion inhibitor and from dyeing and tanning industries. Chromium is used in the manufacture of refractory bricks to line metallurgical furnaces, chrome steels and alloys, and in plating operations. Other uses of Cr include topical antiseptics and astringents, defoliants for certain crops and photographic emulsions (Page, 1974). Chromium is widely distributed in soils, water, and biological materials. The range of Cr in native soils is 1-1000 ppm with an average concentration of 100 ppm Cr (Bowen, 1966). Soils derived from serpentine rocks are very high in Cr and nickel.

The Cr in most industrial wastes is present in the +6 oxidation state as chromate (CrO_4^{-2}) or as dichromate ($Cr_2O_7^{-2}$). In this +6 or hexavalent form, Cr is toxic and quite mobile in soil. Under acid conditions there is a conversion from chromate to dichromate. Soluble salts of Cr, such as sulfate and nitrate, are more toxic than insoluble salts of Cr such as oxides and phosphates. This toxicity becomes more important as the acidity of the soil is increased (Aubert and Pinta, 1977). Overcash and Pal (1979) state that in an aerobic acid soil, hexavalent Cr is quickly converted to the less toxic trivalent Cr or chromic, which is quite immobile; they consider the trivalent form to be relatively inert in soils. The oxidation of trivalent to hexavalent Cr has not been documented in field studies but does warrant further consideration because of the extreme toxicity and mobility of the hexavalent form.

Downward transport of Cr will be more rapid in coarse-textured soils than in fine textured soils because of the larger pores, less clay and faster downward movement of water. Chromium (III) readily precipitates with carbonates, hydroxides and sulfides, and would likely be a means of reducing leaching (Murrmann and Koutz, 1972). These precipitation reactions are also favored by a pH>6. Data from Wentink and Edzel (1972) show that three different soils were capable of almost 100% retention of Cr(III).

Chromium has been shown to be toxic to plants and animals, and recent studies indicate it may also be toxic to soil microorganisms. Ross et al. (1981) found that levels as low as 7.5 ppm in the growth media were toxic to gram negative bacteria including Pseudomonas and Nocardia. This indicates that soil microbial transformations such as nitrification and hydrocarbon degradation may be adversely affected by Cr. Rudolfs (1950) reviewed the literature on metals in sewage sludge and recommended a 5 ppm limit for Cr+6 in sewage sludge which is land treated. Mutations in bacterial populations have also been observed in bacteria grown in the presence of Cr+6 (Petrilli and De Flora, 1977).

Many investigators have found that Cr is toxic to plants. Dichromate is apparently more phytotoxic than chromate (Pratt, 1966b) and that both of these tetravalent forms are more toxic than the trivalent state (Hewitt, 1953). Application of 75 ppm Cr to soil is not toxic to sweet-orange (Citrus sinensis) seedlings, but additions of 150 ppm Cr are toxic. In sand cultures, 5 ppm Cr as chromate ion was toxic to tobacco (Nicotiana tabacum) and 10 ppm was toxic to corn (Zea mays) (Pratt, 1966b). Plants affected by Cr toxicity are stunted and frequently have narrow, discolored and necrotic leaves (Hunter and Vergnano, 1953).

There is some indication that Cr is accumulated in plant roots. The influence of plant Cr concentration on plant growth is given in Table 6.29 which indicates that some plants experience decreased yield at soil concentrations as low as 0.5 ppm Cr. These data indicate that the phytotoxic concentration is greater than 10 ppm. Soane and Saunder (1959) found the Cr content of tobacco roots to be twenty times higher than in the leaves of plants showing symptoms of Cr toxicity. They found only slightly higher Cr levels in the leaves of plants showing toxic symptoms than in leaves of healthy plants. Therefore, translocation of Cr from roots to the plant tops apparently is not a serious problem. This does not, however, eliminate Cr as a toxic element since it has a definite toxic effect on roots.

TABLE 6.29 PLANT RESPONSE TO CHROMIUM IN SOIL AND SOLUTION CULTURE

Amount of Cr (ppm)	Media	Species	Effect	Reference
.01	Silt soil	Fescue (Festuca clatior) & alfalfa (Medicago sativa)	No increase in plant Cr	Stucky & Newman (1977)
0.5	Solution	Soybean (Glycine max)	Reduced yield	Turner and Rust (1971)
4.8	Sand	Mustard	Decreased yield	Gemmell (1972)
5.2	Solution	Cotton (Gossypium hirsutum)	83% yield reduction	Rehab and Wallace (1978b)
10	Pot experiments	Mustard	Toxic	Andrziewski (1971)
10	Solution	Oat (Avena sativa)	Iron clorosis	Hewitt (1953)
10	Soil	Soybean (G. max)	Reduced yield	Turner and Rust (1971)
25	Pot experiments	Mustard	Toxic	Andrziewski (1971)
30-60	Solution	Soybean (G. max)	Toxic	Turner and Rust (1971)
52	Pot experiments	Potato (Solanum tuberosum) seedlings	Threshold of toxicity	Mukherji and Roy (1977)
55	Sandy loam	Rye (Secale cereale)	No increase in plant Cr	Kelling et al. (1977)
100-200	Yolo loam	Bush bean (Phaseolus limensis)	Decreased yield	Wallace et al. (1976)
128-640	Sand & peat	Mustard	Reduced yield	Gemmell (1972)
150	Soil	Sweet orange (Citrus sinensis)	Toxic	Pratt (1966b)
400	Submerged soil	Rice (Oryza sativa)	Slight yield reduction	Kamada and Doki (1977)
300-500	Soil	Rice (O. sativa)	No effect	Silva and Beghi (1979)

Chromium is essential for glucose metabolism in animals and its activity is closely tied to that of insulin (Scott, 1972). Although Cr is highly toxic to many invertebrates, it is only moderately toxic to higher animals, and most mammals can tolerate up to 1000 ppm Cr in their diets. In animals, however, experimental data have shown conclusively that Cr in the hexavalent form can cause cancer (Hernberg, 1977). The predilection of workers in Cr plants to respiratory cancer has been thoroughly documented in several studies and has been reviewed by Enterline (1974).

The use of irrigation water containing the upper limit of the acceptable concentration of Cr recommended by the National Academy of Sciences and National Academy of Engineering (1972) is equivalent to an accumulation of 1000 ppm Cr in the soil. Information obtained from this study indicates that the phytotoxic level of Cr in soil is highly variable, depending on the soil type and plant species, but can be as low as 25 ppm. Therefore, a more suitable criteria on which to base loading rates would be the amount of Cr immobilized by the soil as determined from demonstration of treatability tests.

6.1.6.9 Cobalt (Co)

Cobalt is used in the production of high grade steel, alloys, super-alloys and magnetic alloys. It is also used in smaller quantities as a drier in paints, varnishes, enamels and inks. Compounds of Co are also used in the manufacture of pigments and glass (Page, 1974). The concentration of Co in soils ranges from 1–40 ppm with an average of 8 ppm (Aubert and Pinta, 1977). Extensive areas can be found where the Co level in soil is deficient for animal health (Bowen, 1966).

The availability of Co is primarily regulated by pH and is usually found in soils as Co^{2+}. At low pH it is oxidized to Co^{3+} and often found associated with iron (Ermolenko, 1972). Adsorption of Co^{2+} on soil colloids is high between pH 6 and 7 (Leeper, 1978), whereas leaching and plant uptake of Co are enhanced by a lower pH. Cobalt sorbed on soil exchange sites is held more strongly than the common cations and can revert to a more strongly sorbed form over time (Banerjee et al., 1953). Soils naturally rich in Co have a high pH (Aubert and Pinta, 1977). If Co is added to soils containing lime, precipitation of Co with carbonates can be expected (Tiller and Hodgson, 1960).

Cobalt is water soluble when in the form of chloride, nitrate and sulfate salts. At a pH of 7, Co is 50–80% soluble when it is associated with cations such as ammonium, magnesium, calcium, sodium and potassium. At pH 8.5 Co becomes less soluble and cobaltous phosphate, a compound which is relatively insoluble in water, may regulate solubility (Young, 1948). In soils, Co is bound by organic matter and is very strongly sorbed or coprecipitated with manganese oxides (Leeper, 1978).

There is no evidence that Co is essential for the growth and development of higher plants. It is, however, required for the symbiotic fixation of nitrogen by nodulating bacteria associated with legumes (Ahmed and Evans, 1960 & 1961; Delwiche et al., 1961; Reisenauer, 1960). Excessive amounts of Co can be toxic to plants. Symptoms of Co toxicity vary with species but are frequently described as resembling that of iron deficiency (Vanselow, 1966b). In solution cultures, Co concentrations as low as 0.1 ppm produce toxic effects in crop plants. Cobalt applications to soil of 0.2 ppm had no effect on bean (Phaseolus sp.) growth in a study by dos Santos et al. (1979). In greenhouse experiments, Fujimoto and Sherman (1950) found Sudan grass (Sorghum sudanense) to be unaffected by an application rate equivalent to 224 kg/ha which resulted in a Co content in plants of 3-6 ppm. Phytotoxicity from soil Co occurs in plants containing 50-100 ppm and foliar symptoms are apparent at these levels (Hunter and Vergnano, 1953).

A recent study indicates that plants grown in a Co contaminated soil overlain by uncontaminated soil will accumulate large concentrations of the metal as shown in Fig. 6.16 (Pinkerton, 1982). This appears to be due to healthy vigorously growing roots encountering the elevated soil Co as opposed to having to develop in the high Co soil. This research implies that proper mixing of the Co waste and the soil is essential to preventing excessive plant accumulation of Co.

Most plants growing in soils with native Co concentrations do not accumulate Co and values exceeding 1 ppm are rare. Yet when growing in Co enriched media, these same species may accumulate the element and show yield reductions (Table 6.30). Yamagata and Murakami (1958) found 600 ppm Co in alder (Alnus sp.) leaves, while white oak (Quercus alba), chestnut, saxifrage and dogwood (Cornus florida) growing in the same area had 2-5 ppm Co in leaf ash. Swamp blackgum (Nyssa sylvatica) has also been found to contain a higher concentration of Co than grasses growing in the same area (Vanselow, 1966b). Blackgum is such a good indicator of Co status in a soil that Kubota et al. (1960) consider an area to be Co deficient for grazing animals when the concentration of Co in blackgum trees is less than 5 ppm; this method may be used to indicate soils suitable for amendment with Co-rich waste. The level of Co in cucumbers (Cumcumis sativus) and tomatoes (Lycopersicon esculentum) is increased by increasing the Co additions in nutrient solution (Coic and Lesaint, 1978), yet applications of 0.5-2 kg Co/ha had no effect on the Co concentration of the metal in red clover (Trifolium pratense) hay (Krotkikh and Repnikov, 1976).

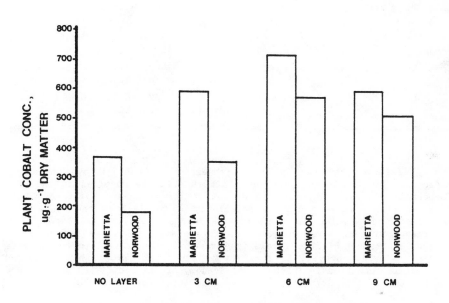

Figure 6.16.   Cobalt concentrations in tall fescue grown
in Marietta and Norwood soils at 400 mg Co
kg^{-1} (added as Co(NO$_3$)$_2$ · 6H$_2$O) with varying layer thicknesses of uncontaminated soil
overlying the cobalt amended soil
(Pinkerton, 1982).

TABLE 6.30   PLANT RESPONSE TO COBALT IN SOIL AND SOLUTION CULTURE

Co Concentration (ppm)	Media	Species	Effect	Reference
5	Solution	Cabbage (Brassica oleracea)	50% yield reduction	Hara et al. (1976)
25	Soil	Corn seedlings (Zea mays)	Top injury	Young (1979)
40	Soil	Oats (Avena sativa)	Toxic	Young (1979)
100	Soil	General	Threshold toxicity	Allaway (1968)
400	Solution	White bean (Phaseolus sp.)	34% yield reduction	Rauser (1978)

Cobalt is required by animals because it is the central atom in vitamin B_{12} (Rickles et al., 1948).   Although vitamin B_{12} is synthesized by microorganisms in the ruminant gut, Co must still be supplied in the diet (Sauchelli, 1969).   Since Co is essential for ruminants, pasture plants deficient in it cause a dietary deficiency of Co which is the cause of a progressive emaciation of ruminants (McKenzie, 1975).   Areas where Co deficiency in animals was observed had forage which contained less than 2.5 ppm Co. Extremely high Co levels in forage can also result in toxicity to grazing animals; however, Co toxicity in livestock has not been reported under field conditions. The National Academy of Science (1980) established 100 ppm Co in plant dry matter as the acute level for ruminants.

The use of irrigation water that contains the upper limit of the acceptable concentration of Co recommended by the National Academy of Sciences and National Academy of Engineering (1972) is equivalent to an accumulation of 500 ppm Co in the upper 15 cm of soil.   However plant toxicity results at soil concentrations well below this value, depending on plant species. Animal health is affected by plants containing 100 ppm Co., therefore loading rates should be based on soil concentrations which produce plants with Co concentrations less than 100 ppm.   A conservative value for cumulative Co of 200 ppm in the soil is suggested to immobilize the element as well as to avoid excess plant uptake.

6.1.6.10 Copper (Cu)

Significant amounts of Cu are produced in wastes from textile mills, cosmetics manufacturing, and sludge from hardboard production. Soil Cu contents range from 2–100 ppm with an average around 30 ppm (Bowen, 1966).

The abundance of Cu enrichment to the environment has prompted studies of the behavior of the element in relation to soil properties. Copper retention in soils is dependent on pH; sorption of Cu increases with increasing pH. In kaolinitic soils where clay surfaces have a net negative charge with increasing pH, the amount of Cu desorbed increased as the pH was lowered from 6 to 2 (Kishk and Hassan, 1973). The lack of adsorption of Cu at a low pH may be due to competition from Mg^{2+}, Fe^{3+}, H^+ and Al^{3+} for sorption sites. Soils selected to represent a broad range of mineral and organic contents were found to have a specific adsorption maximum at pH 5.5 of between 340 and 5780 ppm Cu in soil (McLaren and Crawford, 1973). Land treated Cu waste should be limed if necessary to maintain a pH of 6.5 or greater to ensure the predominance of insoluble forms of Cu, $Cu(OH)_2$ and $Cu(OH)_3$ (Hodgson et al., 1966 and Younts and Patterson, 1964).

Soil organic matter forms very stable complexes with Cu. Carboxyl and phenolic groups are important in the organic complexing of Cu in soils (Lewis and Broadbent, 1961). Sorption of Cu to organic matter occurs at relatively high rates when the concentrations of iron and manganese oxides in the soil are low. There is some evidence that Cu bound to organic matter is not readily available to plants (Purvis and MacKenzie, 1973). Organic matter may provide nonspecific sorption sites for Cu; however, the loss of organic matter through decomposition causes a significant decrease in this retention mechanism.

Clay mineralogy also plays a significant role in determining the amount of Cu sorbed. Experiments have shown that Cu^{2+} is sorbed appreciably by quartz and even more strongly by clays. The adsorption capacity of clays increases in the order kaloninte to illite to montmorillonite (Krauskopf, 1972). The strength of Cu sorption of soil constituents are in the following order:

manganese oxides < organic matter < iron oxides < clay minerals.

A column study by Emmerich et al. (1982) indicated that Cu applied as sewage sludge to a concentration of 512 ppm essentialy did not move below the zone of incorporation and that 94% of that applied was recovered from the soil. This soil had a pH between 5.2 and 6.7 and a CEC of 4.4 to 9.7 meq/100 g. Soil components which are less significant in Cu attenuation include free phosphates, iron salts, and clay-size aluminosilicate minerals.

Cation exchange capacity is a soil property indirectly related to mineralogy which may influence metal loading. Overcash and Pal (1979) have suggested that loading rates based on CEC only be used as a suggestion of the buffering capacity of the soil and critical cumulative limits have been adjusted to soil CEC (0-5 meq/100 g, 125 kg/ha; 5-15 meq/100 g, 250 kg/ha; 15 meq/100 g, 500 kg/ha).

Since the normal Cu concentration in plants (4 to 15 ppm) is lower than Cu levels found in most soils, the soil Cu content appears to be the most important factor in controlling plant levels of Cu. Management practices must be developed considering the chemistry of Cu in soils and Cu toxicity to plants and animals. The data of Gupta (1979) indicate that the toxic range of Cu in the leaves of plants is greater than 20 ppm, depending on species. The influence of soil and solution culture concentration on plant growth are given in Table 6.31, and indicates a soil concentration of over 80 ppm is necessary before most plant growth is adversely affected.

Copper is essential to the metabolic processes common to decomposing bacteria, plants and animals. Small quantities of Cu activate enzymes required in respiration, redox-type reactions and protein synthesis. Copper has been shown to be magnified within the food chain and moderate levels of Cu ingested by ruminants may be poisonous unless the effect is alleviated through proper diet supplements of molybdenum or sulfate (Kubota, 1977).

Several researchers have reported a decrease in plant Cu when large amounts of organic matter are present. Goodman and Gemmell (1978) reported successful reclamation of Cu smelter wastes treated with pulverized fly ash, sewage sludge or domestic refuse. In a greenhouse experiment, MacLean and Dekker (1978) eliminated the toxic effects of Cu on corn (Zea mays) by applying sewage sludge. Kornegay et al. (1976) found that additions of hog manure containing 1719 ppm Cu did not affect the Cu content in grain when compared to grain from control experiments. Purvis and MacKenzie (1973) found that the organic form of Cu was not readily taken up by plants when Cu-laden municipal compost was applied to soil at rates from 50 to 100 metric tons sludge/ha.

A study by Mitchell et al. (1978) evaluated Cu uptake by crops grown in acidic and alkaline soils (Table 6.32 and Table 6.33). In this study, wheat and grain growing in an acid soil showed the greatest amount of Cu accumulation. Copper may be strongly chelated in plant roots; consequently, root concentrations are usually greater than leaf concentrations.

TABLE 6.31 PLANT RESPONSE TO COPPER IN SOIL AND SOLUTION CULTURE

Amount of Cu (ppm)	Media	Species	Effect	Reference
.03	Solution	Andropogon scoparius	Root damage	Ehinger and Parker (1979)
1	Solution	Horse bean (Vicia faba)	Growth inhibited	Sekerka (1977)
10	Soil	Barley (Hordeum vulgare)	Stunted growth	Toivonen and Hofstra (1979)
26	Sand	Barley (H. vulgare); pea (Pisim sp.)	Inhibition of shoot growth	Blaschke (1977)
30	Solution	Coffee	Toxicity threshold	Andrade et al. (1976)
50-115	Soil of mining area	Anthoxanthum odoratum	None	Karataglis (1978)
69	Soil	Corn (Zea mays)	Decreased root weight	Klein et al. (1979)
91	Soil	Barley (H. vulgare)	Reduced yield	Davis (1979)
100	Rooting media	Barley (H. vulgare)	Stunted growth	Toivonen and Hofstra (1979)
100	Soil	Green alder (Alnus americana)	Seedling damage	Fessenden & Sutherland (1979)
130	Soil	Barley (H. vulgare)	Accumulated 21 ppm in leaves	Davis (1979)
150	Soil	Black spruce (Picea mariana)	Growth decrease	Fessenden & Sutherland (1979)
400	Yolo loam	Cotton (Gossypium hirsutum)	Leaf yields reduced by 35%	Rehab & Wallace (1978a)
400	Yolo loam	Cotton (G. hirsutum)	Leaf yields reduced by 53%	Rehab & Wallace (1978a)

TABLE 6.32 COPPER CONCENTRATION IN PLANT TISSUE IN RELATION TO COPPER
ADDITION IN AN ACID SOIL (REDDING FINE SANDY LOAM)*

Cu Concentration (ppm)	Plant Portion	Crop	Plant Concentration	Effect
5	Shoots	Lettuce (Lactuca sativa)	6.8	None
5	Leaves	Wheat (Triticum aestivum)	10.7	None
5	Grain	Wheat (T. aestivum)	7.3	None
80	Shoots	Lettuce (L. sativa)	8.9	None
80	Leaves	Wheat (T. aestivum)	10.7	None
320	Shoots	Lettuce (L. sativa)	10.7	60% yield reduction
320	Grain	Wheat (T. aestivum)	12.3	20% yield reduction
640	Shoots	Lettuce (L. sativa)	18.3	90% yield reduction
640	Grain	Wheat (T. aestivum)	33.0	95% yield reduction

* Mitchell et al. (1978).

TABLE 6.33 COPPER CONCENTRATION IN PLANT TISSUE IN RELATION TO COPPER
ADDITION IN A CALCAREOUS SOIL (DOMINO SILT LOAM)*

Cu Concentration (ppm)	Plant Portion	Crop	Plant Concentration	Effect
5	Shoots	Lettuce (Lactuca sativa)	6.4	None
5	Leaves	Wheat (Triticum aestivum)	10.7	None
5	Grain	Wheat (T. aestivum)	6.7	None
80	Shoots	Lettuce (L. sativa)	7.9	None
80	Leaves	Wheat (T. aestivum)	14.8	None
160	Leaves	Lettuce (L. sativa)	8.2	30% yield reduction
160	Grain	Wheat (T. aestivum)	7.9	None
320	Leaves	Wheat (T. aestivum)	15.4	Significant yield reduction
320	Grain	Wheat (T. aestivum)	9.1	20% yield reduction
640	Grain	Wheat (T. aestivum)	9.2	40% yield reduction

* Mitchell et al. (1978).

In summary, the controlling factor in the prevention of toxic levels of Cu in water, plants and animals is the level of Cu in the soil. While Cu tolerance in plants can be explained by certain mineral interactions, the ultimate sites for adsorption of Cu in the environment remain the organic and inorganic colloid fractions in soil. The National Academy of Sciences and National Academy of Engineering (1972) recommend a soil accumulation of 250 ppm Cu in the upper 15 cm of soil. Tables 6.31, 6.32 and 6.33 indicate that the phytotoxic concentration of Cu ranges from about 70 to 640 ppm Cu in the soil for most plants. A conservative recommendation of 250 ppm is given for Cu concentration in soil. However, if treatability tests show immobilization at higher levels without toxicity, then loading rates could be increased.

6.1.6.11 Gallium (Ga)

Gallium concentration in soil is commonly low, averaging 30 ppm (Kirkham, 1979), except where it occurs in coal, oil, and bauxite ore. Since Ga is sorbed by aluminum (Al) in soil, Ga concentrations are likely to be higher in sandy acidic soils with dominant Al mineralogy. Disposal of Ga present in waste streams of smelter or coal processing plants depends on the degree of Ga retention in soils with dominant Al mineralogy.

6.1.6.12 Gold (Au)

Gold is rarely found in waste streams of any industry because it is a precious metal. Since pure Au is quite dense (19 g/cm^3), it is frequently concentrated in deposits called placers. In Mexico and Australia, placers are concentrated by wind; as the lighter minerals are eroded away, the Au remains in the deposit (Flint and Skinner, 1977). The average Au concentration in igneous and sedimentary rocks is 4 ppb. Gold concentrations in fresh water are normally less than 0.06 ppb, and Au is found in sea water at 0.011 ppb as $AuCl_4$.

Gold is not essential to plants or animals. Bowen (1966) ranks Au as scarcely toxic which means that toxic effects rarely appear except in the absence of a related essential nutrient, or at osmotic pressures greater than one atmosphere. Overcash and Pal (1979) list Au as a heavy metal which reacts with cell membranes to alter their permeability and affect other properties. The Au concentration in land plants ranges from 0.3–0.8 ppb. The horsetail, Equisetum, is said to accumulate Au.

The isotope Au-198 is commonly used in medicine. In mammals, Au in the colloidal form can accumulate in the liver. The typical Au concentration in mammalian livers is 0.23 ppb. The mollusc, Unio mancus, was found to contain 0.3–3.0 ppb Au in its shell and 4.0–40 ppb Au in its flesh (Bowen, 1966). It is expected that any Au present in a waste would be recovered before land treatment.

6.1.6.13 Lead (Pb)

The primary source of Pb in hazardous waste is from the manufacture of Pb-acid storage batteries and gasoline additives (tetraethyl Pb). Tetraethyl Pb production alone consumes approximately 264,000 tons of Pb per year in the U.S. (Fishbein, 1978). Lead is also used in the manufacture of ammunition, caulking compounds, solders, pigments, paints, herbicides and insecticides (Page, 1974). The Pb content of sewage sludge averages 0.17%. In coal, Pb content may range from 2-20 ppm (Overcash and Pal, 1979).

A Pb concentration of about 10 ppm is average for surface soils. Some soil types, however, can have a much higher concentration. In soils derived from quartz mica schist, the Pb content may be 80 ppm. The concentration in soil derived from black shale may reach 200 ppm Pb (Barltrop et al., 1974).

Lead is present in soils as Pb^{2+} which may precipitate as Pb sulfates, hydroxides and carbonates. Figure 6.17 illustrates the various Pb compounds present according to soil pH. Below pH of 6, $PbSO_4$ (anglesite) is dominant and $PbCO_3$ is most stable at pH values above 7. The hydroxide $Pb(OH)_2$ controls solubility around pH 8, and lead phosphates, of which there are many forms, may control Pb^{2+} solubility at intermediate pH values. Solubility studies with molybdenum (Mo) show that $PbMoO_4$ is a reaction product and will govern Mo concentrations in the soil solution.

The availability of Pb in soils is related to moisture content, soil pH, organic matter, and the concentration of calcium and phosphates. Under waterlogged conditions, naturally occurring Pb becomes reduced and mobile. Organometallic complexes may be formed with organic matter and these soil organic chelates are of low solubility. Increasing pH and calcium (Ca^{2+}) ions diminish the capacity of plants to absorb Pb, as Ca^{2+} ions compete with the Pb^{2+} for exchange sites on the soil and root surfaces (Fuller, 1977).

The Pb adsorption capacity of Illinois soils has been found to reach several thousand kilograms per hectare (CAST, 1976). In another study, only 3 ppm soluble Pb was found three days after 6,720 kg Pb/ha was added to the soil (Brewer, 1966b). Lead is adsorbed most strongly from aqueous solutions to calcium bentonite (Ermolenko, 1972).

Lead is not an essential element for plant growth. It is, however, taken up by plants in the Pb^{2+} form. The amount taken up decreases as the pH, cation exchange capacity, and available phosphorus of the soil increase. Under conditions of high pH, CEC and available phosphorous, Pb becomes less soluble and is more strongly adsorbed (CAST, 1976). This insolubilization takes time and Pb added in small increments over long time periods is less available to plants than high concentrations added over a short period of time (Overcash and Pal, 1979).

Lead toxicity to plants is uncommon (Table 6.34). Symptoms of Pb toxicity are found only in plants grown on acid soils. In solution cul-

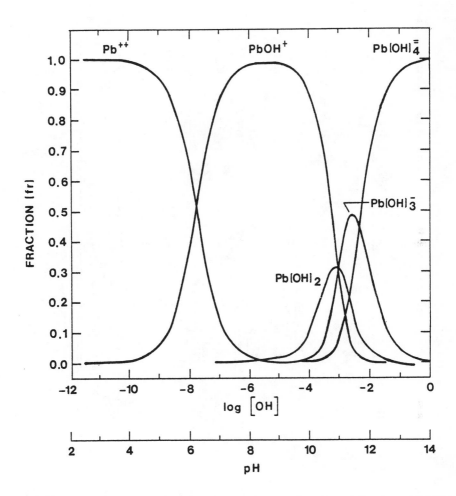

Figure 6.17. Distribution of molecular and ionic species of
divalent lead at different pH values (Fuller,
1977).

TABLE 6.34 PLANT RESPONSE TO LEAD IN SOIL AND SOLUTION CULTURE

Pb Concentration (ppm)	Media	Species	Effect	Reference
0.4	Soil	Eggplant (Solanum melongena)	None	Watanbe and Nakamura (1972)
3.6	Soil	Corn (Zea mays)	None	Sung and Young (1977)
5.0	Solution	Corn (Z. mays)	Reduced root growth	Malone et al. (1978)
21.0	Solution	Sphagnum fimbriatum	None	Simola (1977)
50.0	Solution	Lettuce (Lactuca sativa)	None	John (1977)
66.0	Soil	Loblolly pine (Pinus taeda) & autumn olive	None	Rolfe & Bazzar (1975)
100.0	Solution	Soybean (Glycine max)	None	Malone et al. (1978)
200.0	Sand	Oats (Avena sativa)	Impaired growth	Kovda et al. (1979)
1000.0	Acid Soil	Plantain (Musa paradisiaca)	None	Dikjshoorn et al. (1979)
1000.0	Soil	Red clover (Trifolium pratense)	None	Horak (1979)
1500.0	Soil pH 5.9	Corn (Zea mays)	None	Baumhard and Welch (1972)
1500.0	Solution	Ryegrass (Secale cereale)	None	Jones et al. (1973)
2500.0	Sand	Glyceria maxima	Chlorosis	Raghi-Atri (1978)
3775.0	Sandy clay	Corn (Z. mays) & soybeans (Glycine max)	None	Sung and Young (1977)

ture, root growth of sheep fescue is retarded by 30 ppm and stopped by 100 ppm Pb. Lead content in plants grown on soil with a high Pb level increases only slightly over that of plants grown on soil of average Pb content. Clover tops (Trifolium sp.) show an increase of 7.55 ppm, while kale (Brassica campestris) and lettuce (Lactuca sativa) leaves show an increase of less than 1 ppm. The Pb taken up by plants is rarely translocated since it becomes chelated in the roots. Tops of barley (Hordeum vulgare) grown on a soil extremely high in Pb contained 3 ppm while the roots contained 1,475 ppm Pb (Brewer, 1966b). Translocation of Pb to grain is less than translocation to vegetative parts (Schaeffer et al., 1979). Applied sewage sludge containing 360 ppm Pb resulted in no significant increase in Pb content of corn leaves and grain (Keefer et al., 1979).

Lead poisoning is quite serious and a major human health concern. Perlstein and Attala (1966) estimate that 600,000 children each year in the U.S. suffer from Pb poisoning. Of these, 6,000 have permanent neurological damage and 200 die. One source of elevated Pb in children may be contact with Pb-containing dust (Vostal et al., 1974). In fact, soil Pb content in excess of 10,000 ppm may result in an increase in Pb absorption even by children who do not ingest the contaminated soil (Barltrop et al., 1974). Where high levels of lead are allowed to accumulate, children should be prevented from entering the site throughout the post-closure period.

Cattle and sheep are more resistant to Pb toxicity than horses. There is, however, some tendency for cattle to accumulate Pb in tissues, and Pb can be transferred to milk in concentrations that are toxic to humans (National Academy of Sciences and National Academy of Engineering, 1972). Based on human health considerations, the maximum allowable Pb content in domestic animals is 30 ppm (National Academy of Science, 1980). Cattle ingest large amounts of soil when grazing and may consume up to ten times as much Pb from soil as from forage. Lead poisoning has been reported in cattle grazing in Derbyshire, England, where the soil is naturally high in the element (Barltrop et al., 1974).

The use of irrigation water that contains the upper limit of the acceptable concentration of Pb as recommended by the National Academy of Sciences and National Academy of Engineering (1972) is equivalent to an accumulation of 1,000 ppm of lead in the upper 15 cm of soil. Table 6.34 indicates Pb is generally not toxic to plants and the element does not readily translocate to leaves or seeds. Growth of root crops should be avoided and grazing animals should be excluded from the site to avoid Pb toxicity to animals and humans. If demonstration of treatability experiments verify immobilization of Pb at high concentrations, 1000 ppm total Pb could be safely allowed to accumulate in the soil without phytotoxicity.

6.1.6.14 Lithium (Li)

Lithium normally occurs in saline and alkaline soils and is usually associated with carbonates in soils derived from calcareous parent materials. The average Li content of soils is 20 ppm. Because the concentration of total and soluble Li is not related to depth in the profile, clay content or organic carbon content (Shukla and Prasad, 1973; Gupta et al., 1974), it is expected that Li is not fixed selectively in soil except by precipitation after liming.

The usual Li concentration in plants and animals is low, but levels of 1,000 ppm in plant tissues, which are sometimes reached in plants grown on mineral enriched soils, do not appear to be very phytotoxic. The data provided by the present review indicate that the toxic range of Li in the leaves of plants varies from 80 to 700 ppm depending on species (Table 6.35). At low levels in a nutritive solution, Li stimulates phosphorylase activity in tuber storage of beets (Beta vulgaris), while growth in corn (Zea mays), wheat (Triticum aestivum) and fescue (Festuca sp.) is limited as a result of Li substitution for Na in cellular functions. Tables 6.35 and 6.36 list plant concentrations of Li and crop responses to those concentrations, respectively. Lithium poses little threat to the food chain since it is only slightly toxic to animals.

TABLE 6.35   THE INFLUENCE OF LEAF LITHIUM CONCENTRATION ON PLANTS

Li Concentration (ppm)	Portion of plant	Species	Effect	Reference
26	Leaf	Mean of 200	None	Romney et al. (1975)
45	Leaf	Cotton (Gossypium hirsutum)	None	Rahab & Wallace (1978c)
80	Leaf	Tomato (Lycopersicon esculentum)	Threshold of toxicity	Wallihan et al. (1978)
220	Leaf	Bean (Phaseolus sp.)	Yield reduction	Wallace et al. (1977)
600	Leaf	Bean (Phaseolus sp.)	Severe	Wallace et al. (1977)
700	Leaf	Cabbage (Brassica oleracea)	50% Yield reduction	Hara et al. (1977)

TABLE 6.36   THE INFLUENCE OF SOLUTION CULTURE AND SOIL CONCENTRATION OF
LITHIUM ON PLANT GROWTH AND YIELD

Amount of Li (ppm)	Media	Species	Effect	Reference
2	Solution	Tomato (Lycopersicon esculentum)	Toxicity	Wallihan, et al. (1978)
	Sand	Wheat (Triticum aestivum)	No influence	
8	Solution	Barley (Hordeum vulgare)	No seedlings	Gupta (1974)
50	Loam	Bean (Phaseolus sp.)	Severe injury	Wallace et al. (1977)
50	Yolo loam	Cotton (Gossypium hirsutum)	None	Rehab & Wallace (1978c)
100	Soil	Wheat (T. aestivum)	No influence	Gupta (1974)
		Barley (H. vulgare)		
587	Loam	Cotton (G. hirsutum)	None	Wallace et al. (1977)
1000	Loam	Barley (H. vulgare)	Severe	Wallace et al. (1977)

The use of irrigation water that contains the upper limit of the acceptable concentration of Li as recommended by the National Academy of Sciences and National Academy of Engineering (1972) is equivalent to an accumulation of 250 ppm of Li in the upper 15 cm of soil. Information in Tables 6.35 and 6.36 indicates that the phytotoxic level of Li in soil ranges from 50 to 1000 ppm. An acceptable estimate for cumulative Li in the soil appears to be 250 ppm. However, if treatability tests show that higher concentrations are immobilized without toxicity, then loading rates could be increased.

6.1.6.15 Manganese (Mn)

The major sources of Mn bearing wastes are the iron and steel industries. Other sources of Mn include disinfectants, paint and fertilizers (Page, 1974). Manganese dioxide is found in wastes from the production of alkaline batteries, glass, paints and drying industries.

Concentrations of Mn in mineral soils range from 20–3000 ppm, though 600 ppm is average (Lindsay, 1979). When Mn is released from primary rocks by weathering, secondary minerals such as pyrolusite (MnO_2) and manganite [$MnO(OH)$] are formed. The most common forms of Mn found in soil are the divalent cation (Mn^{2+}) which is soluble, mobile, and easily available, and the tetravalent cation (Mn^{4+}) which is practically insoluble, non-mobile, and unavailable (Aubert and Pinta, 1977). The trivalent cation Mn^{3+}, as Mn_2O_3, is unstable in solution. The tetravalent cation usually appears in well oxidized soils at a very low pH. Under reduced conditions found in water saturated soils, Mn^{2+} is the stable compound, and this divalent ion is adsorbed to clay minerals and organic matter. In strongly oxidized environments, the most stable compound is the tetravalent Mn dioxide, MnO_2.

Manganese availability is high in acid soils and Mn^{2+} solubility decreases 100-fold for each unit increase in pH. (Lindsay, 1972) At pH values of 5.0 or less, Mn is rendered very soluble and excessive Mn accumulation in plants can result. At pH values of 8 or above, precipitation of $Mn(OH)_2$ results in Mn removal from the soil solution.

Reduced conditions in the soil increase Mn solubility and produce Mn^{2+} in solution. Oxidation of Mn occurs at a low redox potential in an alkaline solution (Krauskopf, 1972). Under oxidizing conditions, several Mn compounds may be formed including $(MnSi)_2O_3$, BaMn(II), MnOOH, and the stable product of complete oxidation, pyrolusite (MnO_2).

When the pH of the soil is greater than 7, manganese (Mn^{2+}) is rendered less available by adsorption onto organic matter colloids. Thus, soils of high pH with large organic matter reserves are particularly prone to Mn deficiency. However, the affinity of Mn^{2+} for synthetic chelates is comparatively low, and chelated Mn^{2+} can be easily exchanged by Zn^{2+} or Ca^{2+}.

Interactions of Mn with other elements have been noted in soil adsorption and plant uptake. The formation of manganese oxides in soils appears to regulate the levels of cobalt (Co) in soil solution and hence Co cobalt availability to plants. Bowen (1966) reported that plant uptake of Mn was greater in the absence of calcium and that Mn adsorption was reduced in the presence of iron, copper, sodium, and potassium.

Concentrations of Mn in plant leaves generally range from 15–150 ppm. The suggested maximum concentration value for plants is given at 300 ppm (Melsted, 1973), however the data of the National Research Council (1973) indicate that the toxic range of Mn in leaves is 500 to 2,000 ppm, depending on plant species. Vaccinium myrtillus plants appear healthy when the foliage contains as high as 2431 ppm Mn and Lupinus luteus and Ornithopus sativus are both Mn tolerant (Lohris, 1960). Young plants are generally rich in Mn and the element can be translocated to meristematic tissues. Tables 6.37 and 6.38 list various Mn concentrations in the soil that produce toxic symptoms in plants.

TABLE 6.37   THE INFLUENCE OF LEAF MANGANESE CONCENTRATION ON PLANTS*

Plant Concentration (ppm)	Media	Portion of Plant	Species	Effect
15–84	Solution	Leaves	Soybeans (Glycine max)	None
49–150	Solution	Roots	Soybeans (G. max)	Toxic
70–131	Solution	Tops	Lespedeza (Lespedeza sp.)	None
160	Field	Leaves	Tobacco (Nicotiana tabacum)	None
173–999	Solution	Leaves	Soybeans (G. max)	Toxic
207–1340	Soil	Whole plant	Bean (Phaseolus sp.)	None
300–500	Soil	Leaves	Orange (Citrus sp.)	None
400–500	Field	Tops	Lespedeza (Lespedeza sp.)	Toxic
770–1000	Solution	Tops	Barley (Hordeum vulgare)	Toxic
993–1130	Pots	Whole plant	Tobacco (N. tabacum)	Toxic
1000	Soil	Leaves	Orange (Citrus sp.)	Toxic
1000–3000	Soil	Tops	Bean (Phaseolus sp.)	Toxic
3170	Soil	Roots	Tobacco (N. tabacum)	Toxic
4000–11,000	Soil	Leaves	Tobacco (N. tabacum)	Toxic

* Chapman (1966).

Manganese is absorbed by plants as the divalent cation Mn^{2+}. Its essential functions in plants include the activation of numerous enzymes concerned with carbohydrate metabolism, phosphorylation reactions, and the citric acid cycle. Magnesium, calcium and iron depress Mn uptake in a variety of plant species (Moore, 1972).

Manganese toxicity in young plants is indicated by brown spotting on leaves. One to four grams of Mn per milliliter of solution may depress yields of lespedeza (Lespedeza sp.), soybeans (Glycine max) and barley (Hordeum vulgare) (Labanauskas, 1966). The threshold of toxicity for tomato (Lycopersicon esculentum) plants grown in soil was observed at a Mn concentration of 450 ppm (Jones and Fox, 1978).

TABLE 6.38   PLANT RESPONSE TO MANGANESE IN SOIL AND SOLUTION CULTURE

Amount of Mn (ppm)	Media	Species	Effect	Reference
2.1	Solution	Legume	Toxicity threshold	Helyar (1978)
4-64	Solution	Weeping lovegrass (Eragrostis curvula) & fescue (Festuca sp.)	No effect	Fleming et al. (1974)
5	Solution	Jacoine (Pinus banksiana) & black spruce (Picea mariana)	Toxic No effect	Lafond & Laflamme (1970) Lafond & Laflamme (1970)
5	Solution	Soybean (Glycine max)	Toxic	Brown & Jones (1977)
15	Solution	Soybean (G. max)	No effect	Heenan & Carter (1976)
20	Sand	Groundnut (Apios americana)	Reduced yield	Benac (1976)
30	Solution	Satsuma orange (Citrus reticulata)	Chlorosis	Otsuka and Morizaki (1969)
40	Sand	Macroptilium atropurpureum	No effect	Hutton et al. (1978)
65	Acid soil	Soybean (G. max)	Toxicity	Franco & Dobereiner (1971)
130	Soil	Subterranean clover (Trifolium subterraneum)	Toxic	Simon et al. (1974)
140-200	Soil	Barley (Hordeum vulgare)	Yield decreased	Prausse et al. (1972)
200	Soil	Tobacco (Nicotiana tabacum)	Reduced yield	Link (1979)
250	Soil	Watermelon (Cucumis sp.)	Toxic	Gomi & Oyagi (1972)
450	Soil	Tomatoes (Lycopersicon esculentum)	Toxicity threshold	Jones and Fox (1978)
1400	Soil	Kidney bean (Phaseolus vulgare)	Toxic	Gomi & Oyagi (1972)
3000	Soil	Peppers (Capsicum sp.)	Toxic	Gomi & Oyagi (1972)
5000	Soil	Eggplant (Solanum melongena) & melons (Cucumis sp.)	Toxic	Gomi & Oyagi (1972)

Manganese is an essential element in animal nutrition for reproduction, growth and skeletal formation. Maximum tolerable levels in animals are cattle, 1000 ppm; sheep, 1000 ppm; swine, 400 ppm; and poultry, 2000 ppm (National Academy of Science, 1980).

In summary, the maintenance of certain conditions in the soil can be used to prevent environmental contamination from land treating of Mn bearing wastes. Manganese sorption is enhanced by organic matter colloids and precipitation of Mn is enhanced by carbonates, silicates and hydroxides at high pH values. The maintenance of a pH of greater than 6.5 is essential to reducing Mn solubility. The use of irrigation water that contains the upper limit of the acceptable concentration of Mn as recommended by the National Academy of Sciences and National Academy of Engineering (1972) is equivalent to an accumulation of 1,000 ppm of Mn in the upper 15 cm of soil. Information obtained from Jones and Fox (1978) and Tables 6.37 and 6.38 indicate that the phytotoxic level of Mn in soil is generally greater than 500 ppm.

6.1.6.16 Mercury (Hg)

Mercury has become widely recognized as one of the most hazardous elements to human health. The potential for Hg contamination exists where disposal practices create conditions conducive for conversion of Hg to toxic forms.

Mercury enters land treatment facilities from electrical apparatus manufacturing, electrolytic production of chlorine and caustic soda, pharmaceuticals, paints, plastics, paper products and Hg batteries. Mercury is used as a catalyst in the manufacture of vinyl chloride and urethane. More than 40% of pesticides containing metal contain Hg. Burning oil and coal increases atmospheric Hg which eventually falls to the earth and enters the soil (Page, 1974). Mineral soils in the U.S. usually contain between 0.01-.3 ppm Hg; the average concentration is 0.03 ppm (Lindsay, 1979).

Transformations in the soil and the forms of Hg resulting from these reactions regulate the environmental impact of land application of mercurical waste. Figure 6.18 illustrates these conversions and the cycling of Hg in the soil. Mercury moves very slowly through soils under field conditions. Divalent Hg is rapidly and strongly complexed by covalent bonding to sulfur-containing organic compounds and inorganic particles. These particles bind as much as 62% of the Hg in surface soils (Walters and Wolery, 1974). Mercury, as Hg^{2+}, is also bound to exchange sites of clays, hydrous oxides of iron and manganese, and fine sands (Reimers and Krenkel, 1974). Sorption of Hg by soil organic matter approaches 100% of the Hg added to an aqueous solution and exceeds sorption of a variety of other metal elements (Kerndorff and Schnitzer, 1980).

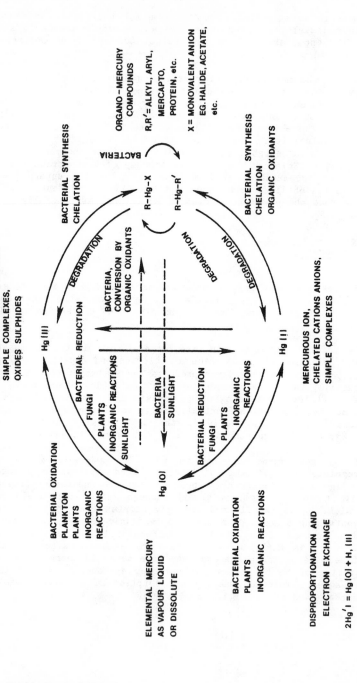

Figure 6.18. The cycle of mercury interconversions in nature (Jonasson and Boyle, 1971). Reprinted by permission of the Royal Society of Canada.

Removal of Hg by adsorption to clay colloids appears to be pH dependent and proportional to the respective CEC value of the clay. A study by
Griffin and Shimp (1978) indicates that 20 to 30% of the observed Hg
removal is due to adsorption by clay, and that Hg removal from the soil
solution is favored by alkaline conditions. The amount of Hg^{2+} removed
from solution by a given clay at a specific pH can be determined as follows:

$$C_R = \frac{(C_I - C_{Eq})V_F}{W} \tag{6.3}$$

where

C_R = amount of Hg^{+2} removed in mg/g clay;
C_I = initial Hg concentration in ppm;
C_{Eq} = equilibrium Hg concentration in ppm;
V_F = total solution volume after pH adjustments in mls;
W = weight of clay in gms.

About two-thirds of the Hg removed by clay is organic Hg, Fig. 6.19 illustrates this removal.

Precipitation of Hg complexes is a means of removing Hg from the
leaching fraction. At pH values above 7, precipitates of $Hg(OH)_2$, $HgSO_4$,
$HgNO_3$, and $Hg(NH_3)_4$ predominate and are very insoluble. Insoluble HgS and
$HgCl_3$ can occur at all pH ranges (Lindsay, 1979).

Organic mercurials associated with soil organic matter or the well-
defined compounds such as phenyl-, alkyl-, and methoxyethyl mercury compounds used as fungicides may be degraded to the metallic form, Hg^o.
This reaction is common in soil when coliform bacteria, or Pseudomonas spp.
are present. This is a detoxication process which produces metallic Hg and
hydrocarbon degradation products; however, the metallic Hg may be
volatilized.

Microbial and biochemical reactions are not only capable of breaking
the link between Hg and carbon in organic mercurials; they may also mediate
the formation of such links. Elemental Hg can be converted to methyl mercury by Methanobacterium omilianskii and also some strains of Clostridium.
These anaerobic microbes are responsible for the formation of toxic Hg
forms, methyl and dimethyl Hg. Both methyl and dimethyl Hg are volatile
and soluble in water, although dimethyl Hg is less soluble and more volatile. The formation of methyl Hg occurs primarily under acidic conditions,
while dimethyl Hg is produced at a near neutral pH (Lagerwerff, 1972). In
addition to being volatile and soluble, methylated forms of Hg are the most
toxic. Methylation of mercury by microbial transformation can be reduced
when nitrate concentrations in the soil are above 250 ppm nitrogen as
KNO_3 (Barker, 1941).

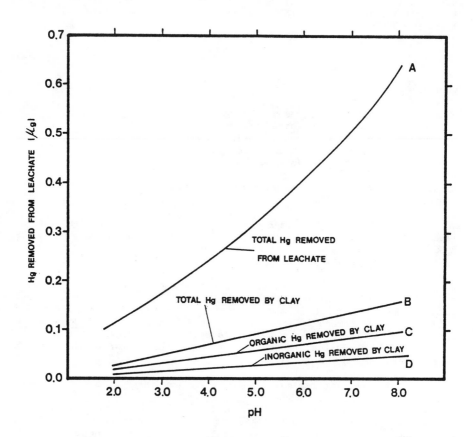

Figure 6.19. Removal of various forms of mercury from
DuPage landfill leachate solutions by
kaolinite, plotted as a function of pH
at 25° C (Griffin and Shimp, 1978).

Methylation of mercury can also occur by a monoenzymatic process involving vitamin B_{12} or one of its analogs, such as methylcobalamine, when CH_3 is transferred from cobalt (Co^{3+}) to Hg^{2+} as shown below:

$$
\begin{array}{c}
CH_3 \\
| \\
2Co^{3+} + Hg^0
\end{array}
\longrightarrow
\begin{cases}
\nearrow CH_3Hg \ + \ CH_4 + 2Co^{2+} \\
\\
\searrow (CH_3)_2Hg \ = \ 2Co^{2+}
\end{cases}
$$

Another method of methylation is facilitated by the fungi Neurospora crassa which can make this conversion aerobically without the mediation of vitamin B_{12} (Lagerwerff, 1972).

Plant content of Hg ranges from 0.001 to 0.01 ppm in plant leaves. Mercury is a nonessential plant element and is taken up by plants in the form of CH_3Hg, Hg^0, and Hg^{2+}. The Hg enters through the roots or by diffusion of gaseous Hg^0 through the stomata. Aquatic plants such as brown algae tend to accumulate Hg relative to its concentration in sea water and contain levels as high as 0.03 ppm (Bowen, 1966). As a result, Hg bioconcentration presents a greater hazard in aquatic food chains than in terrestrial food chains (Chaney, 1973).

The most serious contamination of Hg in the aquatic food chain occurs where Hg exists as methyl mercury. The Hg poisoning in Japan resulted from discharges of Hg containing waste from a plastics factory at concentrations between 1.6 and 3.6 ppb. Local concentrations of Hg were: plankton, 3.5 to 19 ppm; bottom muds, 22 to 59 ppm Hg; and shellfish, 30 to 102 ppm mercury on a dry weight basis (Irukayama, 1966).

No specific concentration of Hg has been shown to be phytotoxic. Applications of 25-37 kg/ha Hg did not reduce yields of wheat, oats, barley, clover or timothy (Overcash and Pal, 1979). The concentration of Hg in soil that is toxic to plants was determined to be greater than 10 ppm by Van Loon (1974). Foliar treatment of rice in Japan has caused Hg concentrations as high as 200 ppb compared with 10 ppb in rice from untreated fields. Mercury levels in tomatoes after application of a Hg containing sludge on an alkaline soil were as high as 12.2 ppm (Van Loon, 1974). Table 6.39 lists the effect of Hg on various plant species and indicates that phytotoxicity does not result from growth in high Hg media.

TABLE 6.39 THE INFLUENCE OF MERCURY ON PLANT GROWTH AND YIELD

Amount of Hg (ppm)	Media	Species	Effect	Reference
.05	Loamy sand	Spring wheat (Triticum aestivum)	Shoots accumulated 5.5 ppm	Findenegg & Havnold (1972)
10	Soil	Alfalfa (Medicago sativa), rape (Brassica sp.), wheat (Triticum aestivum)	No effect	Gracey & Stewart (1977)
10	Solution	Pisum sativum	Toxic	Beauford et al. (1977)
25	Sand	Oat (Avena sativa)	No effect	Kovda et al. (1979)
250	Sand	Oat (A. sativa)	Reduced yield	Sorteberg (1978)
445	Soil	Bentgrass (Agrostis sp.)	No toxic effect	Estes et al. (1973)

Reactions with selenium (Se) and cadmium can decrease Hg toxicity. Methyl Hg readily complexes with Se and when present in equimolar amounts, Se readily detoxifies methyl Hg. Dietary Se protects against the toxic effects of Hg in both rats and quail (El-Begearmi, 1973). It is interesting to note that fish taken from Minimata Bay in Japan had high concentrations of methyl Hg but comparatively low concentrations of Se, with a molar ratio of Se:methyl Hg of about 1:10. Cadmium also seems to react with Hg and has been shown to reduce Hg toxicity in humans and animals (Perry and Yunice, 1965).

In summary, the possibility of methyl mercury reaching the food chain will regulate land treatment waste loading. Uptake of Hg by plant roots can be minimized by maintaining a soil pH above 6.5. Mercury will precipitate as a carbonate or hydroxide at this pH, therefore, maintaining soil pH is a valuable mechanism for attenuating mercury. Adsorption of Hg onto organic matter colloids occurs most readily at a low pH. Mercury is more mobile in soils if it is organically complexed than if it is adsorbed onto clays.

Wastes containing some Se can also alleviate the hazard of Hg toxicity in animals. Application of a waste containing both elements would be less likely to create toxicity problems than a waste that contains only Hg. Sulfur in the waste can also help to attentuate Hg by precipitating HgS which is very insoluble. Chaney (1974) recommends that wastes containing greater than 10 mg/kg Hg not be land applied since extremely low concentrations of Hg are allowed for drinking water. Alternate disposal methods waste containing Hg at these levels should be considered.

6.1.6.17 Molybdenum (Mo)

The largest single use of Mo is in the production of steel and alloys. It is also used in the production of pigments, filaments, lamps and electronic tubes, and is used in small amounts in fertilizers and as a catalyst (Page, 1974). Soils typically have a median Mo concentration of 2 ppm with a range of 0.2 ppm to 5 ppm (Lindsay, 1979). Shale and granite are the major rocks contributing Mo to soils (Goldschmidt, 1954).

At soil pH values above 5, Mo is generally found as the molybdate anion, MoO_4^{2-}. At low pH values (2-4.5) Mo is strongly sorbed by soil colloids and organic matter. However, plants high in Mo are often produced on organic soils, indicating that organic matter is not a major means of rendering Mo unavailable. Sorption of Mo by soil colloids or iron and aluminum oxide coatings on soil colloids appears to be more effective in rendering Mo unavailable for plant uptake. Reisenauer et al. (1962) and Jones (1957) suggest that sorption of Mo by iron and aluminum oxides may be due to the formation of relatively insoluble ferric and aluminum molybdate precipitation at this low pH. Since Mo behaves as an anion at pH values above 2, kaolinite which has a high anion exchange capacity, has been shown to sorb more Mo than montmorillonite (Jones, 1957).

Soil water relationships and their impact on oxidation-reduction relationships also regulate Mo solubility. Kubota et al. (1963) demonstrated this relationship by growing alsike clover on two Nevada soils that contained significant concentrations of Mo. Each soil was held at two moisture levels. One was a wet treatment with the water table maintained 18 cm below the soil surface; another was a dry treatment in which the soil water potential was allowed to decrease to -10 to -15 bars before watering. The clover grown in the wet soil contained greater than 20 ppm Mo, while that grown in the drier regime contained 10 ppm Mo. Therefore, it seems reasonable to suggest that pH measurements alone do not assure a correlation to Mo solubility, and that some soil redox potential measurements should be made.

Molybdenum is an essential plant micronutrient which is required in amounts ranging from 50 to 100 g/ha for agronomic crops (Murphy and Walsh, 1972), and less than 1 ppm in the dry matter (Stout and Meagher, 1948). It is absorbed into the plant as the molybdate anion (MoO_4^{2-}) and is transported to the leaves where it accumulates. The most important functions of Mo in plants is as a component of nitrate reductase and nitrogenase, which are enzymes associated with nitrogen metabolism (Schneider, 1976). Because nitrogenase occurs in bacteria living in the roots of legumes, leguminous plants contain higher amounts of Mo than other plants (Vlek and Lindsay, 1977), and sweetclover (Melilotus offininalis and M. alba) has been termed an accumulator plant.

Plants that accumulate unusually high concentrations of Mo are generally found on high organic matter, alkaline, and poorly drained soils. The element can accumulate in plants to high concentrations without toxicity. Allaway (1975) found plants that contain over 1000 ppm Mo and show no symptoms of toxicity. Molybdenum generally accumulates in the roots and leaves and little enters the seeds. Table 6.40 lists concentrations of Mo found in crops from growth media containing Mo and the data indicate that Mo can accumulate in plants to concentrations well above that contained in the soil.

Interactions between Mo and other elements may also influence the availability of the element for plant uptake. The presence of sulfate reduces the plant availability of Mo, while the presence of ample phosphate has the opposite effect (Stout et al., 1951). Phosphate increases the capacity of subterranean clover (Trifolium subterraneum) to take up Mo by displacing Mo sorbed to soil colloids. Sulfate ions have a similar ionic radius and charge as molybdate ions and compete for the same absorption sites on the root. Manganese decreases Mo solubility and thus uptake by plants, by holding Mo in an insoluble form (Mulder, 1954).

Consumption of high Mo plants by animals may lead to a condition known as molybdenosis, "teart" and "peat scours." Five ppm Mo in forage is considered to be the approximate upper limit tolerated by cattle. Teart pasture grasses usually contain 20 ppm Mo and less than 10 ppm copper (Cu). All cattle are susceptible to molybdenosis, but milking cows and young stock are the most susceptible. Sheep are much less affected and horses are not affected at all (Cunningham, 1950). The high levels of Mo in the

TABLE 6.40 PLANT CONCENTRATION OF MOLYBDENUM FROM GROWING IN MOLYBDENUM AMENDED SOIL

Mo Concentration in the Media (ppm)	Media	Species	Mo Concentration in Leaves (ppm)	Reference
1	Soil	Grass	3.0	Kubota (1977)
2	Organic soil	White clover (Trifolium repens)	6.5	Mulder (1954)
3	Soil	Legume	21.0	Kubota (1977)
	Alkaline soil	Clover Trifolium sp.)	123.0	Barshad (1948)
	Alkaline soil	Rhodesgrass (Chloris gayana)	17.0	Ibid.
4	Organic soil	White clover (T. repens)	13.7	Mulder (1954)
5	Clay	Cotton (Gossypium hirsutum)	320.0	Joham (1953)
	Soil	Alfalfa (Medicago sativa)	2.0	Gutenmann et al. (1979)
	Soil	Bromegrass (Bromus ap.)	1-3.5	Ibid.
	Soil	Orchardgrass (Dactylis glomerata)	2-7	Ibid.
6	Soil	Legume	79.0	Kubota (1977)
6.5	Calcareous clay loam	Bermudagrass (Cynodon dactylon)	177.0	Smith (1982)
13	Clay	Bermudagrass (C. dactylon)	349.0	Ibid.
15	Clay	Cotton (G. hirsutum)	900.0	Joham (1953)
25	Clay	Cotton (G. hirsutum)	1350.0	Ibid.
26	Sandy loam	Bermudagrass (C. dactylon)	449.0	Smith (1982)

digestive tract of ruminants depresses Cu solubility, an essential micro-nutrient, thus Mo toxicity is associated with Cu deficiency. The condition can be successfully treated by adding Cu to the diet to create a Cu:Mo ratio in the diet of the animal of 2:1 or greater. Symptoms of molyb-denosis in ruminants include severe diarrhea, loss of appetite and, in the severest cases, death.

The amount of Mo which can be safely added to the soil depends on the soil mineralogy, pH, the hydrological balance, the crops to be grown, other elements present, and the intended use of the soil. It is evident that additions of Mo are less likely to cause toxicity problems if the soil is acidic and well drained. Establishing vegetation with leguminous plants should be avoided. Care must be taken to assure that leachate does not contain excessive amounts of Mo. If Mo is allowed to leach from the soil, as would occur under alkaline conditions, the loading rate of Mo should be adjusted accordingly.

The use of irrigation water that contains the upper limit of the acceptable concentration of Mo as recommended by the National Academy of Sciences and National Academy of Engineering (1972) is equivalent to an ac-cumulation of 10 ppm of Mo in the upper 15 cm of soil. This recommendation is based on the assumption that plants will accumulate Mo from the soil on a 1:1 relationship, an assumption not always shown to be accurate. Since the relationship between soil concentrations of Mo and plant uptake of the element is difficult to predict, pilot studies are the only accurate means to acquire this data. An estimate of acceptable Mo accumulation is given as 5 ppm Mo in the soil to keep plant concentrations at 10 ppm or less.

6.1.6.18 Nickel (Ni)

The primary uses of Ni are for the production of stainless steel alloys and electroplating. It is also used in the production of storage batteries, magnets, electrical contacts, spark plugs and machinery. Com-pounds of Ni are used as pigments in paints, lacquers, cellulose compounds, and cosmetics (Page, 1974).

The average Ni content in the earth's crust is 100 ppm. In soils, the typical range of Ni is 5-500 ppm (Lindsay, 1979). Soil derived from serpentine may contain as much as 5,000 ppm Ni (Vanselow, 1966c).

Nickel in soil associates with O^{-2} and OH^- ligands and is pre-cipitated as Ni hydroxyoxides at alkaline pH. In an aerobic system, Ni may be reduced to lower oxidation states. Nickel present in the lower oxidation state tends to precipitate as Ni carbonate and Ni sulfide (Bohn et al., 1979).

Nickel sorption by soils has been measured as a function of soil prop-erties and competitive cations. Korte et al. (1975) leached Ni from 10 soils and correlated the amount of metal eluted to various soil properties. The percentage of clay and the CEC values were insignificant to Ni retention. The amount of iron and manganese oxides in the soil was positively correlated to Ni sorption. The magnitude of sorption of three cations to a calcium bentonite was shown to be silver<nickel<lead (Ermolenko, 1972) and sorption to a neutral pH alluvial soil was shown to be lead>copper>zinc> cadmium>zinc (Biddappa et al., 1981). A column study by Emmerich et al. (1982) indicated that when 211 ppm Ni was added as sewage sludge, 94% of the Ni added was recovered from the column indicating essentially no Ni leached below the depth of incorporation. Organic matter has the ability to hold Ni at levels up to 2000 ppm (Leeper, 1978); maximum sorption of Ni by soils is often near 500 ppm (Biddappa et al., 1981). However, other studies show Ni sorption is decreased in the presence of a strong chelating agent such as EDTA, and suggest Ni mobility would be enhanced when present with naturally occurring complexing agents such as sewage sludge (Bowman et al., 1981).

The effects on nitrification and carbon mineralization of adding 10-1000 ppm Ni to a sandy soil were studied by Giashuddin and Cornfield (1978). These researchers found that high levels of the element may decrease both processes by 35 to 68%. This result may imply that high Ni concentrations in an organic waste may inhibit the decomposition of the waste by reducing these processes.

Total Ni content in soil is not a good measure of the availability of the element; exchangeable Ni is more closely correlated to the Ni content of plants. Nickel is not essential to plants and in many species produces toxic effects. Normally the Ni content of plant material is about 0.1-1.0 ppm of the dry matter. Toxic limits of Ni are considered to be 50 ppm in the plant tissue (CAST, 1976). The early stages of Ni toxicity are expressed by stunting in the affected plant.

Liming the soil can greatly reduce the extent of Ni toxicity. Yet, in some cases, plants continue to absorb high amounts of Ni after liming. The effect of lime on Ni toxicity is related to more than just the elevated pH, as illustrated in a case where a small increase in pH from 5.7 to 6.5 resulted in a substantial reduction in Ni toxicity. Apparently, calcium provided by liming is antagonistic to Ni uptake by plants (Leeper, 1978). Potassium application also reduces Ni toxicity; the application of phosphate fertilizers results in increased toxic symptoms (Mengel and Kirkby, 1978).

When corn (Zea mays) was grown on a silt loam soil amended with a sludge containing 20 ppm Ni, a slight increase in plant uptake was observed as the loading rate was increased from 0 to 6.7×10^4 kg/ha; however, there was no significant increase in the Ni content in corn grown on a sandy loam amended with 6.7×10^4 kg/ha of sludge containing 14,150 ppm Ni was a less soluble form. Although Ni was more concentrated in the second sludge, it was less soluble and consequently less available to plants (Keefer et al., 1979). Mitchell et al. (1978) studied Ni toxicity to lettuce (Lactuca sativa) and wheat (Triticum aestivum) plants in an acidic and alkaline soil (Tables 6.41 and 6.42). Nickel uptake and toxicity was found to be much greater in the acidic soil. Solution and soil concentrations of Ni and the response in plants associated with each concentration are given in Table 6.43 which shows a varied response depending on the plant species.

TABLE 6.41   NICKEL CONCENTRATION IN PLANT TISSUE IN RELATION TO NICKEL ADDITION IN A CALCAREOUS SOIL (DOMINO SILT LOAM)*

Concentration Ni (mg/kg)	Plant Portion	Crop	Tissue Concentration (mg/kg)	Effect
5	Shoots	Lettuce (Lactuca sativa)	6.0	None
5	Leaves	Wheat (Triticum aestivum)	3.2	None
5	Grain	Wheat (T. aestivum)	<1.0	None
80	Shoots	Lettuce (L. sativa)	23	20% yield reduction
80	Grain	Wheat (T. aestivum)	<1.0	15% yield reduction
320	Shoots	Lettuce (L. sativa)	61	35% yield reduction
320	Grain	Wheat (T. aestivum)	26	25% yield reduction
640	Shoots	Lettuce (L. sativa)	166	95% yield reduction
640	Grain	Wheat (T. aestivum)	50	65% yield reduction

* Mitchell et al. (1978).

TABLE 6.42   NICKEL CONCENTRATION IN PLANT TISSUE IN RELATION TO NICKEL
             ADDITION IN AN ACID SOIL (REDDING FINE SANDY LOAM)*

Concentration Ni (mg/kg)	Plant Portion	Crop	Tissue Concentration (mg/kg)	Effect
5	Shoots	Lettuce (Lactuca sativa)	6.6	None
5	Leaves	Wheat (Triticum aestivum)	2.6	None
5	Grain	Wheat (T. aestivum)	1.7	None
80	Shoots	Lettuce (L. sativa)	241	25% yield reduction
80	Leaves	Wheat (T. aestivum)	46	Significant yield reduction
80	Grain	Wheat (T. aestivum)	64	20% yield reduction
320	Shoots	Lettuce (L. sativa)	960	90% yield reduction
320	Grain	Wheat (T. aestivum)	247	90% yield reduction
640	Shoots	Lettuce (L. sativa)	1,150	95% yield reduction

* Mitchell et al. (1978).

TABLE 6.43   THE INFLUENCE OF SOLUTION CULTURE AND SOIL CONCENTRATION OF
             NICKEL ON PLANT GROWTH AND YIELD

Amount of Nickel (mg/kg)	Media	Species	Effect	Reference
.8 kg/ha	Soil & sludge	Fescuegrass (Festuca sp.)	7 ppm Ni in grass	King (1981)
2.5	Solution	Tomato (Lycopersicon esculentum)	Yield reduction	Foroughi et al. (1976)
10	Soil	Plantain (Solanum paradisiaca)	Contained 2.5 ppm Ni	Dikjshoorn et al. (1979)
28	Soil & sludge	Ryegrass (Secale cereale)	Contained 3.1 ppm Ni	Davis (1979)
28	Soil & sludge	Barley (Hordeum vulgare)	Contained 3.9 ppm Ni	Davis (1979)
100	Solution	Cotton (Gossypium hirsutum)	90% reduction in plant mass	Rehab and Wallace (1978e)

Grasses growing around Ni smelting complexes have been shown to develop a tolerance for high concentrations of Ni in the growing media, that is, they express no phytotoxic symptoms or yield reductions as a result of the element. These grass species are 10 times more tolerant of Ni than plants growing on a normal soil and have developed this tolerance because selection pressure was high. Attempts are being made to use these metal tolerant strains to revegetate metal contaminated soils, but few tolerant crops are now available commercially. Wild (1970) found Ni accumulators with foliar Ni over 2000 ppm and Ni tolerant excluder plants with low foliar Ni at the same Ni rich site. Where available it seems wiser to introduce excluder type tolerant species and strains to eliminate risk to the food chain. "Merlin" red fescue and the grass Deschampsia cespitosa are considered to be Ni tolerant (Cox and Hutchinson, 1980; Chaney et al., 1981).

There is a possibility that Ni, in trace amounts, has a role in human nutrition. However, there is also a strong possibility that Ni is carcinogenic. Numerous investigations have shown Ni to be carcinogenic to animals when administered by intramuscular, intravenous or respiratory routes (Sundernam and Donnelly, 1965). Occupational exposure to Ni compounds has been shown to significantly increase the incidence of lung and nasal cancer in workmen (Sunderman and Mastromalleo, 1975). In small mammals, the LD_{50} of most forms of nickel is from 100 to 1000 mg/kg body weight. $Ni(CO)_4$ is extremely toxic (Bowen, 1966).

The use of irrigation water that contains the upper limit of the acceptable concentration of Ni as recommended by the National Academy of Sciences and National Academy of Engineering (1972) is equivalent to an accumulation of 100 ppm of Ni in the upper 15 cm of soil. Information obtained from Mitchell et al. (1978) and Tables 6.41-6.43 indicate that the phytotoxic level of Ni in soil ranges from 50 to 200 ppm. A soil accumulation of 100 ppm Ni appears to be acceptable based on phytotoxicity and microbial toxicity. However, if demonstration of treatability tests indicate that higher concentrations of Ni can be safely immobilized without either plant or microbial toxicity, loading rates could be increased.

6.1.6.19 Palladium (Pd)

Palladium is a by-product of platinum extraction. It is used in limited quantities in the manufacture of electrical contacts, dental alloys and jewelry. In 1975 the American automobile industry began installing catalytic converters containing Pd. Various industries use Pd catalysts (Wiester, 1975). The average annual loss of Pd to the environment is 7,596 kg; much of it as innocuous metal or alloys.

Palladium has varying effects on plant and animal life. Palladium chloride ($PdCl_2$) in solution at less than 3 ppm stimulates the growth of Kentucky bluegrass, yet at concentrations above 3 ppm toxic effects appear. Concentrations of 10 ppm or greater are highly toxic. The element was detected in the bluegrass roots but not in the tops (Smith et al., 1978).

Palladium (II) ions are extremely toxic to microorganisms. Palladium is carcinogenic to mice and rats, however, rabbits show no ill effects from dietary Pd. Aquatic life forms, particularly microflora and fish, may suffer ill effects from the discharge of Pd (II) compounds by refineries and small electroplaters (Smith et al., 1978). Palladium toxicity to lower life forms suggests that losses to the environment should be monitored.

6.1.6.20 Radium (Ra)

Radium-226 is a radioactive contaminant of soil and water which often appears in uranium processing wastewaters. Commercial uses of Ra includes manufacture of luminous paints and radiotherapy. The lithosphere contains 1.8×10^{13} g Ra and ocean water contains about 10^{-13} g/l.

Radium is highly mobile in coarsely textured soils and creates a potential for polluting water. The attenuation of Ra is positively correlated with the alkalinity of the soil solution and the retention time in soil, which are governed by the exchangeable calcium content of the soil solution and the soil pore size distribution, respectively (Nathwani and Phillips, 1978). Liming increases Ra retention in soil by the formation of an insoluble calcium-beryllium complex with Ra. The release of organic acids may increase the mobility of Ra in the soil solution. The bound forms of Ra are arranged in the order: acid-soluble>exchangeable>water soluble (Taskayev et al., 1977). Although the forms of Ra have been shown to vary with depth, Ra should be tightly bound in limed soil by the effects of pH and CEC on Ra fixation.

Radium should be prevented from reaching the food chain since it is severely animal toxic and carcinogenic because of its radioactivity. Due to its chemical similarities to calcium, Ra can concentrate in the bone where alpha radiation can breakdown red blood cell production. Radium must be applied so that the leachate does not exceed 20 pCi/day (National Academy of Sciences and National Academy of Engineering, 1972). While the soil may have the capacity to retain large amounts of Ra, the loading rate must be controlled to prevent the Ra concentration in plants and leachate water from reaching unacceptable levels.

6.1.6.21 Rubidium (Rb)

Rubidium concentrations range from 50 to 500 ppm in mineral soils, with an average soil concentration of 10 ppm. Rubidium is typically contained in superphosphate fertilizers at 5 ppm and in coal at 15 ppm (Lisk, 1972).

Most of the information about Rb in soils is derived from plant uptake studies of potassium. Potassium and Rb ions, both monovalent cations in the soil solution, are apparently taken up by the same mechanism in plants. The quantity of Rb absorbed is controlled by pH. Rubidium adsorption by barley roots is greater at pH 5.7 than at 4.1 (Rains et al., 1964). Rubidium has a toxic effect on plants in potassium deficient soils due to increased Rb uptake and blockage of calcium uptake (Richards, 1941).

Average Rb levels in plants range from 1-10 ppm in the Graminae, Leguminosae and Compositae plant families (Borovik-Romanova, 1944). Alten and Goltwick (1933) observed a reduction in tobacco yield when plants were grown in soil containing 80 ppm Rb. Rubidium is rarely phytotoxic in soil that contains sufficient potassium for good plant growth.

6.1.6.22 Selenium (Se)

Selenium is used by the glass, electronics, steel, rubber and photographic industries (Page, 1974). Selenium concentrations in sludges from sixteen U.S. cities ranged from 1.7 to 8.7 ppm (Furr et al., 1976). Fly ash from coal burning power plants can be quite rich in Se when western coals are burned (Furr et al., 1977). The average concentration of Se in soils of the U.S. is between 0.1 and 2 ppm (Aubert and Pinta, 1977).

Most Se in the soil occurs in the form of selenites (+4) and selenates (+6) of sodium and calcium, while some occur as slightly soluble basic salts of iron. Selenium has six electrons in its outer shell (making it a metalloid) and upon addition of two more electrons, Se is transformed into a negative bivalent ion. These anions may combine with metals to form selenides. Selenides formed with mercury, copper and cadmium are very insoluble.

Selenium in soil is least soluble under acid conditions, which is the reverse of most other metals with the exception of Mo. Ferric hydroxides in acidic soils provide an important mechanism of Se precipitation by forming an insoluble ferric oxide selenite. Under reducing conditions that occur in water saturated soils, Se is converted to the elemental form. This conversion provides a mechanism for attenuation since selenate, the form which is taken up by plants, occurs only under well aerated, alkaline conditions. Figure 6.20 illustrates forms of Se at various redox potentials.

Selenium is closely related to sulfate-sulfur both chemically and biologically. Both have six electrons in their outer shell and both ions have an affinity for the same carrier sites for plant uptake. The incorporation of Se into amino acids analogous to that of sulfur has been observed in a number of plant species (Petersen and & Butler, 1962). It is theorized that Se toxicity to plants may be a result of interference with sulfur metabolism.

Little evidence exists to suggest that Se is an essential element for plants, yet plants can serve as carriers of Se to animals for whom the element is essential. Plants will translocate selenate only under aerated alkaline conditions. Plants containing above 5 ppm Se are considered to be accumulator plants since 0.02-2.0 ppm is the normal range of Se in plant

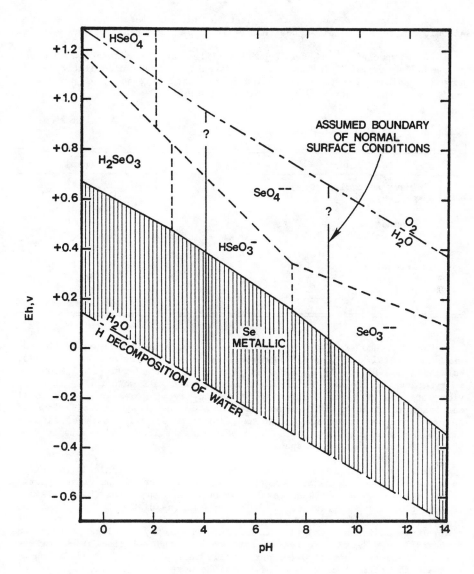

Figure 6.20. Forms of selenium at various redox potentials. (Fuller, 1977).

leaves. A suggested maximum concentration value of Se in plants is given at 3-10 ppm to avoid animal health problems (Melsted, 1973).

Plant species that have been identified as accumulator plants are given in Table 6.44. It has been suggested that these accumulator plants have the ability to synthesize amino acids containing Se, thus preventing toxicity to the plant (Butler and Petersen, 1967).

TABLE 6.44   SELENIUM ACCUMULATOR PLANTS

Plant Genus	Se (ppm)
Primary accumulators:	
Zylorhiza	1400-3490
Stanelya	1200-2490
Oonoposis	1400-4800
Astragalus	1000-15,000
Secondary accumulators:	
Grindelia	38
Atriplex	50
Gutierrezia	60
Astor	70

Excess concentrations of Se in plants result in stunting and chlorosis. The metal can be partially accumulated in growing points in seeds. Watkinson and Dixon (1979) observed plant leaf concentrations of 2500 ppm in ryegrass (Secale cereale) and a reduced growth rate when the Se application rate was 10 kg/ha. Wheat (Triticum aestivum) grown in a sandy soil was tolerant to Se applied as sodium selenate, and phosphorus additions of 50 ppm increased tolerance (Singh and Singh, 1978). The data of Allaway (1968) indicates that the toxic range of Se in the leaves of plants is from 50 to 100 ppm depending on species.

Selenium is an element for which both deficient and toxic levels exist in animals. Selenium as an essential element is part of the enzyme gluta- thione peroxidase which is necessary for metabolic functions in animals and is required in concentrations of 0.05-1 ppm in the diet. Deficiency of Se results in the "white muscle disease" of lambs, calves, chickens and cat- tle. This condition gives rise to muscular dystrophy and loss of hair and feathers. The deficiency can be corrected by the addition of Se in the diet at concentrations of 0.1-1 ppm. Soils that are deficient in Se can be found in the humid Pacific Northwest and the northeastern U.S.

Impacts of Se on aquatic animal species have been noted at concentra- tions of 0.8 mg/l. Selenium toxicity to Daphnia magna, Hyallela azteca, and fathead minnows was reported by Halter (1980) where the LC$_{50}$ value, or the concentration which was lethal to 50% of the population, was .34 to 1.0 mg/l. Toxicity increased with increasing concentration up to 20 mg/l,

at which 100% mortality was exhibited. Runoff containing Se would be
expected to severely impact aquatic life.

At concentrations in excess of 5 ppm in the diet of animals, there is
a danger of Se toxicity. The condition is known as "alkali disease," so
named because alkaline soils have the highest concentrations of available
Se. Animals that are affected by alkali disease eat well but lose weight
and vitality and eventually die. Lesions, lameness and organ degeneration
result from this condition. The minimum lethal dose of Se in cattle is
documented as 6-8 ppm in the diet after 100 days of feeding Se at this
level. Acute toxicity results when animals graze on plants that accumulate
Se. These animals develop "blind staggers" which is characterized by
emaciation, anorexia, paralysis of the throat and tongue, and staggering
(Allaway, 1968).

When land treating a waste high in Se, the quality of leachate and
runoff water from the site and the accumulation of Se in plants should be
considered. If proper precautions are used, Se additions to soils need not
pose environmental problems. Selenium can be concentrated in plants in
concentrations greater than that in the soil solution, so food chain crops
should be avoided and grazing animals excluded from the site. Maintenance
of low pH values to avoid Se solubility seems impractical as almost all
other metals are solubilized at low pH values. The use of irrigation water
that contains the upper limit of the acceptable concentration of Se as
recommended by the National Academy of Sciences and National Academy of
Engineering (1972) is equivalent to an accumulation of 10 ppm of Se in the
upper 15 cm of soil. However, if studies indicate Se is adequately immobilized by the soil so that leaching does not occur and plant concentrations of the element remain below 10 ppm, phytotoxic limits would allow
greater application rates of Se.

6.1.6.23 Silver (Ag)

Silver is found in waste streams of a diverse group of industries,
including photographic, electroplating, and mirror manufacturing. However,
with the increase in the price of Ag, reduction of the element in waste
streams is expected. Berrow and Webber (1972) observed Ag waste amended
soils often contained 5 to 150 ppm Ag. These concentrations are far in
excess of Ag concentrations normally found in soils, indicating that the
soil has a great capacity for retaining Ag from waste streams. Silver is
held on the exchange sites of soil and precipitated with the common soil
anions, chloride, sulfate and carbonates. The solubility of most Ag compounds is greater in acid soil, but even under acidic conditions high
conditions high concentrations of soluble Ag are not taken up by plants
(Aldrich et al., 1955). However, leaching concentrations of .05 mg/l must
be maintained for drinking water standards.

6.1.6.24 Strontium (Sr)

Strontium in soil naturally occurs as two principal ores, celestite ($SrSO_4$) and strontianite ($SrCO_3$), which are often associated with calcium and barium minerals. The sulfate and carbonate forms of Sr are only slightly soluble in water, and it is thought that carbonates or sulfates supplied in fertilizer improve the retention of Sr in soil. On the other hand, calcium (Ca) has been shown to increase Sr movement in soil columns because Ca reacts similarly to Sr in soil and plants (Essington and Nishita, 1966).

Strontium is indiscriminately taken up by higher plants from soil and has no nutritional value to plants. Strontium is able to partially replace Ca in plant tissues and this form of Sr has a low toxicity. However, the artificial isotopes, SR-89 and SR-90 are extremely hazardous. Consumption of forage containing these isotopes can result in the incorporation of Sr in bones and teeth by replacing Ca. Abbazov et al. (1978) report that the uptake of strontium-90 by plants is inversely related to the exchangeable Ca content of soils. Strontium levels exceeding 17,000 ppm are common in the elm (Vanselow, 1966d). In view of the broad range of the Sr to Ca ratio found in plants, liming may have little effect on Sr uptake from soils (Martin et al., 1958).

With the advent of atomic testing, the contamination of soil with Sr originating from atmospheric fallout has become a concern. Strontium-90 is the fission element that is most readily absorbed by plant tissue. Extensive harvesting of grasses has been shown to reduce Sr-90 in soil (Haghiri and Himes, 1974), although this is a very slow process. Some researchers have claimed that Ca and organic matter applications lower Sr-89 uptake from agricultural soils (Mistry and Bhujbal, 1973; 1974). It is not clear whether the applied Ca reduces uptake through precipitation mechanisms or through substitution for Sr in plant tissues. It is known that pH effects in neutral and alkaline soils are minimal, but these effects may become significant in soils with low Ca content.

It is difficult to suggest a management plan for treatment of Sr-90 contaminated soil because Sr uptake by plants or leaching from soil is poorly understood. Strontium exhibited little mobility as a result of leaching from the soil of a 20-year old abandoned strip mine (Lawrey, 1979). Strontium-90 is the most hazardous of the fission products to mammals. Because of its toxicity and the lack of information on Sr attenuation in soils, the loading rate for wastes containing Sr should be equivalent to the loading rate for uranium.

6.1.6.25 Thallium (Tl)

Thallium occurs in the waste streams of diverse industries, including fertilizer and pesticide manufacturing, sulfur and iron refining, and cadmium and zinc processing. Thallium is transported in wastewaters and is fixed in the monovalent form in soils over a broad pH range. Thallium in sulfur ore is probably in the form of Tl sulfate under low pH conditions. Acidic effluents may contain ligands (e.g., chlorine and organics) that stabilize the thallic state and favor oxidation of Tl ions to Tl_2O_3. While Tl^{+3} can be formed in acidic soils under highly oxidized conditions, it is more often fixed in basic soils on hydrous iron oxides. Soluble Tl^+, on the other hand, is removed by precipitation with common soil anions to form sulfides, iodides or chlorides.

Phytotoxic levels of Tl, in excess of 2 ppm, occur in highly mineralized soils. Because of the similarity of Tl chemistry to the group I elements, there are possible interactions with soil and plant alkali minerals which are likely to occur. An imbalance between Tl and potassium (K) on soil exchange sites can impair plant enzymes responsible for respiration and protein synthesis by the substitution of Tl for K. Antimitotic effects attributed to contamination may occur equally in plants as well as in animals.

Plant tolerance to Tl in soil was observed by Spencer (1937) when high concentrations of calcium (Ca), aluminum (Al) and K were present. As a result, the assimilative capacity for Tl may be increased when Ca, K or Al are present.

6.1.6.26 Tin (Sn)

Tin in waste streams originates primarily from the production of tin cans; it is also used in the production of many alloys such as brass and bronze. Tin is used for galvanizing metals and for producing roofing materials, pipe, tubing, solder, collapsible tubes, and foil (Page, 1974). In addition, Sn is a component of superphosphate which typically contains 3.2 – 4.1 ppm Sn.

Tin is concentrated in the nickel-iron core of the earth and appears in the highest concentrations in igneous rocks. The range of Sn in soil is between 2 and 200 ppm, while 10 ppm is considered to be the average value (Bowen, 1966). Casserite (SnO_2), the principal Sn mineral, is found in the veins of granitic rocks.

As a member of group IV, the chemical properties of Sn most closely resemble those of lead, germanium and silicon. The numerous sulfate salts of Sn are very insoluble as are other forms of Sn in soil; thus, their impact on vegetation yield and uptake is slight (Romney et al., 1975). At a lower pH, increased uptake of Sn occurs as a result of increased solubility. The translocation of Sn by plants is reduced by low solubility in soil. Millman (1957) found that Sn concentrations in plants were not related to the concentration in the soil. For soil pH near neutral, 500 ppm Sn had no effect on crops and did not increase foliar Sn. Several studies show little uptake of Sn by plants even when soil Sn was quite high (Millman, 1957; Peterson et al., 1976).

Since there is no substantial evidence that Sn is beneficial or detrimental to plants and since there are no documented cases of animal toxicity due to consumption of Sn-containing plants, loading of a waste containing Sn should pose little environmental hazard. The insolubility of Sn at a neutral to alkaline pH range prevents plant uptake and subsequent food chain contamination.

6.1.6.27 Titanium (Ti)

Titanium is not a trace element by nature and is found in most rocks of the earth's crust in high concentrations (Aubert and Pinta, 1977). The average content of Ti in seventy Australian soils is 0.6%, tropical Queensland soil contains 3.4% (Stace et al., 1968), tropical Hawaiian soil 15% (Sherman, 1952), and up to 25% is found in some lateritic soils (Pratt, 1966c). The average Ti concentration in the soil solution is estimated to be 0.03 ppm.

Soil Ti is a tetravalent cation, usually present as TiO_2. All six common mineral forms of TiO_2 (Hutton, 1977) are studied for their extreme stability in soil environments. Titanium movement in soil is very slow, and thus is used as a measurement of the extent of chemical weathering. Even old, acidic, and highly weathered tropical soils have a Ti content in the soil solution which is near 0.03 ppm. The absolute Ti content is high because as other minerals have weathered the highly stable TiO_2 is left behind. Titanium in soils may be considered essentially immobile and insoluble.

Titanium is rated as slightly plant toxic (Bowen,1966). The toxicity is believed to be due to the highly insoluble nature of Ti phosphates which may possibly tie up essential phosphorus. The average value in dry plant tissue is 1 ppm (Bowen, 1966). Titanium is so insoluble that no natural uptake of toxic amounts has been reported. Similarly, there are no reported values for toxic or lethal doses of Ti in plants or animals.

The only suggested management for high Ti wastes is to maintain an aerobic environment to ensure rapid conversion to TiO_2. The presence of 25% Ti in tropical soils (Pratt, 1966c) suggests that high loading rates would not pose an environmental hazard. Laboratory studies indicate that Ti may form very insoluble complexes with phosphate. Where Ti wastes are to be applied, the addition of phosphorus could be used to immobilize any Ti and phosphate fertilization to maintain plant health may be necessary.

6.1.6.28 Tungsten (W)

The tungsten concentration in the earth's crust is relatively low. Shales contain 1.8 ppm W, sandstones, 1.6 ppm, and limestones, 0.6 ppm. Soils have an average W concentration of 1 ppm (Bowen, 1966). Radioisotopes of W are the principal source of radioactivity from many of the nuclear cratering tests.

The usual W content of land plants is about 0.07 ppm (Bowen, 1966). Plants grown on ejecta from cratering tests concentrate very high levels of radioactive W through their roots (Bell and Sneed, 1970). Tungsten is moderately toxic to plants, with the effects appearing at 1-100 ppm W in nutrient solution depending on plant species (Bowen, 1966).

Wilson and Cline (1966) studied plant uptake of W in soils. They found that W was taken up readily by barley (Hordeum vulgare). Tungsten uptake was 55 times greater from a slightly alkaline, fine, sandy loam than from a medium acid forest soil. Tungsten is probably taken up by plants as WO_4^{2-}.

There has been no physiological need for W demonstrated in animals, and it is slightly toxic to animals. The LD_{50}, or dose of the element which is lethal to 50% of the animal species, for small mammals is 100-1000 mg/kg body weight (Bowen, 1966). The element is readily absorbed by sheep and swine and concentrated in kidney, bone, brain, and other tissues (Bell and Sneed, 1970).

Tungsten is chemically similar to molybdenum (Mo), therefore its solubility curves and other reactions in soil should resemble those of Mo. Tungsten does not pose animal health risks as does Mo however, therefore loading rates for W could be higher than those for Mo.

6.1.6.29 Uranium (U)

Concentrations of total U in soils range from 0.9 to 9 ppm with 1 ppm as the mean value (Bowen, 1966). Uranium concentrations are also expressed as pica Curies per gram (pCi/g), thus U.S. soils contain from 1.1 to 3.3 pCi/g of U (Russell and Smith, 1966). There appears to be more U in the upper portion of soil profiles. This U occurs naturally as pitchblende (U_3O_8) and is found in Colorado and Utah, and in smaller amounts elsewhere in the U.S.

Wastes generated by U and phosphate mining may contain very high concentrations of U and their disposal represents a problem of long duration as the half-life of U is 4.4×10^9 years. Alpha and gamma radiation are associated with this element.

Uranium is strongly sorbed and retained by the soil when present in the +4 oxidation state and may be bound with organic matter and clay colloids. Uranium concentrations of 100 ppm in water were almost completely adsorbed on several of the soils studied by Yamamoto et al., (1973). Changes in pH values had little or no effect on adsorption. However, U present in the +6 oxidation state is highly mobile, so care should be taken to land apply U water or waste only when it will remain reduced, such as on highly organic soils.

Plant uptake of U from soils naturally high in this element provides the only data available on plant accumulation. Because very high concentrations of U in plants are not phytotoxic, plants containing large amounts of U may provide a food chain link to animals. Yet plant uptake of U is usually rather low since U is so strongly fixed in surface soils.

Uranium and its salts are highly toxic to animals. Dermatitis, kidney damage, acute necrotic arterial lesions, and death have been reported after exposure to concentrations exceeding 0.02mg/kg of body weight. The EPA guidelines for Uranium Surface Mining Discharge (FRL 923-7 Part 440 Subpart E) set the average surface discharge level of 10 pCi/g total and 3 pCi/l dissolved, with daily maximum levels at 30 pCi/l total and 10 pCi/l dissolved.

Wastes containing U should be applied to the soil at a rate that prevents leaching of U to unacceptable levels. Uranium is strongly adsorbed in soils that are high in organic matter, however, U may be mobile when oxidized. Disposal of these wastes should follow guidelines set forth by the Nuclear Regulatory Commission and the EPA.

6.1.6.30 Vanadium (V)

The major industrial uses of V are in steels and nonferrous alloys. Compounds of V are also used as industrial catalysts, driers in paints, developers in photography, mordants in textiles, and in the production of glasses and ceramics. In sewage sludge the total concentration of V varies from 20-400 ppm (Page, 1974).

Vanadium is widely distributed in nature. The average content in the earth's crust is 150 ppm. Soils contain 20-500 ppm V with an average concentration of 100 ppm (Bowen, 1966).

In soils, V can be incorporated into clay minerals and is associated with aluminum (Al) oxides. Vanadium in soils may be present as a divalent cation or an oxidized anion (Barker and Chesnin, 1975). Vanadium may be bound to soil organic matter or organic constituents of waste and also bound to Al and iron oxide coatings on organic molecules.

Vanadium is ubiquitous in plants. The V content of 62 plant materials surveyed ranged from 0.27 to 4.2 ppm with an average of about 1 ppm (Pratt, 1966d) and a survey by Allaway (1968) indicates a range of 0.1 to 10.0 ppm. Root nodules of legumes contain 3-4 ppm V and some researchers feel that V may be interchangeable with molybdenum as a catalyst in nitrogen fixation. Although V has not been proven to be essential to higher plants, it is required for photosynthesis in green algae (Arnon, 1958). In addition, low concentrations of V increased the yield of lettuce (Lactuca sativa), asparagus (Asparagus officinalis), barley (Hordeum vulgare), and corn (Zea mays) (Pratt, 1966d).

Vanadium accumulations in plants appear to vary from species to species. Calcium vanadate in solution culture was shown to be toxic to barley at a concentration of 10 ppm, and when the V was added as V chloride, a concentration of 1 ppm produced a toxic response. Yet, rice seedlings showed increased growth when 150 ppm V oxide was applied as ammonium metavanadate. Toxic symptoms appeared when V oxide was applied at a level of 500 ppm, and a concentration of 1,000 ppm killed the rice plants (Pratt, 1966d). The data of Allaway (1968) indicate that the toxic level of V in the leaves of plants is above 10 ppm, depending on species. However, some studies involving application of sewage sludge and fly ash containing V did not result in any change in the plant concentration of the element (Furr, 1977; Chaney et al., 1978).

When V is present in the diet at 10-20 ppm it has been shown to depress growth in chickens (Barker and Chesnin, 1975). In mammals, V may have a role in preventing tooth decay. The element is not very toxic to humans and the main route of toxic contact is through inhalation of V in dust (Overcash and Pal, 1979).

6.1.6.31 Yttrium (Y)

Concentrations of Y in rocks range from 33 ppm in igneous rocks to 4.3 ppm in limestones (Bowen, 1966). Soils contain 3-80 ppm Y (Bohn et al., 1979). In soil, Y, like the other transition metals, associates with O^{2-} and OH^- ligands and tends to precipitate as hydroxyoxides (Bohn et al., 1979).

Yttrium is not an essential element for plant growth. It is found in dry tissue of angiosperms at a concentration of less than 0.6 ppm. Gymnosperms contain only 0.24 ppm or less. Ferns usually contain about 0.77 ppm Y and have been reported to be capable of accumulating this metal (Bowen, 1966).

Yttrium is only moderately toxic to animals. For small mammals, the LD_{50} of Y is 100-1000 mg/kg body weight (Bowen, 1966).

6.1.6.32 Zinc (Zn)

Zinc wastes originate primarily from the production of brass and bronze alloys and the production of galvanized metals for pipes, utensils and buildings. Other products containing Zn include insecticides, fungicides, glues, rubber, inks and glass (Page, 1974).

Most U.S. soils contain between 10-300 ppm Zn, with 50 ppm being the average value (Bohn et al., 1979). Surface soils generally contain more Zn than subsurface horizons. Zinc is abundant where sphalerite and sulfides occur as parent materials for soil (Murrman and Koutz, 1972).

Zinc in the soil can exist as a precipitated salt, it can be adsorbed on exchange sites of clay or organic colloids, or it can be incorporated into the crystalline clay lattice. Zinc can be fixed in clay minerals by isomorphic substitution where Zn^{2+} replaces aluminum (Al^{3+}), iron (Fe^{2+}) or magnesium (Mg^{2+}) in the octahedral layer of clay minerals. Zinc substitution also occurs in ferromagnesium minerals, augite, hornblende and biotite. Zinc bound in these minerals composes the majority of Zn found in many soils.

Zinc interaction with soil organic matter results in the formation of both soluble and insoluble Zn organic complexes. Soluble Zn organic complexes are mainly associated with amino, organic and fulvic acids. Zinc sorbed on organic colloids may be soluble and easily exchangeable. Hodgson et al. (1966) reported an average of 60% of the soluble Zn in soil is present as Zn organic complexes. The insoluble organic complexes are derived from humic acids.

Zinc found on the exchange sites of clay minerals may be absorbed as Zn^{2+}, $Zn(OH)^+$ or $ZnCl^+$. The intensity of this adsorption is increased at elevated pH. It appears that potassium competes with Zn for the clay mineral exchange sites.

When Zn is complexed with chlorides, phosphates, nitrates, sulfates, carbonates and silicates at higher Zn concentrations, slowly soluble precipitates are formed. The relative abundance of these precipitates is governed by pH. On the other hand, the zinc salts, sphalerite ($ZnFeS$), zincate (ZnO) and smithsonite ($ZnCo_3$), are highly soluble and will not persist in soils for any length of time. Zinc sulfate, which is formed under reducing conditions, is relatively insoluble when compared to other zinc salts.

The predominant Zn species in solutions with a pH less than 7.7 is Zn^{2+}, while $ZnOH^+$ predominates at a pH greater than 7.7. Figure 6.21 illustrates the forms of Zn that occur at various pH values. The relatively insoluble $Zn(OH)_2$ predominates at a soil pH between 9 and 11, whereas $Zn(OH)_3-$ and $Zn(OH)_4^{2-}$ predominate at a soil pH greater than 11. The complexes, $ZnSO_4$ and $Zn(OH)_2$, control equilibrium Zn concentrations in soil at a low pH and high pH, respectively (Lindsay, 1972).

Zinc interacts with the plant uptake and absorption of other elements in soils. For example, high levels of phosphorus (P) induce Zn deficiency in plants by lowering the activity of Zn through precipitation of $Zn_3(PO_4)_2$ (Olsen, 1972). Furthermore, Zn uptake is decreased when copper is present by competition for the same plant carrier site. Similar effects of decreased Zn uptake are caused by iron, manganese, magnesium, calcuim, strontium and barium. On the other hand, dietary Zn may decrease the toxicity of cadmium in animals.

The normal range of Zn in leaves of various plants is 15-150 ppm and the maximum suggested concentration in plants is 300 ppm to avoid phytotoxicity (Melsted, 1973). Zinc is an essential plant element necessary for

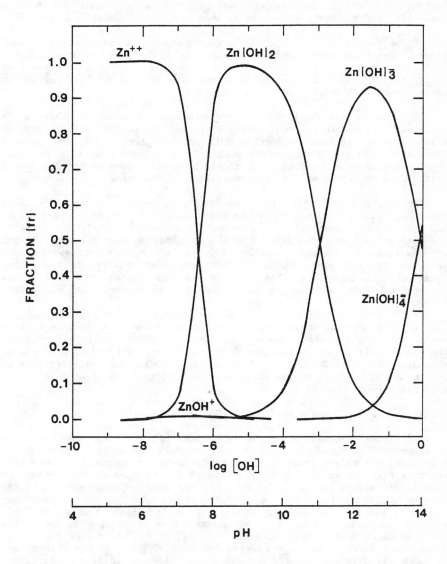

Figure 6.21. Distribution of molecular and ionic species
of divalent zinc at different pH values
(Fuller, 1977).

hormone formulation, protein synthesis, and seed and grain maturation. Table 6.45 lists plant response to various concentrations of Zn.

Toxic levels of Zn occur in areas near Zn ore deposits and spoil heaps. Some plant species, however, tolerate Zn levels of between 600 and 7800 ppm. Agrostis tenuis (bentgrass), Armeria helleri, and Phaseolus vulgaris (bean) have been shown to accumulate as much as 1000 ppm Zn in their leaves (Wainwright and Woolhouse, 1975).

Zinc is an essential element for animals. Animals that have a Zn deficiency are unable to grow healthy skin; poultry produce frizzy, brittle feathers; domestic animals develop dull scraggly fur; and humans develop scaly skin. In addition, animals with a Zn deficiency heal slowly. However, the element may become toxic to microorganisms such as Pseudomonas, a hydrocarbon degrader, at soil concentrations of 500 mg/kg.

Animals are generally protected from Zn poisoning in the food chain since high concentrations of Zn are phytotoxic. Levels of dietary Zn of 500 ppm or more have little adverse effect on animals (Underwood, 1971). The National Academy of Science (1980) recommends maximum tolerable levels of dietary Zn as follows: cattle, 500 ppm; sheep, 300 ppm; swine, 1000 ppm; poultry, 1000 ppm. Aquatic animals are more sensitive to zinc, however; the 96 hour LC_{50} for fathead minnows exposed to Zn(II) was 2.6 ppm and that for rainbow trout is 14.6 ppb (Broderius and Smith, 1979).

Loading rates of Zn bearing wastes can be estimated using a Zn equivalent. However, the use of a Zn equivalent is often unsatisfactory since the equation developed by Chumbley (1971) neglects any toxic effects due to elements other than Zn, nickel (Ni) and copper (Cu). The concentrations of Cu, Zn and Ni (in ppm) in the waste are weighted in terms of Zn to give the zinc equivalence (Z.E.):

$$\text{Z.E. ppm} = Zn^{2+} \text{ ppm} + 2Cu^{2+} \text{ ppm} + 8\ Ni^{2+} \text{ ppm}$$

If proper precautions are used, Zn additions to soils need not pose environmental problems since Zn is rendered insoluble in soils where the pH values are maintained above 6.5. Plants rarely accumulate Zn levels that would be toxic to grazing animals, although Zn can accumulate in plants to high levels before becoming phytotoxic. The use of irrigation water containing the upper limit of the acceptable concentration of Zn as recommended by the National Academy of Sciences and National Academy of Engineering (1972) is equivalent to an accumulation of 500 ppm of Zn in the upper 15 cm of soil. Information in this review indicates that the phytotoxic level of Zn in soil ranges from 500 to 2000 ppm. If the element can be immobilized in soils and excessive plant uptake avoided, concentrations over 500 ppm Zn can be land treated. This concentration (500 ppm) is suggested as a conservative cumulative level.

TABLE 6.45 PLANT RESPONSE TO ZINC IN SOIL

Zn soil concentration (ppm)	Species	Comment	Plant Response	Reference
2-4	Wheat (Triticum aestivum)	(ZnSo$_4$)	Decreased yield in acid soils	Teakle and Thomas (1939)
2-6	Corn (Zea mays) & Oats (Avena sativa)	Control soil was Zn deficient (ZnSO$_4$)	Yield increase, earlier maturation	Barnette and Camp (1936)
2.7	Wheat (T. aestivum)	Highly alkaline soils (ZnSO$_4$)	Reduced Zn deficiency die back	Millikan (1946)
3-5	Wheat (T. aestivum) & Oats (A. sativa)	Counteracted root fungi (ZnSO$_4$)	Superior growth relative to control	Millikan (1938)
11	Corn (Z. mays)	Soil	Toxic, plant leaf level 81 ppm	Takkar and Mann (1978)
27-49	Rye (Secale cereale)	Sewage sludge limed to pH 6.8 rye grown from seed immediately after spreading	Little yield reduction relative to control	Lagerwerff et al. (1977)
40	Rice (Orzya sativa)	Loam soil pH 9.2 sewage sludge limed to pH 6.8	Slight yield reduction	Brar and Sekhou (1979)
49-237	Rye (S. cereale) & Wheat (T. aestivum)	Rye grown from seed, 7 weeks prior to planting	Little yield reduction	Lagerwerff et al. (1977)
89	Wheat (T. aestivum)	As ZnPO$_4$, Zn(NO$_3$)$_2$, Zn(CO$_3$)$_2$	No effect on yield	Voelcker (1913)
140	Alfalfa (Medicago sativa) & fescue (Festuca sp.)	Sewage sludge	Yield increase due to additional macronutrients	Stucky and Newman (1977)

—continued—

TABLE 6.45 (continued)

Zn soil concentration (ppm)	Species	Comment	Plant Response	Reference
156-313	Oats (Avena sativa)	Zn from ore roasting stack gases	Good yields relative to control when crop nutrient added	Lundegardh (1927)
179	Wheat (T. aestivum)	Loamy soil pH 6.7 ($ZnSO_4$)	Promoted growth	Tokuoka and Gyo, (1940)
223	Cowpeas (Vigna unguiculata)	Norfolk fine sand ($ZnSO_4$)	Toxic effect above this level	Gall (1936)
248-971	Corn (Z. mays)	Sewage sludge	No yield effect	Clapp et al. (1976)
300	Sorghum (Sorghum bicolor)	Alkalai soil, Zn concentration in tops, 697 ppm	47% yield reduction	Boawn and Rasmussen (1971)
300	Barley (Hordeum vulgare)	Alkalai soil, Zn concentration in tops, 910 ppm	42% yield reduction	Boawn and Rasmussen (1971)
313	Corn (Z. mays)	Norfolk fine sand ($ZnSO_4$)	Toxic effect above this level	Gall (1936)
480	Lettuce (L. sativa)	Clay soil pH 6.5	No effect	MacLean and Dekker (1978)
500	Corn (Z. mays)	Alkalai soil, Zn concentration in tops, 738 ppm	45% yield reduction	Boawn and Rasmussen (1971)
500	Wheat (T. aestivum)	Alkalai soil, Zn concentration in tops, 909 ppm	45% yield reduction	Boawn and Rasmussen (1971)
500	Beans (Phaseolus sp.)	Alkalai soil, Zn concentration in tops, 235 ppm	Not significant	Boawn and Rasmussen (1971)

--continued--

TABLE 6.45 (continued)

Zn soil concentration (ppm)	Species	Comment	Plant Response	Reference
500	Alfalfa (M. sativa)	Alkalai soil, Zn concentration in tops, 345 ppm	22% yield reduction	Boawn and Rasmussen (1971)
500	Spinach (Spinacia oleracea)	Alkalai soil, Zn concentraion in tops, 945 ppm	40% yield reduction	Boawn and Rasmussen (1971)
500	Potato (Solanum tuberosum)	Alkalai soil, Zn concentration in tops, 336 ppm	Not significant	Boawn and Rasmussen (1971)
500	Sugarbeet (Beta vulgaris)	Alkalai soil, Zn concentration in tops, 1076 ppm	40% yield reduction	Boawn and Rasmussen (1971)
500	Tomato (Lycopersicon esculentum)	Alkalai soil, Zn	26% yield	Boawn and Rasmussen (1971)
535.7 (14 exchangeable)	Wheat (T. aestivum)	Foundry waste, (pH 7.3)	Good yields	Knowles (1945)
620.5	Corn (Z. mays) & wheat (T. aestivum)	Acid & alkaline soils	No effect evident	Chesnin (1967)
640	Lettuce (L. sativa)	Applied to acid soil with sewage sludge	50% yield reduction	Mitchell et al. (1978)
640	Wheat (T. aestivum)	Applied to cal-careous soil	70% yield reduction	Mitchell et al. (1978)
893	Rice (O. sativa) & wheat (T. aestivum)		Toxic action evident	Tokuoka and Gyo (1940)
925	Corn (Z. mays)	Alkaline soil	No effect	Murphy and Walsh (1972)

--continued--

TABLE 6.45 (continued)

Zn soil concentration (ppm)	Species	Comment	Plant Response	Reference
1161	Grass	Galvanized metal contamination (ZnO)	Toxic response	Meijer and Goldenwaagen (1940)
1200	Chard (Beta vulgaris var. Cicla)		No toxicity	Chaney et al. (1982)
1500	Tomatoes (L. esculentum)		Damage	Patterson (1971)
2000	Rice (O. sativa)	Grown on paddy soil	No toxic symptoms	Ito and Iimura (1976)
2143-3571	Oats (A. sativa)	(ZnO) silt loam neutral pH	No adverse effect	Lott (1938)
3839	Vegetable crops	Naturally occuring high Zn peat	Nonproductive soil	Staker (1942)

6.1.6.33 Zirconium (Zr)

Zirconium is not a major constituent of most materials usually associated with pollution of soil and air. The Zr concentration in superphosphate fertilizer is typically 50 ppm and the range in coal is from 7-250 ppm. Sewage sludge usually contains 0.001-0.009% Zr. The average concentration of Zr in urban air is 0.004g per cubic meter (Overcash and Pal, 1979). The principal Zr mineral in nature is zircon ($ZrSiO_4$) which is very common in rocks, sediments and soils (Hutton, 1977). Sandstones are particularly high in Zr with a concentration of 220 ppm. Igneous rocks contain 165 ppm Zr; shales, 160 ppm Zr; and limestones, 19 ppm Zr. The average concentration of Zr in soil is 300 ppm. The immobility of the element in soils makes it useful as an indicator of the amount of parent material that has weathered to produce a given volume of soil (Bohn et al., 1979).

There is no evidence that Zr is essential for the growth of plants or microorganisms. It is moderately toxic to plants. The symptoms of toxicity appear at concentrations of 1-100 ppm in nutrient solution, depending upon plant species (Bowen, 1966). It is less toxic to microorganisms than nickel, but more toxic than thallium (Overcash and Pal, 1979).

Zirconium is not an essential element for animals and can be slightly toxic. Its LD_{50} for small mammals is 100-1000 mg/kg body weight. The element does not, however, accumulate in plants to a level toxic to animals feeding on the plants (Pratt, 1966e).

6.1.6.34 Metal Interpretations

There is a growing consensus in studies on the fate of metals in soils that the toxic effect of a trace metal is determined predominantly by its chemical form (Florence, 1977, Allen et al., 1980). When a metal waste is land treated, soil characteristics such as pH, redox potential, and mineralogy, as well as the source of the metal present in the waste stream, determine the solubility and thus the speciation of the metal. Identifying the metal form will also establish the expected behavior, thus fate of the metal once it is land treated. Sections 6.1.6.1-6.1.6.33 provide information on the toxicity of particular metal forms to microorganisms, plants and animals, as well as the expected fate of each metal.

In the preceding discussion on individual metals, emphasis was placed on soil properties that control the solubility and plant availability of a metal. Of these properties, pH is probably the most important. The solubility of most metal salts decreases as soil pH increases as indicated by the data summarized in (Fig. 6.22). With the exceptions of antimony, molybdenum, tungsten and selenium, which increase in solubility with increasing pH, the normal recommendation for land treatment units is to maintain the pH above 6.5. This is a valuable approach when the predominant metals decrease in solubility at neutral to high pH values. However,

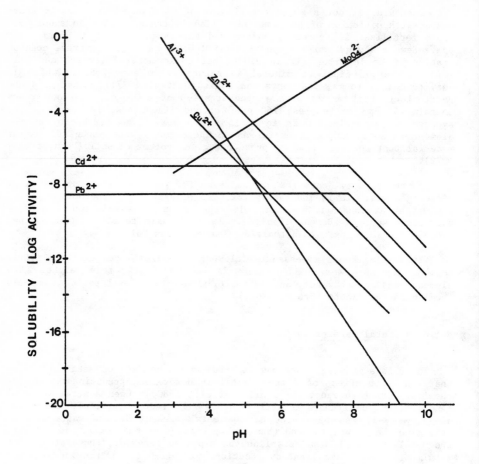

Figure 6.22. Solubilities of some metal species at various
pH values.

for a soil receiving a waste or combination of wastes containing both metals that require a high and low pH, the appropriate pH will need to be carefully determined and maintained to prevent problems. If the pH must be maintained below 6.5, the amounts of metals applied may need to be less than the acceptable levels suggested under each metal section.

It is well known that normally acid soils require repeated lime applications to keep the pH near neutral. While it is expected that pH values will be properly adjusted and maintained during operation and closure, it is likely that following closure, the pH will slowly decrease to the value of the native soil. Therefore, it is possible that some insoluble or sorbed metals will later return in the soil solution. Little information is available on the release of precipitated metals, but when evaluating the long-term impact of land treatment on a normally acidic soil, this possibility should be considered.

There is little evidence that, upon the addition of sludge to soil, significant amounts of metals are permanently held on the cation exchange sites by physical sorption or electrostatic attraction. The soil cation exchange capacity (CEC) has also been shown to make little difference in the amount of metal which is taken up by crops (Hinesly et al., 1982). Most of the metal inactivation in the soil is probably a result of chemical or specific sorption, precipitation and, to a lesser extent, reversion to less available mineral forms, particularly when a soil is calcareous. Chaney (personal communication) suggests that the only reason for considering CEC is to limit the amounts of metals applied to normally acidic soils that have a CEC below 5 meq/100 g since such soils would likely revert to the original pH shortly after liming is discontinued. Consideration of CEC as a measure of the buffering capacity more closely related to the surface area of a soil, rather than as a guide to loading capacity, is the appropriate approach.

The maximum and normal concentrations of metals found in soil are given in Table 6.46. One must be cautious, however, about using the upper limit of the normal range of metal concentrations in soil as an acceptable loading rate. These ranges often include soils that contain naturally high concentrations of metals resulting in toxicity to all but adapted plants.

Table 6.47 is compiled from the National Academy of Science and National Academy of Engineering (1972) irrigation quality standards, sewage sludge loading rates developed by Dowdy et al. (1976), and an extensive review of the literature. National Academy of Science and National Academy of Engineering (1972) recommendations are primarily based on concentrations of metals which can adversely affect sensitive vegetation. The irrigation standards assume a 57.2 cm depth of water applied for 20 years on fine textured soil. Recommendations given by Dowdy et al. (1976) limit application based on the soil CEC. The final column in Table 6.47 is compiled from the literature review in this document and is based on microbial and plant toxicity limits, animal health considerations, and soil chemistry which reflects the ability of the soil to immobilize the metal elements. Although immobilization was considered in developing these recommendations, there is little information in the literature on which to base loading

TABLE 6.46   TRACE ELEMENT CONTENT OF SOILS*

Element	Common Range (ppm)	Average	Element	Common Range (ppm)	Average
Ag	0.01–5	.05	Li	5–200	20
Al	10,000–300,000	71,000	Mg	600–6,000	5,000
As	1–50	5	Mn	20–3,000	600
Au		<1	Mo	0.2–5	2
B	2–100	10	Ni	5–500	40
Ba	100–3,000	430	Pb	2–200	10
Be	0.1–40	6	Ra	8×10^{-5}	
Br	1–10	5	Rb	50–500	10
Cd	0.01–0.7	.06	Sb	2–10	
Cl	20–900	100	Se	0.1–2	.3
Co	1–40	8	Sn	2–200	10
Cr	1–1,000	100	Sr	50–1,000	200
Cs	0.3–25	6	U	0.9–9	1
Cu	2–100	30	V	20–500	100
F	10–4,000	200	W		1
Ga	0.4–300	30	Y	25–250	50
Hg	0.01–0.3	.03	Zn	10–300	50
I	0.1–40	5	Zr	60–2,000	300
La	1–5,000	30			

* Lindsay (1979).

rates and treatability studies may indicate that higher levels are acceptable in a given situation. As is true of any general guideline developed to encompass a large variety of locations and conditions, these suggested metal accumulations could be either increased or decreased depending on the results of the treatment demonstration or the suitability of a particular site.

TABLE 6.47   SUMMARY OF SUGGESTED MAXIMUM METAL ACCUMULATIONS WHERE MATERIALS WILL BE LEFT IN PLACE AT CLOSURE*

Element	Sewage Sludge Loading Rates[†] (mg/kg soil)	Calculated Acceptable Soil Concentrations[#] (mg/kg soil)	(kg/15 cm-ha)	Soil Concentrations Based on Current Literature and Experience[+] (mg/kg)
As		500	1100	300
Be		50	110	50
Cd	10	3	7	3
Co		500	1100	200
Cr		1000	2200	1000
Cu	250	250	560	250
Li		250	560	250
Mn		1000	2200	1000
MO		3	7	5
Ni	100	100	220	100
Pb	1000	1000	2200	1000
Se		3	7	5
V		500	1100	500
Zn	500	500	1100	500

* If materials will be removed at closure and plants will not be used as a part of the operational management plan, metals may be allowed to accumulate above these levels as long as treatability tests show that metals will be immobilized at higher levels and that other treatment processes will not be adversely affected.

† Dowdy et al. (1976); for use only when soil CEC>15 meq/100 g, pH>6.5.

National Academy of Science and National Academy of Engineering (1972) for 20 year irrigation application.

+ See individual metal discussions for basis of these recommendations; if metal tolerant plants will be used to establish a vegetative cover at closure, higher levels may be acceptable if treatability tests support a higher level.

To better understand the impact of metals on the environment, the elements are combined into three groups. Of primary importance are metals which are established carcinogens including arsenic, chromium (as chromate), beryllium and nickel (Norseth, 1977). The second group includes metals such as cadmium, molybdenum, selenium and perhaps nickel and cobalt that are taken up by plants in sufficient quantities to be transmitted up the food chain. Interestingly enough, molybdenum and selenium are also metals that leach from the soil at elevated pH levels if soil properties permit downward movement of solutes. Leaching of metals below the root zone depends on soil physical and chemical properties, climate and the presence or absence of soil horizons of low permeability. Downward transport of metals is generally more rapid in coarse-textured soils than in clays because larger pores allow faster movement of soil water. However, clay soils with cracks have a fairly high leaching potential. Similarly, transport is greater in high rainfall areas. Though coarse textured surface horizons allow greater apparent leaching, an underlying horizon of low permeability, such as an argillic or petrocalcic, will impede further downward movement. If the system can be managed to allow leaching at concentrations that are acceptable to the receiving aquifer, the buildup of these metals may be avoided, thus minimizing contamination of the food chain. The concentration of metals leaching to aquifers should meet drinking water standards; Table 6.48 lists the water quality criteria of interest.

The third group of metals includes those metals that are excluded from the food chain since they are toxic to plants at concentrations that are less than levels toxic to animals. Common concentrations of metals in plants and phytotoxic levels are given in Table 6.49. The upper level of chronic lifetime diet exposure for cattle and swine are given in Table 6.50. A comparison of these data reveals that phytotoxicity would be expected to protect the food chain from arsenic, copper, nickel and zinc. However, some plants take up cobalt and mercury in concentrations that may cause an adverse impact on animals consuming forage containing these elements. Cadmium, molybdenum and selenium are not toxic to plants at fairly high concentrations and are, consequently, accumulated in plants in concentrations that are toxic to animals.

There is a wide range of tolerance among plants for heavy metals. Certain species can withstand much greater metal concentrations in the soil than others. Tolerant plants are often found around outcrops of metal-bearing geological deposits, on spoils from mining activities, or on areas where the soil has been contaminated due to the activities of man. Heavy metal tolerance may be achieved by exclusion of the metal at the root surface or by chelation inside the plant root (Giordano and Mays, 1977).

While metals are taken up by plants, it is generally not possible to use plants to significantly decrease the metal content of soils. Plant uptake typically amounts to less than one percent of the metal content in the soil and thus several hundred years of growth and removal would be needed to result in a significant reduction of the metal content of the soil (Chaney, 1974). However, there are certain species that concentrate selenium, nickel, zinc, copper and cobalt. These plants have internal mechanisms that prevent the metals from reaching the sites of toxic action in the plant. If these plants are grown and harvested, they could possibly decrease metal concentrations to acceptable levels in a reasonable time. Table 6.51 lists several plant genera that have exhibited hyperaccumulation

TABLE 6.48 WATER QUALITY CRITERIA FOR HUMANS AND ANIMALS*

	Standards & Criteria for Drinking Water in mg/l		Quality Criteria for Drinking Water for Farm Animals in mg/l
	EPA	NAS/NAE	
Common Parameters			
PH		5–9	
Total dissolved solids			3000
Common Ions			
Chloride		250	
Flouride	1.4–2.4		
Nitrate (as N)	10	10	
Metals			
Arsenic	0.05	0.1	0.2
Aluminum			5
Barium	1	1	
Boron			5
Cadmium	0.01	0.01	0.05
Chromium	0.05	0.05	1
Cobalt			1
Copper		1	0.5
Cyanide		0.2	
Iron		0.3	
Lead	0.05	0.05	0.1
Mercury			0.01
Molybdenum	.002		
Selenium	0.01	0.01	0.05
Silver	0.05		
Vanadium			0.1
Zinc		5	25

* EPA (1976); National Academy of Sciences and National Academy of Engineering (1972).

TABLE 6.49   NORMAL RANGE AND TOXIC CONCENTRATION OF TRACE ELEMENTS IN PLANTS

Concentrations of Elements in Plant Leaves (ppm Dry Weight)			
Element	Range*	Toxic	Source
As	0.01–1.0	>10	National Academy of Sciences and National Academy of Engineering
B	5–30	>75	Allaway (1968)
Ba	10–100	––	
Be	1–40	>40	Williams and LeRiche (1968)
Cd	0.2–0.8	5–700[†]	
Co	0.01–0.30	200	Pinkerton (1982)
Cr	0.1–1.0	10–20	Table 6.29
Cu	4–15	>20	Gupta (1979)
F	2–20	20–1500	Table 6.20
Fe	20–300	––	
Hg	0.001–0.01	>10	VanLoon (1974)
I	0.1–0.5	>10	Newton and Toth (1952)
Pb	0.1–5.0	Low plant uptake[†]	Table 6.34
Li	0.2–1.0	50–700	Table 6.36 and Table 6.37
Mn	15–150	500–2000	National Research Council (1973)
Mo	1–100	>1000	Joham (1953) and Smith (1982)
Ni	0.1–1.0	50–200	Tables 6.41, 6.42 and 6.43
Se	0.02–2.0	50–100	Allaway (1968)
V	0.1–10.0	>10	Allaway (1968)
Zn	15–150	500	Boawn and Rasmussen (1971)

* Melsted (1973); Bowen (1966); Swaine (1955); Allaway (1968).

[†] Chaney, personal communication.

Note: Toxicity is defined by a 25% reduction in yield.

TABLE 6.50   THE UPPER LEVEL OF CHRONIC DIETARY EXPOSURES TO
ELEMENTS WITHOUT LOSS OF PRODUCTION*

Element	Cattle (ppm)[†]	Swine (ppm)[†]
Al	1,000	200
As	50	50
Ba	20[#]	20
Bs	400[#]	400[#]
B	150	150[#]
Br	200	200
Cd	0.5	0.5
Ca	20,000	10,000
Cr as Cl	1,000[#]	1,000[#]
Cr as oxide	3,000[#]	3,000[#]
Co	10	10
Cu	100	250
F	40	150
I	50	400
Fe	1,000	3,000
Pb	30	30
Mg	5,000	3,000[#]
Mn	1,000	400
Hg	2	2
Mo	10	20
Ni	50	100[#]
P	10,000	15,000
K	30,000	20,000
Se	2[#]	2
Si	2,000	--
Ag	--	100[#]
Sr	2,000	3,000
S	4,000	--
W	20[#]	20[#]
V	50	10[#]
Zn	500	1,000

* National Academy of Sciences (1980).

[†] Concentrations in the diet on a dry weight basis unless indicated otherwise.

[#] Concentration supported by limited data only.

TABLE 6.51   HYPERACCUMULATOR PLANTS

Plant Species	Highest Metal Concentration Recorded (mg/kg)	Reference
Mint family (Labitae)		
Aeolanthus biformifolius	2820 Co	Malaisse et al. (1979)
Haumaniastrum homblei	2010 Co	Ibid.
H. robertii	10200 Cu, 1960 Cu	Brooks (1977)
Legume family (Leguminosae)		
Crotalaria cobalticola	3000 Co	Brooks (1977)
Vigna dolomitica	3000 Co	Brooks et al. (1980)
Figwort family (Scrophulariceae)		
Alectra welwitschii	1590 Co	Brooks et al. (1980)
Buchnera henriquesii	352 Cu, 1510 Co	Ibid.
Lindernia damblonii	100 Co	Malaisse et al. (1979)
Crucifer family (Cruciferae)		
Alyssum alpestre	3640 Ni	Brooks and Radford (1978)
A. corsicum	13000 Ni	Brooks et al. (1979)
A. masmenaeum	15000 Ni	Ibid.
A. syriacum	6200 Ni	Ibid.
A. murale	7000 Ni	Brooks and Radford (1978)
Homaliaceae		
Homalium austrocale donicum	1805 Ni	Brooks et al. (1979)
H. francii	14500 Ni	Brooks et al. (1977)
H. guillianii	6920 Ni	Ibid.
Nod violet family (Hybanthus)		
Hybanthus austrocaledoniaum	13700 Ni	Ibid.
H. floribundus	14000 Ni	Ibid.
Psychatria doyarrei	34000 Ni	Brooks et al. (1979)

--continued--

TABLE 6.51 (continued)

Plant Species	Highest Metal Concentration Recorded (mg/kg)	Reference
Milk vetch family (Astragulus)		
Astragalus beathii	3100 Se	Beath et al. (1941a)
A. crotalaria	2000 Se	Trelease and Beath (1949)
A. osterhoutii	2600 Se	Beath et al. (1941a)
A. racemosa	15000 Se	Beath et al. (1941b)
Atriplex confertifolia	1700 Se	Trelease and Beath (1949)
Catilleja chromosa	1800 Se	Ibid.
Oonopsis condensata	4800 Se	Beath (1949)
Stanleya pinnata	1200 Se	Ibid.
Xylorrhiza parryi	1400 Se	Trelease and Beath (1949)
Achillea millefolium	4100 Zn	Robinson et al. (1947)
Betula grandulosa	22400 Zn	Warren (1972)
Equisetum arvense	7000 Zn	Robinson et al. (1947)
Linaria vulgaris	4500 Zn	Ibid.
Lobelia inflata	4400 Zn	Ibid.
Populus grandidentata	2000 Zn	Ibid.
Trifolium pratense	1300 Zn	Ibid.
Viola sagittata	3500 Zn	Ibid.

of a particular metal. Although commercial propagation of these plants is increasing, their availability at the present time is limited.

Caution should be exercised when evaluating plant toxicity data generated from experiments where large amounts of metal containing sludges were applied at one time to simulate long-term loading. The metals may be bound by the organic fraction of the waste and may not be released for plant uptake until the organic matter degrades. If it is desirable to test metal availability from single large applications, it is best to use waste that has aged naturally or has been aged by composting.

Many industrial wastewater treatment sludges, particularly those from the petroleum industry, have metal concentrations lower than those normally found in sewage sludge. However, the use of specific catalysts or chemicals in certain processes may result in much higher concentrations of one or a few metals. If these metals limit land application, perhaps the waste stream contributing the metal could be isolated and the metal disposed by some other means, or an alternate catalyst or chemical could be found that would allow the reduction of the limiting metal. In many instances, such reductions have allowed the economical land treatment of wastes which would otherwise not be acceptable.

Table 6.52 lists acceptable levels of metals for which less data are available. This list is based on limited understanding of the behavior of these metals in the soil and should be used only as a preliminary guide. If a waste which contains excessive levels of these metals is to be disposed, it is advisable to conduct laboratory or field tests to supplement the limited information on their behavior available in the literature.

TABLE 6.52   SUGGESTED METAL LOADINGS FOR METALS WITH LESS WELL-DEFINED INFORMATION

Element	TOTAL kg/ha-30 cm	Element	TOTAL kg/ha-30 cm
Ag	400	Re	4,000
Au	4,000	Rh	2,000
Ba	2,000	Ru	4,000
Bi	2,000	Sb	1,000
Cs	4,000	Sc	2,000
Fr	4,000	Si	4,000
Ge	2,000	Sn	4,000
Hf	4,000	Sr	40
Hg	40	Ta	4,000
Ir	40	Tc	4,000
In	2,000	Te	2,000
La	2,000	Th	2,000
Nb	2,000	Ti	4,000
Os	40	Tl	1,000
Pd	2,000	W	40
Pt	4,000	Y	2,000
Rb	1,000	Zr	4,000

The inclusion of the long list of metals given here should not be taken to mean that any waste should be analyzed for all these metals. Wastes may be analyzed only for the metals that are known to be included in the plant processes, or that are an expected contaminant during storage.

There is little evidence that the rate a metal is added to a soil influences its ultimate availability to plants. Thus, the total accceptable metal loading may be done in a single application if other constituents of the waste are not limiting or the applications may be stretched over a 10 or 20-year period. The net result would be similar levels of available metals once the summation of the periodic application equals the amount that had been applied in a single application.

6.2   ORGANIC CONSTITUENTS

To determine the suitability of a waste for land treatment, it is essential to understand the probable fates of the organic constituents in the land treatment system. Organic constituents are frequently part of a complex mixture of hazardous and nonhazardous organic and inorganic compounds. To simplify the determination of which organic constituents may limit the capacity or rate of waste application, it is helpful to know the feedstocks and industrial unit processes that are involved in generating the waste.

Individual wastes are generated by a combination of feedstocks and catalysts reacting in definable unit processes to give predictable products and by-products. Often, enough can be determined from this readily available information to predict the predominant hazardous organic constituents in a waste. Once these constituents are determined, options can be explored for in-plant process controls and waste pretreatment (Section 5.2) that may either increase the loading rate and capacity or reduce the land area required for an HWLT unit. In addition, knowledge of the predominant organic constituents in a waste greatly reduces the analyses necessary in waste characterization and site monitoring. In the following sections, hazardous organic constituents are defined and the fate of these waste constituents are discussed in terms of fate mechanisms and the fate of organic constituent classes.

6.2.1   Hazardous Organic Constituents

Understanding the probable fate of land treated hazardous organic constituents is simplified if their basic physicochemical properties are known. These include such properties as water solubility, vapor pressure, molecular weight, octanol/water partition coefficient, boiling point and melting point. These values are given in Table 6.53 for the 361 commercial chemical products or manufacturing intermediates that have been identified by the EPA as either an "acute hazardous waste" or a "toxic waste" if they are discarded or intended to be discarded.

TABLE 6.53 PROPERTIES OF HAZARDOUS CONSTITUENTS

Hazardous Constituents	Hazardous Waste #	Density (gm/cm³)*	Molecular Weight	Water Solubility Qualitative	Water Solubility PPM	Octanol/water Partition Coef.	Vapor Pressure (Torr)*	Melting Point °C,760Torr*	Boiling Point °C,760Torr*	CAS #
Acetaldehyde	U001	0.78349 @ 18°C	44.05	soluble	10,000	1.0	740@20°C	-12	20.8	75-07-0
Acetone	U002	0.79792@15°C	58.08	miscible	100,000	$1 \times 10^{-0.38}$	400@31.5°C	-95.4	56.2	67-64-1
Acetonitrile	U003	0.7857	41.1	soluble			74@20°C	-46	81.6	75-05-8
3-(alpha-acetonylbenzyl)-4-hydroxycoumarin and salts	P001									
Acetophenone	U004	1.0281	120.14	insoluble			1@15°C	20.5	202.0	98-86-2
2-Acetylaminofluorene	U005									
Acetyl chloride	U006	1.11	78.5	decomposes in water (reacts violently)		$1 \times 10^{-1.1}$	150@20°C	-112	50.9	75-16-5
1-Acetyl-2-thiourea	P002	0.8410	56.1	soluble	400,000	$1.0^{-9.9}$	215@20°C	-86.95	53.0	107-02-8
Acrolein	P003	1.22	71.1	highly soluble	1 to >1	$1 \times 10^{0.13}$	1.6@61.5°C	84.5	125@25Torr	79-06-1
Acrylamide	U007	1.0511	72.1	miscible		$1 \times 10^{-0.92}$	3.2@20°C	13	142	107-13-1
Acrylic acid	U008	0.8060	53.1	miscible	73,500	$1 \times 10^{-0.14}$	100@228°C	-83.5	77.5	109002
Acrylonitrile	U009			slightly soluble	0.025		2.3×10^{-4}@20°C	-104		107-18-6
Allyl alcohol	P005	0.854	58.08	miscible			100@10.5°C	-129	97	20859-73-8
Aluminum phosphide	P006	2.85@25°C	57.96	miscible						
5-(Aminomethyl)-3-isoxazolol	U010									
4-Aminopyridine	P007									
Amitrole	U011	1.719	94.12	soluble				159	100@127Torr explodes@410	08-89-1
Ammonium picrate	P009	1.02	84.1	soluble	35,000	$1 \times 10^{0.96}$		decomposes		62-53-1
Aniline	U012		246.14			1.0	1@34.8°C	-6.3	184	
Arsenic acid (m-) (o-)	P010	2.0-2.5	93.1	slightly soluble	2300x10⁶ppb @20°C			decomposes @315 315(sublimes)		7778-19-4
Arsenic pentoxide	P011	4.32	150.9	slightly soluble	21x10⁶ ppb@25°C					
Arsenic trioxide	P012	4.09@25°C	229.8						136	1127-53-1
Asbestos	U013		197.8							1332-21-4
Auramine	U014		267.4							
Azaserine	U015									
Barium cyanide	P013		189.4							
Benz[c]acridine	U016	1.29	229.3	insoluble	0.0011	$1 \times 10^{5.61}$	0.3@20°C	-16	214	60448-23-9
Benzal chloride	U017		161.03	practically insoluble				162	415 sublimes	225-51-4
Benz[a]anthracene	U018		220	insoluble						98-87-3
Benzene	U019	0.879	78.11	slightly soluble	1,280 @25°C	$1 \times 10^{2.28}$	95.2@25°C	5.5	80	56-55-3
Benzenesulfonyl chloride	U020	1.3842 @15°C	176.6	slightly soluble	lg in 2,447g@12°C lg in 107g@100°C	$1 \times 10^{1.81}$		-14.8	168.7	71-43-2
Benzenethiol	P014	1.0766	110.2	lg in hot HNO₂				122-128	400@760Torr	98-09-9
Benzidine	U021	1.250	184.23							108-98-5
										92-87-5

TABLE 6.53 (continued)

Hazardous Constituents	Hazardous Waste #	Density (gm/cm³)*	Molecular Weight	Water Solubility — Qualitative	Water Solubility — PPM*	Octanol/water Partition coef.*	Vapor Pressure [Torr]*	Melting Point °C,760Torr*	Boiling Point °C,760Torr*	CAS #
Benzo[a]pyrene	U022		252.3	practically insoluble @25°C	6.0018 @25°C	$1\times10^{6.04}$	7.32×10^{-7} Pa	176.5		50-32-8
Benzotrichloride	U023	1.38@15.5°C	195.46	insoluble		$1\times10^{4.03}$		-5°C	221	12002-48-1
Beryllium (dust)	P015	1.85	4.01	insoluble		$1\times10^{1.26}$		1283	2970	7440-41-7
Bis(2-chloroethoxy) methane	U024		173.1	low solubility		$1\times10^{1.58}$			218.1	111-91-1
Bis(2-chloroethyl) ether	U025	1.2199	143.02	practically insoluble	81,000@25°C / 10,200		<0.1@20°C / 0.71@20°C	-46.8	178	111-44-4
N,N-bis(2-chloroethyl) 2-Naphthylamine	U026		268.2							
Bis(2-chloroisopropyl) ether	U027		171.07	practically insoluble	1,700	$1\times10^{2.58}$	0.85@20°C	-97	189	108-60-1
Bis(chloromethyl) ether	P016	1.328	114.96	immediately hydrolyzes	22,000@22°C	$1\times10^{-0.38}$	30@22°C	-41.5	104	542-88-1
Bis(2-ethylhexyl) phthalate	U028	0.985	391.0	almost insoluble	0.4-1.3@25°C	$1\times10^{5.3}$	2×10^{-7}@20°C	-50	386.9@5Torr	117-81-7
Bromoacetone	P017	1.631@0°C	136.99					-54	136	598-31-2
Bromomethane	U029	1.676@-20°C	94.94		900@20°C	$1\times10^{4.28}$	1400@20°C	-93.6	4.5	74-83-9
4-Bromophenyl phenyl ether	U030		391.0					18.72	310.14	101-55-3
Brucine	P018		394.45					178		
2-Butanoneperoxide	U019	0.811(sp.gr.)	88.1	very soluble	90,000@25°C	$1\times10^{0.88}$	6.5@25°C	-79.9	117.7	71-36-3
n-Butyl alcohol	U031		74.12							
2-sec-Butyl-4,6-dinitrophenol	P020		158.1					decomposes>350		60448-22-8
Calcium chromate	U032		92			100		-110.8	46.5	75-15-0
Calcium cyanide	P021	1.263	66.01					-114	<8>	
Carbon disulfide	P022	1.139@-110°C	147.4	very soluble	2,200@25°C	$1\times10^{-1.41}$	60@20°C	-57.5	97.8	75-87-6
Carbonyl fluoride	U033	1.51					5@20°C			
Chloral	U034	1.67	403.8		14,740	$1\times10^{2.78}$	$\times10^{-5}$	107.0-108.9(Cis)/103.0-105.0(trans)	175@7Torr	12789-03-6
Chlorambucil	U035									
Chlordane (tech.)	U036									
Chloroacetaldehyde	P023	1.19	78.5	very soluble	10,000 / 10,000@20°C	$1\times10^{1.8}$ / $1\times10^{2.34}$ / $1\times10^{2.84}$	100@45°C / 85@1.1°C	-16.3	90.0-100.1	107-20-0
p-Chloroaniline	P024	1.21	127.6	very soluble				72.5	230.5	106-47-8
Chlorobenzene	U037	1.11	112.56	moderately soluble	480@25°C		.015@20°C	-45	132	108-90-7
Chlorobenzilate	U038		325	soluble			14000.00@4Torr			4755-72-0
1-(p-Chlorobenzoyl)-5-methoxy-2-methylindole-3-Acetic acid	P025									
p-Chloro-m-cresol	U039		142.54	slightly soluble	3,850@20°C	$1\times10^{2.95}$		66	235	59-50-7
Chlorodibromomethane	U040									
1-Chloro-2,3-epoxy propane	U041	1.1761	92.52	relatively high solubility	15,000	$1\times10^{1.28}$	0@16.6°C / 6.75@20°C	-57.1	112.9 / 109@740	110-75-8
2-Chloroethyl vinyl ether	U042	1.0525	106.55	slightly soluble	1.1@25°C / 8,200			-70.1		
Chloroethene	U043	0.9106	62.50	slightly soluble		$1\times10^{0.60}$	.660	-153.8	-13.37	75-01-4
Chloroform	U044	1.49	119.4	highly soluble		$1\times10^{1.97}$	50.5@20°C	119.5	61.7	67-66-3

TABLE 6.53 (continued)

Hazardous Constituents	Hazardous Waste #	Density (gm/cm³)	Molecular Weight	Water Solubility Qualitative	Water Solubility PPM	Octanol/Water Partition Coef.	Vapor Pressure (Torr)	Melting Point °C,760Torr	Boiling Point °C,760Torr	CAS #
Chloromethane	U045	0.997(sp-gr.)	50.49	slightly soluble	400@25°C	$1\times10^{0.91}$	3.765@20°C	-97.73	-24.2	74-87-3
Chloromethyl methyl ether	U046		80.52	practically insoluble		$1\times10^{0.91}$				
2-Chloronaphthalene	U047		162.62	almost insoluble	6.74(calc.)	$1\times10^{4.12}$	0.017@20°C(calc.)	61	256	91-58-7
2-Chlorophenol	U048	1.24	128.56	slightly soluble	28.500@20°C	$1\times10^{2.16}$	2.2@20°C(calc.)	8.4	175.6	95-57-8
1-(o-Chlorophenyl)thiourea	P026					$1\times10^{2.49}$				
4-Chloro-o-toluidine hydrochloride	U049								176(decomposes)	542-76-7
3-Chloropropionitrile	P027	1.1163@25°C	89.5	slightly soluble	<1000	$1\times10^{1.33}$	6@50°C	-51	179	100-44-7
alpha-Chlorotoluene	U028	1.1026@18°C	126.58	slightly soluble	0.002@25°C	$1\times10^{5.61}$	11@60°C	-43	440@1.01x10⁻⁵/Pa	218-01-9
Chrysene	U050	1.274	228.28	almost insoluble				256	decomposes before melting	544-92-3
Copper cyanide	P029		115.61	almost insoluble	5	$1\times10^{2.70}$				
Creosote	U051	1.07	94.136	very soluble			1@20°C	11-15	200-250	
Cresol	U052	1.092@25°C	108.15	very soluble		$1\times10^{1.97}$	19@51°C	-76.0	191-203	1319-77-3
Crotonaldehyde	U053	0.851	70.09	almost insoluble	2.4-3.18	$1\times10^{3.75}$	3.2@20°C	10.9-35.5	104	4170-30-3
Cresylic acid	U055	1.034-1.048	108.73	reacts slowly	50@25°C			-96.0	191-203	
Cumene	U055	0.86(sp-gr.)	120.19	reacts slowly					152	98-82-8
Cyanides	P030									
Cyanogen	P031	0.866@17°C	52.04	slightly soluble	2500@25°C	$1\times10^{3.44}$	3800@20°C	-34.4	-21.0	57-12-5
Cyanogen bromide	P031	2.015	105.93	almost insoluble	45@25°C	$1\times10^{3.51}$	100@22.6°C	52	61.1	2074-87-5
Cyanogen chloride	P033	1.186(sp-gr.)	61.48				610@20°C	-6.5	13.1	506-68-3
Cyclohexane	U056	0.78	84.16				77@20°C	6.5	80.7	110-82-7
Cyclohexanone	U057	0.445(liquid)	98.15	soluble	24.000@25°C		10mm@38.7; 5mm@26.4°C	-45.0	115.6	108-94-1
2-Cyclohexyl-4,6-dinitrophenol	P034		266.23	soluble	.02-.1	$1\times10^{5.99}$		41-45		
Cyclophosphamide	U058		261.1			3.48		decomposes@190		20830-81-3
Daunomycin	U059		526.6					112		72548
DDD (p,p')	U060		320	almost insoluble	5.5 ppb@25°C	$1\times10^{5.97}$ (calc.)	10.2×10^{-7}@30°C	108.5-109	185	50/4
DDT (p,p')	U061	2.440@25°C	354.5	slightly soluble	0.0005@25°C		1.5×10^{-7}@25°C	25-30	150@Torr	
Diallate	U062	2.172@25°C	270.2	almost insoluble		$1\times10^{2.09}$	10^{-6}@20°C	270		53-70-3
Dibenz(a,h)anthracene	U063	1.044@25°C	278.36	insoluble			15@10.5°C	281.5	118-122@748Torr	124-48-1
Dibenzo[a,i]pyrene	U064		302	insoluble			17.4@10°C	<-20	196	46-12-8
Dibromochloromethane	U065		208.24	almost insoluble	11@25°C	$1\times10^{5.2}$		9.3	131.4	
1,2-Dibromo-3-chloropropane	U066		236.4	almost insoluble	12@25°C	$1\times10^{3.38}$	0.10@15°C			74-95-3
1,2-Dibromoethane	U067	2.172@25°C	174.88	slightly soluble	125@25°C	$1\times10^{3.4}$	1.9@25°C	-35.0	340	84-74-2
Dibromomethane	U068	1.044@25°C	119	almost insoluble	29@25°C		2.28@25°C	-24.7	160.5	95-50-1
Di-n-butyl phthalate	U069	1.047	278.34	insoluble	42@25°C	$1\times10^{3.02}$ (calc.)	1.18@25°C	53.1	173	541-73-1
o-Dichlorobenzene	U070	1.307	147.01	almost insoluble				132	174	106-46-7
1,3-Dichlorobenzene	U071	1.29	147.01	slightly soluble						
1,4-Dichlorobenzene	U072	1.46	147.01	almost insoluble						91-94-1
3,3'-Dichlorobenzidine	U073		253.12	almost insoluble						

TABLE 6.53 (continued)

Hazardous Constituents	Hazardous Waste #	Density (gm/cm³)*	Molecular Weight	Water Solubility Qualitative	Water Solubility ppm	Octanol/Water Partition Coef.	Vapor Pressure (Torr)*	Melting Point °C,760Torr*	Boiling Point °C,760Torr*	CAS #
1,4-Dichloro-2-butene	U074	1.183@25°C	125	slightly soluble	200@20°C	$1 \times 10^{2.16}$	4.36@20°C	1-3	156	764-41-0
Dichlorodifluoromethane	U075	1.174	120.92	almost insoluble	5.500	$1 \times 10^{2.79}$	100@20°C	-158	-29	75-71-8
1,1-Dichloroethane	U076	1.26	98.96	highly soluble	8,700	$1 \times 10^{1.5}$	61@20°C	-96.98	57.28	75-34-3
1,2-Dichloroethane	U077	1.21)	98.96	insoluble			82.28mmHg@25°C	-35.36	83.47	107-06-2
1,1-Dichloroethylene	U078	1.2743@25°C	97.0	slightly soluble	600@20°C	$1 \times 10^{1.48}$	200@14°C	-122.53	37@101.33kpa	75-35-4
1,2-trans-Dichloroethylene	U079	1.3255(sp.gr.)	96.94	highly soluble	20,000@25°C	$1 \times 10^{1.1}$	380@22°C	-50	47.5	540-59-0
Dichloromethane	U080	1.38	84.9	highly soluble	4,600	$1 \times 10^{1.14}$	0.21@20°C(calc.)	-95	39.75	75-09-2
2,4-Dichlorophenol	U081		163.01	practically insoluble	0.279	$1 \times 10^{2.4}$	165@5°C	45	210	120-83-2
2,6-Dichlorophenol	U082		163.01					68-69	219-220@740Torr	87-65-0
2,4-Dichlorophenoxy acetic acid (2,4-D)	P015	1.57@30°C	221.0	moderately soluble	620@25°C	$1 \times 10^{0.81}$, $1 \times 10^{3.66}$	0.49@160°C	141	160@0.4Torr	94-75-7
Dichlorophenylarsine	P036									
1,2-Dichloropropane	U083	1.20@25°C	112.99	highly soluble	2,700@25°C	$1 \times 10^{2.20}$	42@20°C	-100	96.8	78-87-5
1,3-Dichloropropene	U084	1.22@25°C (sp.gr.)	110.98	highly soluble	2,700(Cis-), 2,800(Trans)	$1 \times 10^{1.98}$	25@20°C		104.3(Cis), 112(Trans)	542-75-6
Diepoxybutane	U085									
Dieldrin	P037	1.75	381	practically insoluble	0.25@25°C	800	1.78×10^{-7}@20°C	150		60-57-1
Diethylarsine	P038		134	almost insoluble	25@room temp.	1			86	1615-80-1
1,2-Diethylhydrazine	U086	0.797@26°C	88.2							
O,O-Diethyl-S-(2-(ethylthio)ethyl) ester of phosphorothioic acid	P039	1.144	274.4				0.0001@20°C		620@0.01Torr	
O,O-Diethyl-S-methyl ester of phosphorodithioic acid	U087			slightly soluble	1,000@32°C 89@25°C					
Diethyl phthalate	U088	1.1175	222.23			$1 \times 10^{3.22}$	0.05@70°C	-40.5	298	84-66-2
O,O-Diethyl-o-(2-pyrazinyl)-phosphorothioate	P040									
O,O-Diethyl phosphoric acid, o-p-Nitrophenyl ester	P041		268.3					169-172		
Diethylstilbestrol	U089	1.0695	164.2						228	
Dihydrosafrole	U090								46	
3,4-Dihydroxy-alpha-(methyl-amino)-methyl benzyl alcohol	P042			moderately soluble				-82		
Diisopropylfluorophosphate	P043	1.07	209.17					50		
Dimethoate	P044		197	miscible			1300@20°C		6.88	
3,3'-Dimethoxybenzidine	U091		244.29			$1 \times 10^{-0.38}$		137-138		119-90-4
Dimethylamine	U092	0.6806.9°C	45.08	insoluble		$1 \times 10^{-0.49}$		-92.19		75-50-3
p-Dimethylaminoazobenzene	U093		229.1	insoluble				114-117		57-97-6
7,12-Dimethylbenz[a]anthracene	U094		256.1					122-123		119-93-7
3,3'-Dimethylbenzidine	U095		212.3							

TABLE 6.53 (continued)

Hazardous Constituents	Hazardous Waste #	Density (gm/cm³)*	Molecular Weight	Water Solubility Qualitative	Water Solubility PPM	Octanol/Water Partition Coef.	Vapor Pressure (Torr)*	Melting Point °C,760Torr*	Boiling Point °C,760Torr*	CAS #
alpha,alpha-Dimethyl benzylhydroperoxide	U096	1.05	152.2						153	
Dimethylcarbamoyl chloride	U097	1.678@20°C	107.6	miscible				-33	165-167	
1,1-Dimethylhydrazine	U098	0.782@25°C	60.1	miscible			157@25°C	-58	63	57-14-7
1,2-Dimethylhydrazine	U099	0.872@20°C	60.1	miscible			100@25°C	-9	81	540-73-8
3,3-Dimethyl-1-(methylthio)-2-butanone-O-(methylamino)-carbonyloxime	P045									
Dimethylnitrosoamine	P100	1.005@20°C	74.08	soluble		$1 \times 10^{0.06}$			151-153	62-75-9
alpha,alpha-Dimethylphenethyl-amine	P046		121.18							
2,4-Dimethylphenol	U101	0.965@20°C	122.16	slightly soluble	17,000@160°C	$1 \times 10^{2.12}$(calc.)	0.062@20°C,1x0.01@20°C	24-54	194.15	105-679
Dimethyl phthalate	U102	1.189@25°C	194.18	slightly soluble	4,320@25°C	$1 \times 10^{2.85}$		0	210.93	131-11-3
Dimethyl sulfate	U103	1.332@20°C	126.13	sparingly soluble	100@20°C	$1 \times 10^{1.53}$		-31.8	283.7	
4,6-Dinitro-o-cresol and salts	U047		198.13	slightly soluble	5,600@18°C			85.8	188	534-521
2,4-Dinitrophenol	U104,P048	1.683@24°C	184.11		270@22°C	$1 \times 10^{2.01}$	0.0013@59°C	114	(sublimes)	51-28-5
2,4-Dinitrotoluene	U105	1.521@15°C	182.14	Insoluble to slightly soluble		$1 \times 10^{2.05}$		70	300 (decomposes)	121-14-2
2,6-Dinitrotoluene	U106	1.283	182.14	Insoluble	3@25°C	$1 \times 10^{3.2}$(calc.)	<0.2@15°C	65	285	606-20-2
Di-n-octyl phthalate	U107	0.978(ap-gr)	391.0	Insoluble		$1 \times 10^{3.82}$	60@25°C	-25	220@4Torr	117-840-0
1,4-Dioxane	U108	1.053@20°C	88.10	slightly soluble	0.252@20°C	$1 \times 10^{1.67}$	100@25°C	111	101.1	
1,2-diphenylhydrazine	U109	0.731(ap-gr.)	184.19	extremely soluble	10,000		30@25°C	-40	239 (decomposes)	122-66-7
Dipropylamine	U110	0.722	101.19			$1 \times 10^{1.31}$			105	142-84-7
Di-n-propylnitrosamine	U111		130.19	nearly insoluble	9,900			181	205	621-64-7
2,4-Dithiobiuret	P049		135.20		60 to 150ppb	$1 \times 10^{1.55}$	9×10^{-5}@80°C	106	decomposes	
Endosulfan	P050	1.523@30°C	406.9		200pb@25°C	$1 \times 10^{1.62}$	1×10^{-3}@25°C	212		115-29-7
Endrin	P051	1.745@20°C	374.			$1 \times 10^{5.6}$(calc.)	2×10^{-7}@25°C	200 (decomposes)		72-20-8
Ethyl acetate	U112	0.896@25°C	100.12		15,000@25°C	$1 \times 10^{1.01}$	100@27°C	-83.6	77-15	141-78-6
Ethyl acrylate	U113	0.933@20°C	55.08				29@20°C	<-72	99.8	
Ethyl cyanide	U114	0.783@21°C						-103.5	97.1	
Ethylenebisdithiocarbamate	P053		78.12(hyd.) 60.1(anhyd.)	extremely soluble	1×10^{6}@25°C	$1 \times 10^{1.2}$		8.5	117.2	107-15-3
Ethyleneamine			43.07				9@20°C			
Ethylene oxide	U115	0.832@20°C	44.05	miscible			160@20°C	-71.5	55-56	151-56-4
Ethylene thiourea		0.711@20°C	102.5	highly soluble	2,000		1.095@20°C	-111.3	10.7	75-21-8
Ethyl ether	U117		74.12		75,000@25°C	$1 \times 10^{0.53}$			34.6	60-29-7
Ethylmethacrylate	U118		114.17	Insoluble@25°C			442@20°C	-116.2	119	
Ethylmethanesulfonate	U119	0.7134(liquid)	124.2					<-95		
Ferric cyanide	P055		214.98	soluble						
Fluoranthene	U120	0.918@25°C	203.26	Insoluble	0.26@25°C	$1 \times 10^{5.33}$(calc.)	$1 \times 10^{-6} - 1 \times 10^{-4}$@20°C	120	367	206-44-0

TABLE 6.5) (continued)

Hazardous Constituents	Hazardous Waste #	Density (gm/cm³)*	Molecular Weight	Water Solubility Qualitative	Water Solubility PPM	Octanol/Water Partition Coef.	Vapor Pressure (Torr)*	Melting Point °C/760Torr*	Boiling Point °C/760Torr*	CAS #
Fluorine	P056	1.14@-200°C	116.5	slightly soluble	1.98@20°C	1x10^4.18(calc;.1)x10^-1-1x10^-2@20°C		-218	-187	7782-41-4
2-Fluoroacetamide	P057		77	soluble				3)		640-19-7
Fluoroacetic acid sodium salt	P058		78.9					-111	165	62-74-8
Fluorotrichloromethane	U121	1.484@17.2°C	137.38	miscible			35@20°C	-92	24.1	75-69-4
Formaldehyde	U122	0.815@-20°C	30.0	miscible		1x10^-0.54	750@10°C	8.2	-17	50-00-0
Formic acid	U123	1.220	46.03			1x10.88-	1@20°C		100.8	64-18-6
Furan	U124	0.94	68.1	very soluble	10,000	1x10.34	758@20°C	-85.65	31.36	110-00-9
Furfural	U125	1.161°C	96.08	highly soluble	83,000	1x10.88-	1@20°C	-36.5	161.7	
Glycidylaldehyde	U126		86.2			1x10.4)				
Heptachlor	U127	1.58	374	almost insoluble	0.056@25°C	8x10.18	1x10^-4@25°C	95-96		76-44-8
Hexachlorobenzene	U127	3.82(mp.gr.)	284.78	almost insoluble	0.035	1x10.74_	1.09x10^-5@20°C	227-230	322-325	118-74-1
Hexachlorobutadiene	U128	1.68@15.5°C(sp.gr.)	260.74	almost insoluble	5pp@20°C	4x10^4	0.15@20°C	-21	215	87-68-3
Hexachlorocyclohexane [alpha][beta][gamma]	U129		291	almost insoluble	1.63@25°C 0.70@25°C	1x10.81 1x10.14 1x10.99	2.15x10^-5@25°C 2.0x10^-7@20°C @25°C	157-158 309	239	319-84-6 319-85-7 319-86-8
Hexachlorocyclopentadiene	U130	1.72@15°C(mp-gr.)	273	slightly soluble	27.3	1x10^7.54	0.08@25°C	9.4	186@77?Torr	77-47-4
Hexachloroethane 1,2,3,4,10,10-Hexachloro-1,4,4a,5,8,8a-hexahydro-1,4:5,8-endo,endo-dimethano-naphthalene	U131	2.04@20°C(mp.gr.)	236.74	slightly soluble	50	1x10^6.83-	0.4@20°C	sublimes		67-72-1
Hexachloropropene Hexachlorophene	P060 U132		406.9	almost insoluble	0.004			166-167	decomposes above 15	1888-71-7
Hexaethyltetraphosphate	P061 P062	1.101@15°C	248.8 506.4	miscible		1x10^7.54- 1x10^6.83-		-40	113.5	102-01-2
Hydrazine	U133	0.687(mp.gr.)	32.05	miscible	1x10^6	1x10^-0.25	14.4@25°C 400@48°C	14	25.7	302-01-2
Hydrocyanic acid	P063	0.99 liquid	27	miscible	1x10^6	1	400@25°C	-13.2	19.54	74-90-8
Hydrofluoric acid	U134	(mp.gr.@11.6°C)	19.91					-83.1		7664-39-3
Hydrogen sulfide	U135	1.539g/180°C	34.08	very soluble	66.7		15,200@25°C	-85.5	-60.4	7783-06-4
Hydroxydimethyl arsine oxide	U136	1.95	138.0	slightly soluble		1	10-10@20°C	192		193-39-5
Indeno(1,2,3-c,d)pyrene	U137		276.34	insoluble		1x10^7.66(calc.)	400@25.3°C	162.5-164	42.5	74-88-4
Iodomethane	U138	2.279°C	141.95					-66.4		9004-66-4
Iron Dextran	U139		100,000	soluble						
Isobutyl alcohol	U140	0.748@25°C(sp.gr.)	74.1	soluble	95,000@18°C	1x10^0.88	12.2@25°C	-108	108.3	78-83-1
Isosafrole (trans-)	U141	1.180°C	162.2	very soluble				-86 6.2	21.3 253	624-81-9

TABLE 6.53 (continued)

Hazardous Constituents	Hazardous Waste #	Density (gm/cm³)*	Molecular Weight	Water Solubility Qualitative	Water Solubility PPM*	Octanol/Water Partition Coef.	Vapor Pressure (Torr)*	Melting Point °C,760Torr*	Boiling Point °C,760Torr*	CAS #
Kepone	U142		490.7					decomposes@350°C		
Lasiocarpine	U143		411.6							
Lead acetate	U144	2.55	379.35	soluble				75,anhydrous@280		
Lead(-o-)phosphate	U145	6.9-7.3	811.59					1,014		
Lead subacetate	U146									
Maleic anhydride	U147		98.06	very soluble	163,000@30°C	$1 \times 10^{-0.58}$	1@44°C	53	202	108-31-6
Maleic hydrazide	U148	0.734(sp-gr.)	112.1	somewhat soluble miscible		$1 \times 10^{-0.84}$		30.5	220	109-77-3
Malononitrile	U149	1.049@14°C	66.1							
Melphalan	U150		200.61	almost insoluble				-38.87	356-358@20Torr	7439-97-6
Mercury	U151	13.546(sp-gr.)		soluble in hot H₂O	19.2ppb@5°C 81.3ppb@30°C		0.0012@20°C	explodes		
Mercury fulminate	P065	4.42(sp-gr.)	284.7	slightly soluble in hot H₂O		$1 \times 10^{0.29}$	65@25°C C1520@26°C	-36 -123.1	90.3 7.6	74-93-1
Methacrylonitrile	U152	0.805	67.09	miscible		$1 \times 10^{-0.75} - 1 \times 10^{0.71}$	100@21°C	-97.8	64.96	67-56-1
Methanethiol	U153	0.868@20°C	48.10							
Methanol	U154	0.7195(sp-gr.)	32.04	highly soluble	10,000-50,000	2	5×10^{-5}@25°C	78-79	71.4 280@80Torr	16752-75-5 75-55-8
Methapyrilene	U155	1.2946@24°C	261.4					180		56-49-5
Methyl	P066		162.2							
2-Methylaziridine	P067	1.223	58.10	insoluble				-85.9 -86.75	79.57 76.6	101-14-4
Methyl chlorocarbonate	U156	1.28	94.50				71.2@20°C			
3-Methylcholanthrene	U157		268.3							
4,4'-methylene-bis-(2-chloroaniline)	U158	0.805(sp-gr.)	267.2	very soluble	100,000@25°C	1	49.5@25°C 16@20°C	-20.4 -84.7	87.8 116.85	60-34-4
Methylethyl ketone(MEK)	U159		72.1	slightly soluble	19,000@25°C	1				
Methyl ethyl ketone peroxide(R)	P068	0.874	80.1	slightly soluble			28@20°C	-50	101.1	80-62-6
Methyl hydrazine	U161	0.801@25°C(sp-gr.)	46			$1 \times 10^{0.74}$				
Methylisobutyl ketone	U162	0.936(sp-gr.)	100.16	slightly soluble	>20					
2-Methyllactonitrile			100.13							
Methyl methacrylate										
2-Methyl-2-(methylthio)-propionaldehyde-o-(methyl-carbonyl)oxime	P070		190.3							
Methomyl-N'-nitro-N-nitrosoguanidine	U163	1.358	167.1	slightly soluble	55-60@25°C	82	0.97×10⁻⁵@20°C	38		
Methyl parathion	P071		263.2	slightly soluble	34.4@25°C					
Methylthiouracil	U164	1.162	142.2	slightly soluble	30-40	2,300	0.049@20°C	80.55	217.4	91-20-3
Naphthalene	U165		128.19							

TABLE 6.53 (continued)

Hazardous Constituents	Hazardous Waste #	Density (gm/cm³)*	Molecular Weight	Water Solubility Qualitative	Water Solubility PPM*	Octanol/Water Partition Coef.	Vapor Pressure (Torr)*	Melting Point °C, 760Torr*	Boiling Point °C, 760Torr*	CAS #
1,4-Naphthoquinone	U166	1.422	158.16	slightly soluble	>200@25°C	$1 \times 10^{1.74}$		123-126(@100 starts to sublime)		130-15-14
1-Naphthylamine	U167	1.131	143.18	soluble to 0.167%				50	300.8	134-32-1
2-Naphthylamine	U168	1.061@98°C	143.18	slightly soluble				111.5	306.0	91-59-8
1-Naphthyl-2-thiourea	U072	1.318@17°C						-25	43	60?20-56-1
Nickel carbonyl	P073		170.8	slightly soluble	180		400@25.8°C		247.3	557-19-7
Nickel cyanide	P074		110.8							
Nicotine and salts	P075	1.0092	162.23					<-80	-151.18	54-11-5
Nitric oxide	P076	1.3402g/l liquid@-150°C	30.01				1@61.8°C	-161		
p-Nitroaniline	P077	1.424	138.1	slightly soluble	1900@20°C	$1 \times 10^{1.85}$	1@142.4°C	148.5	332.0	100-01-6
Nitrobenzene	U169	1.205@25°C	123.11	slightly soluble	1000@20°C		1@44.4°C	5.6	211	98-95-3
Nitrogen dioxide	P078	1.491@0°C	46				400@80°C	-9.3	21 (decomposes)	10102-44-0
Nitrogen peroxide	P079	1.3402g/l liquid	30.01					-161	-151.18	12031-49-7
Nitrogen tetroxide	P080	1.491@0°C	46				400@80°C	-93	21 (decomposes) explodes@218 decomposes@279	
Nitroglycerine	P081	1.601	227.09	soluble in hot water			1@127°C	13		55-63-0
p-Nitrophenol	U170	1.27	139.11		16,000@25°C	$1 \times 10^{1.91}$	2.2@145°C	113-114		100-02-7
2-Nitropropane	U171	0.992	89.09	moderately soluble			10@15.9°C	-93	12	79-46-9
N-Nitrosodimethylamine	U172		74.1							
N-Nitrosodiethanolamine	U173	0.9422	164.2						152	89-30-6
N-Nitrosodiethylamine	U174		102.2					64-66	81	621-64-7
N-Nitrosodimethylamine	U182	1.005	198.24	insoluble		$1 \times 10^{2.57}$(calc.)				
N-Nitrosodiphenylamine	P083	0.9160	130.19	soluble	9,900@25°C	$1 \times 10^{1.31}$(calc.)				
N-Nitrosodi-n-propylamine	U175		103.1							
N-Nitroso-n-ethylurea	U176		132.2							
N-Nitroso-n-methylurea	U177		114.2							
N-Nitroso-N-methylurethane	U178		100.1							
N-Nitrosomethylvinylamine	P084									
N-Nitrosopiperidine	U179									
N-Nitrosopyrrolidine	U180									
5-Nitro-o-toluidine	U181									
Octamethylpyrophosphoramide	P085	1.137@25°C	286.34				10@26°C	20-21	137-142Torr	
Octyl alcohol condensed with 2 moles ethylene oxide	P086							39.5-41	sublimes@130	20816-12-0
Osmium tetroxide	P087	4.906@22°C	254.20							
7-Oxabicyclo[2.2.1]heptane-2,3-dicarboxylic acid	P088									
Paraldehyde	U182	0.9943(ap-gr.)	132.16	soluble	120,000	$1 \times 10^{1.15}$	25.3@20°C	12.6	124.4	
Parathion	P009	1.267	291.3	slightly soluble	24@25°C	6,400	3.7x10¹ @20°C	375		
Pentachlorobenzene	U183	1.836@17°C(ap-gr.)	250.34	almost insoluble	0.135	154,000		86	277	608-93-5

TABLE 6.53 (continued)

Hazardous Constituents	Hazardous Waste #	Density (gm/cm³)ᵃ	Molecular Weight	Water Solubility Qualitative	Water Solubility PPMᵃ	Octanol/Water Partition Coef.	Vapor Pressure (Torr)ᵃ	Melting Point °C,760Torrᵃ	Boiling Point °C,760Torrᵃ	CAS #
Pentachloroethane	U184	1.671@25°C	202.3	slightly soluble	500	1x10^3.64		-29	162	76-01-7
Pentachlorophenol	U090	1.978	266.35	slightly soluble	14@20°C	1x10^5.01	0.0001@20°C	190	309-310 (decomposes)	87-86-5
Pentachloronitrobenzene	U185	1.718@25°C	295.35	almost insoluble	0.44@20°C	1x10^5.57		146	328	82-68-8
1,3-Pentadiene	U186									504-509
Phenacetin	U187	1.07@25°C(sp-gr.)	179.21	very soluble	67,000-93,000@25°C	1x10^1.5	760@0.1°C	135	181.75	
Phenol	U188	1.654	94.11					40.90	255-275	108-95-2
Phenyldichloroarsine	P091		222.92	slightly soluble			0.021@20°C	-15.4		103-85-5
Phenylmercury acetate	P092	1.3	336.75	slightly soluble		18		149		298-00-2
N-Phenylthiourea	P093		152.2					154		
Phorate	P094	1.156	260.4	slightly soluble	50@room temp.		0.0008@20°C	-15	118-120@0.5Torr	
Phosgene	P095	1.37	98.92	slightly soluble			1180@20°C	-118	6.3	75-44-5
Phosphine	P096	1.52 g/l@0°C	34.04				13200@-3°C	-132.5	-87.5	7803-51-2
Phosphorothioic acid, O,O-diethylester,O-ester with N,N-dimethyl benzene sulfonamide	P097			decomposes		1				
Phosphorous sulfide	P189	2.03	222.24	slightly soluble		0.24	2x10^-4@20°C	131.2	295 (sublimes)	85-44-9
Phthalic anhydride	U190	1.49(sp-gr.)	148.12	soluble			10@24.4°C	-70	129	109-06-8
2-Picoline	U191	0.55@15°C	93.13	soluble				634.5		151-50-8
Potassium cyanide	P098	1.52 @16°C	65.11							506-64-9
Silver cyanide	P099		199.0							
Pronamide	U192	1.0162@25°C	76.11				0.08@20°C		188.2	57-55-6
1,2-Propanediol	P100	0.783@21°C	122.2	miscible			248@20°C	-101.5	97.1	107-12-0
1,3-Propane sultone	P101	1.393	55.08				11.6@20°C	48-49	48-49	107-10-8
Propionitrile	P102	0.9715	56.11	miscible		1x10^0.66	14@20°C	-50	115	107-19-7
n-Propylamine	P102	0.903(sp-gr.)	79.10	slightly soluble in hot H₂O			considerable	-83	115.3	106-51-4
2-Propyn-1-ol	U196	1.318	108.09	insoluble				115.7	(sublimes)	
Pyridine	U197			miscible						
Quinone										
Reserpine	U200	1.285@15°C	608.7				1@08.4°C	264-265,decomposes		108-46-3
Resorcinol	U201	1.0960	110.11				1@63.8°C	110,decomposes	276.5 (sublimes)	
Saccharin	U202	3.004@15°C	183.2				1@56°C	228,decomposes	234.5	
Safrole	U203	3.05@60°C	162.18					11		7446-34-6
Selenious acid	U204		128.98					decomposes	decomposes@118-119	630-10-4
Selenium sulfide	U205							111.03		506-64-9
Selenourea	P103	1.95	133.90					decomposes@320		
Silver cyanide	P104	1.846	65.02				18@17°C	decomposes	1,496	26628-22-8
Sodium azide	P105		49.02					563.7		143-33-9
Sodium cyanide	P106		265.3	decomposes				115		
Streptozotocin	U206	3.07@15°C	119.7							1314-96-1
Strontium sulfide	U207	1.359@18°C	334.40					268	270	57-24-9
Strychnine and salts	P108			almost insoluble	6	1x10^4.99	<0.1@25°C	138	245	
1,2,4,5-Tetrachlorobenzene	U207	1.858@21°C(sp-gr.)	215.9							95-94-3

TABLE 6.53 (continued)

Hazardous Constituents	Hazardous Waste #	Density (gm/cm³)*	Molecular Weight	Water Solubility Qualitative	Water Solubility PPM*	Octanol/Water Partition Coef.	Vapor Pressure (Torr)*	Melting Point °C;760Torr*	Boiling Point °C;760Torr*	CAS #
1,1,1,2-Tetrachloroethane	U208	1.54@25°C(sp.gr.)	167.9	slightly soluble	2600	1x10^4.99	6@25°C	-68	129	630-20-6
1,1,2,2-Tetrachloroethane	U209	1.5953	167.9	slightly soluble	2900	1x10^2.39	5@20°C	-36	146.2	79-34-5
Tetrachloroethene	U210	1.623	165.83	slightly soluble	150-200@20°C	1x10^2.88	14@20°C	-22.7	121	127-18-4
Tetrachloromethane	U211	1.0390@25°C(sp.gr.)	153.82	slightly soluble	1000@25°C	1x10^2.64	90@20°C	-22.9	76.54	56-23-5
2,3,4,6-Tetrachlorophenol	U212		232.0	almost insoluble		1x10^5.08	18@100.0°C	69-70	228	58-90-2
Tetraethyldithiopyrophosphate	P109		3.22	insoluble				125-150	198-202 (decomposes)	
Tetraethyl lead	P110	1.650@18°C	323.5	miscible			1@38.4°C			109-99-9
Tetraethyl pyrophosphate	P111	1.2002	209.2	miscible		1x10^0.46	0.00015@20°C	-108.5	64-65	509-14-8
Tetranitromethane	P112	0.8902	196.04				10@25°C	13	125.7	12251-21-7
Thallic oxide	P113	1.65*@13°C					10@23.7°C	717±15	(-0,)875	
Thallium acetate	U214	9.65@21°C	456.78					110		
Thallium carbonate	U215	3.68	263.43					273		
Thallium chloride	U216	7.11	468.79				105@17°C	430	720	11343-12-2
Thallium nitrate	U217	7.00	239.8					206	430	
Thallium selenite	P114	5.55	266.4						(decomposes)	
Thallium sulfate	P115	6.77	504.84					632	(decomposes)	
Thioacetamide	U218		75.20					113		62-56-6
Thiosemicarbazide	P116		76.1					117		
Thiourea	U219	1.405	92.13							
Thiram	U244									
Toluene	U220	0.866		slightly soluble	470-534.8@25°C	1x10^2.07 / 1x10^2.69	28.7@25°C	-95	110.6	108-88-3
Toluenediamine	U221	1.047	122.17	very soluble			1@106.5°C	99	292	
o-Toluidine hydrochloride	U222	1.22	143.6				0.05@25°C	20-22	251	
Toluene diisocyanate	U223		174.16	almost insoluble	0.4-0.3	825	0.2-0.4@25°C	70-95	decomposes>120	8001-35-2
Toxaphene	P123	1.65	414	slightly soluble	3.0@15°C		10@14°C	8.3	149.5	75-25-2
Tribromomethane	U225	2.840	252.75	slightly soluble	2,190@20°C	1x10^2.2	96.0@20°C	-30.41	74.1	71-55-6
1,1,1-Trichloroethane	U226	1.332	133.41	slightly soluble	4.4x10^3@20	1x10^2.17	19@20°C	-36.5	113.77	79-00-5
1,1,2-Trichloroethane	U227	1.440(sp.gr.)	133.41	slightly soluble	>200	1x10^2.29	57.9@20°C	-73	87	79-01-6
Trichloroethene	U228	1.464@17.2°C	131.34	slightly soluble	1,100@20°C	1x10^2.53	667.4@20°C	-111	23.8	75-69-4
Trichlorofluoromethane	U229		137.4	slightly soluble	1,100					
Trichloromethanethiol	P118									
2,4,5-Trichlorophenol	U230		197.46	slightly soluble	200	1x10^3.72	0.1@25°C	57	252	95-95-4
2,4,6-Trichlorophenol	U231	1.675@25°C(sp.gr.)	197.46	slightly soluble	800@25°C	1x10^3.38	1@76.5°C	69.5	244.5	88-06-2
2,4,5-Trichlorophenoxy-acetic acid(2,4,5-T)	U232	1.662	255.5	slightly soluble	228@25°C	4		151-153		93-76-5

TABLE 6.53 (continued)

Hazardous Constituents	Hazardous Waste #	Density (gm/cm³)*	Molecular Weight	Water Solubility Qualitative	Water Solubility PPM	Octanol/Water Partition Coef.	Vapor Pressure (Torr)*	Melting Point °C, 760Torr*	Boiling Point °C, 760Torr*	CAS #
2,4,5-Trichlorophenoxy-propionic acid(2,4,5-TCPPA)	U233		269.5	slightly soluble	350	$1\times10^{1.37}$		182		
1,3,5-Trinitrobenzene	U234	1.688	213.11	slightly soluble				122	decomposes	99-35-4
Tris(2,3-dibromopropyl)-phosphate	U235	2.27metricton/m³	697.7	soluble			0.025%Pa@25°C	5.5		
Trypan blue	U236		960.8							
Uracil mustard	U237									
Uracil	U238	0.9862	>89.1				10@177.8°C	49	184	
Vanadic acid, ammonium salt	P119									
Vanadium pentoxide(dust)	P120	3.357@18°C	181.90	slightly soluble	175@25°C			690	decomposes@1,750	
Xylene (o-)	U239	0.880@25°C(sp.gr.)	106.2	slightly soluble	130	$1\times10^{2.95}$	10@32.1°C	-25.5	144.4	95-47-6
(m-)		0.868@15°C(sp.gr.)	106.2	slightly soluble	198	$1\times10^{2.26}$	10@28.3°C	-47.9	139	108-38-3
(p-)		0.862@25°C(sp.gr.)	106.2			$1\times10^{3.15}$	10@27.3°C	13-14	138	106-42-3
Zinc cyanide	P121		117.4					decomposes@800		557-21-1
Zinc phosphide	P122	4.55@13°C	285.10					420	1,100	51810-70-9

*Unless otherwise noted; at 20°C unless otherwise noted.

Commercial chemical products or manufacturing intermediates that have been identified as acutely hazardous have been assigned three digit numbers preceded by the letter "P" (i.e., P003 for acrolein). An acutely hazardous waste is defined by the EPA (1980b) as having at least one of the following characteristics:

(1) it has been found to be fatal to humans in low doses;

(2) in the absence of data on human toxicity it has been shown in studies to have an oral LD_{50} toxicity to rats of less than 50 mg/kg;

(3) it has an inhalation LC_{50} toxicity to rats of less than 2 mg/l;

(4) it has a dermal LD_{50} toxicity to rabbits of less than 200 mg/kg; or

(5) it is otherwise capable of causing or significantly contributing to an increase in serious irreversible or incapacitating reversible illness.

Commercial chemical products or manufacturing intermediates that have been identified as toxic have been assigned three digit numbers preceded by the letter "U" (i.e., U0222 for benzo(a)pyrene). A toxic waste is defined by the EPA as having been shown in scientific studies to have toxic, carcinogenic, mutagenic or teratogenic effects on humans or other life forms (EPA, 1980b).

Physicochemical properties listed in Table 6.53 were compiled from the EPA background documents on the identification and listing of hazardous waste (Dawson et al., 1980; Sax, 1979). The table is largely self-explanatory (i.e., highly water soluble compounds may be leachable, and compounds with high vapor pressures may be lost through volatilization), with the possible exception of the octanol/water partition coefficient. This is defined by Dawson et al. (1980) as "the ratio of the chemical's concentration in octanol to that in water when an aqueous solution is intimately mixed with octanol and allowed to separate." Dawson goes on to say that this value reflects the bioaccumulative potential, which he defines as the ratio of the concentration of the compound in an aquatic organism to the concentration of the compound in the water to which the organism is exposed. The octanol/water partition coefficient may also be used to estimate the distribution coefficient (K_d) for organic constituents in a soil/water system (Karickoff et al., 1979) as follows:

$$[K_d]_i = 6.3 \times 10^{-7} \ f_{oc} \ [K_{ow}]_i \qquad (6.4)$$

where

f_{oc} = fraction of organic carbon in the soil (g of organic carbon per g dry soil);
K_{ow} = octanol/water partition coefficient; and
  i = solute index.

It is important to understand the fate of hazardous organic constituents because of their potential impact on human health should they be released from the treatment unit. Consequently, it would be helpful to have a means of obtaining available data on the human health impact of the hazardous constituents in a land treated waste. Table 6.53 lists the Chemical Abstract Service (CAS) Registry numbers which are the primary listing mechanism for a variety of computerized data searching services such as the Dialog computerized listing of Chemical Abstract and Environmental Mutagen Information Center (Oak Ridge, Tennessee). These data bases are continuously updated and can therefore be extremely useful where more information is needed on specific waste constituents.

6.2.2   Fate Mechanisms for Organic Constituents

To be considered suitable for land treatment, all major organic components of a waste applied to soil must degrade at reasonable rates under acceptable application rates and conditions. A reasonable rate of degradation is a rate rapid enough that degradation, rather than volatilization, leaching or runoff, is the controlling loss mechanism within the HWLT unit. The allowable degree of loss by volatilization, leaching and runoff depends on the types of compounds involved. Air and water leaving the site should meet current air and water quality standards. Organic waste constituents that are recalcitrant under land treatment conditions may limit the life of a facility even though they may be present in relatively small concentrations.

There are five primary mechanisms for the removal of organic waste constituents from a treatment site: degradation, volatilization, runoff, leaching, and plant absorption. Each of these mechanisms is examined in the following discussions.

6.2.2.1 Degradation

Degradation is the loss of organic constituents from soil by chemical change induced by either soil microorganisms, photolysis, or reactions catalyzed by soil. While the nonbiological sources of chemical change can play an important role in degradation, the primary mechanism of organic chemical degradation in soil is biological.

While degradation of organic constituents over time may appear to be exponential, it is actually made up of distinct components that will vary in importance with climatic conditions, soil type (Edwards, 1973), and substrate properties. If the approximate half-life of a constituent is known for a given soil-climate regime, it is possible to estimate the amount of the constituent that will accumulate due to repeated applications of the constituent to the treatment soil. For instance, if 5,000 kg/ha/year of a one year half-life constituent is applied to soil, there will still be 2,500 kg/ha left in the soil when the second 5,000 kg is applied.

Consequently, the amount of the substance in the soil immediately after the 2nd, 3rd, 4th, 5th, 6th and 7th yearly application would be approximately 7,500, 8,750, 9,315, 9,688, 9,844 and 9,922 kg/ha. For substances with half-lives of no more than one year, and assuming that the substance is not toxic to soil microbes at the maximum accumulated concentration, no more than twice the amount applied yearly should accumulate in soil (Edwards, 1973; Burnside, 1974). More generally, the accumulation of an organic constituent can be held at twice the amount placed in the soil in one application so long as the applications are separated by the time length of one half-life of the constituent. Degradation of approximately 99% of the substance should be attained within 10 years of the last waste application (Table 6.54). After a 30 year post-closure period, an initial concentration in the soil of 0.5% or 10,000 kg/ha should have been reduced to 0.5 ppb or approximately 1 gm/ha. Methods for evaluating the degradation rate or half-life of organic constituents in a waste are discussed in Section 7.2.1.2.

TABLE 6.54. PERCENT DEGRADATION AFTER 10, 20 AND 30 YEARS FOR ORGANIC CONSTITUENTS WITH VARIOUS HALF-LIVES IN SOIL

Half-Life In Soil	Percentage of Substance Degraded		
	After 10 Years	After 20 Years	After 30 Years
3 months	100		
6 months	99.9999	100	
1 year	99.90	99.9999	100
2 years	96.88	99.90	99.9999
3 years	89.56	98.96	99.90
4 years	81.25	96.88	99.39
5 years	75.0	93.75	98.44
10 years	50.0	75.0	87.5
20 years	25.0	50.0	62.5
30 years	16.6	33.3	50.0

Both the rate and extent of biodegradation of waste in soil depend primarily on the chemical structure of the individual organic constituents in the waste. Other factors that affect biodegradation include the waste loading rate and the degree to which the waste and soil are mixed. If, for instance, an oily waste is applied too frequently or at too high a loading rate, anaerobic conditions may prevail in the soil and decrease biodegradation. If toxic organic constituents are applied at too high a rate, either microbial numbers may be reduced or a soil may even become sterilized (Buddin, 1914). Adequate mixing of waste with soil tends to decrease localized concentrations of toxic waste components while it increases both soil aeration and the area of contact between soil microbes and the waste.

Soil factors that affect biodegradation include texture, structure, temperature, moisture content, oxygen level, nutrient status, pH, and the kind and number of microbes present. In a study that evaluated the effect of soil texture on biodegradation of refinery and petrochemical wastes, a sandy clay soil consistently degraded more waste than a sandy loam soil and two clay soils (Brown et al., 1981). The low degradation rate exhibited by the clay soils was at least partly due to anaerobic conditions (excess water and low oxygen levels) that developed in these soils. This condition might be overcome with time if the waste applied were to impart a more aggregated structure to the soils allowing better drainage and a higher rate of oxygen transfer into the soil.

Soil pH strongly influences biodegradation rate, presumably by affect-ing the availablity of nutrients to the soil microbes. Dibble and Bartha (1979) noted a significantly higher biodegradation rate for oily sludge at soil pH of 7.0 to 7.8 than at pH 5 to 6. In general, however, the availa-bility of most nutrients is optimal in the pH range of 6 to 7. The most common method of increasing soil pH to near 7 is the application of agri-cultural lime. Management of soil pH is discussed in Section 8.6.

Soil temperature for optimal degradation of oily sludge has been reported to be above 20°C but below 40°C (Dibble and Bartha, 1979). Another study found that the biodegradation rate for petrochemical and refinery wastes doubled when soil temperatures increased from 10°C to 30°C, but decreased slightly when temperatures increased from 30°C to 40°C (Brown et al., 1981).

Soil moisture content for optium biodegradation varies with soil type, soil temperature, waste type, and waste application rate. Consequently, the optium moisture level needs to be determined on a case-by-case basis. However, very dry or saturated soils have been reported to exhibit lower biodegradation rates than moist soils (Brown et al., 1981). As a general rule, a soil water content that supports plant growth will also encourage microbial degradation of waste (Huddleston, 1979).

The nutrient status of a soil-waste mixture depends on both the pres-ence and availability of the necessary elements. Adding nitrogen ferti-lizer to soils where oily wastes had been applied increased biodegradation by 50% in one study (Kincannon, 1972), but the increase in biodegradation was substantially less in similar studies (Brown et al., 1981; Raymond et al., 1976). Nitrogen additions have the greatest effect on degradation of wastes that are readily degradable but are nitrogen deficient. For more slowly degradable organic wastes, lower levels of nitrogen are necessary for optimal biodegradation (Huddleston, 1979). The amount of carbon in relation to the amount of nitrogen needed to optimize degradation (the C:N ratio) may be as low as 10:1 or as high as 150:1 (Brown et al., 1981). Care must be taken when applying nitrogen fertilizer to avoid an excess of nitrogen which could contribute to the leaching of nitrates. Fertilization with potassium or phosphorus is usually not necessary unless the receiving soil has a deficiency or large amounts of wastes deficient in these ele-ments are land applied.

Both kind and number of soil microbes determine which and how much of the organic constituents degrade in soil. In native, undisturbed soil, a large variety of microbes are present. After application of waste, the microbes that cannot assimilate the carbon sources present in the waste are rapidly depleted, while microbes that can use these carbon sources tend to flourish. In this manner, the microbial population of the soil is automatically optimized for the applied waste. In some cases, there may be an initially low degradation rate as the number of microbes that can use the waste as a food source multiply. Several studies report substantial increases in total numbers of bacteria soon after addition of hydrocarbons to soils (Dotson et al., 1971; Jobson et al., 1974). The two genera of hydrocarbon-utilizing bacteria most often found to contribute to biodegradation of oily wastes are <u>Pseudomonas</u> and <u>Arthrobacter</u> (Jensen, 1975).

6.2.2.2 Volatilization

Volatilization is the loss of a compound to the atmosphere. Two studies note that soil, as compared to water, decreased volatilization by an order of magnitude (Wilson et al., 1981). Factors affecting volatilization include the properties of the specific compound (vapor pressure, water solubility, and Henry's Law Constant), the soil (air-filled porosity and temperature), interactions between the waste and soil (application method and degree of mixing), and atmospheric conditions (wind velocity, air temperature, and relative humidity). One study found that the highest emission rate of volatile organic components of waste occurred within minutes of application and decreased substantially within one hour (Wetherold et al., 1981).

Compounds of most concern with regard to their potential volatilization include both those that are persistent, toxic, and/or weakly adsorbed to soil and those that exhibit either low water solubility or high vapor pressure. Organic constituents with high vapor pressures are more readily volatilized from soil. Compounds that are not soluble in water tend to be available for volatilization longer because they are less likely to be removed in leachate or runoff water. Persistent organic constituents may similarly be more of a volatilization problem because they tend to be present in the soil longer. In addition, organic compounds are more easily volatilized if they are less strongly adsorbed by soil. Finally, the toxicity of the compound is of concern since the more toxic an organic constituent, the larger the environmental impact per unit of material volatilized.

In a study of volatilization of oily industrial sludges from land treatment, the amount of the total weight of the sludges volatilized within the first 30 minutes after waste application ranged from 0.01 to 3.2% (Wetherold et al., 1981). In this same study, emissions were measured for oily sludges that were subsurface injected at two depths. When the waste was injected to a depth of 7.5 cm, the emissions were relatively high because the sludge bubbled to the surface. Sludge injected to a depth of 15 cm produced no detectable emissions, and no sludge appeared on the surface.

Reduction of waste volume through volatilization is not an acceptable treatment process for organic chemicals. However, it can be a substantial loss mechanism. For instance, Schwendinger (1968) noted that 41, 37 and 36% of a light oil volatilized from soil within 7 weeks when oil application rates were 25, 63 and 100 ml oil/kg soil, respectively. In nine out of ten cases, more oil was lost by volatilization than by biodegradation (Schwendinger, 1968). Methods for evaluating volatilization of waste components from soil are discussed in Section 7.2.3.

6.2.2.3 Runoff

Runoff is that portion of precipitation that does not infiltrate a soil, but rather moves overland toward stream channels or, in the case of HWLT units, to retention ponds. HWLT units should be designed to collect all runoff from the active portion of the facility because this water may be contaminated with various constituents of the waste. Methods for the retention and treatment of runoff are discussed in Section 8.3.3–8.3.5 Factors affecting the loss of organic constituents by runoff include watershed properties, organic constituent properties, waste-soil interactions, and precipitation parameters.

The watershed of an HWLT unit is the area of land that drains into the retention ponds. Since run-on, or surface drainage water from outside the unit must be diverted, runoff will only be generated from the active portion. The amount of the organic constituents removed in runoff is closely tied to how much runoff is generated. Although organic constituents removed in this manner will largely be those that are water soluble, some may be removed through adsorption to suspended solids in the runoff water. Edwards (1973) suggested that insoluble organics that strongly sorb to soil particles could be transported off-site on soil particles in runoff water. Since the amount of suspended solids increases as the rate of runoff increases, removal of organic constituents adsorbed to these solids is also expected to increase as the rate increases. The organic constituents that are adsorbed to suspended solids vary with the nature of the suspended solid and may be considerably different from the constituents dissolved in the runoff water.

Waste-soil interactions that affect the amount of organic constituents released to runoff water are waste loading rate, application timing, and application method. A larger portion of the organic waste constituents can be expected in runoff water as the loading rate is increased beyond the adsorption capacity of the soil. Application timing can also increase the organic constituents in runoff particularly when a large application of waste is made just prior to a heavy rainstorm, or when a large portion of the yearly waste produced is applied to a soil during a rainy season. The release of organic constituents to runoff can be substantially reduced by subsurface injection.

6.2.2.4 Leaching

Leaching of organic chemicals from surface soil to groundwater is a potential problem wherever these chemicals are improperly disposed. Some of the most widely used organic chemicals, halogenated and nonhalogenated solvents, have been found both in groundwater in the U.S. and to a lesser extent in the other industrialized countries (Table 6.55). Though the source of these constituents is not known, most of the synthetic organic compounds found in groundwater are quite volatile, inferring that these compounds were probably leaking from buried wastes rather than wastes applied to soil. If the volatile and slowly degradable halogenated solvents were land treated, the major loss mechanism would probably be volatilization rather than leaching. However, neither volatilization nor leaching is considered an acceptable loss mechanism for these toxic organics. Wastes containing chlorinated solvents should undergo a dehalogenation pretreatment before they are considered land treatable. With a properly managed HWLT unit, numerous studies have shown that at least the nonhalogenated hydrocarbons can be completely degraded before they leach from the soil. Methods for evaluating the constituent mobility are given in Section 7.2.2 and techniques for the collection and treatment of leachate are discussed in Section 8.3.6.

Effective land treatment of readily leachable organics requires an understanding of the soil and organic constituent properties that affect compound leachability. Following are discussions of these properties and how they effect the leachability of organic constituents.

6.2.2.4.1 Soil Properties that Affect Leaching. Soil properties that influence the leaching of organic constituents of land treated waste are texture, structure, horizonation, amount and type of clay present, organic matter content, cation exchange capacity (CEC), and pH. Relative influence of the soil properties can vary with waste composition, application method, loading rate, and climatic conditions. While there are no simple methods for predicting the rate at which a particular organic constituent will leach, an understanding of how soil properties influence leaching can aid in site selection and soil management. Determination of the leachability of individual hazardous organic constituents should be determined by pilot studies (Chapter 7). Discussions of how the soil properties affect leaching of organic constitutents follow.

Soil texture and structure have been shown to have substantial influence on the leachability of organic constituents (Brown and Deuel, 1982; Brown et al., 1982a). Leaching can be substantial from sandy soils due to their low CEC, low clay content, low organic matter content, and relative high number of large pores and resultant high permeability. Clay soils can limit leaching due to their high CEC, high clay content, high organic matter content, and high number of small intraaggregate pores and resultant low permeability. For instance, in one study where industrial wastes were applied to four soils and leachate was collected in field lysimeters, sandy soil allowed the greatest amount of organic constituent leaching (Brown et

TABLE 6.55   TWO CLASSES OF SYNTHETIC ORGANIC CONSTITUENTS WIDELY FOUND IN GROUNDWATER[*]

Organic Constituent	Highest Level Detected in Groundwater (µg/1)	
	USA[†]	Netherlands[#]
HYDROCARBONS		
Cyclohexane	540	30
Benzene	330	100
Toluene	6,400	300
Xylenes	300	1,000
Ethyl benzene	2,000	300
Isopropyl benzene	290	300
HALOGENATED HYDROCARBONS		
Chloroform	490	10
Dichloromethane	3,000	3,000
Carbon tetrachloride	400	30
Dibromochloromethane	55	0.3
1,1-Dichloroethane	400	10
1,2-Dichloroethane	11,330	3
1,1,1-Trichloroethane	5,440	3,000
Dichloroethylenes	860	10
Trichloroethylene	35,000	1,000
Tetrachloroethylene	1,500	30

[*] This list represents some examples of compounds in two classes of organic compounds that have been found several times in groundwater and is in no way a comprehensive list of the leachable constituents in those organic constituent classes.

[†] Burmaster and Harris (1982); Dyksen and Hess (1982).

[#] Zoeteman et al. (1981).

al., 1982a). In another study, deep soil cores were taken from five HWLT units to examine the depth of penetration of land-applied hydrocarbons (Table 6.56). An HWLT unit with a sandy loam soil (site E) that received large amounts of oily waste allowed hydrocarbons to move 180-240 cm in one year. Another HWLT unit with a clay soil (site A) had not allowed detectable quantities of hydrocarbons to penetrate below the treatment zone (top 18 cm) after two years of operation. The potential benefits of horizonation can be seen in site B, where a clay subsoil seems to have minimized the depth to which hydrocarbons penetrated into that soil.

While soil texture can be used to estimate the distribution of pore sizes in sandy soils, the pore size distribution in clay soils can be greatly affected by clay particles clumping into larger aggregate structures. These aggregates tend to allow the formation of larger pores between aggregates, while they contain many small internal or intraaggregate pores. When liquid waste is applied by either spray irrigation or overland flow to structured clay soil, organic constituents may move through the large interaggregate pores without being appreciably adsorbed by the majority of the soil surface present in the intraaggregate pores (Helling, 1971; Davidson and Chang, 1972). However, if organic constituents are dewatered first and then incorporated into a soil surface, water later percolating through the interaggregate pores may not have enough residence time to desorb organic constituents adsorbed on the intraaggregate surfaces. Dekkers and Barbera (1977) found that leachability of organic constituents incorporated into soil decreased as the soil aggregate size increased.

Both amount and type of clay present in a soil have been found to affect the mobility of pesticides (Helling, 1971). Mobility of nonionic pesticides was found to be inversely related to clay content. Soils high in montmorillonitic clays were found to inhibit the movement of cationic pesticides. Anionic or acidic pesticides were relatively more mobile in montmorillonitic soils, suggesting possible negative adsorption. Acidic pesticide mobility was found to be inversely related to nonmontmorillonitic clay content.

Several studies have noted that the movement of organic chemicals in soil is inversely related to the organic matter content of the soil (Helling, 1971; Filonow et al., 1976; Roberts and Valocchi, 1981; Miles et al., 1981; Nathwani and Phillips, 1977). Helling (1971) found that the retardation of organic chemical movement through soils was highly correlated to the adsorption of these organic chemicals by the native soil organic matter.

Cation exchange capacity (CEC), the capacity of soil to adsorb positively charged compounds, decreases the mobility of cationic and nonionic organic constituents and it may increase the mobility of anionic organic constituents (Helling, 1971). CEC can be thought of as the capacity of the negatively charged soil to attract and hold positively charged compounds such as cationic organic constituents. The correlation between CEC and reduced mobility of nonionic compounds is probably due to the component of the CEC represented by native soil organic matter. Organic matter has the

TABLE 6.56 DEPTH OF HYDROCARBON PENETRATION AT FIVE REFINERY LAND TREATMENT UNITS*

Site	Soil Type	Depth of Hydrocarbon Penetration (cm)	Waste Types Applied†	Time Between Last Waste Application and Sampling (Months)	Approximate Application Rate (M³/Ha/Yr#)	Length of Operation (Years)
A	Clay	Less than in untreated area	1,3,8	4	30	2
B	Loamy surface with clay subsoil	23	2,7	16	1-4% oil (one time application)	6
C	Sandy clay loam	30	1,3,4,6	3	25 (one time application)	4
D	Sandy clay loam	91	1,3,6	11	54	6
E	Sandy loam	180-240	1-6	<1	7000	1

* Brown and Deuel (1982).

† Waste types applied were: (1) API separator sludge; (2) DAF sludge; (3) Tank bottoms; (4) Filter clays; (5) ETP sludge; (6) Slop oil emulsion; (7) Treatment pond sludge; and (8) Leaded sludge.

Unless otherwise noted.

capacity to adsorb cationic, nonionic and anionic organic constituents. The increased mobility, or negative adsorption, of anionic organics is due to the electrical repulsion between the negatively charged clay minerals and the anionic organic constituents.

Soil pH has been found to be an important parameter affecting the mobility of organic acids. Helling (1971) noted that as soil pH increased, the mobility of acidic organic constituents increased. Organic acids exist in soil as anions when the soil pH is greater than the dissociation constant (pK_a) of the compounds. As anions, these compounds exhibit negative adsorption and are increasingly mobile in clay soils.

6.2.2.4.2 Organic Constituent Properties that Affect Leaching. The main properties of organic constituents that affect their leaching in soils include water solubility, concentration, strength of adsorption, sign and magnitude of charge, and persistence. Additional organic class-specific information is given in Section 6.2.3.

Only when soil is saturated with oils or solvents will these fluids flow in liquid phase (Davis et al., 1972). In a properly managed HWLT unit, the percolating liquid will be water, and the concentration of organic constituents in the leachate will be limited to the water solubility of the constituent (Evans, 1980). However, many land treated organics, and especially their organic acid decomposition by-products, have unlimited water solubility. Consequently, land treatment units should, if at all possible, be maintained at water contents at or below field capacity. In climatic regions of seasonally high rainfall, an effort should be made to apply wastes only during dry seasons. Where this is not possible, under-drainage may be a workable alternative. Leachate collection systems are discussed in Section 8.3.6.

Generally, the higher the organic constituent concentration in an applied waste, the higher the concentration of these constituents in the leachate. Where substantial quantities of leachate are generated, waste loading rates should not exceed the adsorption capacity of the soil. Adsorption capacity can be considered as the concentration of a constituent in soil above which an unacceptably high concentration of the constituent will enter leachate generated on-site. Ideally, pilot tests should be conducted to assure that the adsorption capacity of the soil for specific hazardous organic constituents will not be exceeded at the planned waste loading rates (Chapter 7). For cationic organic constituents, either increasing valence, or number of positive charges per molecule, will increase the adsorption capacity of the constituent. For anionic organic constituents, the reverse is usually true. That is, the stronger the negative charge on a compound, the stronger will be the negative adsorption and hence, the greater rate of leaching for the compound. As discussed in Section 6.2.2.4.1, by maintaining the soil pH below the pK_a of anionic organic species, the leachability of these species can be minimized. Care should be taken that the pH is not lowered to a point that will decrease degradation rates or increase leachability of heavy metals or other constituents to be immobilized in the treatment zone.

Persistance of organic constituents increases the likelihood that these compounds will be leached by increasing the period of time over which they are exposed to percolating water. Laboratory or field studies can be designed to determine if the half-life of an organic constitutent is too long to allow it to be degraded before it leaches from the treatment zone (Chapter 7). It may be necessary to pretreat certain waste streams before land treatment if the waste contains hazardous organic constituents that are both readily leachable and persistent in the soil environment.

Leaching of trace level organics from a rapid infiltration facility constructed in loamy sand was evaluated in a study by Tomson et al. (1981). By comparing the concentration of various organics in the effluent and in the groundwater underlying the site, it was possible to evaluate leaching in terms of removal efficiency for various organic compound classes. Most classes of compounds had 90-100% removal efficiencies, with low removal achieved for chloroalkanes, alkylphenols, alkanes, phthalates, and amides. Overall removal efficiency for organics was 92%. However, most HWLT units are not designed for rapid infiltration, in part due to the incomplete treatment usually exhibited by these facilities. In addition, the loamy sand soil at the site would provide little attenuation of the applied organics.

HWLT units should not be designed for rapid infiltration of the applied wastes when this would result in significant leaching of hazardous constituents. When waste loading rates are designed to optimize retention of organics in the zone of incorporation (top 30 cm of soil), degradation efficiencies of well over 99% can be achieved (Table 6.54).

6.2.2.5 Plant Uptake

The ability of higher plants to absorb and translocate organic molecules has been recognized for over 70 years. However, only within the past thirty years has this phenomenon received much attention, mostly during trials for possible systemic pesticides. Furthermore, until the relatively recent advent of radioactive labeling techniques studying the uptake of organic compounds was extremely difficult. Recent studies have shown that plant uptake of toxic organic compounds may both pose environmental risk and potentially threaten the quality of human food. Plewa (1978) has reviewed recent studies indicating that various chemicals absorbed by plants may become mutagenic, or that their mutagenic activity may be enhanced through metabolic processes within the plant. Numerous toxic organics, including PCBs, hexachlorobenzene, dimethylnitrosamine, 2,4,5-T, and others, have been observed to be taken up by plant roots (Table 6.57). However, insufficient data currently exist to predict the plant uptake of particular compounds or groups of compounds. Also, the data are insufficient to describe specific mechanisms of uptake and factors that influence uptake. Empirical testing may, therefore, be required to evaluate the absorption, translocation and persistence of toxic organic compounds in higher plants.

TABLE 6.57   ORGANIC CONSTITUENTS ABSORBED BY PLANT ROOTS

Organic Constituent Class and Name	References
Organic Nitrogen Compounds	
α-Alanine	Nissen (1974); Ghosh & Burris (1950)
β-Alanine	Ghosh & Burris (1950)
Arginine	Nissen (1974); Ghosh & Burris (1950)
Asparagine	Ghosh & Burris (1950)
Aspartic Acid	Ibid.
Cystine	Ibid.
Glutamic Acid	Ibid.
Glycine	Nissen (1974); Ghosh & Burris (1950)
Histidine	Ghosh & Burris (1950)
Hydroxyproline	Ibid.
Isolecucine	Ibid.
Leucine	Ibid.
Lysine	Nissen (1974); Ghosh & Burris (1950)
Methionine	Ghosh & Burris (1950)
Phenylalanine	Nissen (1974); Ghosh & Burris (1950)
Proline	Ghosh & Burris (1950)
Serine	Ibid.
Threonine	Ibid.
Tryptophane	Ibid.
Tyrosine	Ibid.
Valine	Ibid.
Glutamine	Ibid.
α-Amino-n-butyric acid	Ibid.
Norleucine	Ibid.
Oxime, α-keto-glutaric acid	Ibid.
Oxime, oxalacetic acid	Ibid.
Oxime, pyruvic acid	Ibid.
Casein hydrozolate	Ibid.
Cysteine	Ibid.
Peptone	Ibid.
Urea	Ibid.
Dimethyl nitrosamine	Dean-Raymond and Alexander (1976)
Cyanide	Wallace et al. (1981)--applied as ^{14}C sodium cyanide; possible absorption as organic cyanide complex.
EDTA	Hill-Cottingham and Lloyd-Jones
EGTA	(1965)--compounds applied as metal
DTPA	chelates.
Chloine Sulfate	Nissen (1974)
Indole acetic Acid	Bollard (1960)
Indole butyric Acid	Ibid.
Indole proprionic Acid	Ibid.

-- continued --

TABLE 6.57 (continued)

Organic Constituent	References
Organic Dyes	Kolosov (1962). Dyes used to study root functions.

    Methylene Blue
    Malachite Green
    Light Green
    Orange I (α-Naphthol)
    Toluidine Blue
    Soluble Indigo
    Aurantia
    Indigo Red

Derivatives of Aromatic Hydrocarbons

Napthalene acetic acid	Bollard (1960)
Phenyl acetic acid	Ibid.
Phenyl proprionic acid	Ibid.
Di-(2-ethylhexyl)phthalate	Kloskowski et al. (1981)

Sugars   Nissen (1974)

    Glucose
    3-0-methyl glucose
    Sucrose
    Fructose

Antibiotics   Bollard (1960)

    Streptomycin
    Clorotetracycline
    Griseofulvin
    Penicillin
    Chloramphenicol
    Cycloheximide
    Oxytetracycline

Organic Sulfur Compounds

Sulfanilamide	Bollard (1960)
Sulfacetamide	Ibid.
Sulfaguanidine	Ibid.
Sulfapyridine	Ibid.
Sulfadiazine	Ibid.
Sulfathiazole	Ibid.
4,4'-Diaminodiphenyl-sulfone	Ibid.

-- continued --

TABLE 6.57 (continued)

Organic Constituent	References

Organic Sulfur Compounds (continued)

N-Dodecylbenzene-sulfonate	Kloskowski (1981)
p-Chlorphenyl-methyl-sulfide	Guenzi et al. (1981)
p-Chlophenyl-methyl-sulfoxide	Ibid.
p-Chlorphenyl-methyl-sulfone	Ibid.

Organochlorine Compounds (excluding pesticides)

Dichlorobiphenyl	Moza et al. (1979)
Trichlorobiphenyl	Moza et al. (1979); Kloskowki et al. (1981)
Tetrachlorobiphenyl	Moza et al. (1979)
Pentachlorobiphenyl	Kloskowski et al. (1981); Weber & Mrozek, 1979
4-Chloroaniline	Kloskowski et al. (1981)
Hexachlorocyclopentadiene	Ibid.
Chloroalkylene-9	Ibid.
Trichloroethylene	Ibid.
Hexachlorobenzene	Kloskowski et al. (1981); Smelt (1981)
Pentachloronitrobenzene	Smelt (1981)
Pentachloroaniline	Dejonckheere et al. (1981)

Insecticides

Bis(dimethylamino)fluoro-phosphine oxide	Bollard (1960)
Sodium fluoroacetate	Ibid.
Schradan	Ibid.
Paraoxon	Ibid.
Parathion	Ibid.
Diethyl chlorovinyl phosphate	Ibid.
Dimethyl-carboxomethoxy-propenyl-phosphate	Ibid.
Demeton	Ibid.
Diethyl-diethylaminoethyl-thiophosphate	Ibid.
Aldrin	Kloskowski et al. (1981)
Dieldrin	Ibid.
Kepone	Ibid.
Heptachlor	Plewa (1978)
Chlordane	Ibid.

-- continued --

TABLE 6.57   (continued)

Organic Constituent	References
Fungicides	
Benomyl	Hock et al. (1970)
N-(trichloromethyl-thio)-4- cyclohexane-1-dicarboximide	Stipes & Oderwald (1971)
Thiabendazole	Ibid.
Pentachloronitrobenzene	Smelt (1981)
Herbicides	
Picloram	O'Donovan and Vanden Born (1981)
Methabenzthiazuron	Fuhr & Mittelstaedt (1981)
2,4-D	Bollard (1960)
2,4,5,-T	Ibid.
Amino-triazole	Ibid.
Propham	Ibid.
Monuron	Ibid.
Trichloroacetic acid	Ibid.
Ammonium sulfamate	Ibid.
Maleic hydrazide	Ibid.
3-hydroxy-1,2,4-triazole	Ibid.
Chlorbis(ethylamino)triazine	Ibid.
Simazine	Walker (1971); Shone et al. (1972)
Atrazine	Walker (1971); Shone et al. (1972)
Linuron	Walker (1971); Shone et al. (1972)
Lenacil	Walker (1971); Shone et al. (1972)
Aziprotryne	Walker (1971); Shone et al. (1972)
S-ethyl-dipropyl-thio- carbamate	Gray & Joo (1978)

N,N-dialyl-1-2,2-dichloro-
  acetamide (herbicide
  antedote)
	Ibid.

Hydroxyatrazine (nonphyto- toxic atrazine)	Shone et al. (1972)
Cyanazine	Plewa (1978)
Procyazine	Ibid.
Eradiacane	Ibid.
Metolachlor	Ibid.

Evidence collected thus far indicates that plants may absorb organic acids, organic bases, and both polar and nonpolar neutral organic compounds. Absorption by roots is believed to be a passive mechanism which is influenced by the rate of transpiration and soil moisture conditions (Walker, 1971). Absorption is also influenced by conditions in the root zone and soil properties. Weber and Mrozek (1979) observed that additions of activiated carbon to a sandy soil inhibited the uptake of PCBs by soybeans (Glycine max) and fescue (Festuca clatior). Hock et al. (1970) noted that absorption of the fungicide benomyl by American Elm (Ulmus americana) seedlings was 1.5 to 2.5 times greater from sand culture than from silt loam soil, and 2 to 6 times greater than from a soil, peat, and perlite mixture. Soil applied surfactants were observed by Stipes and Oderwald (1971) to enhance the absorption of three fungicides by elm trees in the field. Nissen (1974), in a discussion of plant absorption mechanisms, suggested that the absorption of choline sulfate and perhaps other compounds was mediated by bacterial activity in the rhizosphere.

Once an organic molecule is absorbed by a plant, the compound may persist, or be metabolized or removed by some other mechanism. PCB absorption by pine trees in a three year study by Moza et al. (1979) indicated that these compounds were not readily degraded by the plants. Dean-Raymond and Alexander (1976) showed that both spinach (Spinacia oleracea) and lettuce (Lactuca sativa) readily absorbed labeled dimethylnitrosamine to the leaves, but the chemical disappeared over time. Rovira and Davey (1971) noted that foliar applied agricultural chemicals were often exuded by roots into the soil. Factors which influence the metabolism of organic chemicals in plants include plant species, part of the plant in which the chemical locates, maturity of the plant and the plant environment (Rouchaud and Meyer, 1982).

Further research is needed to define both the mechanisms of plant absorption of organics from soil and the fate of these compounds once they are absorbed. Virtually no information exists regarding either phytotoxicity or plant bioaccumulation which might threaten the human food chain. Information is needed both to identify accumulator and nonaccumulator plant species and the compounds that are selectively absorbed. Until adequate research data are available, food chain crops grown on HWLT units that receive toxic organics should be closely scrutinized for plant absorption of toxic chemicals.

6.2.3   Organic Constituent Classes

Land treatability of organic constituents often follows a predictable pattern for similar compound types. For instance, where all other properties are constant, the soil half-life of aromatic hydrocarbons increases with the number of aromatic rings. Since it is beyond the scope of this document to address the fate of each organic compound in soil, the following sections discuss organic waste constituents based on their functional groups or other chemical similarities. Where data are available, examples of representative constituents within each group are used to illustrate the trend of land treatability of that group. Specific information given on the degradation of organic constituents in soil is based partially on extrapolation from studies of compounds in other aerobic systems.

6.2.3.1 Aliphatic Hydrocarbons

Aliphatic hydrocarbons are open chain or cyclic compounds that resemble the open chain compounds. Included in this chemical family are the alkanes, alkenes, alkynes, and their cyclic analogs (Morrison and Boyd, 1975). While only a few are listed as hazardous (Table 6.53), aliphatic compounds can be the rate limiting constituents in many oily wastes generated by the organic chemical, petroleum refining, and petroleum re-refining industries. In addition, a wide variety of industries generate aliphatic solvent wastes. Animal and plant processing generates wastes high in aliphatic compounds, but these waste streams are not usually considered hazardous.

A large portion of the wastes that are currently land treated are oily wastes. These wastes generally range from 1 to 40% oil by weight. Oils in these wastes are generally composed of three main organic constituent classes: aliphatics (10-80%), aromatics (5-50%), and miscellaneous (5-50%). If aliphatics and aromatics contain the pentane and benzene extractable constituents, respectively, the miscellaneous compounds are usually those extractable with polar solvents such as dichloromethane. Examples of the names assigned to the constituents in the miscellaneous include asphaltenes, resins, heterocycles, and polar organics.

Degradation of aliphatic hydrocarbons in soil depends on molecular weight, vapor pressure, water solubility, number of double bonds, degree of branching, and whether the compound is in an open chain or cyclic configuration. Perry and Cerniglia (1973) ranked aliphatic and aromatic hydrocarbons from most to least biodegradable as follows: straight-chain alkanes $(C_{12}-C_{18})$ > gases (C_2-C_4) > straight-chain alkanes (C_5-C_9) > branched alkanes (up to C_{12}) > straight-chain alkenes (C_3-C_{11}) > branched alkenes > aromatics > cycloalkanes. Microbial degradation of straight-chain alkanes proceeds faster than with branched alkanes of the same molecular weight (Humphrey, 1967). Degradation rate decreases with either the number and size of alkyl groups or the number of double bonds present. Straight or branched open chain aliphatics degrade much more rapidly than their cyclic analogs. Degradation of straight chain aliphatics also decreases with the addition of a benzene group. Microbial degradation of alkanes to carbon dioxide and water begins at a terminal carbon and initially produces the corresponding organic acid (Morrill et al., 1982). Other degradation by-products of alkanes include ketones, aldehydes and alcohols, all of which are readily degradable in aerobic soil.

Cycloalkane and its derivatives are remarkably less degradable in soil than other aliphatic hydrocarbons. Haider et al. (1981) obtained no significant biodegradation of cyclohexane after the compound was incubated in a moist loess soil for 10 weeks (see Section 6.2.3.4.1, Table 6.60). Even the penta- and hexa-chlorinated cycloalkanes appeared to biodegrade in soil to a greater extent than cycloalkane.

Moucawi et al. (1981) compared the biodegradation rates of saturated and unsaturated hydrocarbons in soil. Four soils were amended with 2,000 mg/kg of an alkane (octadecane) and the corresponding alkene (1-octadecene). While the percent of the added substrate that degraded varied between soils (16.4-32.3% degradation in 4 weeks), the amount of the alkane and alkene that biodegraded in a given soil was essentially the same In the same study, the effect of chain length on n-alkane biodegradation. was evaluated. Six soils were amended with 2,000 mg/kg of C-19 (nona-decane), C-22 (docosane), C-28 (octacosane) and C-32 (dotriacontane) alkanes and percent degradation for the compounds after 4 weeks incubation in the soils ranged from 7.5 to 54.0%, 4.6 to 50.6%, 1.3 to 39.1%, and 0.6 to 43.3%, respectively. The authors noted a clear difference in the degra-dation rates between acid and non-acid soils. Decomposition of both the short and long chain alkanes was consistently greater in the non-acid soils.

Decomposition of oily wastes high in aliphatics can be accelerated by maintenance of optimal soil moisture, temperature, waste loading and nutri-ent levels (Brown et al., 1981). The relative influence of each factor on decomposition varies from waste to waste. Generally speaking, wastes high in aliphatic hydrocarbons are both nitrogen and phosphorus deficient. Kincannon (1972) found that the addition of nitrogen and phosphorus ferti-lizers could double the decomposition of certain oily wastes. Nitrogen additions have increased the decomposition rate of straight chain alkanes (Jobson et al., 1974) and waxy cake (Gydin and Syratt, 1975). Fedorak and Westlake (1981) incubated crude oil in a soil enriched culture for 27 days with and without nitrogen and phosphorus nutrient additions. They obtained essentially complete degradation of the n-alkane fraction and substantial degradation of the branched alkanes with nutrient additions, but noted only slight degradation of these constituents when nutrients were not added.

While aliphatic hydrocarbons are usually degraded rapidly in a well managed land treatment unit, there may be a long-term accumulation of recalcitrant decomposition by-products. Kincannon (1972) determined that one major by-product of oil decomposition is naphthenic acid, which may degrade slowly in soil (Overcash and Pal, 1979).

Volatilization can be a significant loss mechanism for low molecular weight aliphatics. Wetherold et al. (1981) examined air emissions from simulated land treatment units where hexane and several aliphatic rich (oily) sludges were applied to the soil. Results obtained from the study include the following:

(1) volatility of the material applied to the soil was the most significant factor affecting emission levels;

(2) emission rates increased with increasing ambient air humidity, soil temperature and soil moisture;

(3) emission rates were highest in the first 30 minutes after waste application; and

(4) emission rates decreased with depth of subsurface injection of the waste, with a 7.5-10 cm and 15 cm depth of injection yielding high and undetectable emission levels, respectively.

Volatile aliphatic hydrocarbons (vapor pressure greater than 1) are readily assimilated by soils at low application rates. However, at application rates above the critical soil dose level, volatile compounds temporarily decrease the number and type of microorganisms present (Table 6.58). In either case, where volatile aliphatic hydrocarbons are surface applied, the dominant loss mechanism is volatilization. In addition, the rate of volatilization of nonpolar organic chemicals (such as aliphatic hydrocarbons) would increase with the water content of the soil. This may be due to displacement of the adsorbed nonpolar chemicals from the soil surfaces by water (Spencer and Farmer, 1980).

TABLE 6.58   CRITICAL SOIL DOSE LEVEL (CSDL) FOR FOUR ALIPHATIC SOLVENTS*

Aliphatic Solvent	Vapor Pressure		CSDL (ppm)	Time for Microbial Population to Recover (Days)
	mm H_2O @ 25°C	psi @ 80°F		
Heptane	---	0.9	10,000	24-63
Cyclohexane	99	2.0	840	<38
Hexane	144	3.3	430	<20
Pentane	509	---	7,200	30-53

* Buddin, 1914.

Runoff and leaching of aliphatic hydrocarbons are generally thought to be minimal due to low water solubility (Raymond et al., 1976). It should be noted, however, that large applications of oily wastes will, at least initially, decrease the infiltration rate in soil and thereby both increase runoff volume and decrease leachate volume (Plice, 1948). Within months, the elevated level of microbial activity in oil-treated soil may lead to improved soil structure, increased infiltration and leaching, and decreased runoff volume. However, the increase in leachate volume may be less than the decrease in runoff volume because the moisture holding capacity of the soil often increases when soil structure is improved.

A study of organic constituent leaching in land treatment units indicated the strong influence of both soil texture and soil layering on the depth of hydrocarbon penetration (Table 6.56). The least depth of penetration was obtained in a clay textured soil followed by a soil with a near surface clay subsoil. As might be expected, hydrocarbons penetrated to the greatest depth in the soil with the coarsest texture.

Although plants are known to produce and translocate unsubstituted aliphatic compounds, no references could be found in literature concerning the absorption of aliphatic compounds from soil.

6.2.3.2 Aromatic Hydrocarbons

Aromatic hydrocarbons are cyclic compounds having multiple double bonds and include both mono- and polyaromatic hydrocarbons. Monoaromatic compounds are benzene and substituted benzenes such as nitrobenzene and ethylbenzene. Polyaromatic hydrocarbons are composed of multiple fused benzene rings and include compounds such as naphthalene (2 fused rings) and anthracene (3 fused rings). Chlorinated aromatic compounds are discussed in Section 6.2.3.4.

Aromatic compounds are usually present in oily wastes and wastes generated by petroleum refineries, organic chemical plants, rubber industries, coking plants, and nearly all waste streams associated with combustion processes. These compounds are typically present in native soils as a result of open air refuse burning, vehicle exhaust, volcanoes and the effects of geologic processes on plant residues (Groenewegen and Stolp, 1981; Overcash and Pal, 1979). The accumulation of polyaromatic hydrocarbons in a treatment soil is particularly important because these compounds may be both carcinogenic and resistent to degradation (Brown et al., 1982b).

At very low dose levels, the decomposition rate of aromatic compounds depends more on substance characteristics than on the precise dosage (Medvedev and Davidov, 1981). Furthermore, while general trends in the decomposition rate of aromatics can be related to substance properties, there are nearly always exceptions. One general trend observed for aromatic compounds is that the higher the number of fused rings in the structure, the slower its decomposition rate (Cansfield and Racz, 1978). While aromatic compounds with five or more fused rings are not used as a sole carbon source by microbes, there is evidence that these compounds are slowly co-metabolized in the presence of other organic substrates (Groenewegen and Stolp, 1981).

Another general trend with respect to decomposition rates of aromatic compounds in land treatment soils is that the higher the water solubility of the compound, the more rapidly it degrades in soil. As stated before, there are exceptions to nearly every rule governing the decomposition of aromatic compounds. For instance, the relatively insoluble compound anthracene (75 mg/l) was found in one study (Groenewegen and Stolp, 1981) to degrade more rapidly than the more soluble compound fluoranthene (265 mg/l).

In a soil enriched culture, the aromatic constituents of a crude oil were found to degrade in the following order: naphthalene $\approx$ 2-methylnaphthalene > 1-methylnaphthalene > dimethylnaphthalenes $\approx$ dibenzothiophene $\approx$ phenanthrene > C_3-naphthalenes > methylphenanthrenes > C_2-phenanthrenes (Fedorak and Westlake, 1981). Parent aromatic compounds were generally more readily degraded than their alkyl substituted counterparts.

A number of studies have noted short-term accumulation of aromatic hydrocarbons after land treatment of oily wastes. This is apparently due to the formation of aromatic hydrocarbons as by-products of aliphatic hydrocarbon decomposition (Kincannon, 1972). In a well managed land treatment unit, most of the rapidly degradable aliphatic hydrocarbons of oily wastes will decompose within a few months after application. After that point, aromatic hydrocarbons should decrease at a faster rate since they will no longer be added to the soil as decomposition by-products.

Several of the lower molecular weight aromatic hydrocarbons have been reported in large concentrations as organic constituents contaminating groundwater (Table 6.55). In addition, several polyaromatic hydrocarbons (such as benzo(a)pyrene) have been found at low concentrations in groundwater (Zoeteman et al., 1981). While several of the polyaromatic hydrocarbons are naturally occurring pyrolysis by-products, the fact that they have been found in groundwater contaminated by improperly disposed synthetic organic compounds indicates their potential for leaching if they are improperly disposed.

No references were found to indicate the plant absorption of unsubstituted aromatic hydrocarbons. However, plant absorption has been found to occur with carboxylic acid derivatives of aromatics (Bollard, 1960) and halogenated aromatic compounds (Kloskowski et al., 1981) (See Table 6.57).

6.2.3.3 Organic Acids

Organic acids are organic constituents with phenolic or carboxylic acid functional groups. Where the pH of a soil is above the dissociation constant of an organic acid, the acid will exhibit a net negative charge and, consequently have little adsorption to soil and high water solubility. These factors combine to make organic acids relatively volatile, leachable and able to enter runoff water. Organic acids are components of numerous hazardous wastes, but the primary source in land treatment soil will be from the biodegradation by-products of the other organics present in the waste treated soil. Chlorinated organic acids, including chlorinated phenols, are discussed in Section 6.2.3.4.

Degradation of organic acids in soil can be relatively rapid under favorable environmental conditions. Too high a loading rate of acids can sufficiently lower the soil pH so that biodegradation is inhibited. Martin and Haider (1976) showed that several carboxylic acids would degrade as rapidly as glucose in a sandy soil (Table 6.59). Higher molecular weight carboxylic acids may degrade more slowly. Moucawi et al. (1981) compared the percent degradation of 2 long chain, saturated fatty acids (C-18 stearic acid and C-28 montanic acid) after these acids were incubated in 2 microbially active and 2 acid soils for 4 weeks. Stearic acid underwent substantial degradation in the microbially active soils (23.6-31.2%) but little degradation in the acid soils (3.9-5.1%). The longer chain acid underwent very little degradation in all 4 soils (0-2.1%). An unsaturated C-18 fatty acid (Oleic acid) underwent substantial degradation in both the acid (23.4-24.8%) and microbially active soils (33.0-41.4%).

TABLE 6.59 DECOMPOSITION OF THREE CARBOXYLIC ACIDS AND GLUCOSE IN SANDY SOIL*

| Organic Constituent[†] | % Decomposition | |
	After 7 days	After 84 days
Acetic acid	52-76	71-87
Pyruvic acid	47-83	70-93
Succinic acid	52-89	71-95
Glucose	75	87

* Martin and Haider (1976).

[†] All organics applied to the soil at 1000 ppm.

Phenolic acids are also rapidly degraded in soil at low concentrations but can cause a lag phase of low microbial degradation at higher concentrations. Scott et al. (1982) evaluated the curves representing cumulative adsorbed and microbially degraded phenol with two soils in a batch test using a 1:5 soil to solution concentration and continuous shaking. At concentrations $<10^{-3}$M phenol the curves had the following three characteristic phases:

(1) there was an initial lag phase whose length (of time) increased with increasing phenol concentration;

(2) next, there was an exponential growth phase whose rate of growth decreased with increasing phenol concentration; and

(3) finally, there was a stationary phase where essentially all the phenol that was not adsorbed had been degraded.

In another experiment, repeated applications of phenols to soil first increased and then decreased the rate at which phenol was biodegraded (Medvedev et al., 1981). The initial decomposition rate increase was thought to be due to rapid multiplication of the phenol-decomposing microorganisms, and the subsequent decrease, due to a gradual accumulation of toxic metabolic by-products or the proliferation of another microbe that fed on phenol-decomposing bacteria. Haider et al. (1981) studied the degradation in soil of phenol, benzoic acid, and their chlorinated derivatives (See Section 6.2.3.4.1, Table 6.60).

Four phenolic acids (p-hydroxybenzoic, ferulic, caffeic and vanillic acids) were found to be quickly metabolized when 5 mg of the compound was incorporated into each gram of soil (5,000 ppm). After 4 weeks of incubation, both extractable phenols and soil respiration rates had returned to levels near that of the control soil (Sparling et al., 1981). In another study that examined respiration after soil amendment with phenolic acids, the soil respiration rate decreased substantially by the fourth week of the study (Haider and Martin, 1975). However, less than 60% of carbon-14 labelled caffeic acid had evolved as carbon dioxide (CO_2) in 4 weeks and less than 70% had evolved in 12 weeks. This indicates that a decrease in the respiration rate is not necessarily an indication that all of the phenolic acids have been degraded.

Some phenolic compounds have been found to be relatively resistent to biodegradation because they readily undergo polymerization reactions and the higher molecular weight polymers are only slowly degraded. Martin and Haider (1979) incubated two carbon-14 labelled phenols that readily polymerize (coumaryl alcohol and pyrocatechol) in moist sandy loam and found that only 42% and 24%, respectively, of the ring carbons had evolved as CO_2. When the pyrocatechol was linked into model humic acid-type polymers, evolution of carbon-14 from five soils ranged from 2-9% after 12 weeks. When coumaryl alcohol was incorporated into a model lignin, evolution of carbon-14 from five soils ranged from 7-14% after 12 weeks. In both cases where the phenols were linked into model polymers, the addition of an easily biodegradable carbon source to the treatment soil had little effect on the biodegradation rate of the phenols as measured by carbon-14 evolution.

Leaching and runoff of organic acids can be substantial due to the high water solubility of these compounds. If the pH of the soil is greater than the pK_a of an organic acid, mobility of the acid will be increased in clay soils (Section 6.2.2.4.1).

No information was found on vapor loss of organic acids from soil. Judging from the vapor pressure of these compounds, low molecular weight carboxylic acids may undergo substantial volatilization, while the vapor loss of phenolic compounds would be somewhat less.

Plant uptake of organic acids has been shown in several studies (Table 6.57). Bollard (1960) showed that several carboxylic acid derivatives of aromatic hydrocarbons can be taken up by plants. Ghosh and Burris (1950) found plants can take up several amino acids.

6.2.3.4 Halogenated Organics

Halogenated organics contain one or more halogen atoms (Cl, F, Br, or I) somewhere in their molecular structure. Chlorinated organics comprise the vast majority of halogenated organics found in wastes. A notable exception is the group of brominated biphenyls, which until recently were widely used as flame retardants. Halogenated organics can be further broken down into aliphatics, aromatics, and arenes (molecules that contain both aromatic and aliphatic parts).

Most of the interest in the past few years has been directed toward chlorinated aromatics such as chlorinated biphenyls (PCB), chlorinated benzenes and their phenolic metabolic by-products. Little quantitative data are available on such critical areas as the soil half-life, volatilization or leaching rates from soil, or the ability of plants to absorb these compounds. Land treatment of halogenated organics should be avoided unless preliminary studies have assured that biodegradation (not volatilization or leaching) will be essentially the only loss mechanism for these hazardous constituents. In addition, preliminary studies should determine the soil half-life of the halogenated constituents for the following reasons: (1) to ensure that the loading rate schedule does not cause accumulation of these compounds to the point that the concentration is toxic to the microbial population or that the adsorption capacity of the soil is exceeded causing leaching or volatilization to become significant loss mechanisms; and (2) to ensure that the degree of degradation required for closure is achievable within the operational life span of the HWLT unit.

Many of the halogenated organics can not be expected to be satisfactorily degraded within the 10-30 year life span of HWLT units. The low degradability, high leachability and high volatility of the halogenated solvents make these compounds especially unsuitable for land treatment. Wastes containing these compounds should either undergo some type of dehalogenation pretreatment or be disposed in some other manner.

Halogenated organics span the range of leachability, volatility and degradability. At one end of this range are some of the most toxic and persistant compounds made by man. Many of the light weight chlorinated hydrocarbons are among the most prevalent synthetic organic chemicals found in groundwater (Table 6.55). For these reasons, wastes containing even low concentrations of halogenated organics may require a dehalogenation pretreatment prior to land treatment of the waste. Wastes that may contain halogenated hydrocarbons include textiles, petrochemical, wood preserving, agricultural, and pharmaceutical wastes. Halogenated organics may also be found in the wastes of industries that use halogenated solvents.

Degradation of halogenated organics in soil has been documented. However, the range in degradation rates for these compounds may be anywhere from rapid to extremely slow. As with all organic chemicals, the slower the degradation rate, the more likely it is that the compound would be lost by volatilizing, leaching or entering runoff water rather than through biodegradation.

Chlorinated hydrocarbon insecticides are among the most resistant to biodegradation of all pesticides (Edwards, 1973). Soil half-life of many of the early chlorinated pesticides are measured in years rather than days or weeks. With further research, it was discovered that factors such as position of halogens on a ring structure could significantly alter its degradation rate (Kearney, 1967). Isomers of the same chlorinated compound have been found to have order of magnitude differences in soil half-life (Stewart and Cairns, 1974). Another problem that has been encountered with chlorinated organics is that the terminal residue or metabolic by-products may be either more toxic (Kiigemagi et al., 1958) or more persistent (Smelt, 1981) than the parent compound.

6.2.3.4.1 Chlorinated Benzene Derivatives. Chlorinated aromatics are, as a group, less degradable, volatile and leachable than their chlorinated aliphatic counterparts. In many cases, however, the lower degradation rate makes leaching, volatilization, runoff or plant uptake significant loss mechanisms. Following are discussions of chlorinated benzenes (hexachlorobenzene, pentachlorobenzene, trichlorobenzenes, dichlorobenzenes, and chlorobenzene), and brominated and chlorinated biphenyls, along with several derivatives and metabolic by-products of the chlorinated aromatic compounds.

Hexachlorobenzene (HCB) has been found to be both a by-product of numerous industrial processes and a contaminant in a variety of chlorinated solvents and pesticides (Farmer et al., 1980). Beck and Hansen (1974) found HCB, quintozene (PCNB), and pentachlorothioanisol (PCTA) to have soil half-lives (in days) of approximately 969-2089 (calculated), 213-699, and 194-345, respectively. These three compounds follow the general trend in that the less chlorinated otherwise similar compounds are, the more biodegradable they are likely to be. While the water solubility and vapor pressure of these compounds are relatively low, their extreme persistence makes both leaching and volatilization potential loss mechanisms.

Another problem encountered with HCB and its derivatives has been their absorption and translocation in plants. Since these compounds are relatively immobile in soil (Overcash and Pal, 1979), they may be present near the soil surface for centuries and, consequently, accessible to plant roots. Smelt (1981) found several studies that documented the plant absorption of both HCB and PCNB. The ratio of crop to soil concentration was as high as 29:1 for HCB and 27:1 for PCNB. Plants that were found to accumulate higher concentrations of the chlorinated organics than was present in the soil included lettuce (Lactuca sativa), carrots (Daucus carota), grasses, parsley (Petroselinum crispum), radishes (Raphanus sativus), potatoes (Solanum tuberosum) and tulip (Tulipa sp.) bulbs.

HCB and its derivatives could pose a hazard to grazing animals long after closure of a land treatment unit. Consequently, there is a need for HWLT operators to monitor incoming wastes to be sure that untreated chlorinated wastes are detected and rejected before they pass through the front gates. It should also be noted that in soils where HCB is present, there may also be several HCB metabolites. Smelt (1981) examined soil plots that had previously been treated with compounds containing HCB and found the following related compounds: quintozene (PCNB), pentachlorobenzene (QCB), pentachloroaniline (PCA), and pentachlorothioanisol (PCTA). Since plant absorption has been shown to occur for HCB and PCNB, the potential exists for metabolites of these compounds to be either absorbed by plants or formed in the plant as metabolic by-products of HCB or PCNB. PCA has been found in lettuce (<u>Lactuca</u> <u>sativa</u>) leaves (Dejonckheere et al., 1981) but it could not be determined if it entered lettuce from the soil or formed in the plant from decomposition of the PCNB that was also in the plant tissue. Dejonckheere et al. (1981) pointed out that these compounds, if they were consumed by grazing animals would either concentrate in fatty tissue (HCB) or be passed into the milk of dairy cows (PCNB and PCA).

Trichlorobenzenes (TCB) are constituents of both textile-dying wastes and transformer fluids containing polychlorinated biphenyls (EPA, 1976). Two TCBs (1,2,3- and 1,2,4-TCB) were found to biodegrade very slowly (0.35 and 1.00 nmol/day/20 gms soil, respectively) when these compounds were incubated in a sandy loam soil at concentrations of 50 μg TCB per gram of soil (Marinucci and Bartha, 1979). Neither fertilizer additions nor the addition of other microbial substrates appeared to increase TCB biodegradation rates.

Since anaerobic conditions are known to increase the rate of some dechlorination reactions but may suppress aromatic ring cleavage, weekly alterations of anaerobic and aerobic soil conditions were studied to see if TCB biodegradation could be increased. The authors assumed that, since this cycling of soil conditions failed to increase biodegradation, the kinetics of TCB mineralization suggested rate-limiting initial reactions. The only factor found to increase TCB biodegradation was increased temperature (28°C or above). Maximum biodegradation rate for the compounds was obtained at TCB concentrations between 10-25 μg per gram of soil and this rate was found to markedly decrease above that concentration range.

A mixture of dichlorobenzene has been shown to degrade in soil much slower than benzene, chlorobenzene, or a mixture of trichlorobenzenes (Haider et al., 1981). After incubation in a moist loess soil for 10 weeks, only 6.3% of the original 20 ppm carbon-14 labeled dichlorobenzenes had evolved as carbon dioxide. This translates into a soil half-life for these compounds of roughly 2 years. With a 2 year half-life it would take approximately 14 years to achieve 99% degradation. By contrast, the trichlorobenzenes were 33% biodegraded after 10 weeks. At this degradation rate, 99% degradation of the trichlorobenzenes could be achieved in less than 3 years. Chlorobenzene was degraded somewhat slower than the trichlorobenzenes but at four times the degradation rate for the dichlorobenzenes (Table 6.60). While these rates of degradation are somewhat lower than those reported elsewhere, the trends in the data indicate there are significant exceptions to the general rule that "the less chlorinated an organic, the more degradable it is."

TABLE 6.60   DEGRADATION OF CHLORINATED BENZENES, PHENOLS, BENZOIC ACIDS AND
CYCLOHEXANES AND THEIR PARENT COMPOUNDS[*][†]

Compounds	3 days	1 week	2 weeks	5 weeks	10 weeks
Benzene	7.5	24	37	44	47
Chlorobenzene	16.2	18.3	20	25	27
Dichlorobenzenes	0.1	1.1	1.2	1.7	6.3
Trichlorobenzenes	3.6	20.3	22	30	33
Phenol	45.5	48	52	60	65
2-Chlorophenol	7.5	13	14.7	21	25
4-Chlorophenol	15.4	22.2	24	31	35
Dichlorophenols	1.4	31.4	35	43	48
Trichlorophenols	1.6	35	38	47	51
Benzoic acid	40	44	49	57	63
3-Chlorobenzoic acid	21	28	32	38	59
Cyclohexane	<0.02	0.1	0.2	0.3	0.3
γ-Hexachlorocyclohexane	0.05	0.3	0.7	1.8	2.6
γ-Pentachlorocyclohexane	0.01	0.3	0.8	2.3	3.5

[*] Haider et al. (1981).

[†] Degradation was measured by the release of marked CO_2 from the
carbon-14 labeled organic compounds. Values given in the table are sum
values in % of added radioactivity.

Metabolic by-products of chlorinated benzenes include chlorinated
phenols and carboxylic acids. Degradation of phenol, benzoic acid, and
some of their chlorinated derivatives are given in Table 6.60. While the
chlorinated derivatives of these acids are generally less degradable in
soil than their nonchlorinated counterparts, they are usually more degradable than their parent chlorinated benzene derivatives.

Baker and Mayfield (1980) studied the degradation of phenol and its
chlorinated derivatives in aerobic, anaerobic, sterile and non-sterile soil
(Table 6.61). Phenol, o-chlorophenol, p-chlorophenol, 2,4-dichlorophenol,
2,6-dichlorophenol, and 2,4,6-trichlorophenol were biodegraded rapidly in
the aerobic soil, while m-chlorophenol, 3,4-dichlorophenol, 2,4,5-trichlo-
rophenol, and pentachlorophenol were degraded more slowly. The most slowly
degraded compounds under aerobic conditions were 3,4,5-trichlorophenol and
2,3,4,5-tetrachlorophenol. While nonbiological degradation occurred in
both the aerobic and anaerobic soil, no biological degradation of any of
the chlorophenols was indicated for the anaerobic soils.

6.2.3.4.2 Halogenated Biphenyls. Halogenated biphenyls are no longer produced in the U.S., but the extreme recalcitrance of these compounds and
their past widespread use in chemical industries indicates that they will

TABLE 6.61 AEROBIC AND ANAEROBIC DEGRADATION OF PHENOL AND ITS CHLORINATED DERIVATIVES IN SOIL*

Compounds	Aerobic Degradation				Anaerobic Degradation			
	Non-sterile		Sterile		Non-sterile		Sterile	
	Days	% Degraded	Days	% Degraded	Days	% Degraded	Days	% Degraded
Phenol	5.00	100	40	15	40	20	40	7
o-chlorophenol	1.50	100	40	67	80	78	80	82
m-chlorophenol	160.00	87	160	31	160	37	160	15
p-chlorophenol	20.00	83	20	5	40	13	40	17
2,4-dichlorophenol	40.00	81	40	31	80	62	80	59
2,6-dichlorophenol	0.75	100	40	55	80	82	80	81
3,4-dichlorophenol	160.00	88	160	21	160	-4	160	-3
2,4,6-trichlorophenol	3.00	95	80	27	80	28	80	25
2,4,5-trichlorophenol	160.00	72	160	9	80	8	80	5
3,4,5-trichlorophenol	160.00	17	160	0	80	-2	80	4
2,3,4,5-tetrachlorophenol	160.00	31	160	-1	80	5	80	7
Pentachlorophenol	160.00	80	160	20	160	7	160	5

* Baker and Mayfield (1980).

be an important concern of the waste disposal community for at least several decades. Polychlorinated biphenyls (PCB) are still in widespread use in transformers and capacitors around the world (Griffin and Chian, 1980). Polybrominated biphenyls (PBB) were produced for use as flame retardants in business machines, electrical housings, and textiles (Griffin and Chou, 1982).

Degradation of PCBs has been found to be affected by the nature of the chlorine (Cl) substituents as follows (Morrill et al., 1982; Kensuke et al., 1978):

(1) degradation decreased as amount of Cl substitution increased;

(2) PCBs with two Cl atoms in the ortho position on one or both rings had very low degradability; and

(3) PCBs with only one chlorinated ring degraded more rapidly than PCBs with a similar number of Cl atoms but with these divided between the two rings.

In many cases, the mono-, di-, and tri-chlorinated biphenyls have been found to be degradable by mixed microbial populations (Furukawa and Matsumura, 1976; Metcalf et al., 1975). Most reports on the degradability of tetra-, penta-, and hexachlorobiphenyls indicate that these compounds degrade extremely slowly in most environments (Metcalf et al., 1975; Nissen, 1981).

Nissen (1981) investigated the degradability of Arochlor 1254 (a mixture of PCBs with from 4 to 7 chlorine substituents) in moist, warm soil with nutrients added. No biodegradation was evident after 60 days of incubation in the soil. Moein et al. (1975) returned to the site of a two year old spill of Archlor 1254 on soil and found that no perceptible degradation of the PCBs had occurred over that time period. In another study, Iwata et al. (1973) found that the lower chlorinated biphenyls exhibited significant degradation in 12 months on five California soils.

A study by Wallnofer et al. (1981) indicated that PCBs were absorbed by the lipid rich epidermal cells on carrots (Daucus carota) and to a lesser extent by radish (Raphanus sativus) roots. Moza et al. (1976), however, found a phenolic metabolic by-product of 2,2'-dichlorobiphenyl in carrot leaves. Mrozek et al. (1982) demonstrated that salt marsh cordgrass has the capacity to accumulate PCBs above the level of these compounds in the soil. PCBs were taken up by the plant from sand and an organic mud soil. Furthermore, the PCBs were translocated throughout the plant. While PCBs are strongly adsorbed by organic matter in soils, they have been found to be largely associated with the partially decomposed plant litter rather than humic substances (Scharpenseel et al., 1978). These plant remnants are readily taken up by soil fauna thereby providing a means for the PCBs to enter the food chain. Several other studies that noted the plant uptake of various chlorinated biphenyls are listed in Table 6.57.

Polybrominated biphenyls (PBB) were found to be strongly adsorbed by soils and not leached by water by Griffin and Chou (1982). Similar results were obtained by Filonow et al. (1976). Jacobs et al. (1976) found that PBBs were only very slowly degradable in soil and taken up in very small quantities by plants. From all available data it would appear that PBB contaminated soil will pose little threat to groundwater or crop purity, with the possible exception of root crops. There is, however, no information available concerning the toxicity, degradability, leachability or ability for plants to take up metabolites of PBB (Getty et al., 1977).

6.2.3.5 Surface-active Agents

Surface-active agents (surfactants) are organic compounds with two distinct parts to each molecule. One part is hydrophilic or water soluble (such as a sulfonate, sulfate, quarternary amine or polyoxyethylene) and the other part is hydrophobic or water-insoluble (such as an aliphatic or aromatic group) (Huddleston and Allred, 1967). It is the presence of these two different groups on the same molecule that causes these molecules to concentrate at surfaces or interfaces. The presence of these molecules at interfaces reduces the surface tension of liquids. Surfactants are commonly found in industrial wastes as a result of their use in various industries as detergents, wetting agents, penetrants, emulsifiers spreading agents and dispersants. Industries that use large quantities of surfactants include textile, cosmetic, pharmaceutical, metal, paint, leather, paper, rubber, and agricultural chemical industries. The three main types of surfactants produced are cationics, nonionics and anionics. These surfactants accounted for 6, 28 and 65%, respectively, of the total surfactant production in the U.S. in 1978 (Land and Johnson, 1979).

Most cationic surfactants are salts of either a quarternary ammonium or an amine group (with an aromatic or aliphatic side chain) and either a halogen or hydroxide. Many of these surfactants can cause problems due to their strong antimicrobial action.

Nonionic surfactants are so named because they do not ionize in water. Two main types are alkyl polyoxyethylenes and alkylphenol polyoxyethylenes. The former has been found to be readily biodegradable, but decreasingly so as the polyoxyethylene chain is lengthened (Huddleston and Allred, 1967). Half-life of an alkyl polyoxyethylene surfactant in a moist (28% H_2O) sandy loam soil was found to be approximately 60, 90, 120 and 160 days when the surfactant was applied at 250, 1,000, 5,000 and 10,000 ppm, respectively (Valoras et al., 1976). Although the study did not extend long enough to achieve 50% degradation of higher dosage levels extrapolation of the data indicated that when applied to this soil at 20,000 ppm, the half-life of the surfactant may have approached 1 year.

Anionic surfactants are negatively charged ions when in solution. The three major forms are alkyl sulfates, alkylbenzene sulfonates and carboxylates. Alkylbenzene sulfonates are the most widely used surfactants, accounting for 35% of all surfactants produced in the U.S. in 1978 (Land

and Johnson, 1979). Most widely used surfactants of this type are the linear alkyl benzenes (LAS), which are composed of a benzene ring with both a sulfonate and a roughly linear alkyl chain attached. Major factors influencing the biodegradation rate for the LAS type surfactants are as follows (Huddleston and Allred, 1967).

(1) the position of the sulfonate group relative to the alkyl chain;

(2) the alkyl chain length and point of attachment of the benzene ring; and

(3) the degree of branching along the length of the alkyl chain.

Another type of alkylbenzene sulfonate called ABS is a mixture of branched chain isomers of sodium dodecylbenzene sulfonate. While LAS and ABS have both been found to inhibit nitrification activities, LAS is apparently biodegraded more quickly in soil (Vandoni and Goldberg, 1981). Neither of these surfactants is likely to volatilize from the soil surface, but both can be mobile in soils when they are in an ionic state. There is some evidence that these and other surfactants may increase the leachability of other organic constituents and some microorganisms under saturated flow conditions. A discussion of the effects of anionic surfactants on plants has been published by Overcash and Pal (1979).

Surfactants can have strong influences on the chemical, physical and biological properties of a soil. If the hydrophilic portion of a surfactant adsorbs to soil particles, the hydrophobic portion would extend outwards, imparting to soil particles a hydrophobic surface. Under these conditions, the saturated flow (flow due to gravity) increases while the unsaturated flow (flow due to capillary forces) decreases (Sebastiani et al., 1981). Luzzati (1981) found that applying the equivalent of 3,200 kg/ha of nonionic and anionic surfactants to test plots slightly improved soil structure but substantially inhibited soil enzyme activity. Vandoni and Goldberg (1981) found that anionic surfactants significantly inhibited nitrification (metabolism of ammonium in soil) while nonionic surfactants seemed to slightly stimulate nitrification. Letey et al. (1975) showed that infiltration rates were increased with soil application of nonionic surfactants. Aggregation, aeration and water holding capacity of a soil can be increased by surfactant applications to soil (Batyuk and Samochvalenko, 1981). However, Cardinali and Stoppini (1981) found that while anionic surfactant dosages of 16-80 ppm improved the structural stability of some soils, at dosages over 400 ppm the structural stability of the soils always significantly decreased. When calculating the loading rates for biodegradable surfactants, both the half-life and effect on soil properties of these constituents should be carefully considered.

CHAPTER 6 REFERENCES

Abbazov, M.A., I. D. Dergunov, and R. Mikulin. 1978. Effect of soil properties on the accumulation of strontium-90 and cesium-137 in crops. Soviet Soil Sci. 10(1):52-56.

Adams, D. F., J. W. Hendrix, and H. G. Applegate. 1957. Relation among exposure periods, foliar burn, and fluorine content of plants exposed to hydrogen fluoride. J. Agr. Food Chem. 5:108-116.

Adriano, D. C., L. T. Novak, A. E. Erickson, A. R. Wolcott, and B. G. Ellis. 1975. Effect of long term land disposal by spray irrigation of food processing wastes on some chemical properties of the soil and subsurface water. J. Environ. Qual. 4:242-248.

Ahmed, S. and H. J. Evans. 1960. Cobalt: a micronutrient element for the growth of soybean plants under symbiotic conditions. Soil Sci. 90:205-210.

Ahmed, S. and H. J. Evans. 1961. The essentiality of cobalt for soybean plants grown under bymbiotic conditions. Proc. Natl. Acad. Sci. U.S.A. 47:24-36.

Aldrich, D. G., J. R. Buchanan, and G. R. Bradford. 1955. Effect of soil acidification on vegetative growth and leaf composition of lemon trees in pot culture. Soil Sci. 79:427-439.

Allaway, W. H. 1968. Agronomic controls over the environmental cycling of trace elements. Adv. Agron. 20:235-271.

Allaway, W. H. 1975. Soil and plant aspects of cycling of chromium, molybdenum and selenium. In T. C. Hutchinson (ed.) Proc. of International Conference on Heavy Metals in the Environment, Vol. 1. Toronto, Canada.

Allen, H. E., R. H. Hall, and T. D. Brisbin. 1980. Metal Speciation. Effects on aquatic toxicity. Environ. Sci. Technol. 14:441-443.

Alten, F. and R. Goltwick. 1933. A contribution to the question of the substitution of Rb and Cs for K in plant nutrition. Ernahr Planze. 29:393-399.

Anastasia, F. B. and W. J. Kender. 1973. The influence of soil arsenic on the growth of lowbush blueberries. J. Environ. Qual. 2:335-337.

Anderson, A. and K. O. Nilsson. 1974. Ambio 3:198-200.

Andrziewski, M. 1971. Effect of chromium application on yields of several plant species and on the chromium content of soil. Rocznik Nauk Rolniczych. Agricultural College, Pozlon, Poland.

Arnon, D. I. 1958. The role of micronutrients in plant nutrition with special reference to photosynthesis and nitrogen assimilation. pp. 1-32. In C. A. Lamb, O. E. Bently, and J. M. Beate (ed.) Trace elements. Academic Press, New York.

Aston, S. R. and P. H. Brazier. 1979. Endemic goitre, the factors controlling iodine deficiency in soils. The science of the total environment. 11:99-104.

Aubert, H. and M. Pinta. 1977. Trace elements in soils. Elsevier Sci. Publ. Co., New York. 395 p.

Baker, M. D. and C. I. Mayfield. 1980. Microbial and non-biological decomposition of chlorophenols and phenol in soil. Water, Air and Soil Pollution 13:411-424.

Ballard, R. and J. G. A. Fiskell. 1974. Phosphorus retention in coastal plain forest soils: I. Relationship to soil properties. Soil Sci. Soc. Am. J. 38:250-255.

Banerjee, D. K., R. H. Bray, and S. W. Melsted. 1953. Some aspects of the chemistry of cobalt in soils. Soil Sci. 75:421-431.

Barber, S. A., J. M. Walker, and E. M. Vasey. 1963. Principles of ion movement through the soil to the plant root. pp. 121-124. Trans. Joint Meeting, Comm. IV and V. Intern. Soc. Soil Sci., New Zealand, 1962.

Barker, D. E. and L. Chesnin. 1975. Chemical monitoring of soils for environmental quality and animals and human health. Adv. Agron. 27:306-374.

Barker, H. A. 1941. Studies on methane fermentation. V. J. Biol. Chem. 127:153-167.

Barltrop, D., C. D. Strehlow, I. Thornton, and S. S. Webb. 1974. Significance of high soil lead concentrations for childhood lead burdens. Environ. Health Perspec. 7:75-82.

Barnette, R. M. and J. P. Camp. 1936. Chlorosis in corn plants and other field crop plants. Fla. Agr. Exp. Sta. Ann. Rep. 45.

Barshad, I. 1948. Molybdenum content of pasture plants in relation to toxicity to cattle. Soil Science 66:187-195.

Bartlett, R. and B. James. 1979. Behavior of chromium in soils: III. Oxidation. J. Environ. Qual. 8:31-34.

Batyuk, V. P. and S. K. Samochvalenko. 1981. The influence of surface-active substances on the water-physical characteristics of soil and the physiological-biochemical characteristics of plants. pp. 407-418. In M. R. Overcash (ed.) Decomposition of toxic and nontoxic organic chemicals in soils. Ann Arbor Science Publ., Inc. Ann Arbor, Michigan.

Baumhard, G. R. and L. F. Welch. 1972. Lead uptake and corn growth with soil applied lead. J. Environ. Qual. 1(1):92-94.

Beath, O. A. 1949. Economic potential and botanic limitations of some selenium bearing plants. Wyoming Agr. Exp. Sta. Bull. 360.

Beath, O. A., C. S. Gilbert, and H. F. Eppson. 1941a. The use of indicator plants in locating seleniferous areas in the western United States. I. General. Am. J. Bot. 26:257.

Beath, O. A., C. S. Gilbert, and H. F. Eppson. 1941b. The use of indicator plants in locating seleniferous areas of the western United States. II. Correlation. Am. J. Bot. 28:887.

Beauchamp, E. G. and J. Moyer. 1974. Nitrogen transformation and uptake. In Proc. of Sludge Handling and Disposal Seminar, Ontario Ministry of the Environment. Toronto, Ontario.

Beauford, W., J. Barber, and A. R. Barringer. 1977. Uptake and distribution of mercury within higher plants. Physiologia Plantarium 39:261-265.

Beck. J. and K. E. Hansen. 1974. The degradation of quintozene, pentachlorobenzene, hexachlorobenzene, and pentachloroaniline in soil. Pestic. Sci. 5:41-48.

Bell, M. C. and N. N. Sneed. 1970. Metabolism of tungsten by sheep and swine. pp. 70-71 In C. F. Mills (ed.) Trace elements metabolism in animals. E and S Livingston, Edinburgh.

Benac, R. 1976. Effect of the manganese concentration in the nutrient solution on groundnuts. Oleagineaux 31:539-543.

Benson, N. R., R. P. Covey, and W. Hagland. 1978. The apple replant problem in Washington state. J. Amer. Soc. Hort. Sci. 103:156-158.

Bernstein, L. 1974. Crop growth and salinity In J. Van Schilfgaarde (ed.) Drainage for agriculture. Am. Soc. Agron. Madison, Wisconsin.

Berrow, M. L. and J. Webber. 1972. Trace elements in sewage sludges. J. Sci. Food Agr. 23:93.

Biddappa, C. C., M. Chino, and K. Kumazawa. 1981. Adsorption, desorption, potential and selective distribution of heavy metals in selected soils of Japan. J. Environ. Sci. Health L.

Bielby, D. G., M. H. Miller, and L. R. Webber 1973. Nitrate content of percolates from manured lysimeters. J. Soil and Water Cons. 28(3):124-126.

Bingham, F. T. 1973. Boron in cultivated soils and irrigation waters. pp. 130-143. In Trace elements in the environment. Am. Chem. Soc., Washington, D.C.

Bingham, F. T. 1979. Bioavailability of Cd to food crops in relation to heavy metal content of sludge amended soil. Environ. Health Perspec. 28:3943.

Blaschke, H. 1977. The influence of increased copper supply on some crop plants and their root development. Zeitschrift fur Ackerund Pflanzenbau 144:222-229.

Boawn, L. C. and P. E. Rasmussen. 1971. Crop response to excessive Zn fertilization of an alkaline soil. Agron. J. 63: 874-876.

Boggess, S. F., S. Willavize, and D. E. Koeppe. 1978. Differential response of soybean varieties to soil cadmium. Agron. J. 70(5):756-760.

Bohn, H. L., B. L. McNeal, and G. A. O'Connor. 1979. Soil chemistry. John Wiley and Sons, Inc., New York. 329 p.

Bollard, E. G. 1960. Transport in the xylem. Ann. Rev. Plant Physiology 11:141-166.

Borges, A. C. and A. G. Wollum. 1981. Effect of cadmium on symbiotic soybean plants. J. Environ. Qual. 10:216-221.

Borovik-Romonova, T. F. 1944. The content of Rb in Plants. Dokl. Aka. Nauk. SSR 44:313-16.

Bouwer, H. and R. L. Chaney. 1974. Land treatment of wastewater. Adv. Agron. 26:133-76.

Bowen, H. J. M. 1966. Trace elements in biochemistry. Academic Press, New York. 241 p.

Bowman, R. S., M. E. Essington, and G. A. O'Connor. 1981. Soil sorption of nickel: influence of solution composition. Soil Sci. Soc. Am. J. 45:860-865.

Bradford, G. R., 1966. Boron. p. 33. In Chapman, H. D. (ed.), Diagnostic criteria for plants and soils. University of Calif., Div. Agr. Sci. Riverside, California.

Brady, N. C. 1974. The nature and properties of soils. 8th ed. MacMillan Publ. Co., Inc., New York. 639 p.

Brar, M. S. and G. S. Sekhon. 1979. Effect of applied zinc and iron on the growth of rice. Riso 28(2):125-131.

Bremner, J. M. and K. Shaw. 1958. Denitrification in soils II. Factors affecting denitrification. J. Agr. Sci. 51(1):22-52.

Bresler, E., B. L. McNeal, and D. L. Carter. 1982. Saline and sodic soils. Springer-Verlag Inc., New York, New York. 236 p.

Brewer, R. F. 1966a. Fluorine. In H. D. Chapman (ed.) Diagnostic criteria for plants and soils. University of California, Riverside.

Brewer, R. F. 1966b. Lead. pp. 213-217, In H. D. Chapman (ed.) Diagnostic criteria for plants and soils. University of California, Riverside.

Broadbent, F. E., K. B. Tyler, and G. N. Hill. 1957. Nitrification of ammonical fertilizers in some California soils. Hilgardia 27:247-267.

Broderius, S. and L. Smith. 1979. Lethal and sublethal effects of binary mixtures of cyanide and hexavalent chromium, zinc or ammonia to the fathead minnow and rainbow trout. J. Kish. Res. Board Can. 36:164.

Brooks, R. R. 1977. Copper and cobalt uptake by Haumaniastrum species. Plant and Soil 48:541-544.

Brooks, R. R., J. Lee, R. D. Reeves, and T. Jaffre. 1977. Detection of nickeliferous rocks by analysis of herbarium specimens of indicator plants. J. of Geochem. Exploration 7:49-57.

Brooks, R. R., R. S. Morrison, R. D. Reeves, T. R. Dudley, and Y. Akman. 1979. Hyperaccumulation of nickel by Alyssum Linnaeus (Cruciferae). Proc. R. Soc. Lond., B 203:387-403.

Brooks, R. R. and C. C. Radford. 1978. Nickel accumulation by European species of the genus Alyssum. Proc. R. Soc. Lond., B 200:217-224.

Brooks, R. R., R. D. Reeves, R. S. Morrison, and F. Malaisse. 1980. Hyperaccumulation of copper and cobalt--a review. Bull. Soc. Roy. Bot. Belg. 113:166-172.

Brown, J. C. and W. E. Jones. 1977. Manganese and iron toxicities dependent on soybean variety. Comm. in Soil Sci. and Plant Analysis 8:1-15.

Brown, K. W. and L. E. Deuel, Jr. 1982. Evaluation of subsurface landfarm contamination after long-term use. Final Report of a Cooperative Study for the American Petroleum Institute and the U.S. Environmental Protection Agency. 116 p.

Brown, K. W., L. E. Deuel, Jr. and J. C. Thomas. 1982a. Soil disposal of API pit wastes. Final Report of a Study for the Environmental Protection Agency (Grant No. R805474013). 209 p.

Brown, K. W., K. C. Donnelly, and B. Scott. 1982b. The fate of mutagenic compounds when hazardous wastes are land treated. pp. 383-397. In David W. Shultz (ed.) Proceedings at the Eighth Annual Research Symposium at Ft. Mitchell, Kentucky. EPA-600-9-82-002b.

Brown, K. W., K. C. Donnelly, J. C. Thomas, and L. E. Deuel, Jr. 1981.
Factors influencing the biodegradation of API separator sludges applied to
soils. pp. 188-199. In Proceedings of the Seventh Annual Research Symposium
at Philadelphia, Pennsylvania, March 16-18. U.S. EPA. Cincinnati, Ohio.
EPA-600/9-81-002B.

Buddin, W. 1914. Partial sterilization of soil by volatile and nonvolatile
antiseptics. J. Agric. Sci. 6:417-451.

Burmaster, D. E. and R. H. Harris. 1982. Groundwater contamination: an
emerging threat. Technology Review 35(5):50-62.

Burnside, O. V. 1974. Prevention and detoxification of pesticide residues
in soils. pp. 387-412. In W. D. Guenzi. (ed.) Pesticides in soil and water.
Soil Sci. Soc. of Amer., Inc. Madison, Wisconsin.

Butler, G. W. and P. J. Peterson. 1967. Uptake and metabolism of inorganic
forms of selenium-75 by Spiradela oligorrhiza. Australian J. Biol. Sci.
20:77-86.

Cansfield, P. E. and G. R. Racz. 1978. Degradation of hydrocarbon sludges
in the soil. Canadian Jour. of Soil Sci. 58:339-345.

Cardinali, A. and Z. Stoppini. 1981. Detergent polluted waters and
hydrologic characteristics of agrarian soils. II. Research on Structural
Stability. pp. 419-433. In M. R. Overcash (ed.) Decomposition of toxic and
nontoxic organic chemicals in soils. Ann Arbor Science Publ., Inc. Ann
Arbor, Michigan.

CAST. 1976. Applications of sewage sludge to cropland: appraisal of
potential hazards of the heavy metals to plants and animals. Office of
Water Program Operations, EPA. Washington, D. C. Report No. 64. EPA
420/19-76-013.

CAST. 1980. Residual effects of sewage sludge on the cadmium and zinc
content of crops. Prepared by a Task Force for the Council for Agriculture
Science Technology. CAST Report No. 83. 250 Memorial Union. Ames, Iowa.

Chaney, R. L. 1973. Crop and food chain effects of toxic elements in
sludges and effluents. pp. 129-141 In Recycling municipal sludges and
effluents on land. National Assoc. State Univ. and Land-Grant Colleges.
Washington, D.C.

Chaney, R. L. 1974. Recommendations for management of potentially toxic
elements in agriculture and municipal wastes. pp. 97-120. In Factors
involved in land application of agricultural and municipal wastes. USDA.
Beltsville, Maryland.

Chaney, R. L., D. D. Kaufman, S. B. H. Jornick, J. F. Parr, L. J. Sikora, W. D. Burge, P. B. March, G. B. Willson, and R. H. Fisher. 1981. Review of information relevant to land treatment of hazardous wastes. Solid and Hazardous Waste Research Divison, Office of Research and Development. US. EPA. Draft.

Chaney, R. L., G. S. Stoewsand, A. K. Furr, C. A. Bache, and D. J. Lisk. 1978. Elemental content of tissue of guinea pigs fed swiss chard growth on muncipal sewage sludge--amended soil. J. Agr. Food Chem. 26:994-997.

Chaney, W. R., R. C. Strickland, and R. J. Lamoreaux. 1977. Phytotoxicity of cadmium inhibited by lime. Plant and Soil 47(1):275-278.

Chao, T. T., M. E. Harward, and S. C. Fang. 1963. Cationic effects on sulfate adsorption by soils. Soil Sci. Soc. Am. Proc. 27:35-38.

Chapman, H. D. (ed.) 1966. Diagnostic criteria for plants and soils. Univ. of California, Division of Agr. Sci. Riverside, California. p. 484.

Chesnin, L. 1967. Corn, soybeans, and other great plains crops. Micronutr. Manual Farm Tech. 23(6).

Chessmore, R. A. 1979. Profitable pasture management. Interstate Printers and Publishers, Inc., Danville, Illinois. 424 p.

Chrenekova, C. 1977. Dynamics of arsenic compounds in soil Rostlinna Vyroba 23:1021-1030.

Chrenekova, E. 1973. Intensity of toxic effects of arsenate ion in relation to method of application. Pol nohospodarstvo 19:921-929.

Chumbley, C. G. 1971. Permissible levels of toxic metals in sewage used on agricultural land. A.D.A.S. Advisory Paper 10. p. 12.

Clapp, C. E., R. H. Dowdy, and W. E. Larcon. 1976. Unpublished data. Agr. Res. Ser., USDA. St. Paul, Minnesota.

Coic, Y. and C. Lesaint. 1978. Microelement levels in tomatoes and cucumbers grown in soilless culture. Supplementation with solutions of microelements essential for man. Comptes Rendus des Seances de l'Academie d'Agriculture de France 64(10):787-792.

Coombs, T. D. 1979. Cadmium in aquatic organisms. pp. 93-139. In The chemistry, biochemistry and biology of cadmium. Elsevier. New York.

Cox, R. M. and T. C. Hutchinson. 1980. Multiple metal tolerances in the grass Deschampsia cespitosa from the Sudbury smelting area. New Phytol. 84:631-647.

Cruz, A. D., H. P. Haag and J. R. Sarruge. 1967. Effect of aluminum on wheat (Triticum vulgare v. Piratini) grown in nutrient solution. Anals. Esc. Sup. Agric. Luiz Queiroz 24:107-117.

Cunningham, J. D., D. R. Kenney, and J. A. Ryan. 1975. Yield and metal composition of corn and rye grown on sewage sludge-amended soil. J. Environ. Qual. 4:448-454.

Cunningham, J. J. 1950. Copper and molybdenum in relation to diseases of cattle and sheep in New Zealand. In Copper metabolism, A Symposium on Animal, Plant, and Soil Relationships. John Hopkins Press. New York.

Davidson, J. M. and R. K. Chang. 1972. Transport of Picloram in relation to soil physical conditions and pore-water velocity. Soil Sci. Soc. Am. Proc. 36:257-261.

Davis, J. B., V. E. Farmer, R. E. Kreider, A. E. Straub, and K. M. Reese. 1972. The migration of petroleum products in soil and groundwater: principles and countermeasures. American Petroleum Institute Publication No. 4149. Washington, D.C.

Davis, R. D. 1979. Uptake of copper, nickel and zinc by crops growing in contaminated soils. J. Sci. Food Agric. 30:937-947.

Dawson, G. W., C. J. English, and S. E. Petty. 1980. Physical and chemical properties of hazardous waste constituents. Attachment 1, Appendix B in background document (RCRA subtitle C) identification and listing of hazardous waste. Office of Solid Waste, EPA. May 2, 1980.

Dean-Raymond, D. and M. Alexander. 1976. Plant uptake and leaching of dimethylnitrosamine. Nature 262:394-396.

Dejonckheere, W., W. Steurbaut, and R. H. Kips. 1981. Problems caused by quintozene and hexachlorobenzene residues in lettuce and chicory cultivator. II. pp. 3-13. In M. R. Overcash (ed.) Decomposition of toxic and nontoxic organic compounds in soils. Ann Arbor Science Publ., Inc. Ann Arbor, Michigan.

Dekkers, W. A. and F. Barbera. 1977. Effect of aggregate size on leaching of herbicides in soil columns. Water Research 17:315-39.

Delahay, P., M. Pourbaix, and P. Van Rysselberghe. 1952. Diagrammes d'equilibre potential-pH de quelques elements. Compt. Rend., Reunion Comite Thermodynam. Cinet Electtrochim 1951:15-29.

Delwiche, C. C., C. M. Johnson, and H. M. Reisenauer. 1961. Influence of cobalt on nitrogen fixation by Medicago. Plant Physiol. 36:73-78.

DeMarco, J., J. Jurbiel, J. M. Symons, and G. G. Robeck. 1967. Influence of environmental factors on the nitrogen cycle in water. J. Am. Water Works. Assn. 59(2):580-592.

Deuel, L. E. and A. R. Swoboda. 1972. Arsenic solubility in a reduced environment. Soil Sci. Soc. Am. Pro. 36:276-278.

Dibble, J. T. and R. Bartha. 1979. Effect of environmental parameters on biodegradation of oil sludge. Applied Environ. Microbiol. 37:729-738.

Dijkshoorn, W., L. W. van Broekhoven, and J. E. M. Lampe. 1979. Phytoxicity of zinc, nickel, cadmium, lead, copper, and chromium in three pasture plant species. Neth. J. Agric. Sci. 27:241-253.

dos Santos, A. B., C. Viera, E. G. Loures, J. M. Braga, and J. T. L. Thiebalt. 1979. The response at Phaseolus vulgaris beans to molybdenum and cobalt in soils from Vicosa dn Paula Candido, Minas Gerias. Revista Ceres 26(143):92-101.

Dotson, G. K., R. B. Dean, B. A. Kenner, and W. B. Cooke. 1971. Land-spreading, a conserving and non-polluting method of disposing of oily wastes. Proc. of the 5th Int. Water Pollution Conference Vol. 1, Sec. II. 36/1-36/15.

Dowdy, R. H., R. E. Larson, and E. Epstein. 1976. Sewage sludge and effluent use in agriculture. In Land application of waste materials. Soil Conservation Society of America. Ankeny, Iowa.

Downing, A. L., H. A. Painter, and C. Knowles. 1964. Nitrification in the activated sludge process. Journal and Proc. of the Inst. of Sewage Purification, Part 2.

Dyksen, J. E. and A. F. Hess, III. 1982. Alternatives for controlling organics in groundwater supplies. J. Amer. Water Works Assoc. 74(8):394-403.

Dutt, G. R., R. W. Terkeltouk, and R. S. Rauschkolb. 1972. Prediction of gypsum and leaching requirements for sodium-affected soils. Soil Science 114(2):93-103.

Eaton, F. M. 1942. Toxicity and accumulation of chloride and sulfate salts in plants. J. Agr. Res. 64:357-359.

Eaton, F. M. 1966. Chlorine. pp. 98-135. In H. D. Chapman (ed.) Diagnostic criteria for plants and soils. University of California, Riverside.

Edwards, C. A. 1973. Persistent pesticides in the environment. Critical Reviews in Environmental Control, Vol. 1, No. 1. CRC Press, Cleveland, Ohio.

Edwards, J. H., B. D. Horton, and H. C. Kirkpatrick. 1976. Aluminum toxicity symptoms in peach seedlings. J. Am. Soc. Hort. Sci. 101:139-142.

Ehinger, L. H. and G. R. Parker. 1979. Tolerance of Andopogon scoparius to copper and zinc. New Phytologist 83:175-180.

El-Bergearmi, M. M. 1973. Attempts to quantitate the protective effect of Se against mercury toxicity using Japanese quail. Fed. Proc. 32:886.

Embleton, T. W., W. W. Jones, and R. G. Platt. 1978. Tissue and soil analyses as an index of yield. University of California. Riverside, California.

Emmerich, W. E., L. J. Jund, A. L. Page, and A. C. Chang. 1982. Movement of heavy metals in sewage sludge-treated soils. J. Environ. Qual. 11:174-181.

Enterline, P. E. 1974. Respiratory cancer among chromate workers. J. Occup. Med. 16:523-526.

EPA. 1976a. National Interim Primary Drinking Water Regulations. EPA 570/9-76-003.

EPA. 1976b. Study of the distribution and fate of polychlorinated biphenyls and benzenes after a spill of transformer fluid. EPA No. 904-9-76-D14. U.S. EPA. Washington, D.C.

EPA. 1980a. Subtitle C resource and conservation recovery act of 1976. Listing of hazardous waste, background document for RCRA, section 261.31 and 261.32. Office of Solid Waste, Washington, D.C.

EPA. 1980b. Interim status standards for owners and operators of hazardous waste treatment, storage, and disposal facilities. Federal Register Vol. 45, No. 98, pp. 33154-33259.

EPA. 1982. Hazardous waste management system; permitting requirements for land disposal facilities. Federal Register Vol. 47, No. 143, pp. 32274-32388. July 26, 1982.

Ermolenko, N. F. 1972. Trace elements and colloids in soils. Jerusalem: Israel Program for Scientific Translation. National Technical Information Service. Washington, D.C.

Essington, E. H. and H. Nishita. 1966. Effect of chelates on the movement of fission products through soil columns. Plant and Soil 24:1-23.

Estes, G. O., W. E. Knoop, and F. D. Houghton. 1973. Soil-plant response to surface applied mercury. J. Environ. Qual. 2:451-452.

Evans, G. E. 1980. The mobility of water soluble organic compounds in soils from the land application of petroleum waste sludge. M.S. Thesis. Texas A&M University. College Station, Texas.

Farmer, W. J., M. Yang, J. Letey, and W. F. Spencer. 1980. Land disposal of hexachlorobenzene wastes: controlling vapor movement in soil. EPA/600-7-80-119. U.S. EPA. Cincinnati, Ohio.

Fedorak, P. M. and D. W. S. Westlake. 1981. Degradation of aromatics and saturates in crude oil by soil enrichments. Water, Air and Soil Pollution. 16:367-375.

Filonow, A. B., L. W. Jacobs, and M. M. Mortland. 1976. Fate of polybrominated biphenyls (PBB's) in soils. Retention of hexabromobiphenyl in four Michigan soils. J. Agric. Food. Chem. 24(6):1201-1204.

Findenegg, G. R. and E. Havnold. 1972. Uptake of mercury by spring wheat from different soils. Bodenkultur (Austria) 23:252-255.

Fishbein, L. 1978. Sources of synthetic mutagens. p. 257-348. In W. G. Flamm and M. A. Mehlman (ed.) Advances in modern toxicology. Mutagenesis Vol. 5.

Fleming, A. L., J. W. Schwartz, and C. D. Foy. 1974. Chemical factors controlling the adaptation of weeping lovegrass and tall fescue to acid mine spoils. Agron. J. 6(6):715-19.

Flint, R. F. and B. J. Skinner. 1977. Physical Geology. 2nd ed. John Wiley and Sons, New York. 679 p.

Florence, T. M. 1977. Trace metal species in fresh waters. Water Res. 11:681-687.

Foroughi, M., G. Hoffman, K. Teicher, and F. Venter. 1976. The effect of increasing levels of cadmium, chromium, and nickel on tomatoes in nutrient solution. Stand and Leistung Agrikulturchemischer und Agrar biologischer Forschung XXX. Kongressband (1975) Landwirt schastlicke Forschung. Sonderhest 32(1):37-48.

Fox, R. H. 1968. The effect of calcium and pH on boron uptake from high concentrations of boron by cotton and alfalfa. Soil Sci. 106:435-439.

Franco, A. A. and J. Dobereiner. 1971. Effect of Mn toxicity on soybean Rhizobium symbiosis in an acid soil. Pesquisa Agropecuaria Brasileria 6:57-66.

Franklin, W. T. Unpublished information.

Friberg, L., M. Piscator, G. Nordberg, and T. Kjellstrom. 1974. Cadmium in the environment. 2nd Ed. CRC Press, Cleveland, OH.

Frost, D. V. 1967. Arsenicals in biology--retrospect and prospect. Fed. Proc. 26(1):194-208.

Fuhr, F. and W. Mittelstaedt. 1981. The behavior of methabenzthiazuron in soil and plants after application of methabenzthiazuron [Benzone Ring-U-^{14}C] on spring wheat under field conditions during the year of application and subsequent cultivation. pp. 125-134. In M. R. Overcash (ed.) Decomposition of toxic and nontoxic organic compounds in soils. Ann Arbor Sci. Publ., Inc. Ann Arbor, Michigan.

Fujimoto, G. and G. D. Sherman. 1950. Cobalt content of typical soils and plants of the Hawaiian Islands. Agron. J. 42:577-581.

Fuller, W. H. 1977. Movement of selected metals, asbestos and cyanide in soil: applications to waste disposal problems. EPA 600/2-77-020. PB 266-905.

Furr, A. K., A. W. Lawrence, S. C. Tong, M. C. Grandolfo, R. A. Hofstader, C. A. Bache, W. H. Gutenmann, and D. J. Lisk. 1976. Multi-element and chlorinated hydrocarbon analyses of municipal sewage sludges of American cities. Envir. Sci. Technol. 10:683-687.

Furr, A. K., T. F. Parkinson, R. A. Hinrichs, D. R. VanCampen, C. A. Bache, W. H. Gutenmann, L. E. St John, I. S. Pakkala, and D. J. Lisk. 1977. National Survey of elements and radioactivity in fly ashes. Environ. Sci. Technol. 11:1194-1201

Furukawa K. and F. Matsumura. 1976. Microbial metabolism of polychlorinated biphenyls. J. Agric. Food. Chem. 24(2):251-256.

Gall, O. E. 1936. Zinc sulfate studies in the soil. Citrus Ind. 17(1):20-21.

Gamble, A. W. and D. W. Fisher. 1964. Occurrence of sulfate and nitrate in rainfall. J. Geophysical Res. 69(20):4203-4210.

Gemmell, R. P. 1972. Use of waste materials for revegetation of chromate smelter waste. Nature (London) 240:569-571.

Getty, S. M., D. E. Rickett, and A. L. Trapp. 1977. Polybrominated biphenyl (PBB) toxicosis: an environmental accident. CRC Critical Reviews in Environmental Control. 7(4):309-323.

Ghosh, B. P. and R. H. Burris. 1950. Utilization of nitrogenous compounds by plants. Soil Science 70(3):187-203.

Giashuddin, M. and A. H. Cornfield. 1978. Incubation study on effects of adding varying levels of nickel on nitrogen and carbon mineralization in soil. Environ. Poll. 15:231-234.

Gilpin, L. and A. H. Johnson. 1980. Fluorine in agricultural soils of southeastern Pennsylvania. Soil Sci. Soc. Am. J. 44(2):255-258.

Giordano, P. M. and D. A. Mays. 1977. Effect of land disposal applications of municipal wastes on crop yields and heavy metal uptake. National Fertilizer Development Center, Tennessee Valley Authority. Muscle Shoals, Alabama. EPA 600/12-77-014. PB 266-649/3BE.

Goldschmidt. V. M. 1954. Geochemistry. Clairdon Press, Oxford, United Kingdom.

Goodman, G. T. and R. P. Gemmell. 1978. The maintenance of grassland on smelter wastes in the lower Swansee Valley II. Copper smelter waste. J. Appl. Ecol. 15:875-883.

Gomi, K. and T. Oyagi. 1972. Manganese nutrition of vegetable crops. Bulletin, Faculty of Agriculture, Miyazaki University 19:493-503.

Goring, C. A., D. A. Laskowski, J. W. Hamaker, and R. W. Meikle. 1975. Principles of pesticide degradation in soil. p. 135-172. In R. Hague and V. H. Freed (eds.) Environmental dynamics of pesticides. Plenum Press, New York.

Gracey, H. I. and J. W. B. Stewart. 1977. The fate of applied mercury in soil. pp. 97-103. In Proc. Int. Conf. Land for Waste Management, Ottawa, Can. 1973. Dept. Environ./Nat. Research Council.

Gray, R. A. and G. K. Joo. 1978. Site of uptake and action of thiocarbonate herbicides and herbicide antidotes in corn seedlings. pp. 67-84. In F. M. Pallos and J. E. Casidia (eds.) Chemistry and Action of Herbicide Antidotes. Academic Press, New York.

Griffin, R. A. and E. S. K. Chian. 1980. Attenuation of water-soluble polychlorinated biphenyls by earth materials. EPA No. 600/2-80-027. U.S. EPA. Cincinnati, Ohio.

Griffin, R. A. and S. F. J. Chou. 1982. Attenuation of polybrominated biphenyls and hexachlorobenzene by earth materials. U.S. EPA. Cincinnati, Ohio. PB 82-107558.

Griffin, R. A. and N. F. Shimp. 1978. Attenuation of pollutant in municipal landfill leachate by clay minerals. EPA 600/2-78-157. PB 287-140.

Groenewegen, D. and H. Stolp. 1981. Microbial breakdown of polycyclic aromatic hydrocarbons. pp. 233-240. In M. R. Overcash (ed.) Decomposition of toxic and nontoxic organic compounds in soils. Ann Arbor Science Publ., Inc. Ann Arbor, Michigan.

Guenzi, W. D., W. E. Beard, R. A. Bowman, and S. R. Olsen. 1981. Plant uptake and growth responses from p-chlorophenyl methyl sulfide,--sulfoxide, and--sulfone in soil. J. of Env. Quality 10(4):532-536.

Gudin, C. and W. J. Syratt. 1975. Biological aspects of land rehabilitation following hydrocarbon contamination. Environ. Poll. 8(2):107-112.

Gupta, I. C. 1974. Lithium tolerance of wheat, barley, rice, and grain at germination and seedling stage. Indian J. Agr. Res. 8:103-107.

Gupta, I. C., S. V. Singh, and G. P. Bhargava. 1974. Distribution of lithium in some salt affected soil profiles. J. Indian Soc. Soil Sci. 22:88-89.

Gupta, U. C. 1979. Copper in agricultural crops. pp. 255-288. In J. O. Nriagv (ed.) Copper in the environment. John Wiley and Sons, Inc. New York.

Gutenmann, W. H., I. S. Pakkala, D. J. Churey, W. C. Kelly, and D. J. Lisk. 1979. Arsenic, boron, molybdenum and selenium in successive cuttings of forage crops field grown on fly ash amended soil. J. Agric. Food Chem. 27:1393-1395.

Haghiri, F. 1974. Plant uptake of cadmium as influenced by cation exchange capacity, organic matter, zinc and soil temperature. J. Environ. Qual. 3:180-183.

Haghiri, F. and F. L. Himes. 1974. Removal of radiostrontium by leaching runoff and plant uptake as influenced by soil and crop management practices. Dept. of Agronomy, Ohio Agricultural Research and Development Center. Wooster, Ohio. Bull. 1072, 1453.

Halter, M. 1980. Selenium toxicity to Daphnia magna, Hyallea azteca amd fathead minnow in hard water. Bull. Environ. Contam. Toxicol 24:102.

Haider, K., G. Jagnow, R. Kohnen, and S. U. Lim. 1981. Degradation of chlorinated benzenes, phenols, and cyclohexane derivatives by benzene and phenol-utilizing soil bacteria under aerobic conditions. XXIII. pp. 207-223. In M. R. Overcash (ed.) Decomposition of toxic and nontoxic organic compounds in soils. Ann Arbor Publ., Inc. Ann Arbor, Michigan.

Haider, K. and J. P. Martin. 1975. Decomposition of specifically carbon-14 labelled benzoic and cinnamic acid derivatives in soil. Soil Sci. Soc. Am. Proc. 39(4):657-662.

Hara, T., Y. Sonada, and I. Iawi. 1976. Growth response of cabbage plants to transitional elements under water culture conditions. II. Cobalt, nickel, copper, zinc, and molybdenum. Soil Sci. Plant Nutr. 22:317-325.

Hara, T., Y. Sonoda, and I. Iawi. 1977. Growth response of cabbage plants to arsenic and antimony under water culture conditions. Soil Sci. Plant Nutr. 23(2):253-255.

Hart, R. H. 1974. Crop selection and management. In Factors involved in land application of agricultural and municipal wastes. Agr. Res. Ser., USDA. Washington, D.C.

Hassett, J. J., J. E. Miller, and D. E. Koepe. 1976. Interaction of lead and cadmium on maize root growth and uptake of lead and cadmium by roots. Environ. Poll. 11(4):297-302.

Heath, M. E., D. S. Metcalfe, and R. F. Barnes. 1978. Forages: the science of grassland agriculture (3rd ed.). Iowa State Univ. Press, Ames, Iowa. 755 p.

Helling, C. S. 1971. Pesticide mobility in soil III. Influence of soil properties. Soil Sci. Soc. Am. Proc. 35(5):743-748.

Helyar, K. R. 1978. Effects of aluminum and manganese toxicities on legume growth. pp. 207–231. Mineral nutrition of legumes in tropical and subtropical soils. Proc. of a workshop held at CSIRO. Cunningham Lab. Brisbane, Australia.

Hernberg, S. 1977. Cancer and metal exposure. pp. 147–157. In H. H. Hiatt, J. D. Watson, and J. A. Winston (eds.) Origins of human cancer. Book A. Incidence of cancer in humans. Cold Spring Harbor Laboratory.

Hess, R. E. and R. W. Blanchar. 1977. Arsenic determination and As, Pb and Cu content of Missouri soils. Research Bulletin No. 1020 University of Missouri.

Hewitt, E. J. 1953. Metal interrelationships in plant nutrition. I. Effects of some metal toxicities on sugar beet, tomato, oat, potato, and marrowstem kale grown in sand culture. J. Exptl. Botany (London) 4:59–64.

Hill-Cottingham, P. G. and C. P. Lloyd-Jones. 1965. The behavior of iron chelating agents with plants. J. Exp. Botany 16:233–242.

Hinesly, T. D., R. L. Jones, and E. L. Ziegler. 1972. Effects on corn by applications of heated anaerobically digested sludge. Compost Sci. 13(4):26–30.

Hinesly, T. D., K. E. Redborg, E. L. Ziegler, and J. D. Alexander. 1982. Effect of soil cation exchange capacity on the uptake of cadmium by corn. Soil Sci. Am. J. 46:490–497.

Hock, W. K., L. R. Schreiber, and B. R. Roberts. 1970. Factors affecting uptake concentration and persistence of benomyl in american elm seedlings. Phytopathology 60(11):1619–1622.

Hodgson, J. F., W. L. Lindsay, and J. F. Trierweiler. 1966. Micronutrient cation complexing in soil solution II. Soil Sci. Soc. Am. Proc. 30:723–726.

Horak, O. 1979. Investigations on lead uptake by plants. Bodenkultur 30:120–126.

Hsu, Pa Ho. 1977. Aluminum hydroxides and oxyhydroxides. pp. 99–143. In J. B. Dixon and S. B. Weed (eds.) Minerals in the soil environment. Soil Sci. Soc. Amer., Madison, Wis.

Huddleston, R. L. 1979. Solid-waste disposal: landfarming. Chemical Engineering, Feb. 26. pp. 119–124.

Huddleston, R. L. and R. C. Allred. 1967. Surface-active agents: biodegradability of detergents. pp. 343–370. In A. D. McLaren and G. H. Peterson (eds.) Soil Biochemistry, Vol. 1. Marcel Dekker, Inc. New York, New York.

Humphrey, A. E. 1967. A critical review of hydrocarbon fermentations and their industrial utilization. Biotech. Bioeng. 9:3–24.

Hunter, J. V. and T. H. Kotalik. 1976. Chemical and biological quality of treated sewage effluents. pp. 6-25. In W. E. Sopper and L. T. Kardos (eds.) Recycling treated municipal wastewater and sludge through forest and cropland. Penn. State Univ. Press, Univ. Park, Pennsylvania.

Hunter, J. G. and O. Vergnano. 1953. Trace element toxicities in oat plants. Anal. Appl. Biol. 40:761-777.

Hurd-Karrer, A. M. 1950. Comparative fluorine uptake by plants in turned and unturned soil. Soil Sci. 70:153-159.

Hutchinson, F. E. and A. S. Hunter. 1970. Exchangeable Al levels in two soils as related to lime treatment and growth of six crop species. Agron. J. 62:702-04.

Hutton, E. M., W. T. Williams, and C. S. Andrews. 1978. Differential tolerance to manganese introduced and breed lines of Macroptilium atropurpureum. Aust. J. Agric. Res. 29:67-79.

Hutton, J. T. 1977. Titanium and zirconium minerals. pp. 673-688. In J. B. Dixon and S. B. Weed (eds.) Minerals in soil environments. Soil Sci. Soc. Am. Madison, Wisconsin.

Irukayama, K. 1966. The pollution of Minamata Bay and Minamata disease. Third International Conference on Water Pollution Research. Section 3, Paper 8. 13 p.

Ito, H. and K. Iimura. 1976. Absorption of zinc and cadmium by rice plants, and their influence on plant growth 1. Effect of zinc. J. Sci. Soil Manure. Japan. 47(2):21-24.

Iwata, Y., W. E. Westlake, and F. A. Gunther. 1973. Varying persistence of polychlorinated biphenyls in six California soils under laboratory conditions. Bull. Environ. Contam. and Toxicol. 9:204.

Jacobs, L. W., S. F. J. Chou, and J. M. Tiedje. 1976. Fate of polybrominated biphenyls (PBBs) in soils: persistence and plant uptake. J. Agric. Food Chem. 24:1198-1201.

Jacobs, L. W. and D. R. Keeney. 1970. Arsenic-phosphorus interactions of corn. Commun. Soil Sci. Pl. Analysis 1:85-93.

Jacobs, L. W., J. K. Syers, and D. R. Keeney. 1970. Arsenic adsorption by soils. Soil Sci. Soc. Am. Proc. 34:750-753.

Jain, V. K. 1978. Studies on the effects of cadmium on the growth pattern of Phaseolus aureus varieties. Journal of the Indian Botanical Society 57(Suppl.):84.

Jensen, V. 1975. Bacterial flora of soil after application of oily waste. Oikos 26(2):152-158.

Jobson, A., M. McLaughlin, F. D. Cook, and D. W. S. Westlake. 1974. Effects of amendments on the microbial utilization of oil applied to soil. Applied Microbiology 27(1):166-171.

Joham, H. E. 1953. Accumulation and distribution of molybdenum in the cotton plant. Plant Physiol. 28:275-280.

John, M. K. 1974. Wastewater renovation through soil percolation. Water, Air, and Soil Pollution 3:3-10.

John, M. K. 1977. Varietal response to lead by lettuce. Water, Air, and Soil Pollution. 8(2):133-144.

Jonasson, I. R. and R. W. Boyle. 1971. Geochemistry of mercury. pp. 5-21. In Mercury in man's environment. Proc. Roy. Soc. Can. Symp., Ottawa, Canada. February 15, 1971.

Jones, J. P. and R. L. Fox. 1978. Phosphorus nutrition of plants influenced by manganese and aluminum uptake from an Oxisol. Soil. Sci. 126(4):230-236.

Jones, L. H. P. 1957. The solubility of molybdenum in simplified systems and aqueous soil suspensions. J. Soil Sci. 8:313-327.

Jones, L. P. H., C. R. Clement, and M. J. Hopper. 1973. Lead uptake from solution by perennial ryegrass and its transport from roots to shoots. Plant and Soil 38:403-414.

Kamada, K. and K. Doki. 1977. Movement of Cr in submerged soil and growth of rice plants. 2. Effect of two different soils and the addition of Fe (II) on injury caused by Cr (IV). J. Sci. Soil Manure. (Japan). 48(11-12):457-465.

Karataglis, S. S. 1978. Studies on heavy metal tolerance in populations of Anthoxanthum odoratum. Berichte der Deutchen Botanischen Gesellschaft 91:205-216.

Kardos, L. T., W. E. Sopper, E. A. Meyers, R. R. Parizek, and J. B. Nesbitt. 1974. Renovation of secondary effluent for use as a water resource. EPA 660/2-74-016. PB 234-176/6BA.

Karickoff, S. W., D. S. Brown, and T. A. Scott. 1979. Sorption of hydrophobic pollutants on natural sediments. Water Res. 13:241-248.

Kearney, P. C. 1967. Influence of physiochemical properties of biodegradability of phenylcarbonate herbicides. J. Agr. Food Chem. 15:568-571.

Keaton, C. M. and L. T. Kardos. 1940. Oxidation-reduction potentials of arsenate-arsenite systems in sand and soil medium. Soil Sci. 50:189-207.

Keefer, F. F., R. N. Singh, D. J. Horvath, and A. R. Khawaja. 1979. Heavy metal availability to plants from sludge application. Compost Sci./Land Utilization (May/June 1979). p. 31-34.

Kelling, K. A., D. R. Keeney, L. M. Walsh, and J. A. Ryan. 1977. A field study of the agricultural use of sewage sludge. III. Effect on uptake and extractability of sludge borne metals. J. Environ. Qual. 4(6):352-358.

Kelly, J. M., G. R. Parker, and W. W. McFee. 1979. Heavy metal accumulation and growth of seedlings of five forest species as influenced by soil cadmium level. J. Environ. Qual. 8(3):361-364.

Kensuke, F., K. Tonomura, and A. Kamebayashi. 1978. Effect of chlorine sub-stitution on the biodgradability of polychlorinated biphenyls. Applied Environ. Microbiol. 35:223-227.

Kerndorff, H. and M. Schnitzer. 1980. Sorption of metals on humic acid. Geochimica at Cosmo Chimica Acta 44:1701-1708.

Kerridge, P. C., M. D. Dawson, and D. P. Moore. 1971. Separation of degrees of tolerance in wheat. Agro. J. 63(4):586-591.

Keser, M., B. F. Neubauer, and F. E. Hutchinson. 1975. Influence of aluminum ions on developmental morphology of sugarbeet roots. Agr. J. 67(1):84-88.

Keul, M., R. Andrei, G. Lazar-Keul, and R. Vintila. 1979. Accumulation and effect of lead and cadmium on wheat (Triticum vulgare) and maize (Zea mays). Studii si Cercetari de Biologie, Biologie Vegetala 31(1):49-54.

Khaleel, R., K. R. Reddy, and M. R. Overcash. 1980. Transport of potential pollutants in runoff water from land areas receiving animal waste: a review. Water Res. 14:421-436.

Kiigemagi, U., H. E. Morrison, J. E. Roberts, and W. B. Bollen. 1958. Biological and chemical studies on the decline of soil insecticides. J. Econ. Entomology 51:198-204.

Kincannon, C. B. 1972. Oily waste disposal by soil cultivation process. EPA-R2-72-110, Washington, D.C.

King. L. D. 1981. Swine manure lagoon sludge and municipal sewage sludge on growth, nitrogen recovery and heavy metal content of fescuegrass. J. Environ. Qual. 10:465-472.

Kirkham, M. B. 1979. Trace elements. pp. 571-575. In R. W. Fairbridge and C. W. Finkl, Jr. (eds). The encyclopedia of soil science. I. Dowden, Hutchinson, and Ross, Inc. Stroudsburg, Pennsylvania.

Kishk, F. M. and M. N. Hassan. 1973. Sorption and desorption of copper by and from clay minerals. Plant and Soil 39:497-505.

Kissel, D. E., S. J. Smith, W. L. Hargrove, and D. W. Dillow. 1977. Immobilization of fertilizer nitrate applied to a swelling clay soil in the field. Soil Sci. Soc. Am. J. Vol. 41:346–349.

Klausner, S. D., P. J. Zwerman, and D. R. Coote. 1976. Design parameters for the land application of dairy manure. Environmental Research Laboratory, Office of Research and Development, U.S. EPA, Athens, Georgia. EPA-600/2-76-187.

Klein, H., V. E. Jensch, and H. J. Iager. 1979. Heavy metal uptake by maize plants from soil contaminated with zinc, cadmium and copper oxide. Angewandte Botanik 53:19–30.

Klimashevsky, E. L., Y. A. Markova, M. L. Bernatzkaya, and A. S. Malysheva. 1972. Physiological responses to aluminum toxicity in the root zone of pea varieties. Agrochimia 16:487–96.

Kloke, A. and H. D. Schenke. 1979. The influence of cadmium in soil on the yield of various plant species and their cadmium content. Zeitshrift fur Pflanzenernahrung und Bodenkunde 142(2):131–136.

Kloskowski, R., I. Scheunert, W. Klein, and F. Korte. 1981. Laboratory screening of distribution, conversion, and mineralization of chemicals in the soil-plant-system and comparison to outdoor experimental data. Chemosphere 10(10):1089–1100.

Knowles, F. 1945. The poisoning of plants by zinc. Agri. Prog. 20:16–19.

Kolosov, I. I. 1962. Absorptive activity of root systems of plants. English translation by Indian Scientific Documentation Centre, New Delhi, India.

Konstantinov, G., D. Kovachev, A. Penchev, I. Ermolaev, and M. Mirchev. 1974. Effect of fertilizer application on 137-cesium accumulation in lucerne grown on a leached chernozem. Pochvoznanie: Agrokhimiya 9:34–38.

Kornegay, E. T., J. D. Hedges, D. C. Martens, and C. Y. Kramer. 1976. Effect of soil and plant mineral levels following application of manures of different copper contents. Plant and Soil 45:151–162.

Korte, N. E., J. Skogg, E. E. Niebla, and W. H. Fuller. 1975. A baseline study on tract metal elution from diverse soil types. Water, Air and Soil Pollution 5:149–156.

Kortkikh, T. A. and A. D. Repnikov. 1976. Enrichment of red clover hay with cobalt applied at different rate and by different methods. Trudy, Permskiy Sel'skokhozyaistvennyi Institue 120:25–29.

Kovda, V. A., B. N. Zolotareva, and I. T. Skripnichenko. 1979. The biological reaction of plants to heavy metals in the substrate. Doklady Akademii Nauk SSSR 247(3):766–768.

Krauskopf, K. B. 1972. Geochemistry of microutrients. pp. 7-40. In Micronutrients in agriculture. Soil Sci. Soc. Am.

Kubota, J. 1977. Molybdenum status of U.S. soils and plants. pp. 557-579. In W. R. Chappell and K. K. Peterson (eds.) Molybdenum in the environment. Vol 2. Marcel Dekker, Inc., New York.

Kubota, J., V. A. Lazar, and K. C. Beeson. 1960. The study of cobalt status of soils in Arkansas and Louisiana, using the black gum as the indicator plant. Soil Sci. Am. Proc. 24:527-528.

Kubota, J., E. R. Lemon, and W. H. Allaway. 1963. The effect of soil moisture upon uptake of molybdenum, copper and cobalt by alsike clover. Soil Sci. Soc. Am. Proc. 27:679-683.

Labanauskas, C. K. 1966. Manganese. pp. 264-285. In H. D. Chapman (ed.) Diagnostic criteria for plants and soils. University of California, Riverside, Division of Agricultural Sciences.

Lafond, A. and V. Laflamme. 1970. Relative concentrations of iron and manganese: a factor affecting jack pine regeneration and black spruce succession. pp. 305-312. In C. T. Youngberg and C. B. Davey (eds.) Proc. Third N. Am. Forest Soils. Conf. North Carolina State U. Press.

Lagerwerff, J. V. 1972. Lead, mercury, cadmium as environmental contaminants. pp. 593-636. In J. J. Mortvedt, P. M. Giordano and W. L. Lindsay (eds.) Micronutrients in agriculture. Soil Sci. Soc. Am. Madison, Wisconsin.

Lagerwerff, J. V., G. T. Strickland, R. P. Milberg, and D. L. Brower. 1977. Effects of incubation and liming on yield and heavy metal uptake by rye from sewage sludged soil. J. Environ. Qual. 6(4):427-430.

Lamoreaux, R. J., R. C. Strickland, and W. R. Chaney. 1978. The plastochron index as applied to a cadmium toxicity study. Environ. Poll. 16(4):311-317.

Land, E. and L. Johnson. 1979. Surface active agents. pp. 237-276. In Synthetic Organic Chemicals. U.S. International Trade Commission Publication No. 1001 U.S. Govt. Printing Office, Washington, D.C.

Lawrey, J. D. 1979. Boron, strontium, and barium accumulation in selected plants and loss during leaf litter decomposition in areas influenced by coal strip mining. Can. J. Bot. 57(8):933-940.

Lee, C. R. 1971. Influence of aluminum on plant growth and mineral nutrition in potatoes. Agron. J. 63(4):604-608.

Lee, H. L. 1975. Trace elements in animal production. pp. 39-54. In D. J. D. Nicholas and A. R. Egam (eds.) Trace elements in soil-plant-animal systems. Academic Press, Inc., New York.

Leeper, G. W. 1978. Managing the heavy metals on the land. Marcel Dekker, Inc., New York.

Lees, H. 1951. Isolation of nitrifying organisms from soils. Nature. 167:(335).

Letey, J., J. F. Osborn, and N. Valoras. 1975. Soil water repellency and the use of non-ionic surfactants. Contribution No. 154, California Water Research Center, Riverside, California.

Lewis, T. E. and F. E. Broadbent. 1961. Soil organic matter complexes: 4. Nature and properties of exchange sites. Soil Sci. 91:341-8.

Liebig, G. F. 1966. Arsenic. pp. 13-23. In H. D. Chapman (ed.) Diagnostic criteria for plants and soils. University of California, Riverside.

Lindsay, W. L. 1972. Inorganic phase equilibria of micronutrients in soils. pp. 41-57. In Micronutrients in Agriculture. Soil Sci. Soc. Am.

Lindsay, W. L. 1979. Chemical equilibria in soils. John Wiley and Sons, New York. 449 p.

Link, L. A. 1979. Critical pH for the expression of manganese toxicity on burley tobacco and the effect of liming on growth. Tobacco International 181:50-52.

Lisk, D. J. 1972. Trace metals in soils, plants, and animals. Adv. Agron. 24:267-325.

Loehr, R. C., W. J. Jewell, J. D. Novak, W. W. Clarkson, and G. S. Friedman. 1979a. Land application of wastes. Vol. I. Van Nostrand Reinhold Co., New York. 308 p.

Loehr, R. C., W. J. Jewell, J. D. Novak, W. W. Clarkson, and G. S. Friedman. 1979b. Land application of wastes. Vol. II. Van Nostrand Reinhold Co., New York. 431 p.

Lohris, M. P. 1960. Effect of magnesium and calcium supply on the uptake of manganese by various crop plants. Plant Soil 12:339-376.

Lott, W. L. 1938. The relation of hydrogen ion concentration to the availability of zinc in soil. Soil Sci. Soc. Am. Proc. 3:115-121.

Lucis, O. J., R. Lucis, and K. Aterman. 1972. Tumorigenesis by cadmium. Oncology 26:53-67.

Lund, L. J., A. L. Page, and C. O. Nelson. 1976. Nitrogen and phosphorus levels in soils beneath sewage disposal ponds. J. Environ. Qual. 5:26-30.

Lundegardh, H. 1927. The importance in the development of plants of the quanitites of zinc and lead added to the soil from smoke gases. Meddel. Centalanst. Forsoksv. Jordbruksomsad (Sweden) 326:14.

Luzzati, A. 1981. The effects of surfactants on plants. III. Field experiments with potatoes. pp. 389-394. In M. R. Overcash (ed.) Decomposition of toxic and non toxic compounds in soils. Ann Arbor Sci. Publ. Co., Ann Arbor, Michigan 455 p.

Lyon, T. L. and J. A. Bizzell. 1934. A comparison of several legumes with respect to nitrogen secretion. J. Am. Soc. Agron. 26:651-656.

MacLean, A. J. and A. J. Dekker. 1978. Availability of zinc, copper and nickel to plants grown in sewage-treated soils. Can. J. Soil Sci. 58:381-389.

Maftoun, M. and B. Sheibany. 1979. Effect of fluorine content of irrigation water on the growth of four plant species in relation to soil salinity. Trop. Agr. 56(3):213-218.

Malaisse, F., J. Gregoire, R. R. Brooks, R. S. Morrison, and R. D. Reeves. 1978. Aeolanthus biformifolius: a hyperaccumulator of copper. Science 199:887-888.

Malone, C. P., R. J. Miller, and D. E. Koeppe. 1978. Root growth in corn and soybeans: effects of cadmium and lead on lateral root initiation. Canadian J. Botany 56:277-281.

Mancuso, T. F. 1970. Relation of duration of employment and prior respiratory illness to respiratory cancer among beryllium workers. Environ. Res. 3:251-275.

Marinucci, A. C. and R. Bartha. 1979. Biodegradation of 1,2,3- and 1,2,4-trichlorobenzene in soil and liquid enrichment culture. Applied and Environ. Microbiol. 38(5):811-817.

Martin, J. P. 1966a. Bromine. pp. 62-64. In H. D. Chapman (ed.) Diagnostic criteria for plants and soils. Division of Agricultural Sciences, University of California, Riverside.

Martin, J. P. 1966b. Iodine. pp. 200-202. In H. D. Chapman (ed.) Diagnostic criteria for plants and soils. Division of Agricultural Sciences, University of California, Riverside.

Martin, J. P. and K. Haider. 1976. Decomposition of specifically carbon-labelled ferulic acid free and linked into model humic acid type polymers. Soil Sci. Soc. Am. Jour. 40(3):377-379.

Martin, J. P. and K. Haider. 1979. Biodegradation of ^{14}C-labelled model and cornstalk lignins, phenols, model phenolase humic polymers and fungal melanins as influenced by a readily available carbon source and soil. Applied and Environ. Microbiol. 38:(2):283-289.

Martin, J. P., G. K. Helmkomp, and J. O. Ervin. 1956. Effect of bromide from a soil fumigant and from $CaBr_2$ on the growth and chemical composition of citrus plants. Soil Sci. Soc. Am. Proc. 20:209-212.

Martin, J. P., J. J. Richards, and P. F. Pratt. 1964. Relationship of exchangeable Na percentage at different soil pH levels to hydraulic conductivity. Soil Sci. Soc. Am. Proc. 28:620-622.

Martin, R. P., P. Newbould, and R. S. Russell. 1958. Discrimination between strontium and calcium in plant and soils. Radioisotope Sci. Res. International Conference Proc. Paris, 1957. 4:173-190.

Mathers, A. C., B. A. Stewart, J. D. Thomas, and B. J. Blair. 1973. Effects of cattle feedlot manures on crop yields and soil conditions. Agr. Res. Ser. USDA. Bushland, Texas. Tech. Report 11.

McCalla, T. M. and T. J. Army. 1961. Stubble mulch farming. Adv. Agron. 13:125-196.

McCaskey, T. A., G. H. Rollins, and J. A. Little. 1971. Water quality of runoff from grassland applied with liquid, semiliquid and dairy dry waste. pp. 44-47. In Livestock waste management and pollution abatement. ASAE Proc. 271.

McKee, J. E. and H. W. Wolf. 1963. Water Quality Criteria. State Water Quality Publication, No. 3-1. California.

McKenzie, R. M. 1975. The mineralogy and chemistry of soil cobalt. pp. 83-93. In D. J. D. Nicholas and A. R. Egan (eds.) Trace Elements in soil-plant-animal systems. Academic Press, Inc., New York.

McLaren, R. G. and D. V. Crawford. 1973. Studies on soil copper II. The specific adsorption of copper by soils. J. Soil Sci. 24:443-452.

Medvedev, V. A. and V. D. Davidov. 1981. The transformation of various coke industry products in chernozem soil. pp. 245-254. In M. R. Overcash (ed.) Decomposition of toxic and nontoxic organic compounds in soils. Ann Arbor Science Publ., Inc. Ann Arbor, Michigan.

Medvedev, V. A, V. D. Davidov, and S. G. Mavrodii. 1981. Decomposition of high doses of phenol and indole by a chernozem soil. pp. 201-205. In M. R. Overcash (ed.) Decomposition of toxic and nontoxic organic compounds in soils. Ann Arbor Science Publ., Inc. Ann Arbor, Michigan.

Meijer, C. and M. A. J. Goldewaagen. 1940. A case of zinc poisoning due to galvanized from sheeting. Landbouwk. Tkidschr. 52:17-19.

Melsted, S. W. 1973. Soil-plant relationships p. 121-129. In Proc. recycling municipal sludges and effluents on land. Proceedings from conference in Champaign, Illinois. EPA, Washington, D.C.

Mengel, S. K. and E. A. Kirkby. 1978. Principles of plant nutrition. International Potash Institute, Bern.

Metcalf, R., J. R. Sanborn, P. Lu, and D. Nye. 1975. Laboratory model ecosystem studies of the degradation and fate of radiolabeled tri-, tetra-, and penta-chlorobiphenyl compared to DDE. Archives of Environ. Contam. and Toxicology. 3(2):151-165.

Miles, J. R. W., B. T. Bowman, and C. R. Harris. 1981. Adsorption, desorption, soil mobility and aqueous persistence of fensulfothion and its sulfide and sulfone metabolites. J. Env. Sci. Health B16(3):309-324.

Miles, L. J. and G. R. Parker. 1979. The effect of soil-added cadmium on several plant species. J. Environ. Qual. 8(2):229-232.

Miller, J. E., J. J. Hassett, and D. E. Koeppe. 1976. Uptake of cadmium by soybeans as influenced by soil cation exchange capacity, pH, and available phosphorus. J. Environ. Qual. 5(2):157-160.

Miller, J. E., J. J. Hassett, and D. E. Koeppe. 1977. Interaction of lead and cadmium on metal uptake and growth of corn plants. J. Environ. Qual. 6:18-20.

Millikan, C. R. 1938. A preliminary note on the relation of zinc to disease in cereals. J. Dept. Agr. Victoria. 36:409-416.

Millikan, C. R. 1946. Zinc deficiency in flax. J. Dept. Agr. Victoria. 44:69-88.

Millman, A. P. 1957. Biogeochemical investigations in areas of copper-tin mineralization in southwest England. Geochim. Commochim. Acta. 12:85-93.

Mills, H. A., A. V. Barker, and D. N. Maynard. 1974. Ammonia volatilization from soil. Agron. J. 66:355-358.

Milner, J. E. 1969. The effect of ingested arsenic on methylcholanthre induced skin tumors in mice. Arch. Environ. Health 18:7-11.

Minshall, N. E., S. A. Witzel, and M. S. Nichols. 1970. Stream enrichment from farm operations. Am. Soc. Civil Eng. J. Sanitary Eng. Div. 96(SA2):513-524.

Mistry, K. B. and B. M. Bhujbal. 1974. Effect of calcium and organic matter addition on the uptake or radiostrontium and radium by plants from Indian soils. Agrochemica 18:173-183.

Mitchell, G. A., F. T. Bingham, and A. L. Page. 1978. Yield and metal composition of lettuce and wheat grown on soils amended with sewage sludge enriched with cadmium, copper, nickel, and zinc. J. Environ. Qual. 7(2):165-171.

Miyamoto, S. 1979. Fundamentals of salt management: recent development and challenges. pp. 181-202. In Proceedings of the interamerican conference on salinity and water management technology. Texas A&M University Research Center, El Paso, Texas. December 1979.

Moein, G. J., A. J. Smith, Jr., and P. L. Stewart. 1976. Follow up study of distribution and fate of PCB's and benzenes in soil and groundwater samples after an accidental spill of transformer fluid. pp. 368-372. In Proceedings of 1976 National Conference on Control of Hazardous Material Spills. Rockville, Maryland.

Moore, A. W. 1966. Non-symbiotic nitrogen fixation in soil and soil-plant systems. Soils and Fertilizers 29:113-128.

Moore, D. P. 1972. Mechanisms of micronutrient uptake by plants. pp. 171-198. In J. J. Mortvedt, P. M. Giordano and W. L. Lindsay (eds.) Micronutrients in agriculture. Soil Sci. Soc. Am., Madison, Wis.

Morrill, L. G., B. C. Mahilum, and S. H. Mohiuddin. 1982. Organic compounds in soil: sorption, degradation and persistence. Ann Arbor Science Publ. Inc. Ann Arbor, Michigan.

Morrison, R. T. and R. N. Boyd. 1975. Organic chemistry. Allyn and Bacon, Inc., Boston.

Moucawi J., E. Fustec, P. Jambu, A. Amblfs, and R. Jacquesy. 1981. Biooxidation of added and natural hydrocarbons in soils: effects of iron. Soil Biol. Biochem. 13:335-342.

Moza, P., I. Weisgerber, and W. Klein. 1976. Fate of 2,2'-dichlorodiphenyl-^{14}C in carrots, sugarbeets and soil under outdoor conditions. J. Agric. Food Chem. 24(4):881-885.

Moza, P. N., I. Scheunert, W. Klein, and F. Korte. 1979. Long-term uptake of lower chlorinated biphenyls and their conversion products by spruce trees from soil treated with sewage sludge. Chemosphere 6:373-375.

Mrozek, E. Jr., E. D. Seneca and L. L. Hobbs. 1982. Polychlorinated biphenyl uptake and translation by Spartina Alterniflora Loisel. Water, Air and Soil Pollution. 17:3-15.

Mukherji, S. and B. K. Roy. 1977. Toxic effects of chromium on germinating seedlings and potato tuber slices. Biochemie and Physiologie der Pfanzen. 171(3):235-238.

Mulder, E. G. 1954. Mo in relation to growth of higher plants and microorganisms. Plant and Soil 5:368-415.

Murphy, L. S. and L. M. Walsh. 1972. Correction of micronutrient deficiencies with fertilizers. pp. 347-381. In J. J. Mortvedt, P. M. Giordano and W. L. Lindsay (eds.) Micronutrients in Agriculture. Soil Sci. Soc. Am., Madison, Wis.

Murrmann, R. P. and F. R. Kountz. 1972. Role of soil chemical processes in reclamation of wastewater applied to land. pp. 48-74. In Wastewater management by disposal on land. Spec. Rept. No. 171. U.S. Army Cold Regions Research and Engineering Lab, Hanover, New Hampshire.

Nathwani, J. S. and C. R. Phillips. 1977. Adsorption-desorption of selected hydrocarbons in crude oil on soils. Chemosphere 6(4):157-162.

Nathwani, J. S. and C. R. Phillips. 1978. Rates of leaching for radium from contaminated soils: an experimental investigation of radium bearing soils from Port Hope, Ontario. Water, Air and Soil Pollution 9:453-465.

National Academy of Sciences. 1980. Mineral tolerance of domestic animals. By Subcommitte on Animal Nutrition, Board on Agriculture and Renewable Resources. Commission on Natural Resources., National Research Council. Washington, D.C.

National Academy of Sciences and National Academy of Engineering. 1972. Water quality and criteria. A report of the Committee on Water Quality Criteria, Environmental Studies Board. EPA R3-73-033. March, 1973. PB 236-199/6BA.

National Research Council. 1973. Medical and biological effects of environmental pollutants, manganese. National Academy of Sciences. Washington, D.C. p. 1-191.

Newton, H. P. and S. J. Toth. 1952. Response of crop plants to I and Br. Soil Sci. 72:127-134.

Nissen, P. 1974. Uptake mechanisms: inorganic and organic. Ann. Rev. of Plant Physiology 25:53-79.

Nissen, T. V. 1981. Stability of PCB in the soil. VIII. pp. 79-87. In M. R. Overcash (ed.) Decomposition of toxic and nontoxic organic compounds in soils. Ann Arbor Science Publ., Inc. Ann Arbor, Michigan.

Nommik, H. 1965. Ammonium fixation and other reactions involving a nonenzymatic immobilization of mineral nitrogen in soil. pp. 198-258. In Soil Nitrogen. Am. Soc. Agron. Madison, Wisconsin.

Norseth, T. 1977. Industrial viewpoints on cancer caused by metals as an occupational disease. pp. 159-167. In H. H. Hiatt, J. D. Watson, and J. A. Winsten (eds.) Origins of human cancer. Book A. Incidence of cancer in humans. Colo. Spring Harbor Laboratory.

O'Donovan, J. T. and W. H. Vanden Born. 1981. A microauto radiographic study of ^{14}C-labelled picloram distribution in soybean following uptake. Canadian Jour. of Botany 59(19):1928-1931.

Olsen, S. R. 1972. Micronutrient interactions. p. 243-264. In Micronutrients in agriculture Ed. Soil Sci. Soc. Amer. Inc., Madison, Wisconsin.

Otsuka, K. and T. Morizaki. 1969. Aluminum and manganese toxicities of plants. J. Sci. Soil and Manure (Japan) 40:250-254.

Overcash, M. R. and D. Pal. 1979. Design of land treatment systems for industrial wastes--theory and practice. Ann Arbor Science. Ann Arbor, Michigan.

Page, A. L. 1974. Fate and effects of trace elements in sewage sludge when applied to agricultural lands. National Environ. Res. Center. Cincinnati, Ohio. EPA 670/2-74-005. PB 231-171/0BA.

Page, A. L., F. T. Bingham, and C. Nelson. 1972. Cadmium adsorption and growth of various plant species as influenced by solution cadmium concentration. J. Environ. Qual. 1:288-291.

Paivoke, A. 1979. The effects of lead and arsenate on the growth and acid phosphatase activity of pea seedlings. Annales Botanici Fennici 16:18-27.

Patterson, J. B. E. 1971. Metal toxicities arising from industry. pp. 193-207. In Trace elements in soils and crops. Great Brit. Ministry of Agric., Fisheries, and Food Tech. Bull. 21. Her Majesty's Stationery Office, London.

Pearson, G. A. 1960. Tolerance of crops to exchangeable sodium. USDA Agriculture Information Bulletin No. 216. Washington, D.C.

Perlstein, M. A. and R. Attala. 1966. Neurological sequelae of plumbism in children. Clin. Pediatr. 5:292.

Perry, H. M. and A. Yunice. 1965. Acute pressor effects of intraarterial cadmium and mercuric ions in anesthesized rats. Proc. Soc. Exp. Biol. Med. 120:805.

Perry, J. J. and C. E. Cerniglia. 1973. Studies on the degradation of petroleum hydrocarbons by filamentous fungi. In D. G. Ahearn and S. P. Meyers (eds.) Microbial degradation of oil pollutants. Center for Wetland Resources, Louisiana State University, Baton Rouge, Louisiana.

Peterson P. J. and B. J. Alloway. 1979. Cadmium in soils and vegetation. pp. 45-92. In M. Webb (ed.) the chemistry, biochemistry and biology of cadmium. Elsevier/North Holland Biomedical Press. Amsterdam.

Peterson, P. J. and G. W. Butler. 1962. The uptake and assimilation of selenite by higher plants. Aust. J. Biol. Sci. 15:126-146.

Peterson, P. J., M. A. S. Burton, M. Gregson, S. M. Nye, and E. K. Porter. 1976. Tin in plants and surface waters in Malaysian ecosystems. Trace Subst. Environ. Health 10:123-132.

Petrilli, F. L. and S. DeFlora. 1977. Toxicity and mutagenicity of hexavalent chromium on Salmonella typhimurium. Appl. Environ. Microbiol. 33:805-809.

Pinkerton, B. W. 1982. Plant accumulation of cobalt from soils with artificially elevated cobalt levels. Ph.D. Dissertation. Texas A&M Univ., Univ. Microfilms. College Station, Texas.

Plewa, M. J. 1978. Activiation of chemicals into mutagens by green plants: a preliminary discussion. Environ. Health Perspec. 27:45-50.

Plice, M. J. 1948. Some effects of crude petroleum on soil fertility. Soil Science Soc. of Am. Proc. 13:413-416.

Pound, C. E. and R. W. Curtis. 1973. Characteristics of municipal effluents. pp. 49-61. In Recycling municipal sludges and effluents on land. Nat. Assoc. State Univ. and Land Grant Colleges. Washington, D.C.

Pratt, P. F. 1966a. Aluminum. pp. 3-12. In H. D .Chapman (ed.) Diagnostic criteria for plants and soils. University of California, Riverside.

Pratt, P. F. 1966b. Chromium. pp. 136-141. In H. D. Chapman (ed.) Diagnostic criteria for plants and soils. University of California, Riverside.

Pratt, P. F. 1966c. Titanium. pp. 448-449. In H. D. Chapman (ed.) Diagnostic criteria for plants and soils. University of California, Riverside.

Pratt, P. F. 1966d. Vanadium. pp. 480-483. In H. D. Chapman (ed.) Diagnostic criteria for plants and soils. University of California, Riverside.

Pratt, P. F. 1966e. Zirconium. pp. 400. In H. D. Chapman, Diagnostic criteria for plants and soils. University of California, Riverside.

Pratt, P. F., F. E. Broadbent, and J. P. Martin. 1973. Using organic wastes as nitrogen fertilizers. Calif. Agr. 27(6):10-13.

Pratt, P. F. and S. Davis. Unpublished data. University of California and USDA-ARS, Riverside, California.

Prausse, A., K. Schmidt, and W. Bergmann. 1972. Excess Mn in Thuringian acid soils and its effect on the yield and quality of potatoes and spring barley. Archiv fur Bodenfruchtbarkeit und Pflanzen produktion 16:483-494.

Purvis, D. and E. J. MacKenzie. 1973. Effects of application of municipal compost on uptake of copper, zinc, and boron by garden vegetables. Plant and Soil 35:361-371.

Quimby, P. C., Jr., K. E. Frick, R. D. Wauchope, and S. H. Kay. 1979. Effects of cadmium on two biocontrol insects and their host weeds. Bull. Environ. Contam. Toxicol. 22(3):371-378.

Raghi-Atri, F. 1978. Influence of heavy metals in the substate on Glyceria maxima. Blei Angewandte Botanik 52:185-192.

Rains, D. W., W. E. Schmid, and R. Epstein. 1964. Adsorption of cations by roots. Effects of hydrogen ions and essential role of calcium. Plant Phsyiol. 39:274-278.

Rauser, W. E. 1978. Early effects of phytotoxic burdens of cadmium, cobalt, nickel, and zinc in white beans. Can. J. Bot. 56:1744-1749.

Raymond, R. L., J. P. Hudson, and V. W. Jamison. 1976. Oil degradation in soil. Applied Environ. Microbiol. 31(4):522-535.

Reddy, K. R., R. Khaleel, M. R. Overcash, and P. W. Westerman. 1979. Phosphorous – a potential nonpoint source pollution problem in the land areas receiving long-term application of wastes. In Best management practices for agriculture and silviculture. Ann Arbor Science, Inc., Ann Arbor, Michigan.

Reeves, A. L. and A. J. Vorwald. 1967. Beryllium carcinogenesis. II. Pulmonary deposition and clearance of inhaled beryllium sulfate in the rat. Cancer Res. 27:446-451.

Rehab, F. I. and A. Wallace. 1978a. Excess trace element effects on cotton. 2. Copper, zinc, cobalt and manganese in Yolo loam soil. Comm. Soil Sci. Plant Anal. 9:519-527.

Rehab, F. I. and A. Wallace. 1978b. Excess trace metal effect on cotton: 3. Chromium and lithium in solution culture. Comm. Soil Sci. Plant Anal. 9:637-644.

Rehab, F. I. and A. Wallace. 1978c. Excess trace metal effects on cotton: 4. Chromium and lithium in Yolo loam soil. Comm. Soil Sci. Plant Anal. 9:645-651.

Rehab, F. I. and A. Wallace. 1978d. Excess trace metal effects on cotton: 5. Nickel and cadmium in solution culture. Comm. Soil Sci. Plant Anal. 9:771-778.

Rehab, F. I. and A. Wallace. 1978e. Excess trace metal effects on cotton: 6. Nickel and cadmium in Yolo loam soil. Comm. Soil Sci. Plant Anal. 9:779-784.

Reimers, R. S. and P. A. Krenkel. 1974. Sediment sorption phenomena. CRC Critical Rev. Environ. Control 4:265.

Reisenauer, H. M. 1960. Cobalt in nitrogen fixation in a legume. Nature 186:375-376.

Reisenauer, H. M., A. A. Tabikh, and P. R. Stout. 1962. Molybdenum reactions with soils and the hydrous oxides of iron, aluminum and titanium. Soil Sci. Soc. Am. Proc. 26:23-27.

Rhodes, J. D. 1974. Drainage for salinity control. In J. Van Schilfgaarde (ed.) Drainage for agriculture. ASA No. 17 Am. Soc. of Agron. Madison, Wisconsin.

Richards, F. J. 1941. Physiological studies in plant nutrient. Ann. Bot. (N.S.) 5:263–296.

Rickles, E. L., N. G. Brink, F. R. Koniuszy, I. R. Wood, and K. Folkers. 1948. Crystalline vitamin B_{12}. Science 107:396–397.

Roberts, P. V. and A. J. Valocchi. 1981. Principles of Organic Contaminant Behavior During Artificial Recharge. The Science of the Total Environment 21:161–172.

Robinson, W. O., H. W. Lakin, and L. E. Reichen. 1947. Zinc content of plants on Friedensville zinc slime ponds in relation to biochemical prospecting. Econ. Geol. 42:572–583.

Rolfe, G. T. and F. A. Bazzar. 1975. Effect of lead contamination on transpiration and photosynthesis of loblolly pine and autumn olive. Forest Sci. 21:33–35.

Romney, E. M. and J. D. Childress. 1965. Effects of beryllium in plants and soil. Soil Sci. 100:210–217.

Romney, E. M., A. Wallace, and G. V. Alexander. 1975. Response of bush bean and barley to tin applied to soil and to solution culture. Plant and Soil 42:585–589.

Ross, D. S., R. E. Sjogren, and R. J. Bartlett. 1981. Behavior of chromium in soils. IV Toxicity to microorganisms. J. Environ. Qual. 10:145–148.

Rouchaud, J. and J. A. Meyer. 1982. New trends in the studies about the metabolism of pesticides in plants. Residue Reviews 82:1–35.

Rovira, A. D. and C. B. Davey. 1971. Biology of the rhizosphere. In E. W. Carson (ed.) The plant root and its environment. University Press of Virginia, Charlottesville, Virginia.

Rudolfs, W. (ed.). 1950. Review of literature on toxic materials affecting sewage treatment processes, streams and B.O.D. determinations. Sew. Ind. Wastes 22:1157–1191.

Russell, E. W. 1973. Soil conditions and plant growth, 10th ed. Longman Group Limited, London. 849 p.

Russell, R. S. and K. A. Smith. 1966. Naturally occurring radioactive substances: the uranium and thorium series. In R. S. Russell and K. A. Smith (eds.) Radioactivity and human diet. Pergamon Press, Oxford.

Russo, F. and E. Brennan. 1979. Phytotoxicity and distribution of cadmium in pin oak seedlings determined by mode of entry. Forest Science 25(2):328-332.

Ryan, J. A. and D. R. Keeney. 1975. Ammonia volatilization from surface applied waste water sludge. J. Water Poll. Control Fed. 47:386-393.

Ryden, J. C. and P. F. Pratt. 1980. Phosphorus removal from wastewater applied to land. Hilgardia 48:1-36.

Ryden, J. C. and J. K. Syers. 1975. Use of tephra for the removal of dissolved inorganic phosphate from sewage effluent. New Zeal. J. Sci. 18:3-16.

Saito, Y. and K. Takahashi. 1975. Studies in heavy metal pollution in agricultural land. 4. Cadmium adsorption and translocation in rice seedlings as affected by the nutritional status and co-existence of the other heavy metal. Bull. Shikoku Agr. Exp. Sta. No. 31:111-131.

Sauchelli, V. 1969. Trace elements in agriculture. Van Nostrand Reinhold Co., New York. 248 p.

Sawhney, B. L. and D. E. Hill. 1975. Phosphate sorption characteristics of soils treated with domestic waste water. J. Environ. Qual. 4:342-346.

Sax, N. I. 1979. Dangerous properties of industrial materials. Van Nostrand Reinhold Co., New York.

Schaeffer, C. C., A. M. Decker, and R. L. Chaney. 1975. Effects of soil temperature and sludge application on the heavy metal content of corn. Agron. Abstr. (In Press).

Schaeffer, C. C., A. M. Decker, R. L. Chaney, and L. W. Douglass. 1979. Soil temperature and sewage sludge effects on metals in crop tissue and soils. J. Environ. Qual. 8(4):455-459.

Scharpenseel, H. W., B. K. G. Theng and S. Stephan. 1978. Polychlorinated biphenyls (^{14}C) in soils: adsorption, infiltration, translocation and decomposition. Chapter 50. In W. E. Krumbein (ed.) Environmental biogeochemistry and geomicrobiology, Vol. 2: The terrestrial environment. Ann Arbor Science Publ. Inc., Ann Arbor, Michigan.

Schneider, P. W. 1976. The chemistry and biological nitrogen fixation. pp. 197-204. In H. Sigel (ed.) Metal ions in biological systems. John Wiley and Sons, N.Y.

Schwendinger, R. B. 1968. Reclamation of soil contaminated with oil. J. Institute Petrol. 54:182-192.

Scott, M. L. 1972. Trace elements in animal nutrient. pp. 555-592. In J. J. Mortvedt, P. M. Giordano, and W. L. Lindsay (eds.) Micronutrients in agriculture. Soil Sci. Soc. Am. Inc., Madison, Wisconsin.

Scott, H. D., D. C. Wolf, and T. L. Lavy. 1982. Apparent adsorption and microbial degradation of phenol by soil. J. Env. Qual. 11(1):107-112.

Sebastiani, L. A., A. Borgioli, and A. D. Simonetti. 1981. Behavior of synthetic detergents in the soil. Movement of hard detergents in the soil. pp. 333-347. In M. R. Overcash (ed.) Decomposition of toxic and nontoxic organic chemicals in soils.

Sekerka, V. 1977. Influence of copper on growth and mitotic activity on the cells in the apical meristems of the horse bean (Vicia faba L.). Acta Facultis Rerum Naterallium Universitis Comenianae, Physiologia Plantarum 14:1-16.

Sherman, G. D. 1952. The titanium content of Hawaiian soils and its significance. Soil Sci. Soc. Am. Proc. 16:15-18.

Shone, M. G. T., D. T. Clarkson, J. Sanderson, and A. V. Wood. 1972. A comparison of the uptake of some organic molecules and ions in higher plants. pp. 571-583. In W. P. Anderson (ed.) Ion Transport in Plants. Academic Press, New York.

Shukla, V. C. and K. G. Prasad. 1973. Forms and distribution of lithium, boron, and fluorine in some chernozem soils of Haryana. Indian J. Hort. Sci. 43:934-937.

Siegel, F. R. 1974. Applied geochemistry. Wiley & Sons, New York.

Silva, S. and B. Beghi. 1979. The use of chromium containing organic manures in rice fields. Riso 28(2):105-113.

Simola, L. K. 1977. The tolerance of Sphagnum fimbriatum towards lead and cadmium. Annales Botanici Fennia 14:1-5.

Simon, A., F. W. Cradock, and A. W. Hudson. 1974. The development of manganese toxicity in pasture legumes under extreme climatic conditions. Plant and Soil 41:129-140.

Singh, M. and N. Singh. 1978. Selenium toxicity on plants and its detoxification by phosphorus. Soil Sci. 125(5):255-262.

Smelt, J. H. 1981. Behavior of quintozene and hexachlorobenzene in the soil and their absorption in crops. pp.3-14. In M. R. Overcash (ed.) Decomposition of toxic and nontoxic organic compounds in soils. Ann Arbor Science Publ., Inc., Ann Arbor, Michigan.

Smith, C. 1982. The influence of anions on soil sorption and plant uptake of molybdenum. M.S. Thesis. Texas A&M Univ., Univ. Microfilms, College Station, Texas.

Smith, I. C., B. L. Carson, and T. L. Ferguson. 1978. Trace metals in the environment, vol. 4 - palladium and osmium. Ann Arbor Sci. Publ. Ann Arbor, Michigan.

Sparling, G. P., B. G. Ord, and D. Vaughan. 1981. Changes in microbial biomass and activity in soils amended with phenolic acids. Soil Biol. Biochem. 13:455-460.

Spencer, W. F. and W. J. Farmer. 1980. Assessment of the vapor behavior of toxic organic chemicals. pp. 143-161. In Rizwanul Haque (ed.) Dynamics, exposure, and hazard assessment of toxic chemicals. Ann Arbor Science Publ., Inc. Ann Arbor, Michigan.

Soane, B. D. and D. H. Saunders. 1959. Nickel and chromium toxicity of serpentine soils in southern Rhodesia. Soil Sci. 88:322-330.

Sommers, L. E. 1977. Chemical composition of sewages and analysis of their potential use as fertilizers. J. Environ. Qual. 6(2):225-232.

Sommers, L. E. and D. W. Nelson. 1976. Analyses and their interpretation for sludge application to agricultural land. pp. 3.1-3.7. In B. D. Knezek and R. H. Miller (eds.) Application of sludges and wastewaters on agricultural land; a planning educational guide. N. Ctr. Reg. Res. Pub. 235. Ohio Agr. Res. Division Ctr. Reg. Bull. 1090. Wooster, Ohio.

Sorteberg, A. 1978. Effects of some heavy metals on oats in pot experiments with three different soil types. J. Sci. Ag. Soc. Finland 50:317-334.

Spencer, E. L. 1937. Renching of tobacco and thallium toxicity. Am. J. Bot. 24:16-24.

Stace, H. C. T., G. D. Hubble, R. Brewer, K. H. Northcote, J. R. Sleeman, M. J. Mulcahy, and E. G. Hallsworth. 1968. A handbook of Australian soils. Rellim Technical Publications. Adelaide, South Australia.

Staker, E. V. 1942. Progress report on the control of Zn toxicity in peat soils. Soil Sci. Soc. Am. Proc. 7:387-392.

Steevens, D. R., L. M. Walsh, and D. R. Keeney. 1972. Arsenic residues in soil and potatoes from Wisconsin potato fields. Pesticides Monitoring J. 6:89-90.

Stelmach, Z. 1958. Bromine retention in some soils and uptake of bromine by plants after soil fumigation. Soil Sci. 88:61-66.

Stewart, D. K. R. and K. G. Cairns. 1974. Endosulfan persistence in soil and uptake by potato tubers. J. Agr. Food Chem., Vol. 22. pp. 984-986.

Stipes, R. J. and D. R. Oderwald. 1971. Dutch elm disease: control with soil-injected fungicides and surfactants. (Abstract) Phytopathology 61(8): 913.

Stout, P. R. and W. R. Meagher. 1948. Studies of Mo nutrition of plants with radioactive molybdenum. Science 108:471-473.

Stout, P. R., W. R. Meagher, G. A. Pearson, and C. M. Johnson. 1951. Molybdenum nutrition in croplands. Plant and Soil 3:51-87.

Stover, R. L., L. E. Sommers, and D. T. Silviera. 1976. Evaluation of metals in wastewater sludge. J. Water Poll. Control Fed. 48:2165-2175.

Street, J. J., W. L. Lindsay and B. R. Sabey. 1977. J. Environ. Qual. 6:72-77.

Stucky, D. J. and T. S. Newman. 1977. Effect of dried anaerobically digested sewage sludge on yield and element accumulation in tall fescue and alfalfa. J. Environ. Qual. 6(3):271-273.

Subbarao, Y. V. and R. Ellis. 1977. Determination of kinetics of phosphorus mineralizaiton in soils under oxidizing conditions. Ada, Oklahoma. EPA 600/2-77-180. PB 272-594/3BE.

Sunderman, F. W. 1977. Metal carcinogenesis. pp. 257-296. In R. A. Goyer and M. A. Mehlman (eds.) Toxicology of trace elements. John Wiley and Sons, New York.

Sunderman, F. W. and A. J. Donnelly. 1965. Studies of nickel carcinogenesis. Metastasizing pulmonary tumors in rats induced by inhalation of nickel carbonyl. Amer. J. Path. 46:1027-1041.

Sunderman, F. W. and E. Mastromatteo. 1975. Nickel carcinogenesis. In F. W. Sunderman, F. Coulston, G. L. Eichorn, J. A. Fellows, E. Mastromatteo, H. T. Reno, and M. H. Sanuitz (eds.) Nickel. National Academy of Sciences, Washington, D.C.

Sung, M. W. and H. J. Young. 1977. Effects of various anions on absorption and toxicity of lead in plants. Korean J. Botany 20:7-14.

Swaine, D. J. 1955. Trace element content of soils. Commonwealth Bur. Soil Sci. Tech. Comm. No. 48. Herald Printing, York, England.

Takkar, P. N. and M. S. Mann. 1978. Toxic levels of soil and plant zinc for maize and wheat. Plant and Soil 49(3):667-669.

Taskayev, A. I., V. Y. Ovchevov, R. M. Neksakhin, and I. I. Shok Tomora. 1977. Uptake of Ra^{226} by plants and changes in its state in the soil-plant top-litterfall system. Soviet Soil Sci. 9:79-85.

Taylor, S. A. and G. L. Ashcroft. 1972. Physical edaphology. W. H. Freeman and Co., San Francisco, California.

Teakle, L. J. H. and I. Thomas. 1939. Recent experiments with 'minor' elements in Western Australia. IV. The effect of 'minor' elements on growth of wheat in other parts of the state. J. Dept. Agr. W. Australia 16:143-147.

Tepper, L. B. 1972. Beryllium. CRC critial reviews in toxicology 1:235.

Tiller, K. G. and J. R. Hodgson. 1960. The specific sorption of cobalt and zinc by layer silicates. Clays and Clay Min. 11:393-403.

Tisdale, S. L. and W. L. Nelson. 1975. Soil fertility and fertilizers. 3rd ed. MacMillan Publ. Co., New York.

Tofflemire, T. J. and M. Chen. 1977. Phosphate removal by sands and soils. p. 154. In R. C. Loehr (ed.) Land as a waste management alternative. Proc. of the 1976 Cornell Agricultural Waste Management Conf. Ann Arbor Science Publ. Inc., Ann Arbor, Michigan.

Toivonen, P. M. A. and G. Hofstra. 1979. The interaction of copper and sulfur dioxide in plant injury. Canad. J. Plant. Sci. 59:475-479.

Tokuoka, M. and S. Gyo. 1940. The effect of zinc on the growth of wheat. J. Sci. Soil Manure, Nippon 14:587-596.

Tomson, M. B., J. Dauchy, S. Hutchins, C. Curran, C. J. Cook, and C. H. Ward. 1981. Groundwater contamination by trace level organics from a rapid infiltration site. Water Research 15:1109-1116.

Travis, C. C. and E. L. Etnier. 1981. A survey of sorption relationships for reactive solutes in soil. J. Environ. Qual. 10:8-17.

Trelease, S. F. and O. A. Beath. 1949. Selenium. Published by authors. New York.

Tsuchiya, K. 1978. Cadmium studies in Japan: A review. Elsevier/North-Holland Biomedical Press, N.Y.

Turner, M. A. and R. H. Rust. 1971. Effects of chromium on growth and mineral nutrition of soya beans. Soil Sci. Soc. Am. Proc. 35:755-758.

Turner, W. W. (ed.) 1965. Drugs and poisons. Jurisprudence Publishers, Inc. Rochester, New York.

Underwood, E. J. 1971. Trace elements in human and animal nutrition. 3rd ed. Academic Press, New York.

USDA. 1954. Diagnosis and improvement of saline and alkali soils. L. A. Richards (ed.) USDA Handbook No. 60. Washington, D.C.

Vallentine, J. F. 1971. Range development and improvements. Brigham Young Univ. Press. Provo, Utah. 516 p.

Valoras N., J. Letey, J. P. Martin, and J. Osborn. 1976. Degradation of a nonionic surfactant in soils and peat. Soil Sci. Soc. Am. Jour. 40(1):60-63.

Van Beekom, C. W. C., C. van den Berg, T. A. deBoer, W. H. van den Mulen, B. Verhoeven, J. J. Westerhof, and A. J. Zuvr. 1953. Reclaiming land flooded with salt water. Neth. J. Agr. Sci. 1(3):153-163, 1(4):225-244.

Van Dam, J. G. C. 1955 Examination of soils and crops after the innundation of February 1953. Neth. J. Agr. Sci. 2:242-253.

Vandoni M. V. and F. L. Goldberg. 1981. Synthetic detergents behavior in agricultural soil. Chapter 41. In M. R. Overcash (ed.) Decomposition of toxic and nontoxic organic compounds in soils. Ann Arbor Science Publ., Inc. Ann Arbor, Michigan.

Van Loon, J. C. 1974. Mercury contamination of vegetation due to the application of sewage sludge as a fertilizer. Environ. Letters 6:211-218.

Vanselow, A. P. 1966a. Barium pp. 142-156. In H. D. Chapman (ed.) Diagnostic criteria for plants and soils. University of California, Riverside.

Vanselow, A. P. 1966b. Cobalt. pp. 142-156. In H. D. Chapman (ed.) Diagnostic criteria for plants and soils. University of California, Riverside.

Vanselow, A. P. 1966c. Nickel. pp. 302-309. In H. D. Chapman (ed.) Diagnostic criteria for plants and soils. University of California, Riverside.

Vanselow, A. P. 1966d. Strontium. p. 763. In H. D. Chapman (ed.) Diagnostic criteria for plants and soils. University of California, Riverside.

Velly, J. 1974. Observations on the acidification of some soils in Madagascar. Agronomie Tropicale 29(12):1248-1262.

Vlek, P. L. G. and W. L. Lindsay. 1977. Molybdenum contamination in Colorado pastures. In W. R. Chappell and K. K. Peterson (eds.) Molybdenum in the environment. Marcel Dekker, Inc., New York.

Voelcker, J. A. 1913. Pot culture experiments. J. Royal Agr. Soc. (England) 74:411-422.

Vollanweider, R. H. 1968. Scientific fundamentals of the eutrophication of lakes and flowing waters, with particular reference to nitrogen and phosphorus as factors in eutrophication. OED, DAS/CS 1-68-27. p. 159.

Vostal, J. J., E. Taves, J. W. Sayre, and E. Charney. 1974. Lead analysis of house dust: a method for the detection of another source of lead exposure in inner city children. Environ. Health Perpect. 7:91-97.

Wainwright, S. J. and H. W. Woolhouse. 1975. Physiological mechanisms of heavy metal tolerance in plants. pp. 231-257. In M. J. Chadwick and G. T. Goodman (eds.). The ecology of resource degradation and renewal. Blackwell, Oxford.

Walker, A. 1971. Effects of soil moisture content on the availability of soil applied herbicides to plants. Pesticide Science 2:56-59.

Walker, T. W. 1956. Fate of labeled nitrate and ammonium nitrogen when applied to grass and clover grown separately and together. Soil Science 81:339-352.

Wallace, A., E. M. Romney, J. W. Cha, and F. M. Chaudhry. 1977. Lithium toxicity in plants. Comm. Soil Sci. Plant Anal. 8:773-780.

Wallace, A., E. M. Romney, R. T. Mueller, Sr., and S. M. Soufi. 1981. Plant uptake and transport of ^{241}Am. Soil Science 132(1):114-119.

Wallace, A., S. M. Soufi, J. W. Cha, and E. M. Romney. 1976. Some effects of Cr toxicity on bush bean plants grown in soil. Plant and Soil. 2(44):471-473.

Wallihan, E. F., R. G. Sharpless, and W. L. Printy. 1978. Cumulative toxic effects of boron, lithium, and sodium in water used for hydroponic production of tomatoes. J. Am. Soc. Hort. Sci. 103:14-16.

Wallnofer, P., M. Koniger, and G. Engelhardt. 1981. The behavior of xenobiotic chlorinated hydrocarbons (HCB and PCBs) in cultivated plants. XI. pp. 99-109. In M. R. Overcash (ed.) Decomposition of toxic and non-toxic organic compounds in soils. Ann Arbor Science Publ., Inc. Ann Arbor, Michigan.

Walsh, L. M., M. E. Sumner, and R. B. Corey. 1976. Consideration of soils for accepting plant nutrients and potentially toxic nonessential elements. pp. 22-45. In Land application of waste materials. Soil Conser. Soc. Am.

Walsh, T., M. Neenan, and L. B. O'Moore. 1953. High Mo levels in herbage on acid soils. Nature 171:1120.

Walters, L. T. and T. J. Wolery. 1974. Transfer of heavy metals pollutants from Lake Erie bottom sediment to the overlying water. Ohio State Univ. Columbus, Ohio.

Warren, H. V. 1972. Biogeochemistry in Canada. Endeavor 31:46-49.

Watanbe, T. and H. Nakamura. 1972. Lead absorbed by eggplant from soil and its translocation. Agricultural Chemical Inspection Sta. Bull. No. 12:105-106.

Watkinson, J. H. and G. M. Dixon. 1979. Effect of applied selenate on ryegrass and on larvae of soldier fly, Inopus rubriceps Macquart. New Zealand J. Exptl. Agr. 7(3):321-325.

Weaver, C. M., N. D. Harris, and L. R. Fox. 1981. Accumulation of Sr and Cs by kale as a function of age of plant. J. Environ. Qual. 10-95-98.

Weber, J. B. and E. Mrozek, Jr. 1979. Polychlorinated biphenyls: phytotoxicity, absorption and translocation by plants and inactivation by activated carbon. Bull. Environ. Contam. Toxicol. 23:412-417.

Wentinik, G. R. and J. E. Etzel. 1972. Removal of metal ions by soil. J. Water Poll. Control Fed. 44:1561-1574.

Wetherold, R. G., D. D. Rosebrook, and E. W. Cunningham. 1981. Assessment of hydrocarbon emissions from land treatment of oily sludges. pp. 213-233. In David Shultz (ed.) Proceedings of the Seventh Annual Research Symposium at Philadelphia, Pennsylvania. EPA-600-9-81-002b.

White, M. C. and R. L. Chaney. 1980. Zinc, cadmium and manganese uptake by soybean from two zinc and cadmium-amended coastal plains soils. Soil Sci. Soc. Am. J. 44:308-313.

Whitehead, D. C. 1975. Uptake by perennial ryegrass of iodide, elemental iodine, and iodate added to soil as influenced by various amendments. J. Sci. Food Agric. 26:361-367.

Wiester, M. J. 1975. Cardiovascular actions of palladium compounds in the unanesthesized rat. Environ. Health Perspect. 12:41-44.

Williams, R. J. B. and H. H. LeRiche. 1968. The effect of traces of beryllium on the growth of kale, grass and mustard. Plant and Soil 29:317-326.

Williams, S. E. and A. G. Wollum. Effect of cadmium on soil bacteria and actinomycetes. J. Environ. Qual. 10:142-144.

Wilson, D. O. and J. F. Cline. 1966. Removal of plutonium-239, tungsten-185, and lead-210 from soils. Nature 209:941-942.

Wilson, J. T., C. G. Enfield, W. J. Dunlap, R. L. Cosby, D. A. Foster, and L. B. Baskin. 1981. Transport and fate of selected organic pollutants in a sandy soil. J. Env. Qual. 10(4):501-506.

Woolson, E. A. 1977. Fate of arsenicals in different environmental substrates. Environ. Health Perspect. 19:73-81.

Yamagata, N. and Y. Murakami. 1958. A cobalt accumulator plant, Clethra barbinervis. Nature 181:1808-1809.

Yamamoto, T., E. Yonoki, M. Yamakawa, and M. Shimizu. 1973. Studies on environmental contamination by uranium: 2. Adsorption of uranium on soil and its desorption. J. Radiat. Res. 14:219-224.

Yeh, S. 1973. Skin cancer in chronic arsenicalism. Human Path. 4:469-485.

Young, R. A. 1948. Some factors affecting the solubility of cobalt. Soil Sci. Soc. Amer. Proc. 13:122-126.

Young, R. S. 1979. Cobalt in plants. pp. 50-80. In R. S. Young (ed.) Biology and biochemistry. Academic Press, New York.

Younts, S. E. and R. P. Patterson. 1964. Copper-lime interactions in field experiments with wheat yield and chemical composition data. Agron. J. 56:229-232.

Zoeteman, R. C. J., E. De Greef, and F. J. J. Brinkman. 1981. Persistency of organic contaminants in groundwater-lessons from soil pollution incidents in the Netherlands. The Science of the Total Environment 21:187-202.

Zwerman, P. J., Klausner, S. D., and D. F. Ellis. 1974. Land disposal parameters for dairy manure. pp. 211-221. In Processing and management of agricultural wastes. Proc. Cornell Agricultural Waste Management Conference. Rochester, New York.

PRELIMINARY TESTS AND PILOT STUDIES ON WASTE-SITE INTERACTIONS

Gordon B. Evans, Jr.
K. W. Brown
K. C. Donnelly

The study of waste-site interactions is the key to demonstrating that land treatment of a given waste at a specific site will render the applied waste less hazardous or nonhazardous by degradation, transformation and/or immobilization of hazardous constituents (Appendix B). These interactions also determine the potential for off-site contamination. To understand waste-site interactions, information gathered during the individual assessments of site, soil and wastes must be integrated and used to plan preliminary tests and pilot studies that will provide data on the interaction of system components. Laboratory, greenhouse and field studies provide more valuable information than theoretical models because of the wide range of complex variables involved.

In the flow chart presented in Chapter 2 (Fig. 2.1), Chapter 7 is indicated as a decision point in the evaluation and design process for HWLT. In many ways information gained from the testing procedure outlined in this chapter is the key to decision-making for both the permit evaluator and the facility designer. This chapter discusses a set of preliminary tests and pilot studies used to determine whether a particular HWLT system will meet the goal of rendering the applied wastes less hazardous or nonhazardous. The permit writer must decide whether a unit meets this goal after evaluating test results and other information submitted by the permit applicant. During the design of an HWLT unit, results from testing discussed in this chapter will be used to predict whether the goal of HWLT will be met and will form the basis for many operational and management decisions.

The topics to be discussed in this chapter are illustrated in Fig. 7.1. Sections 7.2 through 7.4 describe a comprehensive experimental approach that considers all of the important treatment parameters, environmental hazards, and potential contaminant migration pathways. The currently available battery of tests, listed in Table 7.1, outlines one

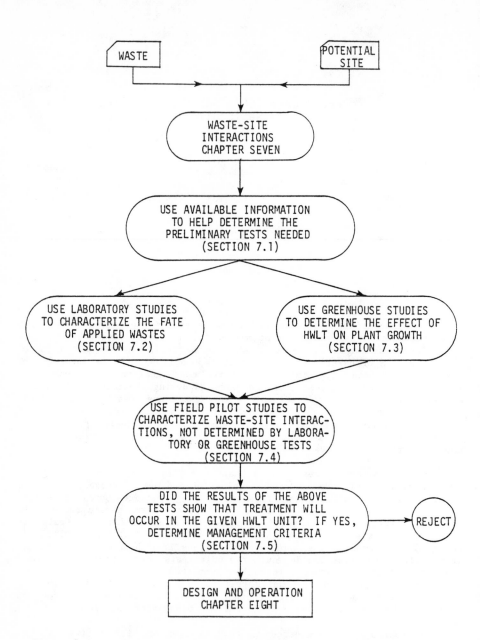

Figure 7.1. Topics to be addressed to evaluate waste-site interactions for HWLT systems.

possible experimental framework that would provide the data to understand the treatment processes at a given HWLT system. As new and more efficient tests are developed, it is expected that new testing procedures will replace those listed in the table. All tests conducted should include an experimental design based on statistical principles so that useful results are obtained. Section 7.5 discusses the interpretation of test results. Results from preliminary testing are used to establish the following:

(1) the ultimate fate of the hazardous constituents of the waste;

(2) the identity of the waste fraction that controls the yearly loading rate, referred to as the rate limiting constituent (RLC);

(3) the identity of the waste constituent that limits the amount of waste that can be applied in a single dose, referred to as the application limiting constituent (ALC);

(4) the identity of the waste fraction that limits the total quantity of waste that can be treated at a given site, referred to as the capacity limiting constituent (CLC);

(5) the criteria for management;

(6) the parameters that should be monitored to indicate contaminant migration into groundwater, surface water, air, and cover crops; and

(7) the land area required to treat a given quantity of waste.

A discussion of the basis for labeling a given waste fraction as either rate, application, or capacity limiting is included in Section 7.5.

TABLE 7.1   CONSIDERATIONS IN A COMPREHENSIVE TESTING PROGRAM FOR EVALUATING
WASTE-SITE INTERACTIONS.

Waste-Site Interactions	Test Method	Manual Reference
Degradation of waste	Respirometry	7.2.1.1
	Field studies by soil testing	7.4.1
Accumulation in soil of nondegradables	Waste analysis (inorganics)	5.3.2.3.1
	Respirometry (organics)	7.2.1.1
Leaching hazards	Soil thin layer chromatography	7.2.2.1
	Soil leaching columns	7.2.2.2
	Field soil leachate testing	7.4.2
Volatilization hazards	Environmental chamber	7.2.3
	Field air testing	7.4.4
Acute toxicity	Respirometry (soil biota)	7.2.1.1
	Beckman Microtox™ System	7.2.4.1.1
	Greenhouse studies (plants)	7.3.2
Chronic toxicity	Microbiological mutagenicity assays	5.3.2.4
Plant uptake	Greenhouse studies	7.3
Pretreatment	Assessment of processes generating waste	5.2

7.1   REVIEW OF AVAILABLE INFORMATION

Although pilot studies are often needed to supplement existing data or
to answer questions posed by unique situations, a review of pertinent
literature and available data from similar HWLT units may reduce the need
for extensive demonstration studies. From this review valuable information
may be found on soils, waste characteristics, and general data for predicting the fate of waste constituents. This information may alert the permit
reviewer and the facility designer to potential problems with recalcitrant
or toxic compounds and provide data for assessing the potential of a particular waste to be land treated. A thorough review of the literature and
other available information, such as monitoring data, may considerably
reduce the amount of testing required and will provide guidelines for
developing an experimental design that will adequately address waste-site
interactions for the particular HWLT unit.

7.2   LABORATORY STUDIES

A series of laboratory studies should be initiated as the first phase
of the waste-site interaction assessment. The major advantages of labora-

tory or bench scale studies are that one may better standardize the methodology and have better control over the important parameters. Laboratory techniques also act as rapid screening techniques by allowing the investigator to look at extremes and individual treatment effects within a reasonable time frame. However, some extrapolations to field conditions may be difficult since bench scale studies involve small, disturbed systems which cannot easily account for time series of events. Therefore, although some definite conclusions can be drawn from laboratory results, field plot and/or field lysimeter studies are usually necessary to verify laboratory results and extrapolations to determine the treatability of a waste. The following suggestions for conducting a comprehensive laboratory evaluation are intended as a general guide and should be adapted to the given situation.

7.2.1   Degradability

The complex nature of a hazardous waste makes it necessary to determine the degradation rate of waste constituents in a laboratory study rather than through theoretical models. The half-life of specific waste constituents cannot be applied to the waste as a whole because of the synergistic, additive, or antagonistic effects of various waste-soil interactions which may significantly alter the overall degradation rate. In circumstances where an equivalent waste has been handled at an equivalent HWLT unit, full-scale laboratory studies may not be necessary. Laboratory studies can be used to define waste loading rates, and to determine if reactions in the soil are producing an acceptable degradation rate for the hazardous organic waste constituents.

Before land applying any waste material, it is necessary to determine to what extent the soil may be loaded with the waste before the microbial activity of the soil is inhibited to the extent that waste degradation falls below acceptable levels. Land treatment of hazardous waste should be designed to utilize the diverse microbial population of the soil to enhance the rate of waste degradation. When environmental parameters are maintained at optimum conditions for microbial activity, efficient use is made of the land treatment site and the environmental impact is minimized. The environmental parameters which can most easily be adjusted at the HWLT unit include application rate and frequency, and the rate of addition of nutrients. To adjust these parameters to optimal levels, waste degradation must be monitored, and the effects of the various parameters on degradation evaluated. An evaluation of waste degradation should include the estimation of microbial populations, the monitoring of microbial activity, and the measurement of waste decomposition products.

The soil respirometer method which is discussed in detail in the following sections is one of the available methods for evaluating the degradation of a complex waste-soil mixture. Use of the soil respirometer requires only a limited amount of laboratory equipment. It is a method that can be quickly set up in most laboratories and can be used to evaluate a large number of parameters. While it does not provide a means for tracing the degradation of the individual components of a complex mixture, unless coupled with chemical analysis, it is a relatively simple and inexpensive method for evaluating the effect of environmental parameters on waste degradation in soil. Other methods which have been used to measure respiration from organic material include infrared gas analysis, gas chromatography, and the Gilson respirometer (Van Cleve et al., 1979). In addition, Osborne et al. (1980) discuss a method for studying microbial activity in intact soil cores.

7.2.1.1 Soil Respirometry

One method to evaluate environmental parameters before field application of waste is to monitor carbon dioxide (CO_2) evolution from waste amended soils in a soil respirometer. The soil respirometer consists of a temperature controlled incubation chamber containing a series of sealed flasks into which various treatments of waste and soil are placed (Fig. 7.2). The respirometer is an apparatus which allows temperature and moisture to be kept at a constant level while other parameters, such as waste application rate and frequency, are varied. A stream of humidified CO_2-free air is passed through the flasks and the evolved CO_2 from the flasks is collected in columns containing 0.1N NaOH. The air stream is purified in a scrubber system consisting of a pump and a series of flasks: one contains concentrated H_2SO_4; two parallel flasks contain 4N NaOH; and a pair of flasks in series contain CO_2-free water. The two flasks of 4N NaOH are placed parallel so that the air stream may be switched to a fresh solution without interrupting the flow of air. Between the scrubber and each flask is a manifold which distributes the air to the flasks through equal length capillary tubes, thus providing an equal flow rate for each flask. Each incubation chamber should include two empty flasks which serve to monitor impurities in the air stream. The air leaving each flask is passed through a 12 mm coarse Pyrex gas dispersion tube which is positioned near the bottom of a 25 x 250 mm culture tube containing 50 ml of CO_2-free 0.1N NaOH. The NaOH solutions are replaced approximately three times a week, depending on CO_2 evolution, and are titrated with 1.0N NCl following precipitation of evolved CO_2 with 3N $BaCl_2$ (Stotzky, 1965) to phenolphthalein end-point. The amount of CO_2 evolved can be determined (Section 7.2.1.1.2.4).

The rate of CO_2 evolution is used as an indication of microbial activity and relative waste decomposition (Stotzky, 1965). Upon termination of the experiment, subsamples may be taken from each flask to determine the residual hydrocarbon content (Section 5.3.2.3.2), and for an estimation of the microbial population (Section 7.2.4.1.1). The data from these tests can provide guidance on the appropriate application rate and frequency to use, the optimum rate of nutrient addition, and the rate of waste degradation in different soil types or at different temperatures. Careful study of these parameters before field application can prevent an accidental overload of the system and unnecessary additions of nutrients.

7.2.1.1.1 <u>Sample Collection</u>. Each hazardous waste stream may possess a variety of compounds that are toxic or recalcitrant, and a unique ratio and concentration of mineral nutrients. Therefore, to begin a laboratory degradation study representative samples of the waste and soil must be collected. Soil collected from the field for the respiration study should be maintained at field capacity (about 1/3 bar moisture tension) and stored at room temperature under a fixed relative humidity to preserve the soil microorganisms. Soil collected where water content is above field capacity should be air dried to reach field capacity, and soil which is collected below field capacity should be wetted with distilled water to field capacity. Since many wastes will require a diverse range of microorganisms to degrade waste constituents, care must be taken in the handling and storage of soil samples. The collection of a truly representative waste sample is also critical to obtaining valid data from the laboratory. Although few, if any, waste streams exist as homogeneous mixtures or have uniform composition. Over time, there are methods of obtaining representative samples; a more complete discussion of waste and soil sampling is presented in Section 5.3.2.1 and Chapter 9, respectively.

7.2.1.1.2 <u>Experimental Procedure</u>. The respiration experiment is begun by equilibrating the respiration chamber (Fig. 7.2) to the desired temperature and starting the scrubber system at least 24 hours before adding the soil to the flasks. Two days prior to waste addition, the soil is brought to the desired moisture content by air drying or wetting with distilled water. A soil sample equivalent to 100 g of dry soil is placed on a glass plate and crushed to reduce the largest aggregates to approximately 1/2 cm. The crushed and weighed soil sample is placed into a preweighed 500 ml Erlenmeyer flask, which is then connected to the CO_2-free air stream and to a column containing 0.1N NaOH. The flow of air through the chamber should be adjusted so that neither stimulation of microbial activity nor inhibition occurs. A flow rate of 20 ml per minute of CO_2-free air per 100 gm of soil appears to provide an adequate supply of oxygen while not affecting the rate of respiration. After the soil has been placed in the respirometer and allowed to equilibrate for at least two days, a 20-40 gram subsample of soil is removed from the flask and placed in an evaporating dish. The desired amount of waste is then mixed with the soil. After mixing, the waste-soil subsample is mixed with the total soil sample from the flask and the mixture is returned to the flask and then put back in the respirometer. This mixing procedure may also be used to add water, or to reapply the waste during the respiration experiment.

7.2.1.1.2.1 <u>Soil moisture</u> is a parameter which may be difficult to adjust in the field. All HWLT units have runoff collection systems and some may have leachate recycling pumps or irrigation systems that can be used to increase the moisture content of dry soil. The optimum range of soil moisture for microbial activity appears to be between the wilting point (about 15 bars moisture tension) and field capacity (1/3 bars moisture tension) of the soil. This range of moisture is also optimum for waste degradation since excess moisture reduces available oxygen and most organics are degraded by an oxidative pathway. In a laboratory, flasks containing the

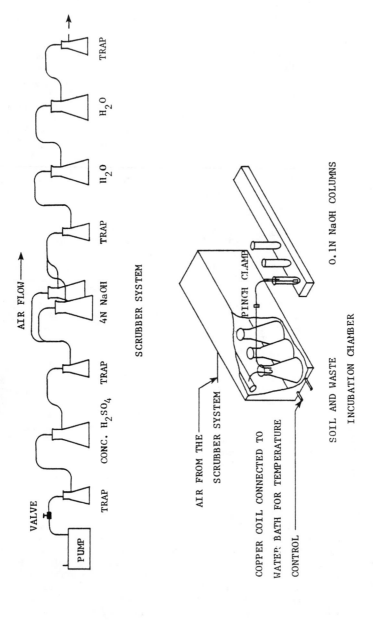

Figure 7.2.   Schematic diagram of respirometer.

soil-waste mixture should be removed and weighed periodically so that the moisture content of the soil can be adjusted. If the moisture content of the soil becomes substantially above field capacity or below the wilting point, the rate of degradation may be significantly altered, and the data should be interpreted with caution.

7.2.1.1.2.2 The temperature of the initial respiration studies may be conducted at $20 \pm 5°C$. This allows the experiment to be carried out at room temperature without requiring temperature control, and provides information on waste treatability. For warmer climates, additional degradation experiments may be performed at 30°C are appropriate. When studying waste degradation in a cold climate the respirometer temperature may need to be regulated to as low as 5°C. Studies at different temperatures provide additional information that can be useful in determining seasonal application rates and frequencies.

7.2.1.1.2.3 Nutrient additions may help stimulate biodegradation. Carbon is used by most bacteria as an energy source and is present in most wastes at much greater concentrations than nitrogen. The addition of large amounts of carbon to the soil will stimulate excess bacterial growth, which will cause nitrogen to be depleted unless nutrient additions are made. The optimum carbon:nitrogen:phosphorus (C:N:P) ratio in a waste-soil mixture is about 50:2:1. However, this ratio should be used only as a guide, and optimum fertilizer rates for individual HWLT units should be determined along with other site-specific parameters. The timing of nutrient additions is important to waste degradation. In some cases it may be more effective to add nutrients after waste degradation has begun and the more susceptible substrates have already been utilized by the microorganisms. In addition to mineral nutrients, lime may be required to maintain the soil pH between 6.5 and 8.5.

7.2.1.1.2.4 Titration of the NaOH solutions are used to determine the amount of CO_2 evolved to indicate the rate of waste degradation. Approximately three times per week the NaOH solutions are replaced to determine the amount of CO_2 absorbed from the air passing through each treatment flask. The frequency of sampling and titration may be reduced or increased as the rate of CO_2 evolution requires. If it is determined that the NaOH solution is nearing saturation, the sampling frequency should be increased, and if the volume of acid required to titrate the treated sample is almost equal to that required to titrate the blank samples, the sampling frequency should be decreased.

The accumulated CO_2 is determined by titrating the NaOH solution with 1.0N HCl following precipitation of evolved CO_2 with 3N $BaCl_2$ (Stotzky, 1965). All titrations are carried to a phenolphthalein endpoint. The amount of CO_2 evolved is determined by the following equation:

$$(B - V)NE = mg\ CO_2 \qquad\qquad (7.1)$$

where

      B = average volume of HCl required to titrate the NaOH from blank treatments;

      V = volume required to titrate the NaOH from the specific treatment;

      N = the normality of the acid; and

      E = the equivalent weight of the carbon dioxide.

Each time the NaOH solutions are replaced, the spent solutions should be titrated and the amount of evolved carbon dioxide determined.

7.2.1.1.2.5 <u>Application rate and frequency</u> are interdependent and depend on climatic conditions, including temperature and rainfall variations. Optimum degradation rates are often achieved when small waste applications are made at frequent intervals. A laboratory study may be used to determine the application rate and frequency that yields the most rapid rate of waste decomposition in a given period of time at a constant temperature and moisture. It is easiest to determine the optimum application rate and then to evaluate the application frequency. Experimental application rate should be varied over a 100-fold range, using a minimum of four treatments with different application rates. One additional flask containing soil to which no waste has been applied should be used as a control. All treatments are conducted in duplicate so that the results can be properly evaluated. Once the optimum application rate is determined for a specific waste stream, the application frequency can be evaluated, using a minimum of three alternate schedules. For example, if it is determined in the rate study that the best compromise between efficiency of land use and biodegradation is achieved when the waste is applied at a rate of 5% (wt/wt), the frequency study would then evaluate the degradation rate of four 1.25% applications, two 2.5% applications, and one 5% application during the same time period. Chemical and biological analyses of the treatments, when evaluated with the cumulative CO_2 data, will indicate the treatment rate and frequency that provide the most efficient degradation rate.

7.2.1.2 <u>Data Analysis</u>

    The data provided by a laboratory respiration experiment may be used to evaluate the potential of a waste to be adequately treated in the land treatment system and to determine the half-life of the organic fraction of the waste. Half-life is defined as the time required for a 50% disappearance of applied carbon. The decision process for determining if a waste is amenable to land treatment is outlined in Fig. 7.3. The first step in this process is to determine how the waste will affect microbial activity when mixed with the soil. If waste application inhibits microbial activity, the following options are available to improve the treatability of the waste:

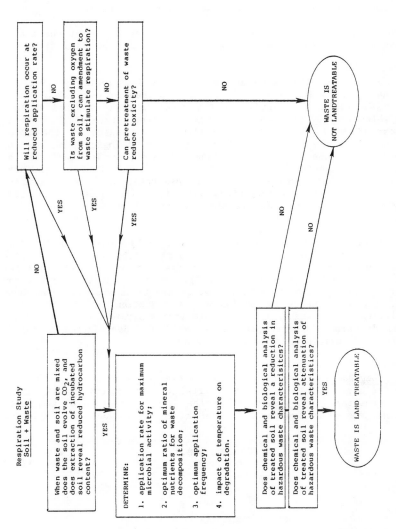

Respiration Study
Soil + Waste

When waste and soil are mixed
does the soil evolve CO_2, and
does extraction of incubated
soil reveal reduced hydrocarbon
content?

Will respiration occur at
reduced application rate?

Is waste excluding oxygen
from soil, can amendment to
waste stimulate respiration?

Can pretreatment of waste
reduce toxicity?

DETERMINE:

1. application rate for maximum
   microbial activity;

2. optimum ratio of mineral
   nutrients for waste
   decomposition;

3. optimum application
   frequency;

4. impact of temperature on
   degradation.

Does chemical and biological analysis
of treated soil reveal a reduction in
hazardous waste characterisitcs?

Does chemical and biological analysis
of treated soil reveal attenuation of
hazardous waste characteristics?

WASTE IS
NOT LANDTREATABLE

WASTE IS LAND TREATABLE

NO

YES

YES

YES

YES

NO

NO

NO

NO

NO

YES

FIGURE 7.3   The information needed to determine if a waste may be land treated.

(1) reducing waste application rates;

(2) pretreating a hydrophobic waste by drying or mixing with a bulking agent to improve the penetration of oxygen into the soil;

(3) pretreating the waste by chemical, physical, or biological means (Section 5.2) to reduce its toxicity; and

(4) making in-plant process changes to alter the waste.

If these options fail and the soil microorganisms cannot alter the nature of the waste, it will not be adequately treated in the land treatment system.

If, after mixing the waste and soil elevated microbial activity is observed the waste is land treatable and the optimum parameters for waste degradation should be determined. If the waste is to be applied at temperatures which vary by more than 10°C from the temperature of the initial respirometer study (20 ± 5°C), the half-life of the waste at the other temperatures should be determined. Chemical and biological analyses of treated soils from the respirometer flasks after incubation indicate the effect of land treatment on the hazardous waste constituents. If these analyses indicate that a waste is rendered less hazardous by incorporation into the soil, half-life calculations (yr) from laboratory application rates (kg/ha) may be used to determine acceptable yearly waste loading rates.

The initial waste loading rate is determined by calculating the time required to degrade 50% of the applied waste constituents. Half-life determinations can be made for the organic fraction of the waste and for each subfraction (acid, base, and neutral). While chemical analysis can define decomposition rates of specific waste fractions and hazardous constituents, the only means of evaluating a reduction in the hazardous characteristics of a waste is through biological analysis (Sections 5.3.2.4 and 7.2.4) or through a previous knowledge of the degradation pathways, by-products, and toxicities of waste conponents.

7.2.1.2.1 Degradation Rate. In most laboratory studies the waste is incubated for a period of six months. After the laboratory experiment is terminated, the rate of degradation for the organic fraction of the waste should be determined by two methods. The first method uses the following equation:

$$D_t = \frac{(CO_2w - CO_2s)0.27}{C} \qquad (7.2)$$

where

$\qquad D_t$ = fraction of total carbon degraded over time;
$\qquad CO_2w$ = cumulative CO_2 evolved by waste amended soil;
$\qquad CO_2s$ = cumulative CO_2 evolved by unamended soil; and
$\qquad C_a$ = carbon applied.

The second method used to determine the rate of degradation requires the extraction of the organic fraction from the soil (Section 5.3.2.3.2). The percent of organic degradation is determined as follows:

$$D_{to} = \frac{C_{ao} - (C_{ro} - C_s)}{C_{ao}} \qquad (7.3)$$

where

D_{to} = fraction of organic carbon degraded over time;
C_{ao} = the amount of carbon applied in the organic fraction of the waste;
C_{ro} = the amount of residual carbon in the organic fraction of waste amended soil; and
C_s = the amount of organic carbon which can be extracted from unamended soil.

To determine the degradation rate of individual organic subfractions the following equation is used:

$$D_{ti} = \frac{C_{ai} - (C_{ri} - C_{si})}{C_{ai}} \qquad (7.4)$$

where

D_{ti} = fraction of carbon degraded in subfraction i;
C_{ai} = carbon applied from subfraction i in the waste;
C_{ri} = carbon residual in subfraction i in waste amended soil; and
C_{si} = the amount of carbon present in an unamended soil from subfraction i.

The clarity of separation of all subfractions should be verified by gas chromatography.

7.2.1.2.2 Half-life Determination. The half-life of the waste may then be calculated for the waste as follows:

$$t_{1/2} = \frac{0.50}{D_t} t \qquad (7.5)$$

where

t = time in days that the waste was degraded to generate the data used in equations 7.2-7.4;
$t_{1/2}$ = half-life of waste organics in soil (days); and
D_t = fraction of carbon degraded in t days.

An optional method that may be used to calculate half-lives is to plot cumulative percent carbon degraded as a function of time on a semi-log scale graph. The point in time where 50% of the waste has been degraded may then be read directly.

Of the half-lives determined by the above methods, the longest half-life should be used as the half-life for the organic fraction of the waste. This half-life is then used to calculate the initial loading rate which will produce maximum microbial activity in the soil. Because of the great number of variables influencing waste biodegradation in soil, it will be difficult to predict the rate of degradation of wastes in the field by using an equation. The preceding equations use zero order kinetics and are designed to make the most efficient use of the land treatment area. Laskowski et al. (1980) suggests that the degradation process for relatively poorly sorbed chemicals appears to follow zero order kinetics at high application rates. Data resulting from both laboratory and field studies are compared in Section 7.5.3.1.4; this comparison indicates that variables not accounted for in laboratory studies may result in an overestimation of the actual waste half-life.

In most cases the rate of degradation of the individual subfractions will vary. In any case, the fraction that degrades at the slowest rate controls waste loading rates. The waste should be applied at a rate that will stimulate microbial activity while not reaching toxic levels of any specific fraction. The degradation of the more resistant fractions will occur after the preferred substrate has been degraded. Gas chromotography can be used to scan the waste after degradation in soil to determine if a specific compound is degrading at a slower rate than the calculated half-life of the other waste fractions. If such a compound is identified, then the half-life of the compound should be used to adjust loading rates. The half-life of the most resistant fraction or compound will restrict loading rates if the compound is mobile in the soil or will remain at an unacceptable concentration far beyond the time when waste applications cease.

7.2.1.2.3 <u>Consideration for Field Studies of Degradation</u>. These calculations are used to provide guidance for establishing design loading rates and developing appropriate field studies. Once the first waste application has been made, waste degradation in the field pilot study should be monitored by periodic soil sampling and subsequent analysis for hydrocarbon and subfraction content (Section 7.4.1). Half-lives determined from experimental field data generally provide a more realistic evaluation of waste decomposition rates. However, the amount of information required from the results of field studies depends on laboratory study results. If, from the laboratory study, it is determined that all waste fractions degrade at equal rates and there is no specific compound which is less susceptible to degradation than the organic fraction as a whole, then the soil sampling need only monitor the removal of the total organics. However, if a particular compound or fraction is evidently resistant to degradation, then this particular compound or fraction should be monitored in the field.

7.2.2   <u>Sorption and Mobility</u>

The potential for organic contamination of surface runoff and leachate from land treatment sites depends on the erosion potential of the soil, the concentration of water soluble constituents in the waste, the adsorptive capacity of the soil, the kinetics of soil water movement, and the degradability of the potentially mobile waste constituents and their degradation products. Proper erosion control and runoff water treatment practices will effectively eliminate the runoff hazard to surface waters. Degradability is discussed in Section 7.2.1 and the results of waste degradation experiments should be integrated with the mobility findings. Therefore, a suitable method for evaluating mobility should account for waste solubility, adsorption, and soil water kinetics. Transport mechanisms or potential leachability may be assessed by soil thin-layer chromatography and column leaching techniques. Where a hazardous waste constituent is demonstrated to be leachable and only slowly degradable, field studies will be necessary to determine the leachate concentrations of the mobile constituents for establishing the maximum safe waste loading rate (Section 7.5.3.1.2). Since the mobility of degradates is often important, laboratory studies may include analyses of aged waste-soil mixtures.

Several modes of transport can be described for the movement of hazardous organic compounds through the soil. As a continuous phase, oil can move as a fluid governed by the same parameters as those which determine soil water movement. Alternatively, water soluble or miscible compounds can be transported by soil water. A small amount of movement might also occur by diffusion, however, diffusion would not occur at a level that would cause a leaching hazard. Sorption and/or degradation account for the attenuation of leachable hazardous constituents. Adsorption capacity is directly related to soil colloidal content and chemical nature of the waste constituents (Bailey et al. 1968; Castro and Belser, 1966; Youngson and Goring, 1962). Soil organic matter is perhaps most responsible for adsorption of nonionic compounds, while polar constituents which are potentially solubilized in water may have a greater affinity to the mineral fraction of soil. Precipitation to less soluble forms and complexation also immobilize and thus attenuate, some waste constituents.

The primary objective of a laboratory leaching study is to evaluate leaching potential rather than to assess actual mobility of a given compound in soil. A disturbed soil can be tested to indicate extremes, but the kinetics of water and solute movement in a bench scale test do not ordinarily approximate field conditions, where precipitation is intermittent and the intact soil profile retains its unique physical characteristics. Soils chosen for leaching studies should be sampled from each horizon in the zone of aeration where adequate microbial populations are ordinarily present for waste degradation. By testing for the mobility of waste constituents in the lower soil horizons, one can establish whether the rapid movement of a waste constituent through a less adsorptive surface soil may be impeded by a more adsorptive subsoil to the extent that the soil biota can adequately decompose the compound(s). Once an organic compound has leached below the zone of abundant microbial activity, however, it has been shown that degradative attenuation is extremely slow (Duffy et al. 1977; Van Der Linden and Thijsse, 1965).

7.2.2.1 Soil Thin-layer Chromatography

The relative mobility of organic fraction components may be determined by the technique of Helling and Turner (1968) and Helling (1971). This technique is similar to conventional preparative thin-layer chromatography (TLC) except that soil is used as the stationary phase rather than materials such as silica gel or alumina. Mobility of a given substance can be expressed by a relative measure, R_F, which describes the distance traversed by a compound divided by the distance traversed by the wetting front. The following description outlines the important aspects of the procedure:

(1) Soil materials used are those passing through a 500 mm sieve for sandy clays and coarser textured soils, or 250 μm sieve for fine loams and clay soils.

(2) Plates are air-dried before use. A smooth, moderately fluid slurry is made of water and sieved soil material and spread on clean glass plates to uniform thicknesses of 500–750 μm for fine textured soils, and 750–1000 μm for the coarser textured soils.

(3) A horizontal line is etched 11.5 cm above the the base. Samples are spotted at 1.5 cm, providing a total leaching distance of 10 cm.

(4) The atmosphere of the developing chamber is allowed to saturate and equilibrate prior to plate development.

(5) Plates are developed in a vertical position in approximately 0.5 cm water. The bottom 1 cm may be covered with a filter paper strip to reduce soil sloughing and maintain the soil-water contact. Development continues until water has risen to the scribed line at 11.5 cm.

(6) Movement is determined by either radioautograms for radioactive materials or scraping and eluting segments of soil from the 10 cm development distance. Scraped materials can be easily eluted with small volumes of solvent by using capillary pipettes as elution columns.

(7) R_F values are computed and correlated to soil properties.

Some drawbacks of soil TLC include the following:

(1) soil particles are oriented in two dimensions;

(2) waste-soil contact is maximized, most closely simulating intraaggregate flow and negating the attenuating effects of soil aggregation; and

(3) flow is rapid and closer to steady state conditions thus minimizing adsorption-desorption kinetics effects.

Soil TLC is a useful rapid screening technique, but where waste constituents are mobile as indicated by R_F values, soil column leaching and field pilot studies will better quantify mobility. Soil column leaching and field pilot studies will provide more accurate predictive data since conditions of these studies more closely resemble conditions in the actual land treatment system.

7.2.2.2 Column Leaching

Column leaching is an approximation of mobility under saturated conditions. It, like the soil TLC method provides a relative index of the potential for leaching. The choice of soils to be tested should be the same as that used for soil TLC. At a minimum, duplicate columns and a control should be used for each waste/soil mixture listed. The general procedure is as follows:

(1) Glass columns (2-3 cm I.D.) are filled with 20 cm air-dry soil previously ground and passed through a 2 mm mesh sieve. Columns should be constructed of glass or other nonreactive material which does not interfere with the analyses.

(2) Columns are filled slowly with soil and tamped to a bulk density approximating that in the field to reduce solution movement by direct channel transport and to more closely resemble field conditions.

(3) Applications of waste are made by mixing waste with a small amount of soil and applying the mix to the soil surface. Alternatively, the organic fraction of the waste may be applied in a minimum amount of solvent to the top of the soil in the column.

(4) Glass wool or a filter pad is placed on the soil surface and leaching is begun by adding at least one column volume of water at a controlled rate no faster than 1 ml/min.

(5) Effluents are analyzed along with the soil extruded and segmented at 2 cm intervals to evaluate depth of penetration as a function of the effective volume partitioned. The volume partitioned can be assumed to be the volume of water retained by the soil at field capacity. Thus an effluent volume equal to the volume of water retained at 1/3 atmosphere soil moisture tension approximates 1 pore volume.

(6) Concentrations of materials in effluent are determined and plotted against cumulative drainage volume.

A soil column offers a better approximation to a natural system than does soil TLC since the column provides a larger soil volume, larger aggregates, and a more random particle orientation. Soil column leaching tests, however, lack the methodological standardization of soil TLC.

The potential leaching hazard of a given waste in a particular soil can be estimated from consideration of the following:

(1) the mobility of waste constituents relative to water;

(2) the concentrations of constituents observed in the leachate and soil;

(3) the degradability of mobile compounds;

(4) the flux and depth of soil solution percolate as observed in the field water balance; and

(5) the toxicity of mobile waste constituents as determined using bioassay techniques (Section 5.3.2.4).

Field pilot studies may be needed to correlate and verify laboratory results. They are particularly important when laboratory data reveal a substantial leaching hazard.

7.2.3   Volatilization

Volatilization is mostly important for those compounds with vapor pressures greater than 10^{-3}mm/Hg at room temperature (Weber, 1972). Environmental variables affecting volatility are soil moisture, adsorption, wind speed, turbulence, temperature and time (Farmer et al., 1972; Plice, 1948). One mechanism of volatilization is evaporative transfer from a free liquid surface. The potential of this mechanism is roughly equivalent to the purgable and easily volatilized fractions; however, the impact should be lessened greatly upon waste-soil mixing. An assessment of volatilization should include this aspect of attenuation. Within a soil, chemicals are not at a free liquid surface and vaporization is dependent upon distribution between air, water and solid surfaces.

Volatilization of waste constituents or degradates may be determined empirically by measuring vapor losses from a known soil surface following waste application. Laboratory investigations using a sealed, flow-through system should consider the following:

(1) the effects of application technique and waste loading rates;

(2) several soil moisture contents, including dry and wet soil;

(3) several temperatures, including the maximum expected surface soil temperature;

(4) variations in air flow; and

(5) changes in volatilized fraction composition and flux with time.

Generally, an air stream is passed over the soil surface and through solid sorbents such as Tenax-GC or florisil and analyzed according to Section

5.3.2.3.2. Results are computed in both concentration (mass/m^3) and flux terms (mass/m^3/surface area).

7.2.4   Toxicity

Treatability tests may include a determination of the levels at which the waste becomes toxic to plants or microbes and/or causes genetic damage. These tests provide an additional qualitative measure of treatability. During the operation of a land treatment unit, and after closure, the biological tests may also be used to monitor environmental samples to evaluate waste degradation and to ensure environmental protection. In addition to the tests described here and in Section 5.3.2.4, the procedure of Brown et al. (1979) may be used to evaluate aquatic toxicity prior to the release of runoff or leachate water from the site. All samples collected for biological analysis should be frozen as described in Section 5.3.2.1 and samples should be processed as soon as is possible after collection.

7.2.4.1 Acute Toxicity

Before a hazardous waste is land applied, it is a good idea to determine if the waste will be acutely toxic to indigenous plants and microbes. Microbial toxicity is particularly important when degradation is one of the objectives of treatment. Methods for evaluating toxicity are discussed below and toxicity testing can generally be combined with any other waste-site interaction study.

7.2.4.1.1 Microbial toxicity. The microbial toxicity of a waste-soil mixture can be evaluated using information obtained from a pour plate method which enumerates total viable heterotrophs and hydrocarbon utilizing microorganisms. This involves collecting soil samples for microbial analysis before waste application and following incubation with the waste in the respirometer. One gram of a soil sample is placed in 99 ml of phosphate buffer and mixed on a magnetic stirrer for fifteen minutes. Subsequent dilutions are made by adding 1 ml of the previous dilution to 99 ml of the buffer. Samples should be assayed on four different media to determine the total number of soil microorganisms. Total viable heterotrophs are enumerated using soil extract agar (Odu and Adeoye, 1969) with 10 mg/l of Amphoteracin B. The presence of soil fungi is determined using potato dextrose agar (Difco) or soil extract agar with 30 mg/l of rose bengal and streptomycin. Hydrocarbon utilizing bacteria and fungi may be detected by replacing the carbon source used in soil extract agar with 6.25 g/l silica gel oil as suggested by Baruah et al. (1967). The silica gel oil is prepared for each waste stream by combining 5.0 g of the waste with 1.25 g of fumed silica gel (Cab-o-sil, Cabot Corporation).

In order to retard spreading of mobile organisms, 0.5 ml of each dilution should be added to 2.5 ml of soft agar (0.75% agar), mixed on a vortex mixer, and poured onto the hard agar surface. Plates are incubated for a minimum of two weeks at the temperature at which the soil waste mixture was incubated. All estimations of viability should be assayed in quadruplicate.

A second method for evaluating microbial toxicity developed by Beckman Instruments, Inc. is currently being tested by the EPA to determine if the procedure can be used as a rapid screening tool for assessing the land treatability of a specific hazardous waste and as a method to determine loading rates. The Beckman Microtox™ system measures the light output of a suspension of marine luminescent bacteria before and after a sample of hazardous waste is added. A reduction in light output reflects a deterioration in the health of the organisms which signifies the presence of toxicants in the waste (Beckman Instruments, Inc., 1982).

Using these, or other, methods the acute toxic effects of land treating a hazardous waste on endemic microorganisms can be assessed. By determining the immediate effects of the waste on soil microorganisms, knowledge is obtained which can aid in the determination of the maximum initial loading rate and in the evaluation of the respiration data (Section 7.2.1.2).

7.2.4.1.2 Phytotoxicity. The phytotoxicity of a hazardous waste may be evaluated in a greenhouse study (Section 7.3) for the types of vegetation anticipated at the land treatment unit. The greenhouse study should evaluate the toxic effects of the waste at various stages of growth, including germination, root extension, and establishment. Root extension may be determined for a water extract of the waste which has been degraded by soil bacteria using the procedures of Edwards and Ross-Todd (1980). Plant bioconcentration for chronic toxicity to humans via the food chain may be measured by analyzing an extract from plants grown in waste amended soil in a biological test system. Plant activation of nonmutagenic agents into mutagens has been demonstrated by Plewa and Gentile (1976), Benigni et al. (1979), Reichhart et al. (1980), Matijesevic et al. (1980), Higashi et al. (1981), and Wildeman et al. (1980).

7.2.4.2 Genetic Toxicity

The genetic toxicity of a waste-soil mixture can be measured using selected bioassays and following the same protocols used to determine the genetic toxicity of the waste itself (Section 5.3.2.4.2). It may be desirable to separate the organic extract of the waste into subfractions (Section 5.3) for determining genetic toxicity. Bioassays of samples taken from the treated waste-soil mixture at different time periods and from different waste application rates can be compared to bioassays of the untreated waste. The reduction in hazardous characteristics following treatment provides a qualitative measure of treatment.

Greenhouse studies are designed to observe the effects of waste additions on plant emergence and growth. Moreover, they can be used to assess
the acute and residual toxicity of the wastes to determine optimum loading
rates. Greenhouse experiments may also aid in selecting application frequencies and site management practices.

In many cases, the concentration of one or more constituents in a
waste, rather than the bulk application rate, may control plant responses.
Therefore, research should include a characterization of which waste compounds are phytotoxic and a determination of the residence times of these
compounds in soils. When short-term growth inhibition is caused by a
rapidly degradable phytotoxin, the quantity of waste which can be applied
in a single application is limited. A more resistant substance in the same
waste may potentially accumulate to toxic concentrations if the long-term
loading of this substance exceeds the rate of degradation. Thus, greenhouse studies of plant responses should be designed to assess the acute
toxicity of freshly applied waste and the toxicities and degradation rates
of resistant compounds.

7.3.1   Experimental Procedure

One general approach to assessing plant toxicity in the greenhouse involves planting a given species in pots containing soil mixed with varying
quantities of waste. The choice of plant species should be based on site
characteristics and the species which will probably be used to establish
the permanent vegetative cover as discussed in Section 8.7. Control plantings receiving no waste must be included, and all pots should be fertilized, watered and carefully maintained to ensure that the results observed
are related to the waste additions. Allen et al. (1976) is a good reference on the proper care and management of greenhouse pot experiments.
Since the toxicity effects are greatest before the fresh waste has begun to
decompose, the emergence and growth tests should consist of only one plant-
harvest cycle of short duration (30–45 days). In practice, management at
an HWLT unit is not striving for maximum yields; therefore, a waste concentration is considered to be toxic when yields are reduced to levels between
50 and 75% of the control yields. The toxic concentration of the waste or
waste fraction in soil is termed the "critical concentration" (C_{crit}).

7.3.2   Acute Phytotoxicity

Using the procedures of 7.3.1, fresh wastes are applied to soil in a
range of concentrations in order to determine the critical concentration of
the waste. This C_{crit} value may be used in conjunction with half-life
($t_{1/2}$) determined from respirometer experiments to establish loading rates

(kg/ha/yr) based on the total organic fraction. If all of the organics in the waste degrade at relatively the same rate, the loading rate established in this manner will be valid for design purposes; however, most complex organic mixtures found in hazardous waste streams do not degrade uniformly. If a loading rate derived from the organic fraction half-life is used, there is likely to be an accumulation of resistant organic constituents with half-lives longer than the half-life of the total organic fraction. Regardless of the portion of the organic fraction which is ultimately established as the rate limiting constituent (RLC), expressed in kg/ha/yr, the loading rate determined from the acute toxicity and degradation rate of a fresh waste may still qualify the total organic fraction as the application limiting constituent (ALC), expressed in kg/ha/application.

7.3.3   Residuals Phytotoxicity

Some particularly resistant organics, if they are not toxic, may pose no special problems if they accumulate in soils. If these resistant compounds are toxic when present in large enough concentrations, then they may limit the loading rate, rather than total organic fraction. Gas chromatographic (GC) analyses of applied waste or wastes incubated in respirometers can quantitatively establish the half-lives of individual compounds and can lead to qualitative determinations of resistant compounds by such techniques as GC-mass spectrometry (GC-MS). Phytotoxicity of these compounds in a waste-soil environment can be determined by spiking the raw waste with various concentrations of the pure compound or compounds, and repeating the greenhouse study using the new mixtures.

Spiking simulates the accumulation of the compound in the land treatment system after repeated waste applications, at the rate established by the organic fraction degradation rate. The concentration which elicits toxic responses by plants is the C_{crit} value for that compound. Two possible scenarios are as follows:

(1) First, establish an economical design life (in years) for the unit. If the C_{crit} value for the resistant compound compound would not be reached during this design life after applying waste at the rate established using the organic fraction degradation rate, then no hazard is posed.

(2) If the C_{crit} value is reached before the design life is attained, or if no specific unit life is specified, then the resistant toxic compound is the RLC for the organic fraction.

Therefore, greenhouse toxicity data can be used in conjunction with respirometer waste degradation data to establish safe HWLT unit loading rates (Section 7.5.3.1.4).

Field pilot studies are intended to verify laboratory results, discover any unforeseen methodological or potential environmental problems, and investigate interactions which cannot be adequately assessed in the laboratory. Field testing is the closest approximation to actual operating conditions, and all aspects of the waste-site system can be observed as an integrated system. In addition to verifying of laboratory results, field studies may function as follows:

(1) to evaluate possible odor or vapor problems;

(2) to provide information on the physical problems associated with distribution and soil incorporation of a particular waste;

(3) to evaluate the possibility of applying greater amounts of waste than would appear possible from the available data or from greenhouse, respirometer or column studies;

(4) to evaluate the runoff water quality;

(5) to provide information on the length of time required for the runoff water quality to become acceptable for uncontrolled release;

(6) to evaluate the fate and mobility of a specific organic constituent or combination of constituents for which little data are available; and

(7) to evaluate the compatibility of a new waste applied to a site previously used for a different waste.

Field pilot studies should be kept small and facilities should be available to retain runoff just as they would be for a fully operational HWLT system. The EPA permit regulations contain certain requirements for conducting demonstration studies (EPA, 1982). Typically, plots should not be greater than 500 m^2, although there may occasionally be justification for larger areas where special equipment for waste application or incorporation activities requires additional space. While field tests often provide much better data than laboratory or greenhouse tests, they are often more costly to conduct. Also, fewer variables, such as application rate, frequency or alternate treatments, can be tested. Furthermore, uncontrolled variables, such as temperature, rainfall and wind, make the data more difficult to interpret.

Application rates to be used in pilot studies must be based on the best available information and be developed in accordance with appropriate procedures. If one of the objectives is to test the feasibility of application rates greater than those that were indicated by the laboratory and greenhouse information, it is often advisable to select waste application rates of 2, 4 and possibly 8 times the optimal rate. Precautions must be taken, however, to protect groundwater from mobile waste constituents loaded onto the soil.

7.4.1   Degradation

     Degradation of organic waste materials in the field should be evalu-
ated by determining the residual concentration of these materials in the
treatment zone. The soil should be analyzed for the hazardous constituents
and perhaps for general classes of organics, including total organics as
suggested in Section 5.3.2.3.2. Sampling procedures should be the same as
for functioning HWLT units. Samples should be taken on a schedule that
allows maximum sampling during the period of maximum degradation. Typically, a geometric sampling schedule of 0, 1, 2, 4, 8, 16, etc. weeks after
application is appropriate.

7.4.2   Leachate

     Leachate water should be collected from below the treatment zone as
will be done when monitoring an operating HWLT unit. Samples should be
collected at sufficiently frequent intervals to be representative of the
water leaching below the normal root zone depth. Typical leachate sampling
depths are 1 to 1.5 m below the soil surface. This ensures an adequate
zone of aerated soil for decomposition and plant uptake. Any waste constituents moving below the 1 to 1.5 m depth will usually continue to the
water table since oxygen availability, microbial populations and plant
uptake decrease markedly below this depth.

7.4.3   Runoff

     Runoff water should be collected and analyzed if these data are needed
to evaluate treatability or the potential for release. The water may be
collected from retention areas if this method is appropriate for the site.
If several treatment rates or options are being tested, it may be necessary
to have different retention areas for each treatment or to install devices
that will collect representative samples as they flow from each plot before
they reach the retention basin. Runoff water should be analyzed for the
constituents to be included in the discharge permit, the hazardous
constituents of the waste, and for the biological activity of the water.

7.4.4   Odor and Volatilization

     If the objective of the test is to evaluate odor problems, periodic
field evaluations should be made by an odor panel as described in Section
8.4.2. Panel observations should be scheduled at frequent intervals following waste application and mixing activities. Again, a geometric sampling schedule may be appropriate. If the pilot test is to provide data on
volatilization, the gases emanating from the surface should be collected

and periodically sampled. A more detailed discussion of volatilization is provided in Section 7.2.3

7.4.5   Plant Establishment and Uptake

If the objective of the test is to evaluate revegetation potential and plant uptake, it may be desirable to plant several species and to try both seeds and sprigs for species that can be planted either way. Planting should not be initiated until the waste has been repeatedly mixed and allowed to degrade. If initial plantings fail, the species should be replanted after further mixing and adjustment of nutrients and soil pH. If water is the limiting factor during germination and emergence, it may be desirable to mulch and irrigate the site to assist establishment. If bio-accumulation is a concern, plants should be harvested and analyzed for accumulated waste constituents.

7.5   INTERPRETATION OF RESULTS

Waste-soil interaction studies generate a variety of data that must be carefully interpreted to determine treatment feasibility, acceptable waste loads, special management needs, and monitoring criteria. Since experiments should have been conducted using the bulk waste, synergistic and antagonistic effects have been considered over the short-term and for mobile or degradable species. However, the effect of long-term accumulation of some waste constituents, especially metals, cannot be established from such condensed investigations. Additionally, only scant information exists regarding the joint toxic effects of several accumulated compounds or elements. In any case, the interpretation of results from literature review, experimental work and/or operational experience may safely consider each important waste constituent independently.

7.5.1   Feasibility and Loading Rates

Treatment feasibility and loading rates are closely related and can be tentatively ascertained from data generated from tests described in Sections 7.2 through 7.4. Practically any hazardous waste may be land treated, although allowable waste application rates may require excessive land area commitments. Consequently, feasibility is essentially an economic decision based on allowable loading rates. The loading rates, on the other hand, are established by calculating the acceptable rates for each waste constituent and adopting the most restrictive value.

A central concept to the understanding of waste loading rates is the way in which waste constituents behave in the given land treatment unit. Basically, the behavior of any given constituent at a given site will fall within one of the following categories:

(1) the constituent is readily degradable or mobile and can be applied to soil at such a rate that the concentration approaches some steady state value;

(2) the constituent is very rapidly lost from the soil system, but overloading in a single application may cause acute hazards to human health or the environment; or

(3) the constituent is not degraded appreciably or is relatively immobile and thus, successive waste applications will cause the concentration in soil to increase.

The waste fraction that controls seasonal loading rates (Case 1 above) is referred to as the rate limiting constituent (RLC). Once the RLC is determined, the land area required to treat the given waste can be determined simply by dividing yearly waste receipts (kg/yr) by the acceptable waste loading rate (kg/ha/yr) based on the RLC.

In Case 2 above, where a constituent limits the amount of waste that may be applied in a single dose, yet the constituent is either rapidly decomposed, lost from the system, or immobilized, it is labeled the application limiting constituent (ALC). The ALC sets the minimum number of applications that can be safely made during a given waste application season (see Section 3.3.3 for discussion of waste application season). If the waste contains an ALC, then the minimum number of applications per year is found by dividing the waste loading rate determined using the RLC (kg/ha/yr) by the waste application limit basis on the ALC (kg/ha/application) and rounding to the next higher integer. In some cases, the ALC may be the same as the RLC.

The final parameter (Case 3 above) needed for determining waste application constraints is what is termed the capacity limiting constituent (CLC). This fraction of the waste is a conservative, accumulating species and sets the upper boundary for the total quantity of waste that may be treated at a given site (kg waste/ha). For a waste that contains a large concentration of a given metal, this metal may be both the CLC and the RLC. However, many industrial wastes have a low metals content so that some organic compound, water, or other constituent may control the application rate while a metal may be the CLC. The CLC controls the maximum design life of the land treatment unit unless some arbitrarily shorter life is chosen. Maximum design life is found by dividing the CLC controlled waste loading capacity ($LCAP_{CLC}$) expressed in kg/ha by the design loading rate (LR) based on the RLC and expressed in kg/ha/yr. Section 7.5.4 more clearly defines this relationship.

7.5.2   Management Needs and Monitoring Criteria

During the course of the pilot studies which include the necessary treatment demonstration tests (EPA, 1982), conditions that influence waste treatment are defined and waste consituents that present a significant risk to the treatment process or the environment are identified. Special management needs identified during pilot studies may include application techniques and timing, pH control, fertility control, and soil aeration. Further evidence gained from the treatment demonstration will dictate which of the waste constituents should be monitored and will determine how the operational program may be streamlined or simplified. All hazardous constituents (Appendix B) of the waste must be monitored unless key constituents can be demonstrated to indicate the success of the treatment processes. These indicators are termed principle hazardous constituents or PHCs (EPA, 1982). PHCs to be monitored should definitely include sampling and analysis for the constituents that have been indicated as the ALC, RLC, and CLC. Chemical analyses and the less specific toxicity bioassays are appropriate analytical approaches to monitoring.

7.5.3   Calculating Waste Loads Based on Individual Constituents

As previously noted, results of pilot studies are interpreted considering each waste constituent independently. The following sections deal with the methods and considerations involved when the entire range of waste constituents are evaluated for the design of the HWLT units. Some elements and compounds are discussed specifically while others are addressed by classes according to their similar behavior. The constituents are discussed in order from most concern to least concern for the treatment of hazardous constituents. For example, organics are discussed first since the organic fraction of the waste is often the main reason for choosing HWLT. Where hazardous organics are land treated, waste loading should be designed so that degradation is maximized. Sample calculations for determining waste loading are presented in Appendix E.

7.5.3.1 Organics

Most hazardous waste streams that are land treated contain a sizeable organic fraction and degradation of organics is usually the principal objective for land treating wastes. The range of possible hazards from waste organics can be generally categorized as the acute or chronic toxicity to soil biota, plants and animals, or the immediate danger of fire or explosion. The potential pathways for loss of organics that must be considered include volatilization, leaching, runoff and degradation. Although the pathways are interrelated, they are acted on by different mechanisms and should be considered separately. Waste application rates, both per application (ALC) and per year (RLC), are established by adopting the most restrictive rate calculated from the four pathways; each of which further discussed below. Plant uptake should also be considered if vegetation will be used as a part of the ongoing management plan. Figure 7.4 illustrates the format for assessing organics.

7.5.3.1.1 Volatilization. Volatility experiments can yield information on vapor concentrations in the atmosphere above a soil, as a function of soil moisture, temperature, surface roughness, wind speed, temperature lapse

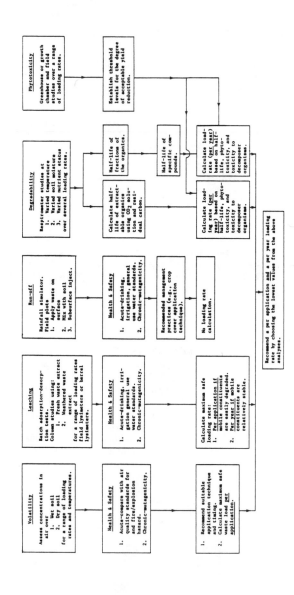

Figure 7.4. A comprehensive testing format for assessing the interactions of organic waste constituents with soil.

rate, waste loading rate, or application technique. The acceptable application rate under a given set of management and environmental conditions may be established using air quality standards, mutagenicity assays, and/or information on concentrations that may cause combustion. If an appreciable quantity of the waste is volatile and hazardous, the quantities of waste per application may be limited and the volatile constituent would be the ALC. The interpretation of test results in this case would specify suitable waste application techniques and timing.

7.5.3.1.2 <u>Leaching</u>. If laboratory leaching tests show the potential for significant movement of some constituents or their metabolites, field lysimeters or leachate samplers beneath an undisturbed soil profile may be used to establish safe waste loading rates. For a mobile hazardous organic compound, loading rates should be controlled to avoid statistically significant increases of the compound in leachate water or soil below the treatment zone. Both the mobility and degradability of an organic compound influence the degree of hazard from leaching. For instance, where a compound is highly mobile, but rapidly degradable in soil, calculations of application limits should be made on a single application basis to reduce the leaching hazard, and the compound is, therefore, a potential ALC. More stable constituents that could potentially leach in the system may limit applications on a yearly basis and may be the RLC.

7.5.3.1.3 <u>Runoff</u>. Since runoff water must be collected and either treated or reapplied, hazards from waste constituents in the runoff do not exert any control on the application rate. For waste fractions which may be eroded by surface water, the emphasis with respect to runoff is to recommend management practices that will minimize erosive waste transport. The degree of management required is, therefore, a function of the degree of hazard presented by mobile waste components. In many cases, the increased management intensity will be more than compensated by decreases in runoff water treatment requirements.

7.5.3.1.4 <u>Degradability</u>. Degradation of organics may be the major objective for land treating a waste; consequently, pilot studies emphasize the characterization of this mechanism by which organics are lost from the HWLT system. Degradability greenhouse and/or field studies should establish the following three facts about the behavior of the waste organics in the given land treatment system:

(1) the quantity of waste that can be applied to a unit of soil in a single application to achieve the best overall system performance;

(2) the half-lives ($t_{1/2}$) of the bulk organics, organic subfractions, or specific organic constituents, leading to a determination of the constituents that are a) most resistant and b) present in significant concentrations in the waste; and

(3)   the threshold concentrations in soil at which these resistant fractions cause unacceptable toxicity to either plants
or, more importantly, waste degrading soil microorganisms.

Given these data, a long-term waste loading rate can be calculated for
the waste based on the organic fraction that is found to be the most
restrictive.   The half-lives for several oily wastes, as determined either
by residual carbon analysis or by monitoring CO_2 evolution, are presented
in Table 7.2.   The results obviously depend on the type of oily waste, the
application rate, and, in some cases, the method of analysis.   The half-
lives, which range from 125-600 days, indicate the need for determinations
on the particular waste proposed for land treatment.   The treatment demonstration should include tests to determine the half-life of the waste under
conditions as near as possible to those expected in the field.   The degradability of the organic fraction of a waste may cause that fraction to be
the RLC.   In addition, toxicity results may further classify some organic
fractions as the ALC.   It should be noted that two entirely different
organic fractions or constituents in the waste may function respectively as
the RLC and the ALC.

The choice of an appropriate half-life is critical to the analysis of
degradability.   Depending on waste characteristics, one of three $t_{1/2}$
values may be chosen.   If degradation is shown to be fairly uniform for all
classes of organics in the waste, the $t_{1/2}$ of the solvent extractables can
be used.   If a given class of compounds which constitutes a large portion
of the waste is particularly resistant to decomposition, the $t_{1/2}$ for that
class can be used.   Finally, if a specific compound is present in a high
concentration and is only slowly degradable, the $t_{1/2}$ for that compound can
be used.

In all three cases, "large" or "high" concentrations of constituents
do not indicate merely a quantitative ranking or comparison.   Instead, the
comparison also considers the relative toxicities of the constituents to
decomposer organisms and, in some cases, plants.   To sustain long-term use
of a land treatment unit, buildup to unacceptably high levels of constituents that are toxic to decomposer organisms should be avoided.   Otherwise,
the system may fall short of the treatment objective.   Where integrated
cover crop management is included in the operating plan, phytotoxicity
should also be determined.   The phytotoxicity threshold is considered to be
the concentration of the waste or constituents that reduce plant yields to
about 50% of controls.   Yield reductions greater than this are an indication that management to provide a protective crop cover will be quite
difficult.

Two types of management plans are described which represent the
extremes of management for HWLT units.   In the first case, the management
plan includes a temporary plant cover over the active treatment area, and
in the second case, a vegetative cover is not established until the
initiation of closure activities (see Section 8.7 for guidance on vegetative management options).   Loading rate calculations for the two plans
would be as follows:

TABLE 7.2 SOIL HALF-LIFE OF SEVERAL OILY WASTES AS DETERMINED BY VARIOUS METHODS

Waste	Application Rate (%)	Half-life (days)	Method of Determination	Reference
Dissolved Air Flotation	10	261	CO_2 evolution	Brown (unpublished data)
Dissolved Air Flotation	20	372	CO_2 evolution	Ibid.
Dissolved Air Flotation	9	125	Residual carbon (field)	Ibid.
API-Separator (refinery)	5	130	CO_2 evolution	Brown, Deuel, & Thomas (1982)
API-Separator (refinery)	5	143	Residual carbon (lab)	Ibid.
API-Separator (petrochemical)	5	600	CO_2 evolution	Ibid.
API-Separator (petrochemical)	5	264	Residual carbon (lab)	Ibid.
Crankcase oil	10	237	Residual carbon (lab)	Raymond, Hudson, & Jamison (1976)
Oil sludge	5	570	CO_2 evolution	Dibble and Bartha (1979)
Oil sludge	5	356	Residual carbon (lab)	Ibid.

(1) When vegetation is a part of ongoing management plan, toxic
     organics, exhibiting either microbial or plant toxicity, may
     limit the loading rate. Assuming that loading rates are
     relatively constant so that the designed area is adequate to
     handle each year's waste production, the following equation
     applies:

$$C_{yr} = \frac{1/2 \ C_{crit}}{t_{1/2}} \qquad\qquad (7.6)$$

where

      C_{yr} = the rate of application of the compound or fraction
              of interest to soil (kg/ha/yr);
      C_{crit} = the critical concentration of the compound or
              fraction in soil at which unacceptable microbial
              toxicity or plant yield reduction occurs (kg/ha);
              and
      $t_{1/2}$ = half-life (yr)

The loading rate is then calculated as follows:

$$LR = \frac{C_{yr}}{C_w} \qquad\qquad (7.7)$$

where

      LR = loading rate (kg/ha/yr); and
      C_w = concentration of the compound or fraction of
           interest in the bulk waste (kg/kg).

If $t_{1/2}$ is less than one year, then the year's loading rate
should be applied in more than one application. To calculate the number of applications let $1/t_{1/2}$ equal the smallest $t_{1/2}$ and use the following equation:

$$NA = 1/t_{1/2} \qquad\qquad (7.8)$$

where

      NA = number of applications/year.

(2) When a vegetated surface is desired only after site closure
     begins, then applications of waste may exceed the phytotox-
     icity threshold value. The only constraints would be that
     the microbial toxicity threshold not be exceeded and that a
     final vegetative cover can be established after a given num-
     ber of years following the beginning of closure. Calcula-
     tions are as follow:

$$C_{max} = C_{crit} \; 2^{(n/t_{1/2})} \qquad\qquad (7.9)$$

where

C_{max} = the maximum allowable concentration of the compound fraction of interest applied to the soil (kg/ha);

n = number of years between final waste application and crop establishment (yr); and

$t_{1/2}$ = half-life (yr).

After C_{max} is determined, loading rates are calculated by applying equations 7.6 and 7.7 substituting C_{max} for C_{crit} in equation 7.6. For wastes with very short half-lives, the resulting loading rate may appear to be excessive; however, assuming that other factors are held constant, a high C_{max} merely indicates that organics will not be limiting. The calculated C_{max} should not be interpreted literally in such cases. Before such high rates of application are reached, some other parameter is likely to be limiting; this possibility will need to be evaluated. For instance, degradation of waste organics may be inhibited at much lower levels than C_{max} due to wetness and the resulting loss of soil aeration.

7.5.3.2 Water

Most land treatable wastes have a high water content, and even fairly viscous sludge may contain greater than 75% water. Therefore, particularly in humid regions, waste water may be the RLC. Using the climatological data on precipitation and evapotranspiration and soil permeability information from Section 4.1.1.5, a water balance model may be developed as discussed in Section 8.3.

The two keys to properly using the water balance models for the given site are first, determining the waste application season (Section 3.3.3) and, second, deciding on a water management scheme (Section 8.3). The waste application season depends on whether cover crops are to be grown during, or only after, active treatment. Determination of the waste application season is essentially the same for both options except that where no cover crop will be grown during the active life of the HWLT unit, phytoxicity need not be considered. The waste can accumulate with little degradation of organics but without presenting a phytotoxicity, leaching, volatilization, or runoff hazard, then the waste application season is based on the period of time when water may be readily applied. If accumulation leads to phytotoxicity or environmental hazards, then the season is based on the time that degradation effectively begins and ends, generally when soil temperature is $\geq 5°C$ and soil moisture can be maintained at or below field capacity. The water balance model can be integrated over the application season to yield the depth of water (H_2O) that may be applied per year to maintain the average soil moisture content at field capacity. The waste analysis shows the percent water by volume and the waste density (kg/liter). Therefore, the waste loading rate on the basis of water content is:

$$LR = \frac{LR_{H_2O}}{F_{H_2O}} \times \rho \qquad (7.10)$$

where

LR = loading rate (kg/ha/yr);
LR_{H_2O} = volumetric H_2O loading rate (l/ha/yr), noting that l cm depth = 10^5 l/ha;
F_{H_2O} = fraction of waste constituted by water; l/l and;
ρ = waste density (kg/l).

Field capacity, defined elsewhere (7.2.1.1.2.1), is chosen because it is the optimum soil moisture content for organics degradation and decreasing the likelihood of pollutant leaching.

7.5.3.3 Metals

Metals management strives to permanently sorb the applied elements within the soil so that no toxicity hazard results. Some elements (e.g., molybdenum and selenium) may cause environmental damage through leaching since these elements occur as anions in the soil system. Leaching of mobile anions should be considered in a manner similar to halide leaching (7.5.3.7). Toxicity assessment should account for phytotoxicity, food chain effects, and direct ingestion of soil by grazing animals. Section 6.1.6 provides background information on metals and suggests maximum concentrations that may be safely added to soils. These amounts are cumulative totals for those metals for which no significant movement occurs. The capacity of a given soil to immobilize a particular element can vary somewhat from the limits suggested in the tables in Section 6.1.6; therefore, in all cases, the associated discussions and literature references should be consulted. At this stage, one must have consciously decided upon a general management plan in order to choose whether metal limits should be based on phytotoxicity or toxicity to decomposer organisms. Many metals are essentially untested at high concentrations in the soil environment simply because, historically, there have been no major cases where these metals have contaminated the soil. However, the increasing uses for various elements in industry indicates that some land treated wastes may contain high concentrations of metals. Therefore, a data base is needed on many elements both from the standpoint of basic research and from observed interactions in natural systems.

Accumulation of metals will often be the factor that controls the total amount of waste that may be treated per unit area. Therefore, even if another waste constituent limits loading rates, a metallic element frequently is the capacity limiting constituent. To compare metals to determine the element potentially limiting total waste applications (potential CLC), one can simply calculate the following ratio for each metal in the waste:

$$\text{Metal loading ratio} = \frac{\text{Metal loading capacity (mg metal/kg soil)}}{\text{Metal content of the waste residual solids}} \quad (7.11)$$
$$\text{(mg metal/kg RS)}$$

Metal loading capacity is determined for each metal from Section 6.1.6 and Table 6.46. The residual solids (RS) determination is found in Section 5.3.2.3.2.2. If the ratio is in all cases less than or equal to 1, then no metal will ever limit the useful life of the land treatment unit. Where one or more of the ratios are greater than 1, then the metal with the largest ratio is the potential CLC.

All of the allowable metal load may be applied during any chosen time frame (e.g., a single application; continuously for ten years; or incrementally over a twenty year period, etc.). However, other constituents in the waste may limit the rate at which the waste is applied.

7.5.3.4 Nitrogen

The following estimates of nitrogen (N) additions and losses from a land treatment unit (Table 7.3), are used to calculate a nitrogen mass balance equation. Actual values for a given site can be estimated using the guidance given in Section 6.1.2.1.

TABLE 7.3 NITROGEN MASS BALANCE

Inputs	Removals
Total N in waste	Denitrification
N in precipitation	Volatilization of ammonia
N fixation	N storage in soil
Mineralization	Leaching
Nitrification	Runoff
	Crop uptake
	Immobilization

Inputs of nitrogen must equal nitrogen removals to maintain acceptable levels of nitrates in runoff or leachate.

The comprehensive equation presented below includes a number of factors in the mass balance calculation. The depth of waste application is computed by taking the sum of the N involved in crop uptake, leaching, volatilization, and denitrification, subtracting the N from rainfall, and then dividing by the N concentration of the waste. When using this equation, estimates of denitrification and volatilization must also be made. The equation is written as follows:

$$LR = 10^5 \frac{10 \ (C + V + D) + (Ld)(Lc) - (Pd)(Pc)}{I + \sum_{t=1}^{n} (M)(O)} \qquad (7.12)$$

where

      LR = waste loading rate (kg/ha/yr);
      C = crop uptake of N (kg/ha/yr);
      V = volatilization (kg/ha/yr);
      D = denitrification (kg/ha/yr);
      L_d = depth of leachate (cm/yr);
      L_c = solute (N) concentration in leachate (mg/1);
      P_d = depth of precipitation (cm/yr);
      P_c = concentration of N in precipitation (mg/1);
      I = concentration of inorganic N in the waste (mg/1 on a wet
          weight basis);
      M = mineralization rate given in Table 6.4;
      O = concentration of organic N in the waste (mg/1 on a wet
          weight basis; if the concentration of N is known on a weight
          basis (mg/kg) then the value of O equals mg/kg x waste
          density in kg/1); and
      t = years after waste application.

The concentration of N in the leachate (L_c) must be chosen with regard to the groundwater quality objectives for the underlying aquifer. A value of 10 mg N/1 is a likely choice since this reflects the primary drinking water standard for NO_3-N. If the land treatment unit does not harvest a crop from the active site, the plant uptake term is removed from the equation. For comparison purposes, nitrogen may qualify as the RLC.

7.5.3.5 Phosphorus

Phosphorus (P) is effectively retained in soil as are the metals, except that the soil has a more easily determined finite P adsorption capacity. This adsorption capacity can be estimated from Langmuir isotherm data. The calculations must include the horizontal area (ha), depth to the water table (cm), and the previous treatment of the soil at the site. It is expected that complete renovation occurs in the root zone, or within a depth of 2 m (Beek and de Haan, 1973). Although the effect of organic matter and long-term precipitation reactions on the P adsorption potential are not well understood, the profile distribution of aluminum, iron, and calcium may greatly influence sorption capacity. It is therefore necessary to calculate the total permissible waste load as a function of the sorption capacity of each soil horizon. The loading capacity can be calculated as follows:

$$LCAP = 10 \sum_{i=1}^{n} d_i \rho (b_{max} - P_{ex}) \qquad (7.13)$$

where

    $LCAP$ = loading capacity (kg P/ha);
      d_i = thickness of the ith horizon;
      ρ = bulk density of soil (g/cm^3);
   b_{max} = apparent sorption capacity estimated from Langmuir
           isotherm (μg/g); and
    P_{ex} = NaHCO$_3$-extractable phosphorus reported on a dry weight
           basis (μg/g).

Total phosphorus application is the sum of the values for all horizons. This total permissible load may be divided at the discretion of the site manager who must consider the life of both the industrial plant and the disposal site. Once this calculated capacity is reached, applied P may leach without attenuation to shallow groundwater, consequently, phosphorus may be the CLC.

7.5.3.6 Inorganic Acids, Bases and Salts

    The accumulation of salts and the associated soil physical and chemical problems, are primary management concerns when land treating acids, bases, and salts or other wastes having significant incidental concentrations of these constituents. Excessive applications of acidic or basic wastes may necessitate mitigation of the adverse affects on soil. For example, lime may be used to control soil pH where waste acids are land treated.

    In any case, no broadly satisfactory method has yet been developed for quantifying salt behavior in soil so that waste loading rates can be determined. Consequently, management of salts must consider two broad cases. In the first case, water inputs or soil drainage are inadequate and salts are conserved and accumulate in the surface soil. Salts would therefore behave as a CLC, where limits are determined based on toxicity to plants or waste decomposer organisms. See Section 6.1.4 for methods of salt measurement and salt tolerance of variuos crops. Total waste loads (kg/ha) would be based on the given management plan. In the second case, adequate site drainage is present or can be artifically provided, salt can be an RLC and some type of model would be needed to calculate loading rates with groundwater quality criteria serving as the limits for leachate quality. Since, as stated in Section 6.1.4, no satisfactory model is currently available, consultation with a soil scientist having salt management experience is recommended. Where a sodium imbalance in the waste could threaten soil structure and cause associated problems, the waste loading rate will still be controlled by salt content, but additional salinity may result from amendments added to control the cation balance.

7.5.3.7 Halides

A halide may qualify as the RLC because loading rates should be controlled to maintain acceptable groundwater quality and these anions will leach readily from the soil. Calculations are similar in many respects to those for the nitrogen model. Determinations may be modified to account for precipitation into less soluble forms, such as CaF_2.

Two halide management cases are possible, depending on the site. Where water inputs or soil drainage are not adequate to remove these anions by leaching, concentrations of available halides will build up in soil. In this case, assuming salt buildup does not physically damage the soil structure, the halide can behave as a CLC, with limits based on toxicity to plants or microbes (see Section 6.1.5). Calculations would be the same as for metals. In the second case, conditions would be favorable for leaching to occur and the given halide would be a potential RLC. A halide will have little interaction with the soil matrix and should therefore leach readily. Additionally, it is assumed that repeated waste applications will allow the system to be approximated by a steady state solution, and the following equation can be used:

$$LR = \frac{(L_d)(L_c) \times 10^5}{I} \tag{7.14}$$

where

$\quad$ LR = waste loading rate (kg/ha/yr);
$\quad$ L_d = depth of leachate (cm/yr);
$\quad$ L_c = solute (halide) concentration in leachate (mg/1); and
$\quad\;$ I = concentration of halide in the waste (mg/1 on a wet weight basis).

The L_c term should be chosen based on water quality standards or other criteria (see Section 6.1.5).

7.5.4 Design Criteria for Waste Application and Required Land Area

Following the independent consideration of each waste constituent which may cause an environmental hazard, a comparison must be made to determine the most limiting constituents. For a given waste and site, the procedure for identifying the ALC and RLC is straightforward once loading rates and capacities have been established for each component of the waste. Information should be organized into a tabular format similar to Table 7.4, where each waste constituent and its associated waste loading rate (based

Constituent	Potential ALC[†]	Potential RLC
Organics	X	X
− Volatiization	X	
− Leaching	X	X
− Degradation		X
Water	X	X
Nitrogen	X	X
Inorganic Acids, Bases, and Salts		X
Halides		X

* The actual comparison should be tabulated similarly, but using calculated
  loading rates in place of the X's. The lowest value under each category
  corresponds to the respective limiting constituent.

[†] Depending upon prevailing site conditions, the ALC may vary seasonally.

on the wet weight of waste) are entered in appropriate columns. Among the
waste components entered under each category, the component having the
smallest calculated rate is chosen as the limiting constituent (ALC or
RLC). After the most limiting constituents are identified, the final
decisions on the required land area (eq. 7.15) and the minimum number of
applications per year (eq. 7.16) are made using the following calculations:

$$A = \frac{PR}{LR_{RLC}} \qquad (7.15)$$

where

A = required treatment area (ha);
PR = waste (wet weight) production rate (kg/yr); and
LR_{RLC} = waste loading rate based on the RLC (kg/ha/yr).

If the value calculated for A is greater than the area available for treatment, then land treatment cannot accommodate all of the waste which is
being produced.

$$NA = \frac{LR_{RLC}}{AL} \qquad (7.16)$$

where

> NA = number of applications per year and is equal to the smallest integer greater than or equal to the actual value calculated;
> LR_{RLC} = waste loading rate based on the RLC (kg/ha/yr); and
> AL = application limit based on the ALC (kg/ha/application).

The land treatment unit life and concomitant choice of a CLC are not predicted in such a straightforward manner. Three classes of potentially conservative constituents have been identified, metals, phosphorus and inorganic acids, bases, and salts. By calculating a unit life based on each, the design unit life and CLC can be chosen to be that constituent which is the most restrictive. Phosphorus is redistributed throughout the treatment zone while salts, if conserved, tend to accumulate near the surface and thus can be described using the following equation:

$$UL = \frac{LCAP_{PS}}{LR_{RLC}} \qquad (7.17)$$

where

> UL = unit life (yr);
> $LCAP_{PS}$ = waste loading capacity beyond which the CLC will exceed allowable accumulations (kg/ha); and
> LR_{RLC} = waste loading rate based on the RLC (kg/ha/yr).

Metals, by contrast, are practically immobile and are mixed in the waste with a heterogeneous matrix of water, degradable organics, mobile constituents and nondegradable residual solids (see Section 5.3.2.3.2.2). Waste application is therefore not merely the addition of a pure element to soil. The residual solids fraction (RS) adds to the original soil mass. Wastes containing high RS concentrations can significantly raise the level of the land treatment unit as well as limit the amount of soil which can be used to dilute the waste. As mentioned under Metals in Section 7.5.3.3, if the concentration of a given metal in the RS of a waste is less than the maximum allowable concentration in soil, then the given metal cannot limit waste application. The metal with the largest ratio greater than one from eq. 7.11 is the possible CLC and unit life is determined as follows:

(1) determine the concentration (c_a) of the metal in the waste residual solids (mg/kg);

(2) calculate the residual solids loading rate from the equation;

$$z_a = \frac{LR_{RLC} \times (\text{weight fraction of residual solids in waste})}{\rho_{BRS}} \times 10^{-5} \qquad (7.18)$$

where

> z_a = volumetric waste loading rate on a residual solids basis (cm/yr);
> ρ_{BRS} = bulk density of residual solids, assumed to be the same as that of the soil after tillage and settling (kg/l); and
> 10^{-5} = conversion factor from l/ha to cm;

(3) choose a tillage or waste-soil mixing method and determine the "plow" depth (z_p) in cm;

(4) from the background soil analysis, obtain the background concentration (mg/kg) of the given metal (c_{po});

(5) from reference to the specific metal in Chapter 6, determine the maximum allowable soil concentration (c_{pn}) of that metal (mg/kg);

(6) using these quantities, solve for n in the following equation (Chapra, unpublished paper) where n is the number of applications which result in the concentration of the surface layer being c_{pn}:

$$n = \frac{z_p}{z_a} \ln \frac{c_{po} - c_a}{c_{pn} - c_a} \qquad (7.19)$$

(7) the corresponding unit life is:

$$UL = nt_a \qquad (7.20)$$

where

> t_a = time between applications.

The equation idealizes the process of application and plowing as a continuous process. To do this, a number of assumptions must be made.

(1) Assume that sludge is applied at equal intervals, t_a in length.

(2) Assume that the sludge always has the same concentration c_a.

(3) Assume that the sludge is always applied at a thickness of z_a.

(4) Assume complete mixing of the surface layer to depth z_p due to plowing.

(5) Assume that the plowed soil and the sludge have equal porosity.

(6) The annually applied waste degrades and dries approximately down to residual solids.

A design unit life (years) is then chosen from among salts, phosphorus and metals. The shortest life of the three is the desired value. For many waste constituents, inadequate information is available to properly assess loading rates. Pilot experiments and basic research are suggested in this document so that an understanding of the fate of various constituents in soil can begin to be developed. Where land treatment is proposed for a waste constituent about which only scant knowledge is available, pilot studies should be conducted to evaluate that constituent, and the loading rate for such a constituent should be conservative to provide a margin of safety.

CHAPTER 7 REFERENCES

Allen, S. E., G. L. Terman, and L. B. Clements. 1976. Greenhouse techniques for soil-plant-fertilizer research. TVA. Muscle Shoals, Alabama. TVA Bull. Y-104.

Bailey, G. W., T. L. White, and T. Rothberg. 1968. Adsorption of organic herbicides by montmorillonite; role of pH and chemical character of adsorbate. Soil Sci. Soc. Am. Proc. 32:222-234.

Baruah, J. N., Y. Alroy, and R. L. Mateles. 1967. The incorporation of liquid hydrocarbons into Agar media. Appl. Microbiol. 15(4):961.

Beckman Instruments, Inc. 1982. Beckman Microtox® system operation manual.

Beek, J. and F. A. M. de Haan. 1973. Phosphorus removal by soil in relation to waste disposal. Proc. of the International Conference on Land for Waste Management. Ottawa, Canada, Oct. 1973.

Benigni, R., M. Bignami, I. Camoni, A. Carere, G. Conti, R. Iachetta, G. Morpurgo, and V. A. Ortali. 1979. A new in vitro method for testing plant metabolism in mutagenicity studies. Jour. of Toxicology and Environ. Health 5:809-819.

Brown, K. W., D. C. Anderson, S. G. Jones, L. E. Deuel, and J. D. Price. 1979. The relative toxicity of four pesticides in tap water from flooded rice paddies. Int. J. Environ. Studies. 14:49-54.

Brown, K. W., L. E. Deuel, and J. C. Thomas. 1982. Final report on soil disposal of API pit wastes. U.S. EPA Grant No. R 805474013. Cincinnati, Ohio.

Castro, C. E. and N. O. Belser. 1966. Hydrolysis of cis- and trans-dichloropropene in wet soil. J. Agr. Food Chem. 14:69-70.

Chapra, S. C. A simple model for predicting concentrations of conservative contaminants at land treatment sites. Unpublished paper.

Dibble, J. T. and R. Bartha. 1979. Effect of environmental parameters on biodegradation of oil sludge. Appl. Environ. Microbiol. 37:729-738.

Duffy, J.J., M.F. Mohtadi, and E. Peake. 1977. Subsurface persistence of crude oil spilled on land and its transport in groundwater. pp. 475-478 In J. O. Ludwigson (ed.) Proc. 1977 Oil Spill Conference. New Orleans, Louisiana. 8-10 March, 1977. Am. Pet. Inst. Washington, D.C.

Edwards, N. T. and B. M. Ross-Todd. 1980. An improved bioassay technique used in solid waste leachate phytotoxicity research. Environ. Exper. Bot. 20:31-38.

EPA. 1982. Hazardous waste management system; permitting requirements for land disposal facilities. Part 264. Federal Register Vol. 47, No. 143. pp. 32274-32388. July 26, 1982.

Farmer, W. J., K. Ique, W. F. Spenser, and J. P. Martin. 1972. Volatility of organochlorine insecticides from soil: effect of concentration, temperature, air flow rate, and vapor pressure. Soil Sci. Soc. Am. Proc. 36:443-447.

Helling, C. S. 1971. Pesticide mobility in soils I. Parameters of thin-layer chromatography. Soil Sci. Soc. Am. Proc. 35:735-737.

Helling, C. S. and B. C. Turner. 1968. Pesticide mobility: determination by soil thin-layer chromatography. Science 162: 562-563.

Higashi, K., K. Nakashima, Y. Karasaki, M. Fukunaga, and Y. Mizuguchi. 1981. Activiation of benzo(a)pyrene by microsomes of higher plant tissues and their mutagenicity. Biochemistry International 2(4):373-380.

Laskowski, D. A., C. A. I. Goring, P. J. McCall, and R. L. Swann. 1980. Terrestrial environmental risk analysis for chemicals. R. A. Conway (ed.). Van Nostrand Reinhold Company, New York.

Matijesevic, Z., Z. Erceg, R. Denic, V. Bacun, and M. Alacenic. 1980. Mutagenicity of herbicide cyanazine plant activation bioassay. Mut. Res. 74(3):212.

Odu, C. T. I. and K. B. Adeoye. 1969. Heterotrophic nitrification in soils - a preliminary investigation. Soil Biol. Biochem. 2:41-45.

Osborne, G. J., N. J. Poole, and E. Drew. 1980. A method for studying microbial activity in intact soil cores. J. Soil Sci. 31:685-687.

Plewa, M. J. and J. M. Gentile. 1976. The mutagenicity of atrazine: a maize-microbe bioassay. Mutat. Res. 38:287-292.

Plice, M. J. 1948. Some effects of crude petroleum on soil fertility. Soil Sci. Soc. Am. Proc. 43:413-416.

Raymond, R. L., J. O. Hudson, and W. W. Jamison. 1976. Oil degradation in soil. App. Environ. Microbiol. 31:522-535.

Reichhart, D., J. P. Salaun, I. Benveniste, and F. Durst. 1980. Time course of induction of cytochrome P-450, NADPH-cytochrome c reductase, and cinnamic acid hydroxylase by phenobarbital, ethanol, herbicides, and manganese in higher plant microsomes. Plant Physiol. 66:600-604.

Stotzky, G. 1965. Microbial respiration. pp. 1550-1572. In C. A. Black (ed.) Methods of soil analysis part 2. Chemical and microbiological properties. Am. Soc. Agron. Madison, Wisconsin.

Tomlinson, C. R. 1980. Effects of pH on the mutagenicity of sodium azide in Neurospora crassa and Salmonella typhimurium. Mutat. Res. 70(2):179-192.

Van Cleve, K., P. I. Coyne, E. Goodwin, C. Johnson, and M. Kelley. 1979. A comparison of four methods for measuring respiration in organic material. Soil Biol. Biochem. 11:237-246.

Van Der Linden, A. C. and G. J. E. Thijsse. 1965. The mechanisms of microbial oxidations of petroleum hydrocarbons. Adv. Enzymology 27:469-546.

Weber, J. B. 1972. Interaction of organic pesticides with particulate matter in aquatic and soil: fate of organic pesticides in the aquatic environment. Am. Chem. Soc. Washington, D.C. pp. 55-120.

Wildeman, A. G., I. A. Rasquinha, and R. N. Nazar. 1980. Effect of plant metabolic activation on the mutagenicity of pesticides. Carcinogenesis, AACR Abstracts 89:357.

Youngson, C. R. and C. A. I. Goring. 1962. Diffusion and nematode control by 1,2 dibromoethane, and 1,2 dibromo-3-chloropropane. Soil Sci. 93:306.

DESIGN AND OPERATION OF HWLT UNITS

D. Craig Kissock
K. W. Brown
James C. Thomas
Gordon B. Evans, Jr.

This chapter discusses the management concerns that are important to the design and effective operation of an HWLT unit. The topics discussed in this chapter (Fig. 8.1) pull together information that has been gathered from waste, soil and site characterizations and from pilot studies of waste-soil interactions. Since system interactions are very site, waste- and soil-specific, the management plan should specify how the design criteria and operational plan address site-specific factors and anticipated operational problems. This chapter considers several options for operating HWLT units in an environmentally sound manner under different general conditions. The specific design and management approach will be established on a case by case basis, however, since each individual unit will have different needs. Permit writers and facility owners or operators should study the principles discussed in this chapter and use those that apply to the specific needs of the HWLT unit being considered.

8.1 DESIGN AND LAYOUT

Actual design and layout depends on the terrain, the number and type of wastes being treated, and the area involved. In laying out a land treatment unit, consideration should be given to minimizing the need to construct terraces to divert water from uphill watersheds. Access roads should be laid out along the top of the grade or on ridges to provide good drainage and minimize traffic problems during wet periods, particularly if waste is to be applied continuously. Disposal areas should be designed so the waste can be easily and efficiently spread by irrigation, by surface or subsurface spreading vehicles, or by graders or dozers after it is dumped. If sludge is to be dumped at one end of an area, spread, and then tilled, plots should be shaped to allow uniform spreading with the available equip-

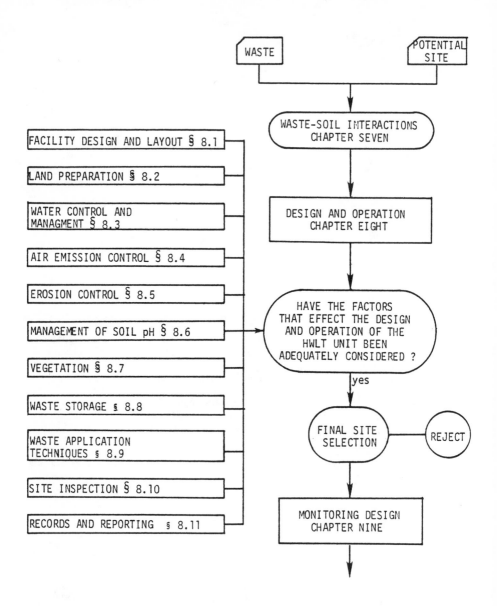

Figure 8.1 Topics to be considered for designing and managing an HWLT.

ment. If equipment will become contaminated during unloading or mixing, a traffic pattern should be established and a wash area or rack constructed so that all equipment can be decontaminated before leaving the confined watershed of the HWLT unit. If equipment remains on-site, a parking facility and possibly a service area should be included in the design.

If erosion is a potential hazard due to climate, topography or soil characteristics, waste should be applied in strips across the slope parallel to terraces or on the contour. Contour application involves alternating freshly treated strips and vegetated areas. Once a vegetative cover is established on the treated strips, applications begin on the previously vegetated buffer strips. This technique serves to reduce the potential for erosion and also provides vegetated areas with better traction for equipment during inclement weather.

While many land treatment units are designed to receive only one type of waste, there is no reason why they cannot be designed and managed to receive any number of wastes which would be rendered less hazardous in the land treatment system. If more than one waste is to be disposed, separate plots can be used for each type; or, it may be possible to dispose several types of waste simultaneously on one plot, if application rates are designed to stay within the constraints of the rate (RLC) and capacity limiting constituents (CLC) of the waste mixture for the particular site. In some cases, it may be beneficial to codispose wastes containing different concentrations of the constituents that limit the application rate. For instance, one waste may contain nitrogen, but be low in phosphorus, zinc and lead, while another waste is deficient in nitrogen but contains significant concentrations of phosphorus, zinc and lead. It should be possible to select application rates for several wastes that achieve the disposal objective without exceeding acceptable leachate concentrations and without accumulating high levels of the constituents involved. Obviously, a more detailed management and record keeping system is needed when several wastes are codisposed. There are other instances where codisposal may be advantageous. Certainly the codisposal of acidic and basic wastes will result in neutralization and can be done provided excessive salts do not result. For such disposal, it is often desirable to first dispose of the basic waste and then apply the acidic waste to prevent the release of immobilized waste constituents such as metals.

When waste characteristics are likely to change in the future, or when it may be desirable to use the land for future disposal of another waste, the site should not be fully loaded with any one constituent which would prevent future addition of that particular constituent. For instance, if there is the possibility that the CLC concentration of the waste may later be reduced or that another waste having a different CLC may also be disposed, it is desirable to cease loading when only a fraction of the allowable capacity has accumulated.

Although the soil is an excellent medium for deactivating and decomposing waste materials, there is the persistent danger at facilities where a variety of wastes are disposed that incompatible wastes could come in contact with each other. Problems can be reduced by thoroughly incorporating wastes that would otherwise be incompatible into the soil as soon as they are received, since the soil will greatly buffer the reactions that take place and can adsorb evolved heat or gases. The greatest dangers occur when wastes come into contact with each other in receiving basins or storage facilities. There have been several instances of deaths resulting from incompatible wastes being mixed together at poorly managed disposal facilities. To avoid such problems, incompatible wastes should be handled separately and precautions should be taken to ensure that pumps and spreading equipment are cleaned before being used for a different waste.

When wastes such as strong acids, strong bases, cyanides, ammonia compounds, chlorine containing compounds, and other compounds that may react with each other to generate toxic gases, or that may cause violent reactions, are received the facility should have a detailed plan for separate handling and the safeguards necessary to prevent mixing. One source of information on the compatibility of binary mixtures of compounds is A Method for Determining the Compatibility of Hazardous Wastes (Hatayama et al., 1980). This is a useful guide for predicting possible reactions resulting from mixing wastes, but this information does not necessarily apply to such mixtures within the soil matrix. Additionally, the information does not address the issues of constituent concentrations or of the heterogeneity or complexity of most waste streams. Lab and field testing may be needed when knowledge about the possible reactions resulting from mixing particular waste streams is insufficient. A list of incompatible wastes is given in Table 8.1 and Fig. 8.2.

8.1.1   Single Plot Configuration

Size and subdivision of the land treatment area depend on the character of the waste involved, including the waste constituents and their behavior in soils (Chapter 6 and 7), the soil characteristics, the amount of waste to be disposed, the disposal schedule, and the climatic conditions of the area. Where applications are made only during one season of the year or, on only a few specific occasions, and the limiting cumulative constituents are present in low concentrations, it may be desirable to spread the waste uniformly over all the available acreage (Fig. 8.3). Such a configuration can be used without subdividing the land treatment area if soils are uniform, provided this procedure does not interfere with establishing a vegetative cover if one is desired.

8.1.2   Progressive Plot Configuration

A controlling factor in the layout of any HWLT unit is the amount of runoff to be collected and options available for disposal of runoff water. Options for runoff are discussed in Section 8.3.5 and include on-site disposal by evaporation and/or reapplication, use of a wastewater treatment plant prior to release, and use of a retention pond to allow settlement of solids and analysis prior to release. In climates where significant

TABLE 8.1   POTENTIALLY INCOMPATIBLE WASTES*

The mixture of a Group A waste with a Group B waste may have the potential consequence as noted.

Group 1-A

Acetylene sludge
Alkaline caustic liquids
Alkaline cleaner
Alkaline corrosive liquids
Alkaline corrosive battery fluid
Caustic wastewater
Lime sludge and other corrosive
  alkalines
Lime wastewater
Lime and water
Spent caustic

Group 1-B

Acid sludge
Acid and water
Battery acid
Chemical cleaners
Electrolyte, acid
Etching acid liquid or solvent
Liquid cleaning compounds
Pickling liquor and other
  corrosive acids
Sludge acid
Spent acid
Spent mixed acid
Spent sulfuric acid

Potential consequences: Heat generation, violent reaction.

Group 2-A

Asbestos waste and other toxic wastes
Beryllium wastes
Unrinsed pesticide containers
Waste pesticides

Group 2-B

Cleaning solvents
Data processing liquid
Obsolete explosives
Petroleum waste
Refinery waste
Retrograde explosives
Solvents
Waste oil and other flammable
  and explosive wastes

Potential consequences: Release of toxic substances in case of fire or explosion.

Group 3-A

Aluminum
Beryllium
Calcium
Lithium
Magnesium
Potassium
Sodium
Zinc powder and other reactive metals
  and metal hydrides

Group 3-B

Any waste in Group in 1-A or 1-B

Potential consequences: Fire or expolsion; generation of flammable hydrogen gas.

--continued--

TABLE 8.1 (continued)

Group 4-A	Group 4-B
Alcohols	Any concentrated waste in
Water	Groups 1-A or 1-B
	Calcium
	Lithium
	Metal hydrides
	Potassium
	Sodium
	SO_2Cl_2, $SOCl_2$, PCl_2,
	CH_3SiCl_3, and other water-
	reactive wastes

Potential consequences: Fire, explosion or heat generation; generation of flammable or toxic gases.

Group 5-A	Group 5-B
Alcohols	Concentrated Group 1-A or 1-B
Aldehydes	wastes
Halogenated hydrocarbons	Group 3-A wastes
Nitrated hydrocarbons and other	
reactive organic compounds and solvents	
Unsaturated hydrocarbons	

Potential consequences: Generation of toxic hydrogen cyanide or hydrogen sulfide gas.

Group 7-A	Group 7-B
Chlorates and other strong	Acetic acid and other organic
oxidizers	acids
Chlorine	Concentrated mineral acids
Chlorites	Group 2-B wastes
Chromic acid	Group 3-A wastes
Hypochlorites	Group 5-A wastes and other
Nitrates	flammable and combustible
Nitric acid, fuming	wastes
Perchlorates	
Permanganatesfuming	
Peroxides	

Potential consequences: Fire, explosion or violent reaction.

* Cheremisinoff et al. (1979).

HAZARDOUS WASTE COMPATIBILITY CHART

REACTIVITY GROUP NO.	REACTIVITY GROUP NAME
1	Acids, Mineral, Non-oxidizing
2	Acids, Mineral, Oxidizing
3	Acids, Organic
4	Alcohols and Glycols
5	Aldehydes
6	Amides
7	Amines, Aliphatic and Aromatic
8	Azo Compounds, Diazo Compounds, and Hydrazines
9	Carbamates
10	Caustics
11	Cyanides
12	Dithiocarbamates
13	Esters
14	Ethers
15	Fluorides, Inorganic
16	Hydrocarbons, Aromatic
17	Halogenated Organics
18	Isocyanates
19	Ketones
20	Mercaptans and Other Organic Sulfides
21	Metals, Alkali and Alkaline Earth, Elemental
22	Metals, Other Elemental & Alloys as Powders, Vapors, or Sponges
23	Metals, Other Elemental & Alloys as Sheets, Rods, Drops, Moldings, etc.
24	Metals and Metal Compounds, Toxic
25	Nitrides
26	Nitriles
27	Nitro Compounds, Organic
28	Hydrocarbons, Aliphatic, Unsaturated
29	Hydrocarbons, Aliphatic, Saturated
30	Peroxides and Hydroperoxides, Organic
31	Phenols and Cresols
32	Organophosphates, Phosphothioates, Phosphodithioates
33	Sulfides, Inorganic
34	Epoxides
101	Combustible and Flammable Materials, Miscellaneous
102	Explosives
103	Polymerizable Compounds
104	Oxidizing Agents, Strong
105	Reducing Agents, Strong
106	Water and Mixtures Containing Water
107	Water Reactive Substances

EXTREMELY REACTIVE! →

(Matrix columns 1 – 13 across bottom of triangular compatibility chart)

--continued--

Figure 8.2.   Hazardous waste compatibility (Hatayama et al., 1980).

Design and Layout   453

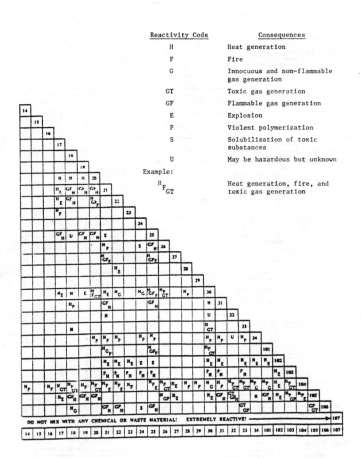

Reactivity Code	Consequences
H	Heat generation
F	Fire
G	Innocuous and non-flammable gas generation
GT	Toxic gas generation
GF	Flammable gas generation
E	Explosion
P	Violent polymerization
S	Solubilization of toxic substances
U	May be hazardous but unknown

Example:

$H_{F_{GT}}$ Heat generation, fire, and toxic gas generation

Figure 8.2.   Continued.

454   Design and Operation

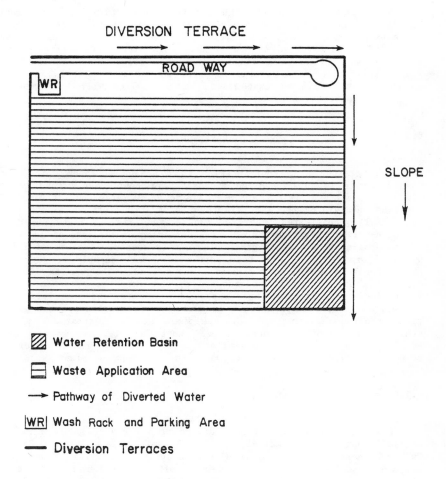

DIVERSION TERRACE

ROAD WAY

WR

SLOPE

▨ Water Retention Basin

⊟ Waste Application Area

→ Pathway of Diverted Water

|WR| Wash Rack and Parking Area

— Diversion Terraces

Figure 8.3. Possible layout of a land treatment unit in a gently sloping uniform terrain when only one plot is used.

volumes of runoff water will be generated, it is particularly important to minimize the acreage from which runoff is generated if on-site disposal will be used.

For some wastes that are high in metals and contain low concentrations of nitrogen and toxic or mobile constituents, it may be possible to load the soil to capacity in a short time. Subsequent waste applications would then need to be diverted to new areas. This situation calls for several small plots rather than a single large area (Fig. 8.4). Following the final application on a particular plot, the closure plan is implemented on the treated cell so that runoff water quality will be improved as quickly as possible.

8.1.3   Rotating Plot Configuration

The rotating plot configuration is a design approach which may be used if waste is to be applied frequently or continuously when the rate limiting constituent (RLC) is low enough to allow large applications. This involves subdividing the land treatment area into plots which are treated sequentially, cultivated, and then revegetated (Figs. 8.5 and 8.6). Following a period of six months or more, depending on the rate of degradation of the applied materials, a given plot can be reused. The use of rotating plots may require 6, 12 or even more plots, each capable of degrading a proportionate fraction of the annual waste load. The use of individual disposal plots offers the advantages of allowing the systematic use of vegetation, minimizing the area exposed to erosion, and maximizing infiltration and evapotranspiration. Enhancement of infiltration and evaporation is often of primary importance where no water treatment plant is available for handling runoff water. Where a water treatment plant is available, the layout may be similar to Fig. 8.6 with runoff water channeled or piped from the retention basin to the treatment plant.

8.1.4   Overland Flow

Overland flow entails the treatment of wastewater as it flows at a shallow depth over a relatively impermeable soil surface with a 2-8% slope. Two treatment options having considerable applicability for industrial use include: using overland flow to treat runoff generated by a land treatment facility or using this method to treat wastewater effluent from industrial processes. Either of these treatment options could be used in conjunction with the treatment alternatives such as a land treatment system. This type of complementary treatment could greatly reduce the cost of treating effluent or runoff water as well as reduce the load on existing wastewater treatment plants.

Overland flow has been effective in removal of nitrogen, biochemical oxygen demand (BOD), total suspended solids (TSS), a variety of metals, and volatile trace organics (Carlson et al., 1974; Jenkins et al., 1981; Martel

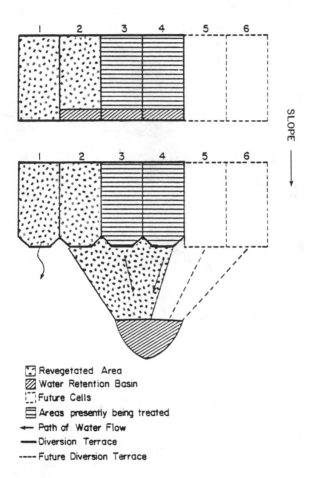

		Revegetated Area
		Water Retention Basin
		Future Cells
		Areas presently being treated
←		Path of Water Flow
		Diversion Terrace
----		Future Diversion Terrace

Figure 8.4.   Possible layout of a land treatment unit in a gently
              sloping uniform terrain when a progressive plot
              configuration is used.

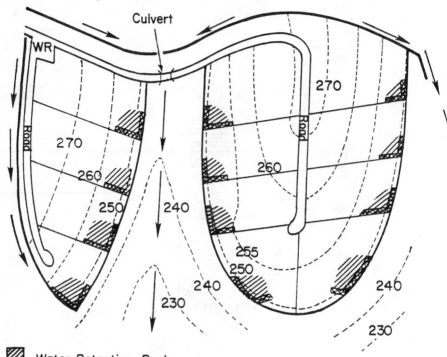

Water Retention Basin

Pathway of Diverted Water

Diversion Terraces

Retention Levees

WR Wash Rack and Equipment Parking

- - - Contour Lines

Figure 8.5.   Possible layout of a land treatment unit in rolling
terrain showing 12 plots and associated runoff retention basins.

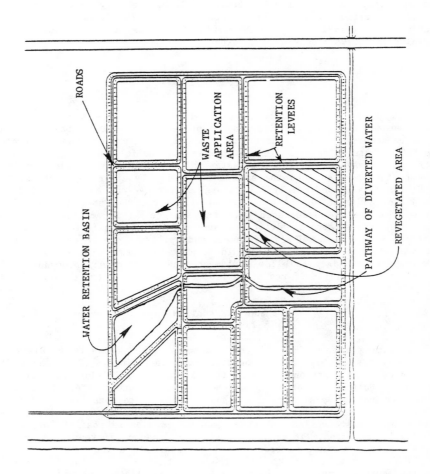

ROADS

WASTE
APPLICATION
AREA

RETENTION
LEVEES

WATER RETENTION BASIN

PATHWAY OF DIVERTED WATER

REVEGETATED AREA

Figure 8.6. Possible layout of a land treatment unit in level terrain.

et al., 1982). Carlson et al. (1974) reports overland flow as being effective in reducing the cadmium, copper, manganese, nickel, lead, and zinc level of secondary effluent. Phosphorus removal by overland flow systems is limited since the exchange sites are used up rather rapidly (Martel, 1982). A more detailed discussion of the topic and the important parameters to be considered during the design phase of an overland flow system can be located in the following sources (Carlson et al. 1974; Hoeppel et al., 1974; Carlson et al., 1974; Peters and Lee, 1978; Thomas et al., 1976; Jenkins et al., 1981; Chen and Patrick, 1981; Dickey and Vanderholm, 1981; Martel et al., 1982; Jenkins and Palazzo, 1981).

8.1.5   Buffer Zones

Land treatment units should be laid out to provide adequate buffer zones between the disposal site and the property boundaries. State regulations concerning required buffer zones should be consulted when designing the HWLT, where no specific regulations exist, the following suggestions on buffer zones may be useful. For wastes which present minimal odor problems and are incorporated into the soil surface shortly after application, the buffer area is needed mainly for diversion terraces and aesthetic reasons. Waste storage areas should be provided with larger buffer zones, particularly if odors are associated with the storage or if aerators are used which may cause aerosol drift. Water retention facilities should be designed and constructed so the levees and spillways can be easily inspected and repaired. Enough area should be provided between the spillways and the property boundary to allow implementation of emergency procedures, if needed, to control runoff resulting from a catastropic storm event.

8.2   LAND PREPARATION

Preparing the surface of the treatment area generally consists of clearing trees or bushes that obstruct the operations. Care should be taken during construction to ensure that design specifications are strictly followed. Surface recontouring may be needed to gather materials to construct external diversion terraces and levees, or to establish grades and internal terraces for water management. If recontouring is required, topsoil should be stockpiled and then respread as soon as possible after regrading is completed. It is often desirable, however, to keep on-site disturbances to a minimum to reduce soil erosion. If a vegetative cover is established, it will tend to hold the soil together and provide traction for the equipment used to spread the initial application of waste. There is no need to plow a field before applying waste if the equipment available for waste incorporation is able to break the turf and incorporate the waste.

Water is the primary means by which pollutants are transported from HWLT units. Hazardous substances may either be dissolved or suspended in water and subsequently carried to off-site land surfaces, surface waters or groundwater. Consequently, water control is of primary importance in land treatment design. When hazardous waste is mixed with the surface soil to achieve the required degradation, almost all water flowing over or through the soil comes into contact with the waste. Water management strives to limit the amount of water contacting treated areas by controlling run-on from untreated areas to reduce the amount of water contaminated. Additionally, runoff from treated areas is collected and either stored, disposed, or treated and released under a permit if the water is shown to be free of contamination.

All water movement on an HWLT unit needs to be carefully planned. When water is directly applied by an irrigation system, it must not be applied at rates above the infiltration capacity of the soil. When intermittent flooding or ridge and furrow irrigation techniques are used, careful timing of applications is needed so applications immediately prior to natural rainfall events are avoided as much as possible. Smaller, more frequent applications are generally better than a few, very large volume applications; however, this consideration should be based on the overall design of the facility. Additionally, all water applications to sloping land should be done in association with some type of erosion control practice such as contour strips, terraces, benches, diversion ditches, or contouring. It may also be desirable on some areas to leave buffer contour strips of undisturbed vegetation to help slow water flow. Any activity that disturbs the soil may decrease the effectiveness of erosion control structures, consequently, these structures should be rebuilt and revegetated as soon after a disturbance as possible.

To provide overall water management, the operator should develop a water balance for the HWLT site and keep a cumulative record of rainfall and available storage volume. To properly manage water at an HWLT, other important climatic parameters may need to be measured, including temperature and pan evaporation. Proper instrument exposures, calibration, and use are essential in order to obtain reliable observations. The National Weather Service establishes the standards for instrumental observations and provides the best source of information on this topic. Additionally, Linsley et al. (1975) provide good discussions of instruments and observations, and list sources of climatic data in their chapter references. Manufacturers of meteorological instruments also provide pamphlets on proper usage. Other useful measurements include wind velocity, soil temperature, soil moisture, and particulate and volatile emissions.

During a wet season, the operator should endeavor to provide sufficient storage capacity for anticipated rainfall runoff during the remainder of the season. For facilities with no discharge permit where runoff water is disposed by evaporation or spray irrigation, reapplication of water should be concentrated during dry periods to reduce the stored volume. The objective of all water management planning and effort is to avoid any release of unpermitted or contaminated water.

8.3.1 Water Balance for the Site

The development of hydrologic information for a site can serve two design purposes, specifying acceptable hydraulic loading rates for liquid-containing wastes and sizing runoff diversion (Section 8.3.2) and retention (Sections 8.3.3 and 8.3.4) structures. Hydraulic loading rates are determined somewhat independently of the natural site water budget while the water budget is the direct means for determining runoff and the associated control structures.

The amount of water which can potentially move into and through the soil profile is primarily a function of the hydraulic conductivity of the most restrictive soil horizon and the site drainage, which may have both vertical and horizontal components available to remove water from the system. Measures for these parameters are described in Section 4.1.1.5. It should be recognized that the waste may dramatically affect the hydraulic conductivity of the surface layer, and measurements obtained from this layer should characterize the waste-soil mixture rather than the unamended native soil. Additionally, the best results are obtained from field measurements taken at enough locations to account for the spatial variability of this parameter.

Once the hydraulic properties of the soils have been characterized, the amount of wastewater that can safely be leached through the system should be determined. This requires knowledge of the climatic setting of the site, the soils, and the mobility of the hazardous constituents to be land treated. In general, as the risk of significant leaching of hazardous constituents increases, the acceptable hydraulic load decreases. At the extremes are the two choices described below. The choice of hydraulic load for intermediate risks should be guided by results of treatment demonstrations (see Chapter 7 for test approaches).

Highly mobile hazardous constituents placed in a groundwater recharge zone would be an example of an extreme case where the leaching risk is great. The objective in this case would be to adjust hydraulic loading so no leaching occurred. In arid regions, this objective may be practically achieved by controlling waste applications to less than the site water deficit. Humid sites may not practically achieve the zero leaching objective, so the unit should be designed so that natural leaching rates would not be significantly increased. At least two approaches may be considered. First, applications can be timed to coincide with dry months, such as summer months in the southeastern U.S. Second, a site can be chosen to include slowly permeable clay soils or soils with shallow clay restrictive layers.

The other extreme is where the mobility of hazardous constituents is minimal and loading rates are based on saturated hydraulic conductivity.

Saturated hydraulic conductivity data should not be used without adjustment, however, because field experience with land treatment of nonhazardous wastewater has shown that practical limits are much lower. Data are very limited, but the USDA and U.S. Army Corps of Engineers (EPA, 1977) indicate that the hydraulic loading rate should be a maximum of 2 to 12% of the saturated hydraulic conductivity for loamy to sandy soils, respectively. Some form of field testing is again necessary to provide an adequate assessment and such information should be developed in conjunction with the waste-site interaction studies (Chapter 7).

8.3.2   Diversion Structures

The primary function of diversion structures on a land treatment unit is to intercept and redirect the flow of surface water. For an HWLT unit to function properly, it needs to be hydrologically isolated. This means that if the treated area lies downslope, all runoff originating above the treatment area should be diverted around the treatment site. Diversion structures must at least be designed to prevent flow onto the treatment zone from the peak discharge of a 25 year storm (EPA, 1982).

Run-on control is normally accomplished by constructing a berm of moderately compacted soil around the site. Excess material from construction of the retention ponds is a good material to use for building these berms. If native topsoil is used, it must be free of residual vegetation that would prevent proper compaction. Berms should run at an angle up the slope so that water moving downslope is intercepted and moved laterally. This design minimizes ponding behind the berm and also allows construction of a smaller berm. If the area draining to the berm is large, a terrace or set of terraces may be needed above the berm to slow the velocity of the water and to assist lateral movement. The terraces and the diversion berms should discharge into a grassed waterway sized to safely divert runoff water without causing serious erosion. For similar reasons, terraces and diversion structures should be vegetated immediately following construction.

Just as diversion structures can be used to prevent run-on from entering the HWLT unit, they can be used to control water on-site. Water flowing from upland portions of the HWLT unit can be carefully channeled to the retention pond to prevent the release of contaminated water. Diversion structures may be useful in some cases to divide the unit into plots.

8.3.3   Runoff Retention

All runoff from an HWLT unit must be controlled; this is usually accomplished by using diversion structures, as previously discussed, to channel water to a retention pond which is normally located in the lowest spot. Another method for controlling runoff is to subdivide the unit and contain the runoff from each smaller area in a separate retention pond.

One advantage to using several ponds is that if water in one pond becomes highly contaminated by waste overloading in one plot, the volume of water to be treated as a hazardous waste is minimized.

Ponds and retention basins must be designed to hold the expected runoff from a 25-year, 24-hour return period storm (Schwab et al., 1971; EPA, 1982). There are two general approaches to meeting this requirement, one is to design a pond for the runoff expected from the specified storm and keep this pond empty and the other approach involves designing a pond to contain rainfall runoff collected from previous storms as well as the runoff for the specified storm. In any case, since the pond cannot be emptied instantaneously, some consideration of accumulated water must be included in the design of runoff retention ponds. If the environmental damage would be extremely high from an inadvertant discharge, the operator may want to use the 50 to 100 year return period storm when sizing the basin. Sizing calculations should take into consideration the potential carryover of water accumulated during previous rainfall events so that the design capacity can be mantained at all times. If a land treatment unit is divided into plots and each plot is equipped with a retention basin designed for a 25-year, 24-hour return period storm, an additional, optional retention basin can be installed to retain any overflow from the smaller ponds. This basin can also be designed to hold the runoff expected to accumulate during a wet season. Retention basins should be lined to comply with regulations concerning surface impoundments (EPA, 1980a; EPA 1982) if the runoff is hazardous. On-site clay materials may be suitable for use in compacted clay liners.

It is imperative that the best available engineering principles be used to design and construct retention basins. Earthern dams should be keyed into the existing soil material whenever possible (Schwab et al., 1971). There are also numerous sources of plastic or other composition liners for sealing industrial storage ponds if clay is unavailable or unsuitable for the given situation. All portions of the dam areas that will not be submerged should be covered with 15 cm or more of topsoil and revegetated with appropriate plant species.

Every pond and retention basin should have an emergency or flood spillway. Whenever possible, ponds should be designed to use an existing ongrade vegetated area as a spillway. Maintenance of a good vegetative cover or riprap in the emergency spillway is needed to hold soil in place and prevent dam failure in the event of an overflow.

8.3.4   Runoff Storage Requirements

Runoff control must be provided to reduce the probability of an uncontrolled release of contaminated runoff water. Obviously, protection against all eventualities (zero probability) is unachievable; consequently, the degree of protection provided should be based on knowledge of the site and the risk associated with an uncontrolled release. The latter is largely based on the characteristics of the waste and the damage which could be caused by those constituents which are likely to be mobilized by runoff water, with erosion control practices, waste application rates and methods, and site management acting as modifiers.

Runoff retention ponds (impoundments) can be likened to multipurpose reservoirs and, as such, can serve two functions, (1) control of normal seasonal fluctuations in rainfall runoff and (2) maintenance of enough reserve capacity to contain stormwater runoff from peak events. Probabilities defining the degree of protection needed should be assigned to each of these functions based on water balance calculations and severe storm records, respectively.

8.3.4.1 Designing for Peak Stormwater Runoff

Consideration of the climatic record for a site includes extreme events, but the effects of these events are usually of little significance to the long-term site water budget. Peak events can, however, have immediate, devastating effects. Therefore, regardless of the general water budget, reserve capacity must be maintained for these singular events. The length of the design storm is usually chosen to be 24 hours since this time increment spans the length of singular storms in most cases while being of short enough duration to be considered practically "instantaneous" in comparison to the climatic record.

A minimum probability which is acceptable for hazardous waste sites is the 25 year, 24 hour storm, which specifies a 4% annual probability that this amount will be equalled or exceeded. Figure 8.7 is a map of the 25 year, 24 hour precipitation for the U.S. Greater values (i.e., lower probabilities), for example the 100 year, 24 hour storm can be used where the given site conditions pose a greater environmental risk. These are precipitation amounts, however, and not runoff. To translate the chosen precipitation value into runoff, a conservative approach would assume that 100% of the precipitation is lost as runoff. Storage volume is simply determined by multiplying the depth of runoff by the total area of the site watershed. For intense storms and high antecedent soil moisture content, this assumption may be acceptable, but some refinement is usually desirable.

Direct runoff from precipitation can be estimated using a procedure, often called the SCS curve number, developed by the Soil Conservation Service (1972). The estimate includes the effect of land management practices, the hydrologic characteristics of the soil, and antecedent soil moisture content on the amount of runoff generated. Although this model is a simple approach to a complex problem, it has an advantage over the more physically realistic models in that the curve number method does not require a great deal of input information and computer time.

To use the curve number method to determine the amount of runoff (i.e., stormwater) retention necessary, first determine the hydrologic group of the soil in the HWLT unit as described in Section 3.4.4. Next,

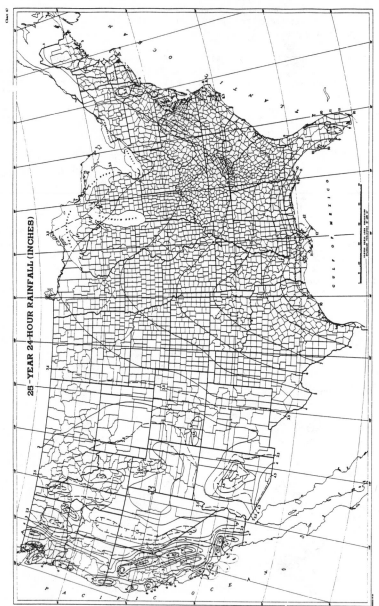

Figure 8.7. 25-Year, 24-Hour rainfall for the United States (Herschfield, 1961).

make an estimate of the rainfall amount which has occurred in the past five days using Table 8.2. However, if fresh waste is applied frequently, the soil may be continually moist and can be classified in antecedent moisture condition (AMC) III. The runoff curve number can now be ascertained from Table 8.3. For example, an HWLT unit planted with pasture grass in fair condition on a soil in hydrologic group C yields a curve number of 79.

TABLE 8.2 SEASONAL RAINFALL LIMITS FOR ANTECEDENT MOISTURE CONDITIONS*

| AMC Group | Total 5-day antecedent rainfall (in inches) | |
	Dormant Season	Growing Season
I	<0.5	<1.4
II	0.5 – 1.1	1.4 – 2.1
III	>1.1	>2.1

* Soil Converstion Service (1972).

The curve number acquired from Table 8.3 represents soils in AMC II and must be converted if the soil is in AMC I or III. In this example, the curve number of 79 is converted to a curve number of 91 using Table 8.4. Figure 8.8 can now be used to estimate runoff amounts. If the 25-year, 24-hour rainfall event is the design parameter, and that equals 7.5 inches, the intersection of 7.5 inches of rainfall with the curve number line of 91 yields direct runoff of 6.4 inches. Multiply this amount of runoff by the acreage of the HWLT watershed, and the total runoff and retention pond size can be calculated in acre-inches.

8.3.4.2 Designing for Normal Seasonal Runoff

Mindful of the two functions of runoff retention ponds, designing ponds to control normal seasonal fluctuations, is more complex. This requires knowledge of numerous independent variables, many simplifying assumptions, and the choice of several management approaches. The complexity of the hydrologic cycle is concomitant with the difficulty of characterizing and measuring the important parameters make the job of predicting the water budget and sizing retention ponds, somewhat of an art, based in part on judgment and experience. Two possible approaches are discussed here, one a relatively straightforward method which can be readily calculated manually and the other a general introduction to a computer modeling approach. Where accumulated rainfall runoff is discharged or otherwise managed so that the storage volume needed for the 25 year, 24 hour storm is maintained, these calculations can be run using the maximum discharge rate for the pump, or wastewater treatment plant used to empty the runoff stor-

TABLE 8.3 RUNOFF CURVE NUMBERS FOR HYDROLOGIC SOIL-COVER COMPLEXES*

(Antecedent moisture condition II, and $I_a = 0.2\ S$)

Land use	Treatment or Practice	Hydrologic condition	A	B	C	D
Fallow	Straight row	----	77	86	91	94
Row crops	Straight row	Poor	72	81	88	91
	Straight row	Good	67	78	85	89
	Contoured	Poor	70	79	84	88
	Contoured	Good	65	75	82	86
	Contoured and terraced	Poor	66	74	80	82
	Contoured and terraced	Good	62	71	78	81
Small grain	Straight row	Poor	65	76	84	88
	Straight row	Good	63	75	83	87
	Contoured	Poor	63	74	82	85
	Contoured	Good	61	73	81	84
	Contoured and terraced	Poor	61	72	79	82
	Contoured and terraced	Good	59	70	78	81
Close-seeded legumes[†] or rotation meadow	Straight row	Poor	66	77	85	89
	Straight row	Good	58	72	81	85
	Contoured	Poor	64	75	83	85
	Contoured	Good	55	69	78	83
	Contoured and terraced	Poor	63	73	80	83
	Contoured and terraced	Good	51	67	76	80
Pasture or range		Poor	68	79	86	89
		Fair	49	69	79	84
		Good	39	61	74	80
	Contoured	Poor	47	67	81	88
	Contoured	Fair	25	59	75	83
	Contoured	Good	6	35	70	79
Meadow		Good	30	58	71	78
Woods		Poor	45	66	77	83
		Fair	36	60	73	79
		Good	25	55	70	77

--continued--

TABLE 8.3 (continued)

	Cover		Hydrologic soil group			
			\(Antecedent moisture condition II, and I_a = 0.2 S\)			
Land use	Treatment or Practice	Hydrologic condition	A	B	C	D
Farmsteads		----	59	74	82	86
Roads (dirt)[#]		----	72	82	87	89
(hard surface)[#]		----	74	84	90	92

* Soil Conservation Service (1972).

[†] Close-dilled or broadcast.

[#] Including right-of-way.

TABLE 8.4   CURVE NUMBERS (CN) AND CONSTANTS FOR THE CASE $I_a = 0.2S$*

1	2	3	4	5	1	2	3	4	5
CN for condition II	CN for conditions I	III	S values†	Curve† starts where P =	CN for Condition II	CN for conditions I	III	S values†	Curve† starts where P =
			(inches)	(inches)				(inches)	(inches)
100	100	100	0	0	60	40	78	6.67	1.33
99	97	100	.101	.02	59	39	77	6.95	1.39
98	94	99	.204	.04	58	38	76	7.24	1.45
97	91	99	.309	.06	57	37	75	7.54	1.51
96	89	99	.417	.08	56	36	75	7.86	1.57
95	87	98	.526	.11	55	35	74	8.18	1.64
94	85	98	.638	.13	54	34	73	8.52	1.70
93	83	98	.753	.15	53	33	72	8.87	1.77
92	81	97	.870	.17	52	32	71	9.23	1.85
91	80	97	.989	.20	51	31	70	9.61	1.92
90	78	96	1.11	.22	50	31	70	10.0	2.00
89	76	96	1.24	.25	49	30	69	10.4	2.08
88	75	95	1.36	.27	48	29	68	10.8	2.16
87	73	95	1.49	.30	47	28	67	11.3	2.26
86	72	94	1.63	.33	46	27	66	11.7	2.34
85	70	94	1.76	.35	45	26	65	12.2	2.44
84	68	93	1.90	.38	44	25	64	12.7	2.54
83	67	93	2.05	.41	43	25	63	13.2	2.64
82	66	92	2.20	.44	42	24	62	13.8	2.76
81	64	92	2.34	.47	41	23	61	14.4	2.88
80	63	91	2.50	.50	40	22	60	15.0	3.00
79	62	91	2.66	.53	39	21	59	15.6	3.12
78	60	90	2.82	.56	38	21	58	16.3	3.26
77	59	89	2.99	.60	37	20	57	17.0	3.40
76	58	89	3.16	.63	36	19	56	17.8	3.56
75	57	88	3.33	.67	35	18	55	18.6	3.72

--continued--

TABLE 8.4 (continued)

1	2	3	4	5
CN for condition II	CN for conditions I	III	S values†	Curve† starts where P =
74	55	88	3.51	.70
73	54	87	3.70	.74
72	53	86	3.89	.78
71	52	86	4.08	.82
70	51	85	4.28	.86
69	50	84	4.49	.90
68	48	84	4.70	.94
67	47	83	4.92	.98
66	46	82	5.15	1.03
65	45	82	5.38	1.08
64	44	81	5.62	1.12
63	43	80	5.87	1.17
62	42	79	6.13	1.23
61	41	78	6.39	1.28

1	2	3	4	5
CN for Condition II	CN for conditions I	III	S values†	Curve† starts where P =
34	18	54	19.4	3.88
33	17	53	20.3	4.06
32	16	52	21.2	4.24
31	16	51	22.2	4.44
30	15	50	23.3	4.66
25	12	43	30.0	6.00
20	9	37	40.0	8.00
15	6	30	56.7	11.34
10	4	22	90.0	18.00
5	2	13	190.0	38.00
0	0	0	infinity	infinity

* Soil Conservation Service (1972).

† For curve number (CN) in Column 1.

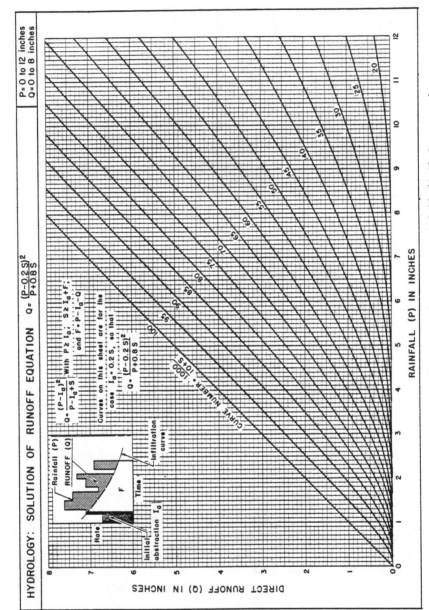

Figure 8.8. Estimating direct runoff amounts from storm rainfall (Soil Conservation Service, 1972).

age pond. These calculations can also help the site manager decide between various discharge rates and pump capacities.

8.3.4.2.1 <u>Monthly Data Approach</u>. An underlying assumption in a water budget for a site must be that, on the average, there is no net change in the volume of runoff stored on a long-term basis. In other words, water management cannot allow a continued increase in water stored because of the "multiplying pond" syndrome (i.e., the need to periodically increase pond capacity). Apart from storage, the means of control for management are enhanced leaching and evaporation and/or discharge under an NPDES permit (Section 8.3.5). Given these considerations, the water budget can be derived from the following basic expression:

$$P + W = EVTS + L + R \tag{8.1}$$

where

$$
\begin{aligned}
P &= \text{precipitation;} \\
W &= \text{water applied in waste;} \\
EVTS &= \text{evapotranspiration;} \\
L &= \text{leachate; and} \\
R &= \text{runoff to be collected.}
\end{aligned}
$$

The equation assumes a negligible change in soil water storage. The runoff (R) term can be broken into two components, storage (S) and discharge (D). Using these terms and rearranging equation 8.1, the expression can be written:

$$S = P + W - EVTS - L - D \tag{8.2}$$

In the long-term, storage will vary approximately sinusoidally with a constant mean.

Choosing a monthly basis as a convenient time increment, to maintain sensitivity while simplifying data requirements, a water budget can be run for the given site by using the climatic record, the watershed properties of the proposed land treatment unit, and the assumption (for the purpose of these calculations), that adequate storage is available to contain all events (i.e. no spillway overtopping). Best results require using a climatic record of at least 20 years. By simulating the entire record, month by month, changes in storage can be seen with time. Appendix E provides an example of how to run the calculations. Arriving at a design storage using this method involves a four step process, as follows:

(1) Assume zero discharge and run the calculations. If there is $\sum_{i=1}^{n} S_i \leq 0$; where S_i = annual change in storage from the previous year), then no discharge is needed;

(2) If $\sum_{i=1}^{n} S_i > 0$; then some discharge and/or enhanced evaporation or leaching is necessary. As a first approximation,

assume that the enhanced water losses equal the average annual storage change (i.e. $\sum_{i=1}^{n} S_i/n$; where n = number of years of record). Now rerun the calculations with the modified values.

(3) Based on risk assessments, choose an acceptable probability of equalling or exceeding the final design storage capacity and then choose a design storage capacity from the record simulated in step (1) or (2) which is equalled or exceeded that portion of the time (e.g., if acceptable probability = 0.1, then design storage should contain runoff all but 10% of the time).

(4) Refinements in the storage capacity determined in step (3) can be made to reflect other considerations. For example, as water loss rates are increased, the storage needed decreases. Therefore, cost considerations might encourage an applicant to treat and discharge more water at some cost to save even higher incremental costs of constructing and maintaining larger retention ponds.

Estimating the input data for the water budget may be a difficult exercise. Monthly precipitation data are relatively easy to locate. Likewise, the amount of water included in the waste is directly ascertainable from waste analyses and projected production rate (volumetric), and converted to a monthly application depth (cm/mo) using the known unit watershed area. In contrast, monthly evapotranspiration and leaching, especially with management modifications, are troublesome parameters to estimate.

The watershed of the HWLT can be divided into areas which behave as free water surfaces (e.g., ponds, ditches, continually wetted plots, and well vegetated plots) and areas of bare soil or poorly cropped surfaces which can vary dramatically in moisture conditions and evaporation rates (e.g., plots, roads and levees). On a monthly basis, the portion of the unit watershed falling in each category should be determined and an estimated evapotranspiration (EVTS) rate determined for each. Free water surface evaporation can be estimated using published monthly Class A pan evaporation data. The assumption may be made that true evaporation equals about 70% of Class A pan evaporation. This assumption may be somewhat inaccurate and can cause an error in estimates since pan coefficients vary widely from month to month, but monthly pan coefficients are not available from any source. If an annual pan coefficient is available for a nearby reservoir, this may be used instead of the 70% figure. Pan evaporation data for the U.S., summarized by Brown and Thompson (1976), is given in Figures 8.9 to 8.20. No data are available for estimating EVTS from a bare soil or the poorly cropped surface of HWLT units.

The only leaching which is of concern here is that which is lost to deep percolation. Perched water having primarily a horizontal component of flow should properly be intercepted by water containment structures and ultimately contribute to the storage or discharge term of the site water

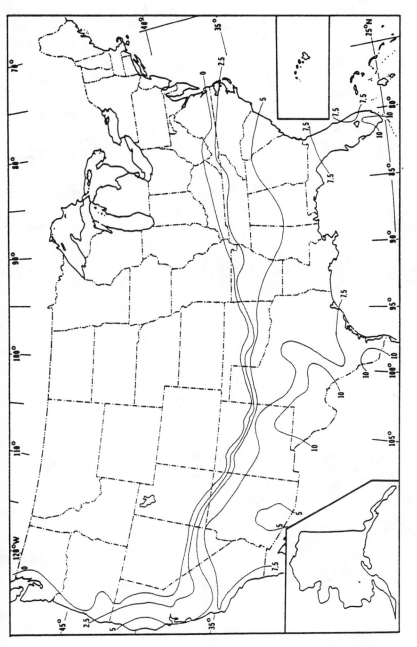

Figure 8.9.   Average pan evaporation (in cm) for the continental United States
for the month of January based on data taken from 1931 to 1960
(Brown and Thompson, 1976).

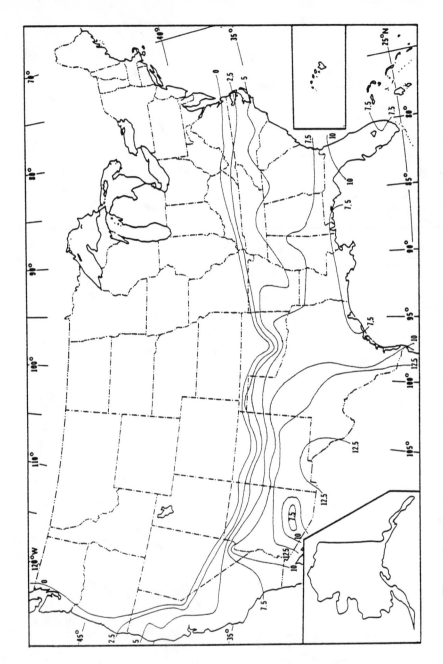

Figure 8.10. Average pan evaporation (in cm) for the continental United States for the month of February based on data taken from 1931 to 1960 (Brown and Thompson, 1976).

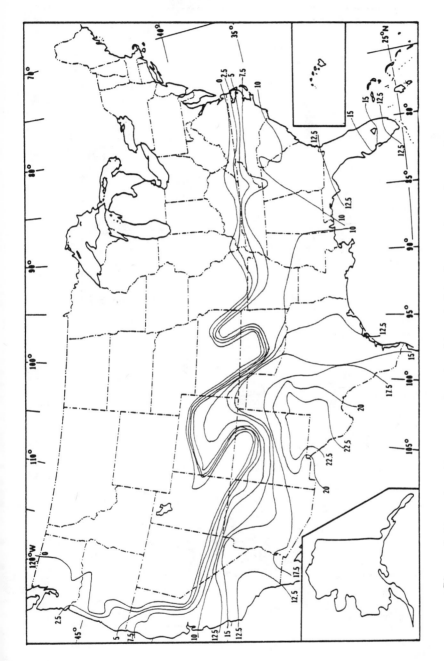

Figure 8.11. Average pan evaporation (in cm) for the continental United States for the month of March based on data taken from 1931 to 1960 (Brown and Thompson, 1976).

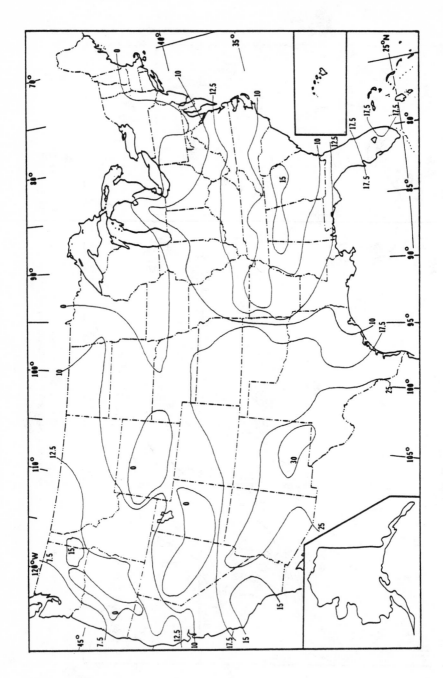

Figure 8.12. Average pan evaporation (in cm) for the continental United States for the month of April based on data taken from 1931 to 1960 (Brown and Thompson, 1976).

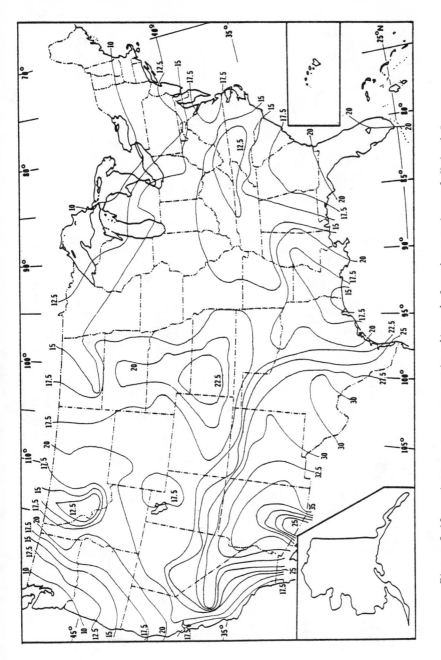

Figure 8.13.　Average pan evaporation (in cm) for the continental United States for the month of May based on data taken from 1931 to 1960 (Brown and Thompson, 1976).

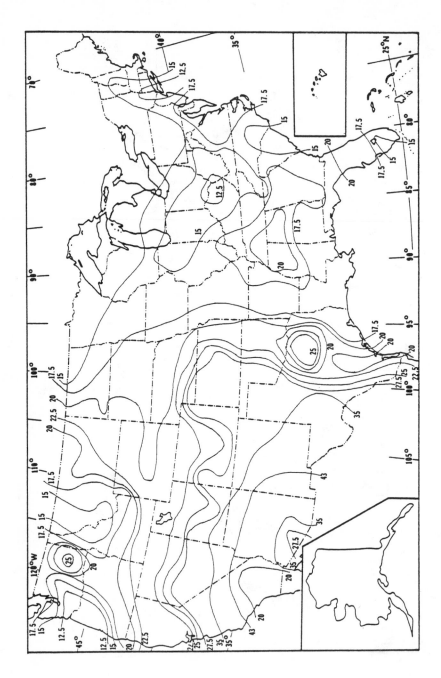

Figure 8.14.   Average pan evaporation (in cm) for the continental United States for the month of June based on data taken from 1931 to 1960 (Brown and Thompson, 1976).

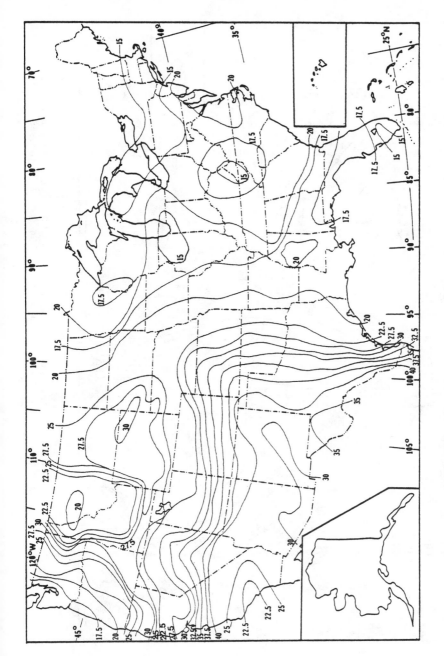

Figure 8.15. Average pan evaporation (in cm) for the continental United States for the month of July based on data taken from 1931 to 1960 (Brown and Thompson, 1976).

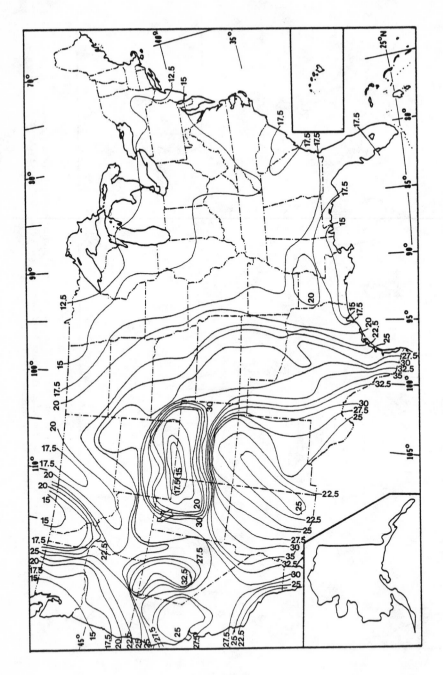

Figure 8.16. Average pan evaporation (in cm) for the continental United States for the month of August based on data taken from 1931 to 1960 (Brown and Thompson, 1976).

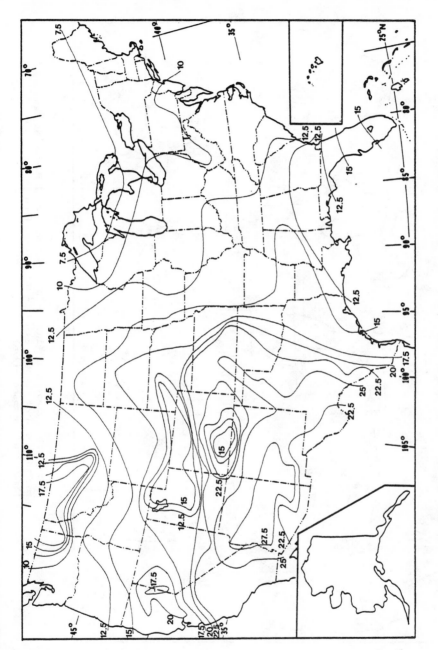

Figure 8.17. Average pan evaporation (in cm) for the continental United States for the month of September based on data taken from 1931 to 1960 (Brown and Thompson, 1976).

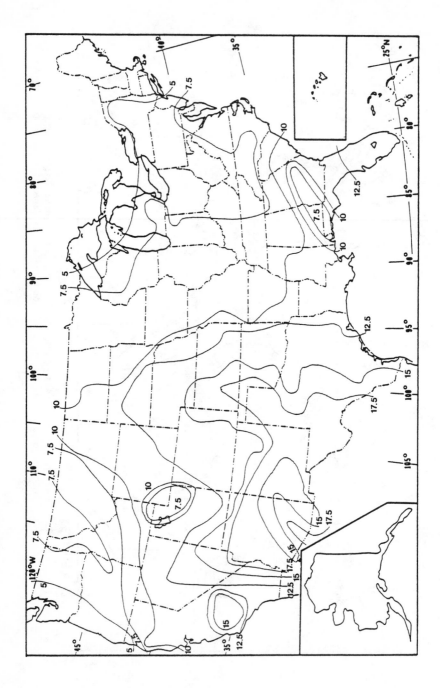

Figure 8.18. Average pan evaporation (in cm) for the continental United States for the month of October based on data taken from 1931 to 1960 (Brown and Thompson, 1976).

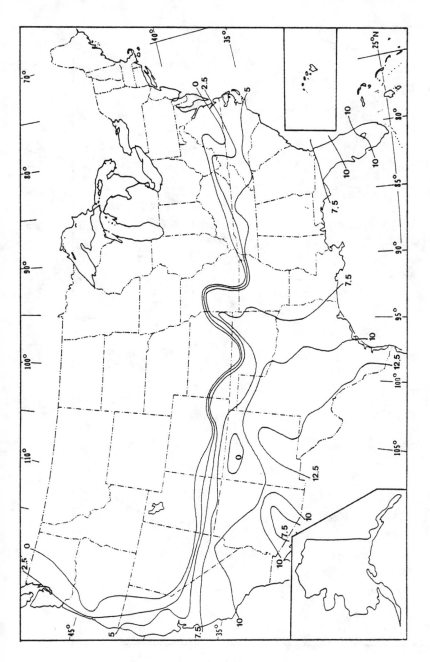

Figure 8.19. Average pan evaporation (in cm) for the continental United States for the month of November based on data taken from 1931 to 1960 (Brown and Thompson, 1976).

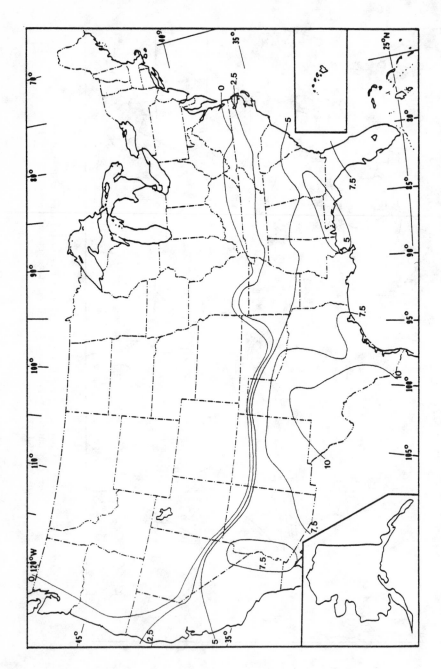

Figure 8.20. Average pan evaporation (in cm) for the continental United States for the month of December based on data taken from 1931 to 1960 (Brown and Thompson, 1976).

budget. A conservative, simplifying assumption which may be acceptable for clay soils or those having shallow, restrictive clay horizons is that leaching is zero. For less restricted conditions, there is unfortunately very little information available for making good leaching estimates. Therefore, unless sound data are provided from field measurements of leaching losses (not hydraulic conductivity), then the conservative strategy is to assume zero leaching or, in cases of heavy hydraulic loading by the waste, use the same approach as previously discussed in Section 8.3.1 for hydraulic loading rates.

8.3.4.2.2 <u>Computer Methods</u>. Computer approaches for water budgets have been designed for a number of special purposes, but none are widely available which can be applied directly for sizing runoff retention ponds. The deterministic model described by Perrier and Gibson (1980) is one useful approach which is readily accessible to practically anyone having access to a computer terminal; however, the model only goes so far as to generate daily runoff data, which must then be manually integrated into a retention pond water budget. Considerations in the manual calculations would be pond evaporation, discharge and enhanced evaporation (EVTS) and leaching (L). The enhanced EVTS and L terms would be handled as a feedback loop in the model by treating them as though they were additional precipitation (an exception is that the quantity reaching the plot must be reduced to account for aerial evaporation losses before the water reaches the ground). There is much need for a package model, possibly incorporating the Perrier and Gibson (1980) model that includes these additional features. Other references on computer modeling are listed and discussed in Fleming (1975).

8.3.4.3 <u>Effects of Sediment Accumulations</u>

One final factor in retention pond sizing is an accounting for decreases in effective capacity because of sediment buildup. Periodic dredging will often be necessary to maintain the designed useful capacity, and some additional capacity must be included to handle sediment buildup prior to dredging. The decision will primarily be based on management and cost factors which are beyond the concern of this document; however, this factor must be included in the pond design calculations.

8.3.4.4 <u>Summary of Retention Pond Sizing</u>

The final pond capacity design must account for the three influences discussed previously: (1) peak storm runoff (8.3.4.1); (2) normal seasonal runoff (8.3.4.2); (3) and sediment accumulations (8.3.4.3). The values for each should be added to obtain a total design pond volume. The storage facility need not be designed to hold all seasonal runoff plus the peak storm runoff if the runoff storage facility will be emptied to maintain the design capacity. However, in practice the storage facility cannot be emptied instantaneously so some additional volume above the 25 year, 24 hour volume will be needed. The design also incidentally specifies design discharge rate (size of water treatment plant, if needed) and/or the quantity of runoff which should be irrigated onto the land treatment unit to provide enhanced evaporation and in some cases leaching. Note that some amount of irrigation is desirable under any circumstances to control wind dispersal of contaminants, provide water for growing cover crops, and sustain optimal soil moisture conditions for organics degradation.

8.3.5   Runoff Treatment Options

Runoff collected in retention basins can be treated or disposed by one of several methods. Water can either be released via a wastewater treatment facility permit, a National Pollutant Discharge Elimination System (NPDES) permit, or treated on-site in a zero discharge system. The method of handling runoff should be considered during the design phase of the facility. If the runoff from the land treatment unit is, itself, a hazardous waste, then it must be handled accordingly. The definition and criteria for identifying a waste as hazardous are found in 40 CFR Part 261 (EPA, 1980b).

If the plant or company that generates the waste owns and operates a wastewater treatment plant, nonhazardous runoff water may be pumped to the plant for treatment and disposal. An analysis of the discharge from the wastewater treatment facility should be performed to determine if existing permit conditions can still be met. Care must be taken to ensure adequate water storage capacity in the runoff retention basins to hold water that exceeds the capacity of the treatment plant.

Where the option of using an existing wastewater treatment facility is not available, application for an NPDES permit may be appropriate if the runoff water is nonhazardous. This would allow direct discharge of the collected runoff water (with or without treatment) after analyses show that the water meets water quality standards set in the permit. Standard engineering principles concerning diversion structures should be followed and care must be taken to keep erosion of drainage ditches to a minimum.

If a company operates an HWLT unit as a zero discharge system, runoff water may be used as a source of irrigation water when soils are dry enough to accept more water. It may also be sprayed into the air above the pond or treatment area to enhance evaporation if no hazard due to volatiles or aerosols would result. When sprinkler irrigation systems are used for reapplication of runoff, the systems should be designed to apply water at a rate not exceeding the soil infiltration rate to minimize runoff. Proper pressure at the nozzles will help spread water uniformly; nozzles that form large droplets are advisable when spray drift and aerosols must be minimized. Collected runoff to be reapplied should be analyzed to determine if it contains nutrients, salts and other constituents important in determining waste loading on the plots. If the water contains significant concentrations of these constituents a record of water applications should be kept and the results used to determine the cumulative loading of the constituents. In most cases, however, collected water contains negligible concentrations of the constituents used in loading calculations when compared to concentrations in the waste.

Regardless of the method used for runoff control, irrigation during dry, hot periods is beneficial to supply adequate moisture to maximize microbial degradation of waste constituents. For this reason, it may be desirable to move the irrigation system around to spread the water over as much of the facility as possible. In some particularly dry seasons or climates, additional irrigation water from off-site may be applied to enhance waste degradation.

8.3.6   Subsurface Drainage

The primary purpose of subsurface drainage from below all or part of an HWLT unit is to lower and maintain the water table below some desired depth, to increase aeration in the surface soil, and to decrease the hazard of groundwater contamination. This may be particularly valuable to help maximize the utility of low lying or poorly drained areas of an HWLT unit. The seasonal high water table should not rise higher than 1 meter (3 feet) below the bottom of the treatment zone (EPA, 1982). If the soil is permeable with a shallow water table, a ditch cut to a specific depth below the water table at the low end of the field may be sufficient to drain the surface soil. Agricultural drainage systems are normally constructed by digging sloped trenches and installing drain tiles or perforated plastic pipe. The top of the pipe is protected by a thin paper or fiberglass covering and the overlying soil is replaced (Luthin, 1957). By controlling the depth of the unsaturated zone using subsurface drainage, a site which would normally remain excessively wet because of a shallow water table might be accessible and usable for land treatment.

Design and spacing of a drainage system can be accomplished using one of several steady state or non-steady state relationships. The decision about which relationship to use is generally based on experience and site conditions. The Soil Conservation Service has historically used the classical Hooghaudt equation (Hooghaudt, 1937; Hillel, 1971), also known as the ellipse steady state drainage equation, which includes a number of simplifying assumptions. The relationship performs well in humid regions where the steady state flow assumption is a fair approximation of site conditions. In the western U.S., however, the Bureau of Reclamation uses a non-steady state approach, particularly the Glover equation (Glover, 1964; Dummn, 1964; Moody, 1966), which accounts for arid conditions where drainage is intermittent. Another non-steady state solution to drain spacing design is the van Schilfgaarde relationship (van Schilfgaarde, 1963; van Schilfgaarde, 1965; Bouwer and van Schilfgaarde, 1963). Additional steady state and non-steady state relationships have been developed based on varying approaches and assumptions, as discussed by Kirkham et al. (1974) and van Schilfgaarde (1974). Two important considerations in choosing and using a suitable relationship are that the explicit assumptions used in the equation fit the particular HWLT site conditions and that the necessary inputs are accurately estimated.

Collection and treatment of the water collected should generally follow guidelines discussed above for runoff water. In general, the water should be collected in a pond or basin. From there it may be discharged to a wastewater treatment plant, directly discharged under an NPDES permit, or used internally for irrigation or other purposes. However, if the water is a hazardous waste, it must be treated and/or disposed as a hazardous waste.

8.4   AIR EMISSION CONTROL

Air quality may be adversely affected by a land treatment operation if hazardous volatiles, odors or particulates are emitted during storage, handling, application and incorporation of waste or during subsequent cultivation. Wind dispersal of contaminants and dust from traffic on facility roads may also present a problem. Management plans should be developed to avoid such emissions as much as possible and to handle these situations if they arise. On an operational basis, wind, atmospheric stability, and temperature are important considerations for timing the application of wastes, especially volatile wastes.

8.4.1   Volatiles

Volatiles may be reduced to an acceptable level through management of loading rates and proper placement of the waste as determined from pilot studies (Section 7.2.3). Wastes containing a significant fraction of easily volatilized constituents should be applied at a depth of 15 cm by subsurface injection. Volatilization losses will effectively be reduced as gases move through the soil profile.

Irrigation of the soil surface may also aid in reducing the net flux into the atmosphere, lessening the impact of volatilized components. Application of wastes containing significant quantities of volatiles should be made when soils are in a moist but friable state. Soils which are too wet are easily puddled by heavy machinery which could reduce aeration and the capacity of the soil to degrade organic waste constituents.

8.4.2   Odor

If a waste contains sufficient easily decomposable organic matter and if oxygen is limited, the waste may develop an undesirable odor. While odors do not indicate that a land treatment system is malfunctioning or that environmental damage is occuring, it has in some cases become a serious enough to prevent the use of land treatment at a site which was otherwise ideally suited. Odors from waste materials often are a result of sulfides, mercaptans, indoles, or amines. Disposal techniques can be designed to avoid the formation and release of these compounds.

The land treatment of waste having potential for emitting an odor generally results in some odor during the period between application and complete incorporation. Little can be done to avoid or circumvent this problem, just as the farmer can do little to avoid odors when he spreads manure. Potential odor problems should be considered when a disposal site is selected, and design should be based on the acceptable limits for odors, volatiles and particulates. Proximity to housing and thoroughfares as well as the prevailing wind direction need to be considered. Frequency and severity of atmospheric inversions that may trap malodorous gases should also be evaluated. Ideally, isolated sites should be selected but, in some cases, this is not possible. When locations adjacent to public areas must be used, certain steps can be taken to minimize odor problems.

Perhaps the best method of odor avoidance is subsurface injection. Soil has a large capacity to absorb gases. If a waste is subsurface injected and does not ooze to the surface, few odor problems are likely to occur. In a properly designed system, the waste application rate depends on the waste degradation rate. Although tilling helps to enhance aeration and degradation, where a significant odor problem exists, tillage may aggravate the odor problem.

If the waste is surface applied, either by dumping or spraying from a vehicle or irrigation system, odor problems can be minimized by quickly incorporating the waste into the soil. Odors often increase when organic wastes are spread or when mixing occurs, particularly when heavy applications are made. It may, therefore, be desirable to spread and incorporate wastes when the wind is from a direction that will minimize complaints. Emission of maladorous vapors can often be reduced substantially by thoroughly mixing the waste with the soil; this can be achieved by repeated discing when the ratio of waste to soil is not too high. In other instances, complete soil cover may be needed to prevent odors. This can be achieved by using turning plows or turning (one-way) discs. Large plows, such as those used for deep plowing, may also be used for covering thick applications of maladorous waste.

Organic wastes that are spread on the land by flooding followed by water decantation are likely to develop odor problems between decantation and incorporation. As long as an adequate layer of water covers the waste, odor is generally not a problem. Consequently, it may be desirable to delay decantation until wind directions are favorble and clear weather is likely. With proper design, including peripheral drainage ditches, it should be possible to rapidly decant excess water so incorporation can begin. While mixing is often desirable to hasten drying and to speed the oxidation of the organic constituents, it may be necessary to minimize mixing after the initial incorporation for situations with potential odor problems, since odor will often occur again when unoxidized material is brought to the surface. Drying and oxidation may be slower, and it may not be possible to repeat applications or establish vegetation as quickly as with more frequent mixing. Therefore, more land may be required for land treatment of a waste having this characteristic and odor might be the application limiting constituent in this situation.

There are many chemicals on the market for odor control. These include: disinfectants which act as biocides; chemical oxidants which act as biocides and also supply oxygen to the microbial population; deodorants which react with odoriferous gases to prevent their release; and masking agents which may impart a more acceptable odor to cover the undesirable odor. Hydrogen peroxide is a commonly used biocide and oxidizing agent. Pountney and Turner (1979) have reported success using hydrogen peroxide to control hydrogen sulfide odors in wastewater treatment facitilites. Strunk (1979) suggests that hydrogen peroxide acts primarily by oxidizing reduced sulfur compounds. Warburton et al. (1979) conducted a study testing the effectiveness of twenty-two commercial odor controlling products including chlorine, mechanical mixing, waste oil, wintergreen oil, and activated charcoal. He found that only mechanical mixing and chlorination signifi-cantly reduced odor from a swine manure. Chlorination may kill the active soil microbes which are important to waste degradation. Thus, while it is possible that some commercial products may be effective in reduction of odors from certain wastes, alternate means including avoidance or oxidizing agents should be considered first.

Odor controlling chemicals have been applied by direct incorporation into the waste prior to application, by manual or solid set spraying along borders or over entire areas, and by point spraying using a manifold mounted on the rear of the machine that spreads or incorporates the waste. Before an odor controlling chemical is employed, testing must demonstrate that it does not inhibit the waste-degrading microbial population.

Presently, there are no instruments available that have the ability to provide an objective determination of odor (Dolan, 1975). Therefore, odor evaluation is accomplished by using a panel of individuals to provide an odor intensity ranking. Experience has shown that an eight member panel, consisting of 50% women yields the most reliable results. Generally, the air sample collected in the field is diluted in varying proportions with fresh air to allow the individuals to establish an odor threshold. The only response that is required from each individual is a yes or no response. Using semilogarithmic paper, the threshold odor concentration is determined from the intersection of the 50% panel response line. From these data the odor emission rate can be computed. A more detailed discus-sion of the odor panel approach is included in the following sources (Dolan, 1975; Dravinicks, 1975).

8.4.3   Dust

Dust problems often occur on access roads used to transport the waste to various plots within an HWLT unit. Occasionally, dust will also be raised during discing or mixing operations when the soil in the treatment zone is dry. The wind dispersal of particulates from the treatment zone must be controlled (EPA, 1982). One method of controlling particulates is to surface apply water. A good source of water for this is often the accumulated runoff. Dust suppressing treatments including oil or calcium chloride may be used on roadways, if desirable, but excessive application should be avoided. Care should be taken in selecting a dust control product to be sure that it does not adversely affect the treatment process or cause environmental damage.

A windbreak may also be planted to help control the dispersal of dust and aerosols. A study of the spread of bacteria from land treating sewage sludge showed that bacteria were recovered 3 m downwind in a dense brushy area and 61 m downwind in a sparsely vegetated area (EPA, 1977). Van Arsdel (1967) and Van Arsdel et al. (1958) have used colored smoke grenades to study the movement of wind around windbreaks and across fields. They found that a spot of dry soil such as a levee or a bare spot in a field produces warmer air which causes an updraft. A windbreak of a single row of trees created a complete circulation cell around the trees. There was an updraft on the sunny side of the tree line and a downdraft on the shady side. The air on the shady side actually moved under the trees and up along the sunny side of the windbreak (Van Arsdel et al., 1958). Although windbreaks may be helpful in certain cases, there effectiveness should be evaluated on a case-by-case basis.

8.5 EROSION CONTROL

Control of wind and water erosion during the active life and closure period for an HWLT unit is needed both to assist in the proper functioning of the unit and to prevent contaminants from moving off-site. Soil conservation methods, developed by the USDA, have been widely used to control erosion on agricultural fields and can readily be adapted for use on HWLT units. Wind erosion may be a particular problem during dry seasons or in arid regions, but maintaining a vegetative cover and moist soil should lessen the problem.

When sloping land is used for an HWLT unit, terraces and grassed waterways should be used to minimize erosion by controlling runoff water. This is essential when large areas are left without vegetation for one or more seasons by repeated waste applications, which may occur with a sludge-type waste disposal operation. Proper conservation terracing is also important if water is applied to a continuously vegetated surface. Terraces slow the flow of intensive storm water, allowing optimal infiltration and putting less strain on retention basins. Furthermore, by decreasing the slope length, less sediment will erode and accumulate in the retention structures. Runoff water quality will be improved before the water enters retention structures; this will reduce the amount of accumulated organics. Improved water quality decreases the load on the wastewater treatment plant and increases the possibility of achieving water quality acceptable for direct discharge.

Terracing is a means of controlling erosion by constructing benches or broad channels across a slope. The original type of bench terrace was designed for slopes of 25 to 30% and resembled a giant stairway. They were very costly and not easily accessible for field equipment. Modern conservation bench terraces, which are adapted to slopes of 6-8% aid in moisture retention as well as erosion prevention (Schwab et al., 1971). The third type of terrace is the broadbase terrace which consists of a water conducting channel and ridge as shown in Fig. 8.21. The general placement of terraces is across the slope with a slight grade toward one or both ends. The collected runoff then drains off the terrace into a waterway.

The number of terraces needed is governed by the slope, soil type and vegetative cover. The vertical interval (VI), defined as the vertical distance between the channels of successive terraces, is calculated as follows:

$$VI = aS + b \hspace{3cm} (8.3)$$

where

VI = vertical interval in feet;
 a = geographic constant (Fig. 8.22);
 b = soil erodibility and cover condition constants (Fig. 8.22); and
 S = slope of the land above the terrace in percent.

This is only an estimate of the amount of terracing needed and can be varied up to 10% in the field without serious danger of failure.

Terraces can be constructed either level or with a grade toward one or both ends. If level, barriers or dams are needed every 120-150 meters to prevent total washout in the event of a break. The advantage to these is that there is no length restriction nor is a grassed waterway needed at the ends. The disadvantage is that the depth needs to be greater to accommodate a rainfall event without overtopping. For graded terraces, with well and poorly drained soils, the minimum grades are 0.1 and 0.2%, respectively. Suggested maximum grades decrease as terrace length increases (Table 8.5). The maximum terrace length is usually considered to be 300 to 550 meters for a one direction terrace and twice that for a terrace draining toward both ends. As slopes increase, terrace width and channel depth increase, resulting in more difficult construction and maintenance (Tables 8.6 and 8.7). The minimum cross sectional area for a sloping terrace is 0.5 to 1 m^2, while for a level terrace 1 m^2 is considered the minimum. Most level terraces are only designed to hold 5 to 10 cm of rain and thus may not be well suited to use at HWLT units in many parts of the country.

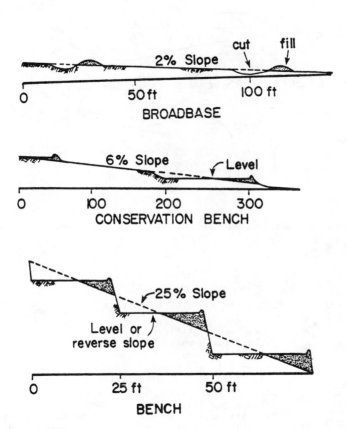

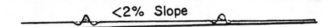

Figure 8.21.   Schematic diagram of general types of terraces
(Schwab et al., 1971). Reprinted by permission
of John Wiley & Sons, Inc.

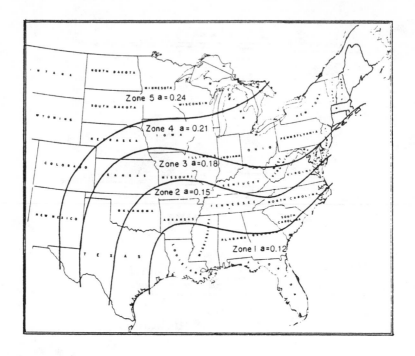

Figure 8.22.   Values of a and b* in terrace spacing equation,
VI = aS + b (ASAE, Terracing Committee, 1980).
Reprinted by permission of ASAE.
*Values for b vary and are influenced by soil
erodibility, cropping systems, and management
systems; in all zones, b will have a value of
0.3, 0.6, 0.9 or 1.2. The low value is applicable to very erodible soils with conventional
tillage and little crop residue; the high value
is applicable to erosion resistant soils where
no-tillage methods are used and a large amount
of crop residue remains on the soil surface.

TABLE 8.5 MAXIMUM TERRACE GRADES*

Terrace length (m) or length from upper end of long terraces	Slope (percent)	
	Erosive soil (Silt loam)	Resistant soil (Gravelly or Rocky)
153 or more	0.35	0.50
153 or less	0.50	0.65
61 or less	1.00	1.50
31 or less	2.00	2.50

* Beasley (1958).

Field layout of terraces may be done along the contours, often result-ing in odd shaped areas, or they may be made parallel, allowing for easier mechanical operations such as waste application, mowing and discing. When parallel terraces are used, it may be necessary to smooth the slope prior to construction. As noted above, variations of the vertical interval can be made up to 10% and some lesser variances in channel grade can be toler-ated.

When the land has a slope of less than 2%, as is the case along much of the Gulf Coast, contour levees similar to those used in rice fields may be used. The vertical interval between levees is typically 6 to 9 cm and the levees are put in along the contour. For proper water management, spillways should be provided to prevent wash out in the event of a heavy storm. Ideally, spillways will conduct water across a grassed area to a retention pond or treatment facility.

Construction is normally accomplished using graders and bulldozers. Allowances of 10-25% must be made for settlement. Any obvious high spots or depressions should be corrected quickly. All traffic on sloped areas should be parallel to the terraces. All terraces should be vegetated as soon as possible using lime and fertilizer as needed. Maintenance should include monthly inspections, annual fertilization, and mowing. Since terraces channel the flow of water, any terrace that is overtopped, washed out, or damaged by equipment should be repaired as soon as conditions per-mit to prevent excessive stress on lower terraces. Without proper mainten-ance and repair, the whole terrace system may be ruined, resulting in the formation of erosion gullies and highly contaminated runoff.

8.5.2   Design Considerations for Vegetated Waterways

A vegetated waterway is a properly proportioned channel, protected by vegetation and designed to absorb runoff water energy without damage to the

TABLE 8.6 TERRACE DIMENSIONS: LEVEL OR RIDGE TERRACE*[†]

Field slope (percent)	Terrace Channel Depth d (cm)	Approximate Slope Ratio[#]		
		CBS	RFS	RBS
2	37	6:1	6:1	6:1
4	37	5:1	6:1	6:1
6	37	5:1	6:1	5:1
8	37	5:1	6:1	5:1
10	37	5:1	5:1	5:1
12	40	4:1	4:1	4:1
15[+]	40	3.5:1	3.5:1	2.5:1

* Soil Conservation Service (1958).

[†] Channel capacity based on retaining 5 cm runoff.

[#] CBS = channel back slope; RFS = ridge front slope; RBS = ridge back slope.

[+] Terrace ridge and RBS to be dept in sod.

TABLE 8.7 TERRACE DIMENSIONS: GRADED OR CHANNEL TERRACE*[†]

Field slope (percent)	Terrace channel depth, d (cm)					Approximate Slope Ratio[#]		
	Terrace length (m)							
	61	122	183	244	305	CBS	RFS	RBS
2	24	27	30	37	37	10:1	10:1	10:1
4[+]	21	27	30	34	34	6:1	8:1	8:1
6	21	24	27	30	30	6:1	8:1	8:1
8	21	24	27	30	30	4:1	6:1	6:1
10	21	24	27	30	30			
12	18	24	27	30	30	4:1	6:1	6:1
15	18	21	27	30	30	4:1	4:1	2.5:1

* Soil Conservation Service (1958).

[†] Channel capacity based on retaining 5 cm runoff.

[#] CBS = channel back slope; RFS = ridge front slope; RBS = ridge back slope.

[+] Terrace ridge and RBS to be kept in sod.

soil. Waterways are used to safely channel runoff from watersheds, terraces, diversion channels and ponds. Thus, in a typical HWLT unit, runoff water from a sloping area is intercepted by either a terrace or diversion channel and flows to a vegetated waterway which directs the water to the retention basin without causing erosion. Emergency spillways for ponds are also frequently designed as vegetated waterways.

The three basic shapes for waterways are trapezoidal, triangular and parabolic. Since many of the waterways at HWLT units flow near a berm, the parabolic shaped waterway will function best with the least danger of eroding the base of the berm. A cross section of a parobolic channel is shown in Fig. 8.23.

When designing a waterway to fit a particular site, the main considerations are vegetation, slope, flow velocity, side slope and flow capacity. Suggested vegetation for use in vegetated waterways is presented in Section 8.7.2 (Table 8.11). The permanent vegetation selected needs to be chosen on the basis of soil type, persistence, growth form, velocity and quantity of runoff, establishment time, availability of seeds or sprigs, and compatability with the waste being applied. Since the area periodically carries large quantities of water, sod forming vegetation is preferred. In many cases, the vegetation being grown on the waste application areas may also be suitable for the waterways.

The design velocity, or flow velocity, is the average velocity within the channel during peak flow. This can be estimated by applying the Manning formula as follows:

$$V = \frac{1.49}{n} \frac{a^{2/3}}{p} S^{1/2} \tag{8.4}$$

where

V = flow velocity in feet/sec (fps);
n = roughness coefficient (0.04 is an estimate for most vegetated areas);
t = design top width of water flow (ft);
d = design depth of flow (ft);
a = cross sectional area in ft^2 calculated as 2/3 td;
p = perimeter calculated as

$$t + \frac{8d^2}{3t}; \text{ and}$$

S = slope of the channel in ft/ft.

Suitable flow velocities for various slopes are given in Table 8.8. The product of flow velocity and cross sectional area of flow gives the flow capacity, which is calculated as follows:

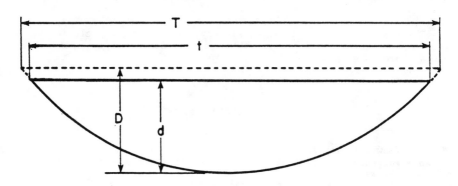

Figure 8.23.   Cross-sectional diagram of a parabolic channel.

$$Q = a\ v \qquad\qquad (8.5)$$

where

Q = flow capacity in ft^3/sec;
a = cross sectional area of flow (ft^2); and
v = velocity in fps.

A properly designed waterway (Fig. 8.23) will carry away runoff from a 25-year, 24-hour storm at velocities equal to or less than the permissible velocity shown Table 8.8. Nomographs such as the one illustrated in Fig. 8.24 are available to determine the channel size needed (Schwab et al., 1971). To use these nomographs, place a mark on the slope scale equal to the channel slope and work the two discharge scales with the designed discharge rate. Using a straight edge, draw a line from the mark on the slope scale through the mark on the nearest discharge scale and extend it until it intersects the top width scale. This is the total construction top width (T). From this point on the top width scale, extend a line through the second discharge scale where marked and extend it until it intersects the total depth scale. This value is the total construction depth (D).

TABLE 8.8 PERMISSIBLE VELOCITIES FOR CHANNELS LINED WITH VEGETATION*

	Permissible velocity (fps)					
	Erosion resistant soils (percent slope)			Easily eroded soils (percent slope)		
Cover	0-5	5-10	Over 10	0-5	5-10	Over 10
Bermuda grass	8	7	6	6	5	4
Buffalo grass Kentucky bluegrass Smooth brome Blue grama Tall fescue	7	6	5	5	4	3
Lespedeza serica Weeping lovegrass Kudzu Alfalfa Crabgrass	3.5	NR[†]	NR	2.5	NR	NR
Grass mixture	5	4	NR	4	3	NR
Annuals for temporary protection	3.5	NR	NR	2.5	NR	NR

* Schwab et al. (1971).

[†] NR = not recommended.

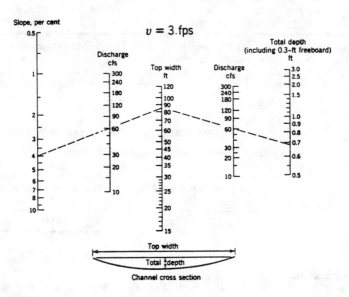

Figure 8.24.   Nomograph for parabolic cross sections with a
velocity of 3 fps (Schwab et al., 1971).
Reprinted by permission of John Wiley & Sons, Inc.

The actual construction of the waterway needs to be done carefully using roadgraders and bulldozers, as necessary. Careful surveying and marking of field areas is needed before beginning earthwork. The entire waterway area should be vegetated as soon as possible after construction and normal agricultural applications of lime and fertilizer used in accordance with site-specific recommendations. Broadcast seeding is the most common practice for planting but drilling, sprigging and sodding are other possible techniques. If drilling or sprigging is used, rows should run diagonally or crosswise to the direction of water flow. Due to the expense, sodding is usually done only on critical areas needing immediate cover.

Maintenance practices for vegetated waterways include periodic mowing to promote sod formation. Annual fertilization is necessary and should be done according to local recommendations. Excess sediment and debris that accumulates in waterways after heavy rains, should be cleaned out to prevent damage to vegetation. A fan shaped accumulation of sediment is likely to form where the waterway joins the retention pond. These deposits need to be removed if they accumulate to a point that interferes with water flow. A more complete discussion of waterway design and construction can be found in Schwab et al. (1971).

In addition to preventing erosion, grassed waterways provide a secondary benefit by improving water quality. In one study, a 24.4 m waterway removed 30% of the 2,4-D that originally entered the waterway (Asmussen et al., 1977). Thus, areas which may potentially carry contaminated runoff water should be vegetated to help improve water quality. Other critical areas that should be vegetated are waterways leading into runoff retention ponds and emergency spillways.

8.6   MANAGEMENT OF SOIL pH

Management of acid or alkaline soils generally requires the addition of some type of chemical amendment for the land treatment unit to operate properly. If a near neutral soil pH is not maintained, plant nutritional problems may develop, soil microorganisms may become less active, and survival of symbiotic nitrogen fixing bacteria may be reduced, resulting in a slower rate of waste degradation. Soil samples should be taken periodically and analyzed for pH. Based on the sample results, the appropriate quantity and type of chemical amendment should be applied.

8.6.1   Management of Acid Soils

Numerous methods exist for measuring soil acidity. The three most common methods are:

(1) titration with base or equilibration with lime;

(2) leaching with a buffered solution followed by analysis of
     the leachate for the amount of base consumed by reaction
     with the soil; and

(3) subtracting the sum of exchangeable bases from CEC (Coleman
     and Thomas, 1967).

Liming of soils refers to the addition of calcium or magnesium compounds that are capable of reducing acidity (Tisdale and Nelson, 1975).
Although the term "lime" is frequently used for material such as $Ca(OH)_2$,
$CaCO_3$, $MgCO_3$, and calcium silicate slags, it correctly refers only to
CaO. The other materials are properly referred to as limestone and liming
agents. When liming agents react with acid soils, calcium or magnesium
replaces hydrogen on the exchange complex (Brady, 1974), as follows:

$$H \diagdown$$
$$\text{Micelle} + Ca(OH)_2 \longrightarrow Ca\text{-Micelle} + 2H_2O$$
$$H \diagup$$

$$H \diagdown$$
$$\text{Micelle} + Ca(HCO_3)_2 \longrightarrow Ca\text{-Micelle} + 2H_2O + 2CO_2$$
$$H \diagup \qquad \text{In solution}$$

$$H \diagdown$$
$$\text{Micelle} + CaCO_3 \longrightarrow Ca\text{-Micelle} + H_2O + CO_2$$
$$H \diagup$$

As the soil pH is raised, plant nutritional problems that accompany
low soil pH are reduced. Soil microorganisms, such as those responsible
for decomposition of plant residues and nitrification, are more active at
pH 5.5-6.5 (Tisdale and Nelson, 1975). Nonsymbiotic nitrogen fixation by
Azotobacter spp. occurs mainly in soils above pH 6.0 (Black, 1968).
Survival of symbiotic nitrogen fixing bacteria, Rhizobium spp., and
nodulation of legume roots is enhanced by liming acid soils (Pohlman,
1966). Many plant diseases caused by fungi are decreased by liming acid
soils. Infection of clover by Sclerotinia trifoliorum was greatly reduced
by liming acid soils in Finland (Black, 1968). It is also desirable to
maintain the pH of the zone of waste incorporation near neutral to minimize
the toxicity and mobility of most metals.

Good management practice requires application of enough liming agent
to raise soil pH to the desired level and addition of sufficient material
every three to five years to maintain that level. Soil sampling and testing should be employed to predict the need for additional liming. The
hydrogen ion concentration of the soil will not reach the desired level
immediately. The change may take six to eight months and, in the case of
added dolomitic limestone, the pH may increase for five years after liming
(Bohn et al., 1979).

8.6.1.1 Liming Materials

Liming agents must contain calcium or magnesium in combination with an anion that reduces the activity of hydrogen, and thus aluminum, in the soil solution (Tisdale and Nelson, 1975). Many materials may be used as liming agents; however, lime (CaO) is the most effective agent since it reacts almost immediately. Thus, lime is useful when very rapid results are needed. Lime is not very practical for common usage because it is caustic, difficult to mix with soil, and quite expensive (Tisdale and Nelson, 1975). The second most effective liming agent is Ca(OH)$_2$, referred to as slaked lime, hydrated lime and builder's lime. Like CaO, it is used only in unusual circumstances since it is expensive and difficult to handle (Tisdale and Nelson, 1975).

Agricultural limestone may be calcitic limestone (CaCO$_3$), dolomite (CaMg(CO$_3$)$_2$), or dolomitic limestone, which is a mixture of the two. Limestone is generally ground and pulverized to pass a specified sieve size. If all the material passes a 10-mesh sieve and at least 50% passes a 100-mesh sieve, it is classified as a fine limestone (Brady, 1974). A fine limestone reacts more quickly than a coarse grade. The neutralizing value of these limestones depends on the amount of impurities, but usually ranges from 65-100% (Tisdale and Nelson, 1975).

In some eastern states, deposits of soft calcium carbonate, known as marl, exist. This material which is usually low in magnesium is occasionally used as a liming agent. Its neutralizing value is usually 70-90% (Barber, 1967). In areas where slags are produced, they are sometimes used as liming agents but their neutralizing value is variable and usually lower than that of marl (Tisdale and Nelson, 1975).

Some waste materials may be suitable as liming agents and can be used when available; but, these materials are generally not as efficient as agricultural limestone. An example of a waste that may be used for liming is blast furnace slag from pig iron production, which is mainly calcium and magnesium aluminosilicates and may also contain other essential micronutrients (Barber, 1967). Basic or Thomas slag, a by-product of the open hearth method of steel production, is high in phosphorus and has a neutralizing value of about 60 to 70% (Tisdale and Nelson, 1975). The composition of slags varies quite a bit, another type of open hearth slag is high in iron and manganese, but has a lower neutralizing value (Barber, 1967). Electric-furnace slag, a by-product of electric-furnace reduction of phosphate rock, is mainly calcium silicate. It contains 0.9-2.3% P$_2$O$_5$ and has a neutralizing value of 65-80% (Tisdale and Nelson, 1975). Miscellaneous wastes such as flue dust from cement plants, refuse lime from sugar beet factories, waste lime from paper mills, and by-product lime from lead mines have been used effectively as liming agents (Barber, 1967). Many fly ashes produced by coal burning power plants are sufficiently alkaline to increase the pH of soil and are frequently used to replace a portion of the lime needed to reclaim mine sites (Capp, 1978).

The lime requirement of a particular soil depends on its buffering capacity and its pH. An equilibrium extraction of the soil with a buffered salt solution followed by determination of exchange acidity is a common method for determining the lime requirement (Peech, 1965b). Many state experiment stations have determined lime requirements for their major soil series and constructed buffer curves. These curves (Fig. 8.25) relate base saturation percentage in the soil to soil pH by expressing milligrams of acidity in soil as a function of soil pH. In addition, lime requirements are expressed in terms of the calcium carbonate equivalent (Table 8.9).

TABLE 8.9 COMPOSITION OF A REPRESENTATIVE COMMERCIAL OXIDE AND HYDROXIDE OF LIME EXPRESSED IN DIFFERENT WAYS*

Forms of Lime	Conventional Oxide Content %	Calcium Oxide Equivalent	Neutralizing Power	Elemental Content %
Commercial oxide	$CaO = 77$ $MgO = 18$	102.0	182.1	$Ca = 55.0$ $Mg = 10.8$
Commercial hydroxide	$CaO = 60$ $MgO = 12$	76.7	136.9	$Ca = 42.8$ $Mg = 7.2$

Brady (1974).

When using $CaCO_3$ as a liming agent, the following formula can be used:

$$\text{Required change in base saturation} \times \text{Soil CEC} \times 1121 = \frac{\text{kg } CaCO_e}{\text{required/ha}} \quad (8.6)$$

Using Fig. 8.25 as an example, to raise the soil pH from 5.5 to 6.0, the base saturation must change from 0.50 to 0.75. Assuming the soil CEC is 17 meg/100 gm, the lime requirement is calculated using equation 8.6 as follows:

$$0.25 \times 17 \times 1121 = 4764 \text{ kg } CaCO_3 \text{ required/ha}$$

When other liming agents are used, a correction factor is added to the equation. This correction factor is the ratio of the equivalent weight of the new liming agent to the equivalent weight of $CaCO_3$. For example, if $CaCO_3$ (equivalent wt = 50) is replaced by $MgCO_3$ (equivalent wt = 42) the lime requirement calculated using equation 8.6 would then be:

$$0.25 \times 17 \times 1121 \times 42/50 = 4287 \text{ kg } MgCO_3 \text{ required/ha}$$

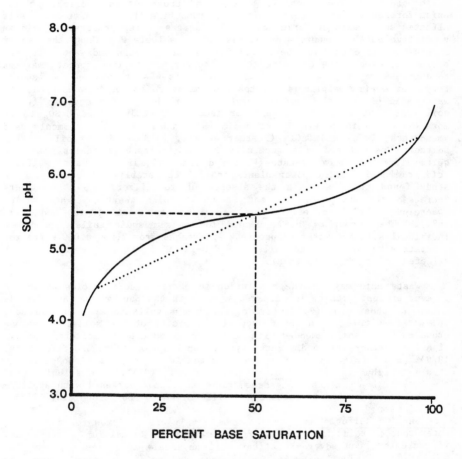

Figure 8.25. General shape for the lime requirement curve for a
sandy loam.

An estimated 4 billion kilograms of waste sulfuric acid are produced each year in the U.S., mainly as a by-product of smelting industries and coal burning power plants (Phung et al., 1978). This acid may have potential for use in the reclamation of salt affected soils. In addition, sulfuric acid could be disposed of by land treating these wastes on saline, saline-sodic, and sodic soils. Using land treatment as a disposal mechanism for these wastes could provide numerous benefits. Land treating salt affected soils with sulfuric acid could increase water penetration, aid in vegetative establishment, and increase water soluble P. Thus, the use of surplus sulfuric acid may be beneficial to both farmers and waste disposal operators. The value of using surplus sulfuric acid from copper smelters to increase water penetration into sodic soils was studied in the laboratory. At optimum application rates equivalent to 12,000-40,000 kg/ha, the waste acid effectively increased water penetration in the sodium-affected soil (Yahia et al., 1975). Another laboratory study showed H_2SO_4 to be more effective in reclaiming of sodic soils than two other commonly used amendments, $CaSO_4$ and $CaCl_2$ (Prather et al., 1978). Mine spoils in the Northern Great Plains are generally saline, calcareous shales that are quite difficult to revegetate (Wali and Sandoval, 1975). Waste sulfuric acid from coal burning power plants could help establish vegetation. One study found that, even in the absence of fertilizer, H_2SO_4 amendments increased the phosphorus content of thick spike wheatgrass and yellow sweetclover (Melilotus officinalis) grown on mine spoil (Safay and Wali, 1979). The amount of H_2SO_4 needed to reclaim sodic soils depends on individual soil and water properties, and ranges from 2,000-6,000 kg/ha for moderately sodium affected soils to 6,000-12,000 kg/ha for severely sodium affected soils (Miyamoto et al., 1975).

Waste acid may provide a solution to nutrient deficiencies which are an ever present problem in calcareous soils in the Southwest. Acid application to phosphorus (P) deficient, calcareous soils in Arizona increased the water soluble P and the P-supplying capacity of the soils. Tomatoes grown on these soils amended with waste acid from copper smelters showed a significant increase in dry matter yield and P uptake (Ryan and Stroehlein, 1979). Spot applications of acid were effectively corrected iron deficiencies in sorghum (Sorghum bicolor) (Ryan et al., 1974). The solubility of the essential nutrients, manganese, zinc and iron, increased with application of sulfuric acid to calcareous soils (Miyamoto and Stroehlein, 1974).

Surplus sulfuric acid may also be a valuable addition for irrigation water that contains high levels of sodium relative to calcium. Such water, if untreated, can adversely affect soil physical properties (Miyamoto et al., 1975). Field studies in Texas showed that acidification of irrigation water reduced the hardness of calcareous soils and lowered the exchangeable sodium percentage of the soils (Christensen and Lyerly, 1965). Acid treatment of ammoniated irrigation waters reduced volatile loss of NH_3 by as much as 50% and also prevented plugging a problem often caused by calcium and bicarbonate (Miyamoto et al., 1975).

Although vegetation is not essential, it may form an important part of the ongoing management plan for the facility. Revegetation is generally required at closure, unless a regulatory variance is granted (EPA, 1982). In all cases, it is desirable to establish a permanent cover following closure to prevent long-term erosional hazards even when not strictly required by the regulations for disposal facilities.

The site manager must be cognizant of the major components required to obtain successful revegetation. The following factors are needed for successful stand establishment and growth:

1) selecting species adapted for the site;

2) preparation of an adequate seedbed;

3) planting during correct season;

4) planting the proper quantity of seed or sprigs;

5) planting seed at the proper depth;

6) allowing sufficient time for plant establishment;

7) implementing a proper fertilization program; and

8) using proper management practices.

Contingency plans should provide for reseeding if the crop does not emerge or fails after emergence.

8.7.1   Management Objectives

The specific objectives of the overall management plan for the HWLT unit are critical to developing a vegetative management plan. Beneficial uses of plants include use to improve site trafficability for waste application or other equipment, to indicate "hot spots" where excessive quantities of waste constituents have accumulated, to minimize wind and water erosion, and to take up excess nitrogen or metals and remove excess water to promote oxidation of organic material. An optional and especially useful function for vegetation at HWLT units is runoff water treatment, where water will be discharged under a permit there are several choices for treating the water. One of these options is to establish a water tolerant species in an overland flow treatment system. The vegetation acts to remove certain types of contaminants from the runoff water through filtering, adsorption, and settling. Other treatment mechanisms are enchanced with increased wastewater detention time. Plants may also be used in land treatment context for aesthetic appeal; since much of the public's perception of a problem or hazard is linked to the visual impression of the facility, a green, healthy crop cover will reassure the public.

One must recognize that there are some limitations associated with using cover crops. Some arguments against a plant cover include the following:

(1) maintaining concentrations of waste in soil which are not phytotoxic may limit the allowable waste application rates to levels far below the capacity of the soil to treat the waste;

(2) where wastes are applied by spray irrigation, hazardous waste constituents may stick to the plant surfaces;

(3) plants may translocate toxins to the food chain; and

(4) a crop cover may filter ultraviolet radiation which could aid in the decomposition of certain compounds.

Table 8.10 presents some of the alternative management techniques that can be used to replace the role of plants in land treatment. The uses of plants at HWLT units are further discussed below.

Where waste is stored and applied only during the warm season and a vegetative cover is desired, the management schedule needs to allow enough time for the establishment of at least a temporary cover crop following waste applications before conditions become unfavorable. In situations where waste is treated year-round, it may be desirable to subdivide the area into plots so the annual waste application can be made within one or two short periods. Following incorporation, surface contouring, or other activities, each plot can be seeded.

If the objective of using vegetation is to take up excess nitrogen, it may be desirable to harvest and remove the crop. The best use of harvested vegetation is as mulch for newly seeded areas. The crop should not be removed from the facility unless a chemical analysis demonstrates that it is acceptable for the specific use. If it is not possible or necessary to harvest the crop, it can be left in place and plowed down when another application of waste is made. In this case, the nitrogen taken up by the crop has not been removed from the system but it has been tied up in an organic form. As the crop residue decomposes, nitrogen will be slowly released. The mineralization rate of nitrogen should be taken into account when determining the nitrogen balance for the site.

For liquid hazardous wastes, it may be possible to use spray irrigation disposal in existing or newly planted forests. With proper design and management, including controlled application rates to match infiltration and storage, it may be possible to minimize direct overland flow of runoff water. Water storage may be necessary to avoid application of waste during unsuitable conditions such as when the site is already saturated. Such systems have been used successfully for treatment of municipal sewage effluent (Myers, 1974; Sopper and Kardos, 1973; Nutter and Schultz, 1975; Overcash and Pal, 1979). The use of such systems when applying hazardous industrial effluents should be fully justified by pilot scale field studies over a sufficient time period to demonstrate their effectiveness. In addi-

TABLE 8.10 ALTERNATIVE MANAGEMENT TECHNIQUES TO REPLACE THE ROLE OF PLANTS
IN A LAND TREATMENT SYSTEM

Plant function	Alternative management
Protective:	
Wind erosion	Maintain a moist soil surface
	Wastes often provide the necessary stability when mixed with the soil.
Water erosion	Minimize slopes and use proper contouring to reduce water flow velocities
	Some wastes, such as oily sludges, repel water and stabilize the soil against water effects.
	Design runoff catchments to account for increased sediment load.
	Runoff water may need some form of treatment before release into waterways.
Cycling:	
Transpiration	Dewater the waste
	Control applications of wastewater to a lower level.
Removal	Plants have only a very minor role in this respect; for organics, manage for enhanced degradation; for inorganics, reduce loading rates.

tion, a method of collecting runoff from this type of system would need to be designed.

At HWLT units where liquid hazardous wastes are spread on the soil surface by irrigation or subsurface injected, it may be desirable to maintain a continuous vegetative cover. Another use of vegetation where wastes are spray irrigated is as a barrier to aerosol drift. In some cases a border of trees may be desirable.

At closure, permanent vegetation is established following the same procedures used for temporary vegetation. In some instances, it is desirable to cover earth structures with 10 to 15 cm of topsoil to assist in establishing vegetation. Lime may need to be added to the final surface, whether it is subsoil or topsoil, to adjust the pH for the species planted. Liming of soils is discussed in Section 8.6.1. Fertilizer and seed may then be applied by the methods described in the following sections. On critical areas, the use of sod or sprigs may be desirable for establishing certain species and mulching may be necessary to prevent erosion. It is generally advisable to use a light disc or cultipacker to anchor the material against displacement by wind and water.

8.7.2   Species Selection

Vegetation should be selected which is easily established, meets the desired management goals, and is relatively insensitive to residual waste constituents. Common residuals occurring at HWLT units include organics, salts, nutrients and possibly excess water. Other important considerations include disease and insect resistance. Grasses are often a good choice because many are relatively tolerant of contaminants, can often be easily established from seed, and can be used to accumulate nitrogen. Various nitrogen accumulating species are discussed in Section 6.1.2.1.4.

Perennial sod crops adapted to the area are often the most desirable surface cover since they provide more protection against erosion and a longer period of ground cover than annual grasses or small grains. In climates where legumes are adapted, it may be desirable to include a grass-legume mixture for the final vegetative cover to provide a low cost nitrogen supply for the grasses. Each species in a mixture will be better adapted to specific site characteristics than other species in that mixture. Rooting habits will vary according to the species planted, thus a mixture of species may allow more efficient use of soil moisture and nutrients at various depths. In cases where a species requires intensive management, it should be planted in a pure stand; many introduced grasses fall into this category.

Water tolerance of vegetation is a concern at many HWLT units because waste dewatering is a common practice. Many perennial grasses can withstand temporary flooding during dormant stages; however, most of the small grains including barley (Hordeum vulgare), oats (Avena sativa), and shallow rooted clovers are very sensitive to flooding. Some relatively tolerant species include Dallisgrass (<u>Paspalum</u> <u>dilatum</u>), switchgrass (<u>Panicum</u> <u>virgatum</u>), bermudagrass (<u>Cynodon</u> <u>dactylon</u>), bahiagrass (<u>Paspalum</u> <u>notatum</u>), Reed canary grass (<u>Phalaris</u> <u>arundinacea</u>), and tall fescue (<u>Festuca</u> <u>arundinacea</u>); however, rice (<u>Oryza</u> <u>sativa</u>) is the most water tolerant plant available. Table 8.11 lists the relative water tolerance of various plants.

Regardless of the specific management objectives, the species selected must be adapted to the climate, topography and soils of the site. Vegetative parameters considered during plant selection include the following:

1) ease of establishment;

2) plant productivity;

3) ability to control erosion;

4) ability to withstand invasion by undesirable plants; and

5) availability of seed at a reasonable price.

Generally, seed of native species should be obtained from local sources or within 200 miles north or south, and 100 miles east or west of the site (Welch and Haferkamp, 1982). Introduced plant materials do not follow these same guidelines; they may be obtained from sources over a relatively broad geographic range. It is highly recommended that certified varieties of either native or introduced plant materials be used when available.

Guidance on species adaptation is given in Table 8.11 and Figs. 8.26 and 8.27. Other sources of information which may be useful are the highway cut revegetation standards available from most state highway departments and recommendations from the Soil Conservation Service, state agricultural extension services, and/or the agronomy departments at state universities. In some instances selected plant materials may be used in climatic zones other than those indicated when special conditions unique to the land treatment unit would permit their use. For example, where irrigation is available, the season for establishment is often longer than indicated in Table 8.11. Thus, Table 8.11 is a general guideline and it is advisable to check selections with local sources because some species are adapted only to certain sites within a given geographic region.

8.7.3   Seedbed Preparation

Prior to seeding, all grading and terracing should be completed and a good seedbed prepared. An ideal seedbed is generally free from live resident vegetation, firm below the seeding depth and has adequate amounts of mulch or plant residue on the soil surface. The most important concerns of seedbed preparation are to reduce existing plant competition and to create a favorable microclimate for developing seedlings or sprigs.

Various methods of seedbed preparation exist; however, plowing is the most common. Use of an offset disc one-way plow, or moldboard plow appears

TABLE 8.11 REGIONAL ADAPTATION OF SELECTED PLANT MATERIALS

Common and scientific names	Pacific Coast	Intermountain	Southwest	Northern Great Plains	Southern Great Plains	Midwest	Southeast	Northeast	Season of growth	Growth habit	Native or introduced	pH	High water	Drought	Cold	Salinity	Sand	Loam	Clay	Lbs. PLS. per acre seeding rate	Special considerations and adaptations
Aeschynomene						x	x	x	W												
Alfalfa (Medicago sativa)	x	x	x	x	x	x	x	x	W	P	I	6.5-7.5	2	2	1-2	2	—	1	2	4.1	M.P.R. 15". Sodformer. Most widely used legume for range and pasture mixtures. Requires well-drained sandy loam to clay soils. Great value as soil improving crop. A fine, mellow, firm seed bed should be prepared. Sensitive to low boron levels. Deep rooted.
Alfileria (Erodium cicutarium)			x						W	A	I						1	1	2		
Alyceclover (Alysicarpus vaginalis)				x	x		x		W	A	I		3	1-2	2-3	2	1	1	1	3.1	M.P.R. 12". Bunchformer.
Bundleflower, Illinois (Desmanthus illinoensis)			x	x	x	a			W	P	N		3	2	3	2	2	1	1	1.5	M.P.R. 16". Bunchforming. Deep rooted. Easily established.
Burclover, california (Medicago hispida)	x						x		C	A	I		3	2-3	3	3	1	1	1	3.0*	Seeding rate based on hulled seed. Prefers moist, well-drained fertile soils. Short season annual which usually re-seeds. Produces less than crimson or arrowleaf clover. Prefer soils high in calcium.
Burclover, southern or spotted (Medicago arabica)									C	A	I		3	2	3	3	2	1	2	3.0*	Seeding rate based on hulled seed. Prefers soils high in calcium.
Burnet, small (Sanguisorba minor)	x	x							C	P	I		3	2	1	3	2	1	2	7.9	Forb with persistent leaves.
Bushsunflower, annual (Simsia exaristata)							a		C	A	N		3	2	3	3	2	1	1		M.P.R. 16". Bunchformer.
Buttonclover (Medicago orbicularis)									C	A	I		3	2	3	3	2	1	2		Prefer soils high in calcium. Commonly used in overseeding of bermudagrass.

TABLE 8.11 (continued)

Common and scientific names	Pacific Coast	Intermountain	Southwest	Northern Great Plains	Southern Great Plains	Midwest	Southeast	Northeast	Season of growth	Growth habit	Native or introduced	pH	High water	Drought	Cold	Salinity	Sand	Loam	Clay	Lbs. PLS. per acre seeding rate	Special considerations and adaptations
Clover, alsike (Trifolium hybridum)	x	x		I		x		x	C	P	I	5.0-7.5	1-2	3	1	2-3	2	1	1	1.9	Noncreeping. Adapted to cool, moist sites. Commonly used in irrigated pasture mixtures. Generally dies after 2 years. Not recommended in areas of South where Ladino clover is adapted. Also produced in many parts the northeast.
Clover, arrowleaf (Trifolium vesiculosum)	a						a		C	A	I	6.0-6.5	3	2	1-2	3	2	1	1	2.5	Seeding rate based on scarified seed. Less tolerant of acidity and low fertility than crimson clover. Should use Pelinor adhesive and arrowleaf clover inoculum (type O). Scarification is beneficial due to hard seed content (>70%).
Clover, ball (Trifolium nigrescens)							x		C	A	I		1-2				2	1	1	1.1*	Tall growth form. Produces growth one month later than crimson clover. Excellent reseeder.
Clover, berseem (Trifolium alexandrinum)						b	b		C	A	I	5.7+		1-2	3		2	1			Produces more forage in winter than most legumes. Erect growth habit.
Clover, crimson (Trifolium incarnatum)	a					x	x	x	C	A	I		3	3	1	2	2	1	2	6.3	M.P.R. 14". Bunchformer. Winter legume. Readily reseeds itself. Tolerant of medium soil acidity. Thrive on both clay and sandy soils. Tolerant of wide range of climatic conditions. Thrives in association with other crops such as coastal bermudagrass. Commonly have 30 to 75% hard seed.
Clover, hop (small) (Trifolium dubium)	x	x					x		C	A	I										Shallow extensive root system. Very competitive with the associated grass. Do not seed alone due to wind damage on young seedlings.
Clover, persian (Trifolium resupinatum)	x						x		C	A	I		1-2	2		3	3	2	1	2.3*	Used for soil improvement.
Clover, red (Trifolium pratense)	x			I		x	x	x	C-W	B	I	6.0-7.0	2				3	2	2	3.2	M.P.R. 19". Bunchformer. Biennial, acts as short-lived perennial but reseeds under mesic conditions. Noncreeping. Prefers fertile, well-drained soils high in lime but will grow on moderately acid soils; often seeded with other legumes and grasses. Susceptable to crown rot, southern anthracnose, and mildew. Hyperaccumulates zinc.
Clover, rose (Trifolium hirtum)	x				x		x		C	A	I		3	1	3	2	2	2	2	6.2	M.P.R. 12". Bunchformer. Widely seeded in California on annual grassland and brush burns. Readily reseeds. Established in Texas. Grows and persists well in areas of limited rainfall (18-25" per year). Northeast Texas growth limited to early spring season. Will grow well in association with summer perennial grasses. Does not do well in poorly drained areas.

TABLE 8.11 (continued)

Common and scientific names	Pacific Coast	Intermountain	Southwest	Northern Great Plains	Southern Great Plains	Midwest	Southeast	Northeast	Season of growth	Growth habit	Native or introduced	pH	High water	Drought	Cold	Salinity	Sand	Loam	Clay	Seeding rate lbs. PLS. per acre	Special considerations and adaptations
Clover, sour (Melilotus indica)	X	1		X					W	A	1		1	3	2	1-2	1	1	1	3.0	Seen in volunteer stands by roadsides. Will tolerate more acid soils than other members of Melilotus genesis.
Clover, strawberry (Trifolium fragiferum)	X						X		C	P	1		2	1-2	2-3	2	1	1	1	13.4	M.P.R. 19". Sodformer. Creeping by rhizomes; low growing. Best use is on wet, salty sites. Very hardy legume.
Clover, subterranean (Trifolium subterraneum)	X					X	X		C	A	1	6.0-7.0	1-2	3	2	2-3	2	1	2	1.5	M.P.R. 16". Sodformer. Well adapted for interseeding mesic annual grasslands in California. Good winter growth. Does best on well-drained, fertile, loam soils with moderate rainfall. Used for erosion control, hay, pasture, soil improvement and seed production. Prostrate growth habit. Tolerant of acid soils.
Clover, white (Ladino) (Trifolium repens)	X	1	1	1	1	X	X	X	C	P	1										M.P.R. 18". Sodformer. Used in pasture mixtures on mesic or irrigated sites. Creeping by stolons. Used in association with grasses and other legumes. Used for soil improvement, erosion control and wildlife. Requires adequate quantities of available phosphorus, potash and calcium. Stand thickness decreases after several years.
Cowpean (Vigna sinensis)			X		X		X		W	A	1		2-3	1-2	1		1	1	2	30.0	One of the most extensive legumes.
Crownvetch (Coronilla varia)	X	X				X	X	X	C	P	1	5.5-7.5	3		1-2		2	2	2	3.0*	M.P.R. 18". Sodforming. Should scarify seeds. Hard seed may be up to 90%. Best adapted to fertile well-drained soils; however, will tolerate some degree of infertility and acidity after established. Excellent for erosion control. Slow to establish but aggressive upon establishment. Commonly seeded with ryegrass.
Field pea (Pisum sativum subsp. arvense)					X		X		C	A			3		1-2		2	1	1		Fall seeding in cotton growing states. Grows well on all soils except wet and poorly drained types. Grown for hay, silage, pasture, seed and green manure.
Flat pea (Lathyrus sylvestris)					X		X		C	P	N	4.0-6.0	2-3	1-2	1		1	2	3	10.0	Seed may be toxic to grazing animals. Slow germination but aggressive upon establishment. Climbing growth form. Maintains a pure stand better than most legumes. Rhizomatous.
Gaillardia, slender (Gaillardia pinnatifida)							X		W	A		5.5-7.0	2-3				1	2	2		M.P.R. 15". Bunchformer. Also adapted to part of Intermountain region.
Indigo, hairy																					Fairly deep rooted and upright.

TABLE 8.11 (continued)

Common and scientific names	Pacific Coast	Intermountain	Southwest	Northern Great Plains	Southern Great Plains	Midwest	Southeast	Northeast	Season of growth	Growth habit	Native or introduced	pH	High water	Drought	Cold	Salinity	Sand	Loam	Clay	Lbs. PLS, per acre	Special considerations and adaptations
Kochi, prostrate or prostrate summercypress (Kochi prostrata)	X	X	X	X	X					P	I		3	1	1	1-2	1	1	1	1.7	M.P.R. 12". Bunchformer. long lived. Extensive root system.
Kudzu (Pueraria lobata)							X		W	P	I		2-3	1-2	3		1	2	2		Plant at 4'x5' spacing. Very little seed produced under southern climatic conditions. Slow to establish, however, grows rapidly after establishment. Will not tolerate close mowing. Other legumes are better adapted in the Southeast since they are easier to establish and more productive.
Lespedeza, bicolor (Lespedeza bicolor)						X	X	X	W	A	I	5.0-6.0	2	1-2	2	3	3	1	1		Grows in low fertility soils. Generally not used for forage.
Lespedeza, common (kobe) (Lespedeza striata)						X	X	X	W	A	I	5.0-6.0	2	2	2		3	1	1	6.3*	Seed rate based on unhulled seeds. Low growing. Better adapted to Texas than Korean lespedeza. Important for pasture, hay and soil improvement. Grown in association with other crops. Neutral to acid soils. Susceptible to bacterial wilt, tar spot, powdery mildew, and southern blight.
Lespedeza, Korean (Lespedeza stipulacea)						X	X	X	W	A	I	5.0-7.0	2	2	2		3	1	1	6.3*	Hard seed 40-60%. Responds to lime and fertilizer applications. Good for soil improvement, hay and seed. Will grow on most soil including poor and acid soils; however, less tolerant of acid soils than common lespedeza. Susceptible to bacterial wilt, tar spot, powdery mildew, and southern blight.
Lespedeza, prostrate (Lespedeza daurica var. schimadai)						X	X	X	W	P	I	5.0-7.0*	2-3	1-2							
Lespedeza, sericea (Lespedeza cuneata)						X	X	X	W	P	I	4.5-7.0	2	2	2	1	2	1	1	6.3*	Seed should be scarified. Seeding rate based on scarified seed since there is usually 75% or more hard seed. Valuable on badly depleted soils as a pioneering legume. Tolerant to low fertility. Should not be mowed in late summer--plant reserve building. Has not performed well in Texas. Bunch-like growth habit.
Medic, black (yellow trefoil) (Medicago lupulina)					X		X		C	A	I	6.0+		2		1	2	2	1	1.5*	Seed scarce (no commercial cultivars). Use alfalfa in oculum. Adapted to lime soils.

TABLE 8.11 (continued)

Common and scientific names	Regional adaptation — Pacific Coast	Intermountain	Southwest	Northern Great Plains	Southern Great Plains	Midwest	Southeast	Northeast	Season of growth	Growth habit	Native or introduced	pH	Tolerance — High water	Drought	Cold	Salinity	Soils — Sand	Loam	Clay	Seeding rate lbs. PLS, per acre	Special considerations and adaptations
Milkvetch, cicer (Astragalus cicer)	x	x		x	x				M	P	I	5.0-6.0	2	2	1-2	2	1	1	2	6.0	M.P.R. 18. Sodformer. Low growing perennial. Fair to good production Rhizomatous. Traits in stand establishment. Non-bloating. Does not accumulate selenium. Hard seed coat. Long-lived.
Penstemon, palmer (Penstemon palmeri)		x								P	N						1	1	1		M.P.R. 15". Sodformer. Short-lived. Also adapted to part of intermountain region.
Penstemon, Rocky Mountain (Penstemon strictus)	x	x								P	N						1	2	3		M.P.R. 15". Bunchformer. Good seedling vigor. Adapted to parts of intermountain and Southwest.
Poppies, gold (Eschscholtzia spp.)	x									A	N						1	2	2		M.P.R. 10". Bunchformer.
Prairieclover, purple (Petalostemum purpureum)				x	x				M	P	N						2	1	1		M.P.R. 15". Bunchformer. Excellent seed producer.
Prairieclover, white (Petalostemum candidum)		x								P	N						1	1	1		M.P.R. 14". Bunchformer.
Sainfoin (Onobrychis viciaefolia)				x	x		a		C	P	I		3	2	2	3	1	1	2	16.8	M.P.R. 16". Bunchformer. Nonbloating legume. Deep rooted species. Well adapted to dry calcareous soils.
Singletary pea (rough) (Lathyrus hirsutus)					x	x	x		M	A	I		1		1		2	2	1	8.8	Should scarify seed. Grows on soils too wet for other winter legumes. Used for hay. Good soil improving crop. Seed is poisonous to animals.
Sunflower, maximilian (Helianthus maximiliana)				x	x	x	x	b	M	P	N			2	1	3	1	1	1	0.3	M.P.R. 18". Sodformer. Does not invade or spread like most sunflowers. Regeneration forms ring around previous years growth. Easily established.
Sweetclover, stiff (Helianthus lactiflorus)				x	x					P	N						1	1	2		M.P.R. 16". Sodformer.
Sweetclover, white (Melilotus alba)	x	x	x	x	x	x	x	b	C	B	I	6.0-8.0	2	1-2	1	1-2	1	1	1	3.4	M.P.R. 16". Bunchformer. Seed of sweetclover should be scarified. Used for green manure more than forage. Excellent seedling vigor. Tall growing. Good soil improving crop due to large tap root. Matures ahead of cotton root-rot infection. Unreliable seed production. Susceptible to sweetclover weevil, root borer and aphid.

TABLE 8.11 (continued)

Common and scientific names	Regional adaptation								Season of growth	Growth habit	Native or introduced	Plant adaptation								Seeding rate lbs. PLS. per acre	Special considerations and adaptations
	Pacific Coast	Intermountain	Southwest	Northern Great Plains	Southern Great Plains	Midwest	Southeast	Northeast				pH	High water	Drought	Cold	Salinity	Sand	Loam	Clay		
Sweetclover, yellow (Melilotus officinalis)	X	X	X	X	X	X	X	p	C	A	I	6.0-8.0	2	1	1	2	2	1	1	3.4	M.P.R. 16". Bunchformer. More tolerant of drought and competition but has a shorter growth period than white sweetclover. Reseeds better than white sweetclover. Acts like biennial if spring seeded. One of the best soil improving crops due to deep tap root. Seeds should be scarified. Unusually susceptible to injury from a number of chemicals used for weed control. Can be established better than white sweetclover in dry conditions. Neutral to alkaline and well drained soils. Susceptible to sweetclover weevil, root borer and aphid.
Trefoil, birdsfoot (Lotus corniculatus)	X			1	X	e	e	X	W	P	I	5.0-7.5	1-2	2-3	2	1-2	2	1	1	2.1	M.P.R. 18". Bunchformer. Does not cause bloat. Rhizomatous. Mostly used in irrigated pastures. May be difficult to establish. Should be planted in mixture with a grass species. New varieties are being developed for the Southeast which are resistant to crown and root diseases. Also adapted to part of Southern Great Plains.
Vetch, American (Vicia americana)	X	X	X	X	X	X			C	P	N										M.P.R. 18". Sodformer.
Vetch, common (Vicia sativa)	e				X		X	X	C	A	I		3	1-2	2		2	1	2	8.7*	Used in combination with small grains--vetch-rye combination; less winter hardy than other vetches. Best adapted to well drained, fertile loam soils.
Vetch, hairy (Vicia villosa)	e			X	X	X	e	X	C	A	I	5.0-7.5	2	2	1-2	1-2	1	1	1	5.6*	M.P.R. 18". Sodformer. Most winter-hardy of cultivated vetches; most widely grown.
Vetch, narrowleaf (Vicia sativa var. nigra)					X		X		C	A			3	2	2		1	1	1	10.0	Often seen in volunteer stands. Prefers well drained soils. Identified by black pods. Limited use.
Vetch, winter (woolly pod) (Vicia dasycarpa)	X		X	X	X	X	X		C	A	I						2	1	2		M.P.R. 12". Bunchformer. Less cold tolerant and more heat tolerant than hairy vetch. Prefers well drained soils.
Zexmenia, orange (Zexmenia hispida)					X					P	N						1	1	1		M.P.R. 18". Bunchformer.

TABLE 8.11 (continued)

Common and scientific names	Pacific Coast	Intermountain	Southwest	Southern Great Plains	Northern Great Plains	Midwest	Southeast	Northeast	Season of growth	Growth habit	Native or introduced	pH	High water	Drought	Cold	Salinity	Sand	Loam	Clay	Seeding rate Lbs. PLS. per acre	Special considerations and adaptations
Bahiagrass (Paspalum notatum and media)							x		W	P	I	4.5-7.5	1	2	3	1	1	1	1	5.2	M.P.R. 30". Sodformer. Rhizomatous. Keep young by mowing.
Barley (Hordeum vulgare)	x	x				x	x	x	C	A	I	5.5-7.8	2-3	2	1	2	1	1	1	3.0*	Commonly springed 17-18" apart. Rhizomatous. Adapted to areas around the Great Lakes and the East Coast to North Carolina. Possible use in gully bottoms.
Beachgrass, American (Ammophila breviligulata)							x	x	C-W	P	I		2-3	1-2	1	1	1	1	1	3	
Bermudagrass (Cynodon dactylon)							x		W	P	I	4.5-7.5	1	1-2	3	3	2	1	2	1.0	M.P.R. 16". Sodforming. Keep young by mowing and ample fertilization. Most varieties must be grown from sprigs at 2x2' spacing; however, common and NK 37 can be seeded. Does best at pH of 5.5 and above.
Bluegrass, big (Poa ampla)	x	x	x		x				C	P	N		1-2	1-2	1	1	2	1	2	1.5	M.P.R. 12". Bunchgrass. Seed in pure stands.
Bluegrass, bulbous (Poa bulbosa)	x		x						C	P	I		3	1	1		2	2	2	1.9	Good erosion control; spreads by aerial bulbils and swollen stem bases. Low yield; unreliable producer.
Bluegrass, Canada (Poa compressa)	x	x			x	a		x	C	P	I	4.5-7.5	2	2	1	3	2	2	2	8.7	Does well on soil too low in nutrients to support good stands of Kentucky bluegrass.
Bluegrass, Canby (Poa canbyi)	x	x			x				C	P	N		2	2	1		3	2	1		M.P.R. 10". Bunchgrass. Adapted to shallow sites.
Bluegrass, Kentucky (Poa pratensis)	x	x			x			x	C	P	I	5.5-7.0	2	2	1	3	1	1	1	0.8	Excellent sod formation. Reproduced by seeds, tillers, and rhizomes. Low production and summer dormancy limit use; however, will grow on disturbed sites. Adapted to Northern Great Plains and Intermountain region where moisture is plentiful. Shallow rooted.
Bluegrass, upland (Poa glaucantha)		x							C	P	-			1-2	2-3	2	1	1	2	1.0	M.P.R. 16". Bunchgrass. Adapted to shallow sites.
Bluestems (Angelton, Gordo, Medio) (Dichanthium aristatum)				x			a		W	P	I		2	2	2	2	3	3	1		M.P.R. 25-30". Bunchgrass.
Bluestem, big (Andropogon gerardii)				x	x	x			W	P	N	5.0-7.5	2	2	1	2	2	1	2	6.0	M.P.R. 25". Bunchgrass. Very productive on mesic sites. Strong, deep rooted. Effective in controlling erosion.

TABLE 8.11 (continued)

Common and scientific names	Pacific Coast	Intermountain	Southwest	Northern Great Plains	Southern Great Plains	Midwest	Southeast	Northeast	Season of growth	Growth habit	Native or introduced	pH	High water	Drought	Cold	Salinity	Sand	Loam	Clay	Lbs. PLS, per acre seeding rate	Special considerations and adaptations
Bluestem, cane (Andropogon barbinodis)			x		x				W	P	N	7.2-8.0	2	2	2	2	2	1	1		M.P.R. 12". Bunchgrass. Adapted to calcareous sites. Seed available in limited quantities.
Bluestem, Caucasian (Bothriochloa caucasica)			x		x		x		W	P	I		2	1	2	2	2	1	1	1.2	M.P.R. 18". Bunchgrass. Generally seeded in pure stand. An "Old World" bluestem.
Bluestem, Kleberg (Dichanthium annulatum)			x		x		p		W	P	I	6.0-8.0	2	2	2	3	2	1	1	1.2	M.P.R. 20". Bunchgrass.
Bluestem, little (Schizachyrium scoparium)			x		x	x			W	P	N		2	2	1	2	1	1	1	3.4	M.P.R. 16-20". Bunchgrass. Dense root system with short rhizomes. More drought tolerant than big bluestem. Good surface protection.
Bluestem, Old World (Dicanthium spp - Bothriochloa spp) (Blend)			x		x		s		W	P	I			1	2	3	1	1	1	1.2	M.P.R. 14".
Bluestem, sand (Andropogon gerardii or hallii var. paucipilus)			x		x				W	P	N		2	2	2	2	2	2	1	6.0	M.P.R. 14-18". Sodformer. Rhizomatous. Very productive on mesic, sandy soil.
Bluestem, yellow (Bothriochloa ischaemum)			x		x				W	P	I		2	2	1	2	2	1	3	1.2	M.P.R. 16". Bunchgrass. Adapted to shallow and calcareous sites.
Bristlegrass, plains (Setaria leucopila or macrostachya)			x						W	P	N		3	1	1	2	1	1	2	3.0	M.P.R. 12". Bunchgrass. Well adapted to disturbed sites. Good seed producer. May produce more than one crop depending on moisture.
Brome, California (Bromus carinatus)	x	x	x						C	A	N	5.5-8.0			1		1	1	3		M.P.R. 14". Bunchgrass. Self seeding.
Brome, meadow (Bromus biebersteinii)		x		x					C		I						2	1	1		M.P.R. 17". Bunchgrass. Rapid establishment.
Brome, mountain (Bromus marginatus)	x	x	x						C	P	N		2	2	1	2	2	1	1	12.4	M.P.R. 18". Bunchgrass. Not commonly used.

TABLE 8.11 (continued)

Common and scientific names	Pacific Coast	Intermountain	Southwest	Northern Great Plains	Southern Great Plains	Midwest	Southeast	Northeast	Season of growth	Growth habit	Native or introduced	pH	High water	Drought	Cold	Salinity	Sand	Loam	Clay	Seeding rate Lbs. PLS. per acre	Special considerations and adaptations
Brome, red (Bromus rubens)	b	x	x						C	A	I	5.5–8.0	2	2	1	2	2	1	1		M.P.R. 12". Bunchgrass. Cultivars are unavailable.
Brome, smooth (Bromus inermis)	x	x	x	x	x	x	x	x	C	P	I	6.0–7.0	2	2	1	2	2	1	1	6.0	M.P.R. 17". Sodforming. Excellent grass for use with alfalfa. Reproduces by seed, tillers and rhizomes.
Bromegrass, field (Bromus arvensis)				x			x	x	C	A	I	6.5–8.0	2	1	1	2	3	1	1	12.0	Bunchgrass. Extensive fibrous root system. Rapid growth and easy to establish.
Buffalograss (Buchloe dactyloides)				x	x	x	a	x	W	P	N		1	1	3		2	1	2	16.0	M.P.R. 15". Sodforming. Seeding rate based on seed in bur. Seeding rate for grain is 3.0 PLS. Low production. Seed only in mixtures. Seeded or transplanted by stolons or rhizomes. Also adapted to part of southwest region.
Buffelgrass (Cenchrus ciliare)			x	x	x		x		W	P	N	5.0–7.5	2	2		2	1	2	1	3.0	M.P.R. 16". Bunchgrass. Mostly rhizomatous. Higgins, Nueces, and Llano can be seeded at 1.5 lb. PLS/A.
Canarygrass, reed (Phalaris arundinacea)	x	x	x	x		x	x	x	W	P	I			3	2–3					2.6	Sodforming. Cut to prevent maturity, seeded, or spread by sod or culm cuttings. Will endure submergence. Seed does not store well.
Carpetgrass (Axonopus compressus)							x		W	P	N				3		2	2	2	5.0	Stoloniferous. Forms a very dense sod.
Centipedegrass (Eremochloa ophiuroides)				b			x		C	P	I										Makes a close turf and is very aggressive. Sod or stolons, no seed available. Easily established, forms a dense turf. Legumes not recommended because of its aggressive nature.
Chess, soft (Bromus mollis)	x								W	A	N										M.P.R. 15". Bunchgrass. Self seeding. Also used in Georgia.
Cottontop, California or Arizona (Digitaria californica, or Trichachne californica)		x	x	x					W	P	N		2–3	2	2	2	2	1	2	1.2	M.P.R. 15". Bunchgrass. Reproduces by seed. Good seed set. Adapted to calcareous sites.
Curlymesquite, common (Hilaria belangeri)			x	x					W	P	N		1	1			2	1	1		M.P.R. 14". Cultivars are unavailable. Stoloniferous.

TABLE 8.11 (continued)

Common and scientific names	Pacific Coast	Intermountain	Southwest	Northern Great Plains	Southern Great Plains	Midwest	Southeast	Northeast	Season of growth	Growth habit	Native or introduced	pH	High water	Drought	Cold	Salinity	Sand	Loam	Clay	Seeding rate Lbs. PLS. per acre	Special considerations and adaptations
Dallisgrass (Paspalum dilatatum)	x		1		1		1		W	P	I		1	2	2	2	2	1	1	4.0	Difficult to establish stand because of low germinating seed. Use in combination with legumes.
Deertongue (Panicum clandestinum)							x	x	W	P	N	3.8-5.0	1-2	2	1		1	1	1	5.0	Bunchgrass with strong fibrous root system. Spreads by rhizomes. Adapted to low fertility soils. Requires 30 days of field stratification, therefore, plant in late fall or very early spring.
Dropseed, giant (Sporobolus giganteus)		x	x						W	P	N						1	2	3		M.P.R. 9". Bunchgrass. Adapted to part of Intermountain region.
Dropseed, mesa (Sporobolus flexuosus)		x	x	x					W	P	N						2	1	3		M.P.R. 8". Bunchgrass. Also adapted to part of Intermountain region. Short-lived.
Dropseed, sand (Sporobolus cryptandrus)		x	x	x	x				W	P	N		2	1	1	2	1	2	2	0.3	M.P.R. 10". Bunchgrass. Adapted to shallow and calcareous sites. Excellent seed producer. Seeded on dry sites where better forages not adapted.
Dropseed, spike (Sporobolus contractus)			x						W	P	N						1	2	3		M.P.R. 10". Bunchgrass. Adapted to shallow sites. Excellent seed producer. Cultivars not available.
Fescue, annual (Festuca megalura)	b		1						C	A	I			1			1	2	1		M.P.R. 10". Bunchgrass. Acid tolerant. Aggressive. Excellent fibrous root system and seedling vigor.
Fescue, Arizona (Festuca arizonica)	x	x	x	x					C	P	N			1	1		1	2	2		M.P.R. 16". Bunchgrass. Adapted to shallow sites.
Fescue, hard (Festuca ovina var. duriuscula)	x	x							C	P	I	5.5-6.5	3	2	1	3	2	1	1	2.3	M.P.R. 16". Bunchgrass. Used mostly in erosion control; robust form.
Fescue, Idaho (Festuca idahoensis)	x	x							C	P	N		3	2	1		3	1	1	1.9	M.P.R. 14". Bunchgrass. Reproduces by seeds. Lack of good seed yields restrict is use.
Fescue, meadow (Festuca elatior)	a					x			C	P	I						3	2	1	4.0	Valuable in Pacific Coast region (La), of limited value elsewhere. Disappearing rather quickly, except on heavy moist soils.
Fescue, red (creeping) (Festuca rubra)	x	x					x	x	C	P	N	5.0-7.5	2	2	1	3	2-3	1	1	4.0	Remains green during summer. Good seeder. Wide adaptation. Slow to establish.
Fescue, sheep (Festuca ovina)		x			x		x	x	C	P	N		2	1	1		1	1	1	10	M.P.R. 10". Bunchgrass.

TABLE 8.11 (continued)

Common and scientific names	Pacific Coast	Intermountain	Southwest	Northern Great Plains	Southern Great Plains	Midwest	Southeast	Northeast	Season of growth	Growth habit	Native or introduced	pH	High water	Drought	Cold	Salinity	Sand	Loam	Clay	Lbs. PLS. per acre seeding rate	Special considerations and adaptations
Fescue, tall (Festuca arundinacea)	x	x	1	x	1	x	x	x	C	P	I	5.0-8.5	1-2	2-3	1-2	1	2	1	1	3.8	M.P.R. 20". Bunchgrass. Generally seeded in pure stands; however, best results will be obtained by planting with an adapted legume. Rapid germination and vigorous seedlings. Easy to establish. Deep rooted.
Fescue, Thurber (Festuca thurberi)	x	x							C	P	N										M.P.R. 16". Bunchgrass.
Fountaingrass (Pennisetum setaceum)	a		x						W	P	I						2	1	1		M.P.R. 8". Bunchgrass. Seed difficult to harvest.
Foxtail, creeping (Alopecurus arundinaceus)	a	x		1		b			C	P	I		1				1	1	2		M.P.R. 19". Sodformer. Acid tolerant. Strong rhizomes.
Foxtail, meadow (Alopecurus pratensis)	x	x		1		x			C	P	I	6.0-8.5	1	2-3	1	3	2	1	1	2.2	M.P.R. 20". Sodformer. Slightly rhizomatous. Very useful in mixture on wet sites.
Galleta, big (Hilaria rigida)	b		x						W	P	N			1		1	2	2	1		M.P.R. 9". Sodforming. Cultivars are not available.
Galleta, common (Hilaria jamesii)		x	x	x	x				W	P	N			1-2	2	3	3	1	1	1.5	M.P.R. 12". Sodformer. Rhizomes. No cultivars are available.
Grama, black (Bouteloua eriopoda)			x	x	x				W	P	N	6.0-8.5	3	1	2	2	2	1	3	1.5	M.P.R. 10". Sodforming. Good quality seed is scarce. May be difficult to establish. Adapted to shallow and calcareous sites.
Grama, blue (Bouteloua gracilis)			x	x	x	x			W	P	N		3	2	1	2	2	1	1	1.5	M.P.R. 10". Bunchgrass. Generally seeded in mixtures. More drought tolerant than sideoats. Extensive root system. Poor seed availability.
Grama, sideoats (Bouteloua curtipendula)			x	x	x	x	a		W	P	N	6.0-7.5	2	2	2	2	2	1	1	5.5	M.P.R. 14". Bunchgrass; rarely forms a sod. Grows well in mixtures of warm-season grasses. Rhizomatous. May be replaced by blue grama in dry areas. Helps control wind erosion. Adapted to shallow and calcareous sites.
Hardinggrass (Phalaris tuberosa var. stenoptera)	x		x		x		x		C	P	I	5.5-7.5	2	2	2	2	3	2	2	2.5	M.P.R. 16". Sodforming. Also adapted to Southwest under irrigated conditions. Primary species for seedling California coastal and inland zones. Rhizomatous.
Indiangrass (Sorghastrum nutans)				x	x	x	x		W	P	N	5.5-7.5	2	3	1	2	1	1	2	4.5	M.P.R. 22". Sodforming. Provides quick ground cover. Rhizomatous. Heavy seed producer.

TABLE 8.11 (continued)

Common and scientific names	Regional adaptation								Season of growth	Growth habit	Native or introduced	Plant adaptation								Lbs. PLS, per acre seeding rate	Special considerations and adaptations
	Pacific Coast	Intermountain	Southwest	Northern Great Plains	Southern Great Plains	Midwest	Southeast	Northeast				pH	High water	Drought	Cold	Salinity	Sand	Loam	Clay		
Johnsongrass (Sorghum halepense)	1		X		X		X		W	P	I		1	1-2	2		1	1	1	7.4	M.P.R. 18". Bunchgrass. Rhizomatous. Difficult to eradicate; therefore, prevent from spreading to cultivated lands. HCN potential. Very productive.
Kleingrass (Panicum coloratum)	X		X		X		X		W	P	I		1	2	3	2	2	1	1	2.0	M.P.R. 20". Bunchgrass. Some varieties are rhizomatous.
Lovegrass, atherstone (Eragrostis atherstonei)			X		X				W	P	I			1	1-2	2	1	1	2		M.P.R. 11". Large vigorous bunchgrass. Generally larger and more productive than either Lehmann or weeping lovegrass. Good seedling vigor.
Lovegrass, Boer (Eragrostis chloromelas)			X		X		X		W	P	I		3	1	2	2	1	1	1	2.0	M.P.R. 10". Bunchgrass. Productive.
Lovegrass, Korean (Eragrostis curvunginea)						b			W	P	I	5.5+									
Lovegrass, Lehmann (Eragrostis lehmanniana) (E. lehmanniana x E. trichophora)	X		X		X				W	P	I		3	1	3	2	1	1	2	2.0	M.P.R. 10". Bunchgrass. Smaller and less cold tolerant than Boer and weeping lovegrass. Reseeds quickly after disturbance. Generally seeded in pure stands. Also adapted to Southern Great Plains (5). Adapted to calcareous sites.
Lovegrass, plains (Eragrostis intermedia)			X	X	X	b			W	P	N	6.0-7.5	3	1	1-2	3	1	2	3	2.0	M.P.R. 16". Bunchgrass.
Lovegrass, sand (Eragrostis trichodes)				X	X				W	P	N		2	2	2	3	1	1	1	2.0	M.P.R. 18". Bunchgrass. Seed in mixtures. Short lived but readily reseeds itself. Fair seed availability. Adapted to calcareous sites.
Lovegrass, weeping (Eragrostis curvula)			X		X		X		W	P	I	4.5-8.0	2	2	2	2	1	1	1	2.0	M.P.R. 16". Bunchgrass. Seeded mostly in southern Great Plains and in pure stands. Adapted to low-fertility sites. Rapid early growth. Good root system. Grows well on infertile soils.
Lovegrass, wilman (Eragrostis superba)			X		X				W		I		2	1	2	2	1	1	2	2.0	M.P.R. 10". Bunchgrass. Adapted to calcareous sites.

TABLE 8.11 (continued)

Common and scientific names	Pacific Coast	Intermountain	Southwest	Northern Great Plains	Southern Great Plains	Midwest	Southeast	Northeast	Season of growth	Growth habit	Native or introduced	pH	High water	Drought	Cold	Salinity	Sand	Loam	Clay	Lbs. PLS. per acre (seeding rate)	Special considerations and adaptations
Millet, browntop (Panicum ramosum)							x		W	A	N	4.5-7.0	2	3							Rapidly growing. Temporary erosion control.
Millet, foxtail (Setaria italica)							x	x	W	A	I	4.5-7.0	1	3						5*	Bunchgrass. Good seedbed preparation important.
Millet, Japanese (Echinochloa crusgalli)							x	x	C	A	I		3	1						5*	Requires good seedbed preparation. Produces large amount of organic material on poor or marginal soils.
Millet, pearl (Pennisetum typhoides)							x		W	A	I									6.3*	Bunchgrass. Proper management is very important.
Millet, proso (Panicum milliaceum)					x		x		W	A	I										
Muhly, bush (Muhlenbergia porteri)			x		x				W	P	N						2	1	1		M.P.R. 9". Bunchgrass. Adapted to part of intermountain region. Adapted to shallow sites. Seed generally unavailable.
Muhly, mountain (Muhlenbergia montana)		x	x						W	P	N						2	1	1		M.P.R. 13". Bunchgrass. Adapted to shallow sites.
Muhly, spike (Muhlenbergia wrightii)		x	x						W	P	N						1	1	2		M.P.R. 13". Bunchgrass.
Natalgrass (Rhynchelytrum roseum)	x						x		W	P	I			1			1	2	3		M.P.R. 19". Bunchgrass. Adapted to shallow sites. Short-lived.
Needle-and-thread (Stipa comata)	x	x		x	x				C	P	N		2	2	1	2	2	1	3		M.P.R. 10". Bunchgrass. Adapted to shallow and calcareous sites. Problem with seed harvesting and availability.
Needlegrass, green (Stipa viridula)		x		x	x				C	P	N	5.0-7.5	2-3	2	2	2	1	1	3	4.8	M.P.R. 15". Bunchgrass. Seeded in mixtures. Low seed quality; delayed germination.
Oatgrass, tall (Arrhenatherum elatius)	x	x				x		x	C	P	I		3	2	2		1	1	1	11.6	Rapid-developing, short-lived bunchgrass adapted to mesic sites. Infrequently used in new seedings. Less heat tolerant than orchardgrass except in Northeast.
Oats (Avena sativa)	x	x					x	x	C	A	I	5.5-7.0	3		2		1	1	1	20*	Requires nitrogen for good growth.

TABLE 8.11 (continued)

Common and scientific names	Regional adaptation Pacific Coast	Intermountain	Southwest	Northern Great Plains	Southern Great Plains	Midwest	Southeast	Northeast	Season of growth	Growth habit	Native or introduced	pH	Tolerance High water	Drought	Cold	Salinity	Soils Sand	Loam	Clay	Seeding rate lbs. PLS per acre	Special considerations and adaptations
Orchardgrass (Dactylis glomerata)	x	i	i	i	i	x	x	x	C	P	I	5.0-7.5	2-3	2-3	2	2-3	2	1	2	2.4	M.P.R. 18". Bunchgrass. Adapted to irrigated or naturally mesic sites. Develops rapidly and is long lived. Seeded in mixtures. Tolerates shade. More summer growth than timothy or bromegrass. Matures early. Tends to be inferior to tall fescue for cover, establishment and persistence.
Paspalagrass (Digitaria decumbens)							x		W	P	I									2.0	Stoloniferous. Well adapted to tropical and subtropical areas. Established vegetatively by fresh stem and stolon cuttings.
Panicgrass, blue (Panicum antidotale)			i		x				W	P	I		3	3	2	2	2	1	1		M.P.R. 20". Sodforming. Rhizomatous. Highly productive on good sites but will produce on droughty infertile soils.
Paragrass (Panicum purpurascens)							p		W	P	I		3	2	1		2	1	1		Propagated by planting pieces of stem or sod. Seed generally available.
Perlagrass or Koleagrass (Phalaris tuberosa v. hirtiglumis)	x								C	P	I		1	3	1		1	1	1		M.P.R. 18". Bunchgrass.
Redtop (Agrostis alba)	i	i	i	i	x	x	x	x	C	P	I	4.0-7.5	2-3	2-3	2	1	2	1	1	0.3	Establishes well from broadcasting on wet soils. Widely adapted to mixtures on soils too wet for other grasses. Spreads by rhizomes.
Reed, common (Phragmites communis australis)		i		i	i		x		C-W	P	N		3	3	1	2	2	1	1		M.P.R. 30". Commonly planted at 1 to 1-1/2 rhizomes (12-18" long) per foot of row. Creeping rhizomes and stolons. Established using vegetative material. Heavy duty shoreline protection.
Reed, giant (Arundo donax)							x		W	P	I										M.P.R. 20". Sodformer. Also adapted to part of Southwest. Established using vegetative materials. Grows to 10' tall.
Rescuegrass (Bromus catharticus or unioloides)			x		x		x		C	A	I		2-3	2-3	1	2	2	1	2	11.0	M.P.R. 25". Bunchgrass. Annual grass under cultivation. Short-lived.
Rhodesgrass (Chloris gayana)	i						x		W	P	I		2	2	3	2	2	1	1	1.0	M.P.R. 20". High sodium tolerance. Also adapted to southern parts of Southwest and southern Great Plains. Most useful in dry portions of South Texas where other grasses are not as well adapted.

TABLE 8.11 (continued)

Common and scientific names	Pacific Coast	Intermountain	Southwest	Northern Great Plains	Southern Great Plains	Midwest	Southeast	Northeast	Season of growth	Growth habit	Native or introduced	pH	High water	Drought	Cold	Salinity	Sand	Loam	Clay	Seeding rate lbs. PLS. per acre	Special considerations and adaptations
Ricegrass, Indian (*Oryzopsis hymenoides*)	x	x		x	x				C	P	N		3	1	-	2	1	1	3	4.6	M.P.R. 7". Bunchgrass. Hard, impermeable seed makes seedling success uncertain. Difficult to establish. Reproduces by seeds.
Rye, winter (*Secale cereale*)		-			-	b	x	x	C	A	I	5.4-7.5	2-3	-	1	2	-	2	2	30	Extensive root system. Generally used as temporary cover. Does not persist more than a year or two out of cultivation.
Ryegrass, annual (*Lolium multiflorum*)	x				1	b	x	x	C	A	I	5.5-7.5	2	3	1	2	1	2	1	3.5	M.P.R. 25". Bunchgrass. Excellent for temporary cover. Can be established under dry and unfavorable conditions. Quick germination, rapid seedling growth.
Ryegrass, perennial (*Lolium perenne*)	x				x				C	P	I	6.0-7.0	2	3	2	2	2	1	1	3.5	M.P.R. 25". Rapid developing, short-lived bunchgrass. Generally used as short term seeding. Easy to establish.
Ryegrass, Wimmera or Swiss (*Lolium rigidum*)	a								C	A	I		2	2	1	1	2	1	1	3.5	M.P.R. 11". Bunchgrass. Short-lived.
Sacaton, alkali (*Sporobolus airoides*)	x	x		x	x				W	P	N		1	1	-	1	3	1	1	1.0	M.P.R. 10". Bunchgrass. Desirable for seeding on saline areas. Seed available from native harvest. Seeds remain viable for many years. Reproduces by seeds and tillers. Cultivars not available.
Saltgrass, inland (*Distichlis stricta*)	x	x		x	x				W	P	N		1-2	-	-	3	2	-	-		M.P.R. 14". Sodforming. Poor seed producer. Seed unavailable.
Sandreed, prairie (*Calamovilfa longifolia*)				x	x		x		W	P	N	6.0-8.0	3	-	3	-	1	2	3	3.2	M.P.R. 11". Sodforming. Seedling limited by inadequate seed supplies and low seed quality. Seed common in native grass seed harvest. Rhizomatous.
Slenderstem (*Digitaria*)	b						x		W	P	N		2	-	3	-	1	-	1		
Smilograss (*Oryzopsis miliacea*)						a			C	P	I		3	1	3	3	2	2	3	1.5	M.P.R. 16". Bunchgrass. Adapted to broadcast seedling after disturbance. Used principally in California. Reproduces by seeds and tillers. Also adapted to portion of Pennsylvania, Maryland and Virginia.
Sorghum almum (*Sorghum almum*)		x		x					W	P	I			2	2	2	2	1	1	15.0	M.P.R. 18". Bunchgrass.
Sprangletop, green (*Leptochloa dubia*)		x		x		b			W	P	N			-	1	2	1	2	2	1.7	M.P.R. 10". Bunchgrass.

TABLE 8.11 (continued)

Common and scientific names	Pacific Coast	Intermountain	Southwest	Northern Great Plains	Southern Great Plains	Midwest	Southeast	Northeast	Season of growth	Growth habit	Native or introduced	pH	High water	Drought	Cold	Salinity	Sand	Loam	Clay	Lbs, PLS, per acre seeding rate	Special considerations and adaptations
Sudangrass (Sorghum sudanense)	x		x	x	x	x	x	x	W	A	I	5.5-7.5	2-3	1			1	1	1	8*	Generally used for temporary cover.
Switchgrass (Panicum virgatum)		x	x	x	x	x	x	x	W	P	N	5.0-7.5	1-2	2	1	2	1	1	1	3.5	M.P.R. 20-25". Sodforming. Seeding rate for Alamo is 2.0. Rhizomatous. Widely seeded in warm season grass mixes on mesic sites. Withstands eroded, acid and low fertility soil. Useful in drainage ways, and terrace outlets.
Timothy (Phleum pratense)	x					x	x	x	C-W	P	I	4.5-8.0	2-3	3	1	1	2	1	1	1.4	Leafy forage. Seeded in mixtures such as alfalfa and clover. Stands are maintained perennially by vegetative reproduction; however, tends to be short-lived. Shallow, fibrous root system.
Tobosa (Hilaria mutica)			1	x					W	P	N						3	2	1		M.P.R. 12". Cultivars are not available.
Trichloris, two flower (Trichloris crinita)		x	x	x					W	P	N			2	2		1	1	2		M.P.R. 8". Bunchgrass. Adapted to shallow and calcareous sites. Seed not commercially available.
Vine-mesquite (Panicum obtusum)			x	x					W	P	N		1	2	2		2	1	1	6.1	Used principally for erosion control. Reproduction by seeds, rhizomes, and stolens.
Wheat, winter (Triticum aestivum)	x	x	x	x					C	A	I	5.0-7.0	3	1-2	2	2-3	2	2	1	30*	Used as temporary cover.
Wheatgrass, beardless (Agropyron inerme)		x	x	x	x				C	P	N		3	1-2	1		2	1	2	6.1	M.P.R. 11". Does well in shallow sites. Bunchgrass.
Wheatgrass, bluebunch (Agropyron spicatum)		x	x	x	x				C	P	N		3	1-2	1		3	1	1	7.4	Bunchgrass. Adaptation and management similar to beardless wheatgrass, but seed less available. Reproduces primarily by seeds. Adapted to shallow and calcareous sites.
Wheatgrass, fairway crested (Agropyron cristatum)	x	x	x	x	x				C	P	I		3	1	1	1-2	2	1	1	4.4	M.P.R. 8". Bunchgrass. Stands thicken sooner and spread more than A. desertorum, also leafier and finer stemmed. Seeds alone or with alfalfa. Best results at altitudes of 1500 m or more. Easily established and extremely long lived. Reproduces by seeds and tillers.
Wheatgrass, intermediate (Agropyron intermedium)	x	x	x	x	x				C	P	I		2	2	1-2	2-3	2	1	1	9.4	M.P.R. 13". Sodformer. Productive on mesic sites and under irrigation. Reproduces by seeds, tillers and rhizomes. Excellent seedling vigor.

TABLE 8.11 (continued)

Common and scientific names	Pacific Coast	Intermountain	Southwest	Northern Great Plains	Southern Great Plains	Midwest	Southeast	Northeast	Season of growth	Growth habit	Native or introduced	pH	High water	Drought	Cold	Salinity	Sand	Loam	Clay	Seeding rate lbs. PLS. per acre	Special considerations and adaptations
Wheatgrass, pubescent (Agropyron trichophorum)	x	x	x	x					C	B	I		2	1-2	1-2	3	1	1	2	9.7	M.P.R. 12". Sodformer. Similar to intermediate wheatgrass but somewhat more drought tolerant.
Wheatgrass, Siberian (Agropyron sibiricum)	x	x	x	x					C	B	I		2-3	1	1	2	1	1	1	4.2	M.P.R. 8". Bunchgrass. Similar to standard crested wheatgrass in adaptation and use but less widely used.
Wheatgrass, slender (Agropyron trachycaulum)		x	x	x					C	B	N		1-2	2	1	1	2	1	1	5.4	M.P.R. 15". Bunchgrass. Short life limits use. Seed in mixtures only. Tends to be stemmy. Reproduces by seeds and tillers.
Wheatgrass, standard crested (Agropyron desertorum)	x	x	x	x					C	B	I		2-3	1	1	2	2	1	1	5.0	M.P.R. 9". Bunchgrass. Refer to Fairway crested wheatgrass, full stands slightly more productive than Fairway.
Wheatgrass, stream bank (Agropyron riparium)		x		x	x				C	B	N		1	1	1	1	1	1	1		M.P.R. 9". Sodformer.
Wheatgrass, tall (Agropyron elongatum)	x	x	x	x	x	D			C	B	I	6.0-8.0	1	2	1	1	2	1	2	11.0	M.P.R. 13". Bunchgrass. High sodium and salinity tolerance. Seed alone rather than in mixtures. Easy to establish.
Wheatgrass, thickspike (Agropyron dasystachym)		x		x					C	B	N		1	1	1	1					M.P.R. 8". Sodformer. Excellent seedling vigor.
Wheatgrass, western (Agropyron smithii)		x	x	x	x	x			C	B	N	4.5-7.0	1	2	1	1	3	1	1	7.0	M.P.R. 16". Sodformer. Seeded in mixtures or in pure stands. Tolerates alkalinity and silting. Rhizomatous. Long lived. Slow germination, spreads rapidly, sodforming. Valuable for erosion control.
Wildrye, Altai (Elymus angustus)		x		x					C	B	I		2	1	1		2	1	1	5.0	Similar to Russian wildrye; deep root system.
Wildrye, basin or giant (Elymus cinereus)	x	x	x	x					C	B	N		1-2	2-3	1	1-2	3	1	1	9.2	M.P.R. 14". Bunchgrass. Vigorous, tall growing bunchgrass. Reproduces by seeds and tillers.
Wildrye, beardless (Elymus triticoides)	x	x	x	x					C	B	N		1	1	1	1	2	1	1		M.P.R. 18". Sodformer. Poor seed production and problems with seed dormancy.
Wildrye, Canada (Elymus canadensis)		x		x		x			C	B	N		2	2	1	2	1	1	1	8.2	Lack of stand maintenance. Reproduces by seeds and tillers.
Wildrye, mammoth (Elymus giganteus)		x		x					C	B	I		2	2	1	1	2	2	3		M.P.R. 10". Sodforming. Established using vegetative material.

TABLE 8.11 (continued)

	Regional adaptation									Season of growth	Growth habit	Native or introduced	Plant adaptation										Lbs. PLS, per acre seeding rate	Special considerations and adaptations
Common and scientific names	Pacific Coast	Intermountain	Southwest	Northern Great Plains	Southern Great Plains	Midwest	Southeast	Northeast					pH	Tolerance					Soils				
														High water	Drought	Cold	Salinity	Sand	Loam	Clay		
Wildrye, Russian (Elymus junceus)	x	x		x					C		I			2	1	1	1	2	1	1	5.0	M.P.R. 13". Bunchgrass. Seed alone or with alfalfa. Early growth. Very hardy once established. Provide a weed-free seedbel.

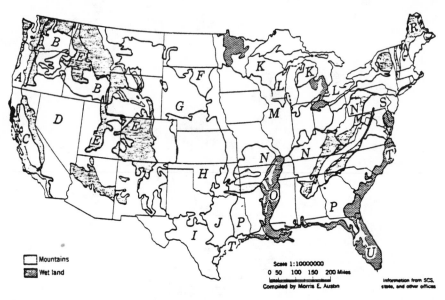

Figure 8.26   Major land resource regions of the United States.
(A) Northwestern forest, forage and specialty
crop region.   (B) Northwestern wheat and range
region.   (C)   California subtropical fruit, truck
and specialty crop region.   (D)   Western range
and irrigated region.   (E)   Rocky Mountain range
and forest region.   (F)   Northern Great Plains
spring region.   (H)   Central Great Plains winter
wheat range region.   (I)   Southwestern plateaus
and plains, range and cotton region.   (J) Southwestern prairies, cotton and forage region.   (K)
Northern lake states forest and forage region.
(L) Lake states fruit, truck and dairy region.
(M) Central feed grains and livestock region.
(N) East and Central general farming and forest
region.   (O) Mississippi Delta cotton and feed
grains region.   (P) South Atlantic and Gulf
Slope cash crop, forest and livestock region.
(R) Northeastern forage and forest region.   (S)
Northern Atlantic Slope truck, fruit and
poultry region.   (T) Atlantic and Gulf Coast
lowlands, forest and truck crop region.   (U)
Florida subtropical fruit, truck crop and range
region (Austin, 1965).

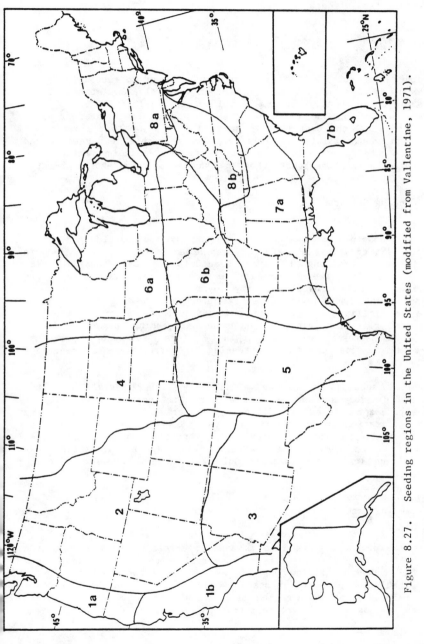

Figure 8.27. Seeding regions in the United States (modified from Vallentine, 1971).

to be the most practical for land treatment. The method selected depends on the waste-soil interactions, present condition of the soil surface and cost-benefit ratios of each method.

8.7.4   Seeding and Establishment

Seeding at the proper time is extremely important to successful stand establishment since it affects the physiological development of the plant. Cool season species usually perform best if seeded in late summer or early fall. Warm season species are normally seeded during late winter or early spring. Generally, the best time to seed is just prior to the period of expected high annual rainfall. This provides favorable temperatures and soil moisture conditions to the developing seedlings. Seeding method, rate and depth also have a direct effect on the success of stand establishment.

8.7.4.1 Seeding Methods

The most commonly used methods of seeding are broadcasting and drilling. Generally, drilling is preferred over broadcasting from an agronomic standpoint because drilling places the seed into the soil, thus improving seed-soil contact and the probability of seedling establishment. With broadcasting, seeds are usually poorly covered with soil which tends to slow stand establishment. Consequently, broadcast seeding is seldom as effective as drilling without some soil disturbance prior to seeding. Better results will be obtained if the broadcast seeding operation is also followed with harrowing or cultipacking. These follow-up operations enhance seed-soil contact, thus increasing the probability for seedling establishment.

Broadcast seeding may be accomplished by either aerial or ground application. Aerial application uses either a helicopter or an airplane equipped with a spreader and a positive type metering device. Broadcasting by ground application may be done by hand using the airstream or exhaust of a farm implement, a rotary spreader, or a fertilizer-spreader type seed box. Ground application tends to be slower than aerial application; however, aerial application is feasible only for large acreages due to the cost involved.

8.7.4.2 Seeding Rate

Using the proper seeding rate is another critical factor to seedling establishment. The actual quantity of seed applied per acre depends on the species, the method of seeding, and the waste-site characteristics. Seeding rates should be adequate for stand establishment without being excessive. When broadcasting seeds, the rates should be increased 50 to 75% since there is less seed-soil contact than is typical for drilling.

The current practice, for calculating seeding rates is based on the quantity (lbs) of seed required to produce 20 live seeds per foot. Pure live seed (PLS) is the percentage of the bulk seed that is considered live, and it can be calculated using the following equation:

$$PLS = (\% \text{ germination} + \% \text{ hardseed}) \text{ X } \% \text{ purity} \qquad (8.9)$$

The tag on the seed bag should contain all the information needed for the various calculations. To determine pounds of available bulk seed needed per acre use the following equation:

$$\text{Lb. PLS/acre} \div \% \text{ PLS of available bulk seed} = \qquad (8.10)$$
$$\text{Lb. of available bulk seed/acre}$$

For seeding mixtures, pounds of PLS needed per acre can be calculated by using the following equation:

$$\text{(decimal equivalent of the percentage for a specific}$$
$$\text{species desired in a mixture) X (lbs. of PLS/acre for} \qquad (8.11)$$
$$\text{a single species seeding)}$$

The quantity of available bulk seed (lbs) needed per acre to obtain the desired mixture can then be calculated using equation (8.10).

8.7.4.3 Seeding Depth

Optimum seeding depth of a particular species depends on seed size and quantity of stored energy and the surface soils at the site. The rule of thumb is to plant seeds at a depth of 4 to 7 times the diameter of the seed (Welch and Haferkamp, 1982). Many seedings fail because seeds are planted too deep and not enough stored energy exists to allow the developing seedlings to reach the soil surface. The major problem with planting seeds at too shallow a depth is the increased potential for desiccation. Seed may safely be planted deeper in light textured soils than in heavy soils.

8.7.4.4 Plant Establishment

Vegetative establishment may require lime, fertilizer, mulch and additional moisture to assure success. Specific cultural practices needed vary according to season and location. Soil tests should be used as a guide to available nutrients and the need for pH adjustment. In most instances, the area will have already been adjusted to a pH of 6.5 or above to obtain optimal waste degradation. Without a proper balance of nitrogen, phosphorus and potassium, plant growth may be poor.

At sites where excessive heat or wind is a problem, a cover crop or mulch can reduce surface soil temperatures, evaporation, crusting and wind erosion. Numerous grasses including various sorghums and millets may be used as mulch; however, it is best to obtain recommendations from local SCS offices or universities. Generally, seed production of a temporary cover crop should be prevented. To accomplish this objective, the species should be planted late in its growing season or cut prior to seed set. Permanent species can then be seeded or sprigged without excessive competition from remnants of the previous cover crop.

8.7.5 Soil Fertility

Soil fertility plays a major role in the ability of plants and microbes to grow and reproduce in a land treatment operation. When vegetation is part of the management plan, nutrient imbalances may adversely affect plant growth. Even if the unit operates without the use of vegetation, nutrient toxicities or deficiencies may deter growth and reproduction of microbes, thus limiting waste degradation.

Numerous macro- and micronutrients are considered essential to plants and microorganisms. A general discussion of this topic is included in Section 4.1.2.3. Micronutrients must be more carefully controlled since there is a narrower range between the quantity of a particular nutrient causing a deficiency or toxicity to plants than with the macronutrients. Attention needs to be given to the total quantity of the nutrient contained in the overall land treatment operation rather than just the quantity present in the treatment medium or the waste alone.

Macronutrients are generally applied in rather large quantities when compared to micronutrients. The three major macronutrients in fertilizer are nitrogen (N), phosphorus (P) and potassium (K). Other macroelements which may need to be applied include calcium, magnesium and sulfur.

Micronutrients include such elements as copper, iron, boron, chloride, molybdenum, zinc and manganese. Other trace elements essential to specific plant groups include sodium, cobalt, aluminum, silicon and selenium (Larcher, 1980). Additions of any one or a combination of micronutrients may be required depending on the characteristics of the treatment medium and the waste.

8.7.5.1 Fertilizer Formulation

Two systems currently exist for reporting composition percentages of fertilizer components. Under the old system, a 13-13-13 fertilizer contained 13% N, 13% P_2O_5 and 13% K_2O; however, under the new system this same fertilizer would contain 13% total N, 30% available P and 16% soluble K. Conversion factors for P and K are as follows:

$$P_2O_5 \times .44 = P \qquad K_2O \times .83 = K$$

$$P \times 2.29 = P_2O_5 \qquad K \times 1.20 = K_2O$$

The average composition of typical fertilizers are given in Table 8.12.

8.7.5.2 Timing Fertilizer Applications

The optimum time to apply fertilizer depends on the amount and distribution of precipitation, the type of fertilizer and the growth characteristics of the plant. Nitrogen is highly mobile in soils, yet phosphorus and potassium move very slowly. Therefore, nitrogen needs to be applied near the period of most active use by the plants, as long as sufficient moisture is present. Phosphorus and potassium can be applied over a longer time frame because precipitation will move them into the active root zone where they eventually can be taken up and used by plants.

8.7.5.3 Method of Application

Two practical fertilizer application methods for land treatment units are broadcasting and sprinkler irrigation. The application method must be compatible with the specific type of fertilizer to be applied. Some fertilizers such as anhydrous ammonia, aqueous ammonia and urea volatize rapidly if they are broadcast so these must be incorporated into the soil shortly after application.

Broadcasting is generally the most cost effective method of application. This method is commonly used when applying granular fertilizers. Minimal surface runoff of fertilizer occurs with this application method since slopes and runoff of land treatment units are restricted.

Sprinkler irrigation may be effective for applying noncorrosive liquid fertilizers. This application method could be easily incorporated into existing land treatment irrigation systems. This method allows frequent uniform applications of fertilizer at lower rates, thus increasing nitrogen utilization by the plants (Vallentine, 1971).

8.8 WASTE STORAGE

Wastes may need to be stored at HWLT units for many reasons, including 1) holding to determine if the waste has the expected concentration of hazardous constituents, 2) equipment breakdown, or 3) climatic restrictions on waste application. If climatic factors will restrict waste application, then sufficient waste storage capacity must be provided for wastes produced during the season when wastes cannot be applied to the HWLT facility.

TABLE 8.12   AVERAGE COMPOSITION OF FERTILIZER MATERIALS*

Fertilizers	% N	% P	% K	% P_2O_5	% K_2O	P solubility in water	% S	$CaCO_3$ Equivalence† Basicity	Acidity
NITROGEN FERTILIZERS									
Ammonia, anhydrous	82								147
Ammonium nitrate	33.5								60
Ammonium phosphate sulfate	16	9		20		Over 75%	16		88
Ammonium sulfate	20						24		110
Di-ammonium phosphate	21	22		50		Over 75%			75
Mono-ammonium phosphate	11	21		48		Over 75%	2.6		58
Potassium nitrate	14		38		46			23	
Urea	45								71
Sodium nitrate	16							28	
PHOSPHATE FERTILIZERS (see also under nitrogen fertilizers)									
Calcium metaphosphate		28		64		Slight		Neutral	
Rock phosphate		15		33		1% or less		Basic	
Superphosphate, single		9		20		Over 75%	12	Neutral	
Superphosphate, triple		20		46		Over 75%	1	Neutral	
Phosphoric acid		24		54		Over 75%			
Mono-potassium phosphate		23	29	52	35	Over 75%		Neutral	110
POTASSIUM FERTILIZERS (see also under nitrogen and phosphorus fertilizers)									
Potassium chloride (muriate of potash)			50		60			Neutral	
Potassium sulfate			44		53		18	Neutral	

--continued--

TABLE 8.12 (continued)

Fertilizers	% N	% P	% K	% P$_2$O$_5$	% K$_2$O	P solubility in water	% S	CaCO$_3$ Equivalence[†] Basicity	Acidity
ORGANIC FERTILIZERS									
Manure, dairy (fresh)	0.7	.13	.54	.30	.65	50%		Slight	
Manure, poultry (fresh)	1.6	.55	.75	1.25	.9	50%		Slight	
Manure, steer (fresh)	2.0	.24	1.59	.54	1.92	40%		Slight	
SULFUR FERTILIZERS									
(see also under nitrogen and phosphorus fertilizers)									
Calcium sulfate (gypsum)							18.6		Acidic
Magnesium sulfate							13		Acidic
Soil sulfur							99		Acidic
Sulfate potash magnesia			21.5		26		18		Acidic
LIMING FERTILIZERS									
Calcium oxide								178	
Dolomite								110	
Limestone, ground								95	
Shell meal								95	

* Vallentine (1971)

† Compared to 100 basicity for CaCO$_3$.

<u>Waste Application Season</u>

The waste application season must be determined to enable the owner or operator to determine the amount of waste storage capacity needed. If accumulation of untreated waste in soil creates no potential toxicity or mobility hazard, waste application will only be limited by freezing temperatures, snow cover and precipitation. Models, developed by Whiting (1976) can be used to determine the waste application season based on various climatic parameters. In the case above, the EPA-1 or EPA-3 model can be applied directly (Whiting, 1976). The climatic data required are the mean daily temperature (°F), snow depth, and daily precipitation for 20-25 years of record.

If accumulation of untreated waste in soil can potentially lead to unacceptable toxicities to plants or soil microbes and/or leaching or volatilization of hazardous waste constituents, then wastes may only be applied when soil temperature is greater than 5°C (41°F) and soil moisture content is less than field capacity. These values are used as thresholds since decomposition of organics and other treatment reactions essentially cease at lower temperatures or greater moisture contents. Soil temperature records are limited, so air temperatures are often used as described in Section 4.1.1.6 to estimate soil temperature. The EPA-1 or EPA-3 models described above may be applied to estimate the waste application season. When the waste application season is limited by cold weather, the nonapplication season for storage volume calculations can be defined as being the last day in fall failing to exceed a minimum daily mean temperature to the first day in spring exceeding the minimum daily mean temperature.

Additional constraints for application of hazardous waste must be evaluated in terms of soil parameters and the 5-year return, month-by-month precipitation for the particular HWLT site. Wetness is restrictive to waste application operations primarily because saturated conditions maximize the potential for pollutant discharge via leachate or runoff and inhibit organic matter degradation. An application season based on periods of excessive wetness can be established in a straightforward manner by applying the EPA-2 model described by Whiting (1976). The required climatic data should be for a 20 to 25-year period of record. Specifically, the required data inputs for the model are as follows:

(1) daily minimum, maximum and mean on-site temperatures (°F);

(2) daily precipitation (inches);

(3) site characteristics and climatic parameters for the station including:

   (a) I, the heat index;

   (b) b, a coefficient dependent on the heat index;

   (c) g, the tangent of the station's latitude;

(d) W, the available water holding capacity of the soil profile (in inches minus 1.0 inch as a safety factor); and

(e) ϕ, the daily solar declination, in radians.

Since the model is driven only by climatic factors, the results should be interpreted carefully; biologic and hydrologic factors should also be considered. The model provides a valuable first estimate of the number of storage days needed. The maximum annual waste storage days for the continental U.S., as estimated by the model are shown in Fig. 8.28. The actual on-site soil profile characteristics including percolation, runoff, profile storage, surface storage, and waste loading rates should be used to determine storage days for a specific HWLT site when the limiting climatic factor is excess precipitation.

8.8.2   Waste Storage Facilities

During the operation of an HWLT unit, there may be periods when waste application is not possible due to wetness, low temperature, equipment failure, or other causes. Suitable facilities must be provided to retain the waste as it is generated until field application can be resumed. The design of the necessary structure depends on the waste material and the actual size of the structure depends on the required waste storage capacity. Waste storage facilities should be sufficient to store the following:

(1) waste generated during extended wet and cold periods as estimated in Section 8.8.1;

(2) waste generated during periods of field work, i.e., plowing, planting, harvesting, etc.;

(3) waste generated during periods of equipment failure;

(4) 25-year, 24-hour return period rainfall over the waste storage structure if it is open; and

(5) waste generated in excess of application capacity due to seasonal fluctuations in the rate of waste production.

Runoff retention areas should not be used to store wastes generated during the above situations; runoff retention areas are designed to retain runoff from the active land treatment areas. Waste storage facilities are discussed below.

8.8.2.1 Liquid Waste Storage

Liquid wastes can be conveniently stored in clay lined ponds or basins. An aeration system may be added to the pond to prevent the liquid

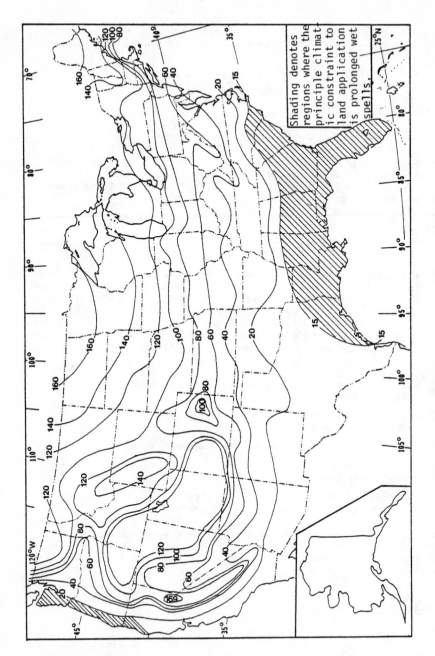

Figure 8.28. Estimated maximum annual waste storage days based on climatic factors (Wischmeier and Smith, 1978).

Shading denotes regions where the principle climatic constraint to land application is prolonged wet spells.

waste from becoming anaerobic. Wastes which are highly flammable or volatile should not be stored in open ponds. Additionally, pond liners must
not be prone to failure. Clay liners and other liner materials may not
acceptable for waste storage if they are chemically incompatiable with the
waste.

A second approach to liquid storage is to construct a tank. The tank
may be either closed or open, is usually made of metal or concrete and can
be equipped with an aerifier. Tanks of this nature are more costly to
construct and require periodic maintenance, but they assure that no waste
is released to percolate through the soil. If differential settling occurs
during storage, some method of remixing the waste may be needed to assure
that the treatment site receives uniform applications. If any of the
liquid wastes being stored are hazardous wastes, the storage facilities for
the wastes must meet specific regulatory requirements for storage (EPA,
1981; EPA, 1982).

8.8.2.2 Sludge Storage

Sludges can be stored in facilities similar to those used for liquids.
Under certain conditions, filling and emptying tanks with sludge may become
a problem. Thus, a properly lined pond or basin may be more appropriate.

8.8.2.3 Solid Waste Storage

The most common method of solid waste storage is to stockpile the
material. If these piles are exposed to the weather, the area should be
bermed sufficiently to contain water from the 24-hour 25-year return period
storm over the storage area, in addition to the waste volume itself. A
buffer factor of at least 20% should be added to the berm to allow for
slumping of the stockpiled waste. The waste application season must,
therefore, be determined to enable the owner or operator to determine the
amount of waste storage capacity needed. Waste piles for hazardous wastes
must meet certain regulatory requirements (EPA, 1982).

8.9   WASTE APPLICATION TECHNIQUES

Waste characteristics such as the total volume and water content,
along with soil properties, topography and climate, need to be considered
to determine the appropriate waste application technique. Liquid wastes
containing between 95% and 100% water with a low volatility hazard may be
successfully applied by sprinkler irrigation; while, relatively dry, volatile and/or toxic materials may require subsurface injection techniques.
Regardless of which application system is chosen, two basic considerations
must be examined. First, the waste application rate chosen should not
exceed the capacity of the soil to degrade, immobilize or transform the

waste constituents. Second, the waste should be applied as uniformly as possible. Waste applications cannot consist of merely pouring or dumping the wastes in one spot. A definite plan must be developed and implemented to uniformly apply the waste to the soil at the design rate over the desired area. There are five basic considerations for choosing an appropriate application system for a given site and waste. They are as follows:

(1) effect on public health and the environment;

(2) operator-waste contact;

(3) ability to handle solids content;

(4) service life; and

(5) cost (capital and operational).

In the following sections, application techniques are discussed with regard to the consistency of the waste as shown in Table 8.13.

TABLE 8.13   WASTE CONSISTENCY CLASSIFICATION

Consistency	Characteristics
Liquid	Less than 8% solids and particle diameter less than 2.5 cm
Semi liquid	3-15% solids or particle diameters over 2.5 cm
Low moisture solids	Greater than 15% solids
Bulky wastes	Solid materials consisting of contaminated lumber, construction materials, plastic, etc.

8.9.1   Liquid Wastes

As a practical definition, a liquid waste is considered to have a solids content of less than 8% and particles with diameters less than 2.5 cm. Handling and transporting many hazardous wastes may be more convenient when the waste is in liquid form. Many wastes are generated in a moist condition and usually require large amounts of energy to dewater them. The cost of transporting a liquid waste from the source to the land treatment unit is a function of distance. Pipelines may be the least costly for short distances, while trucks may be necessary for greater distances.

Applications of liquid wastes are generally accomplished by spraying waste with a sprinkler system or by surface irrigating with flood or furrow irrigation techniques. Liquid wastes should be applied so that direct runoff does not occur. Both techniques may cause air quality problems if the waste applied is highly volatile. Care should be taken when liquid wastes are applied to ensure that leaching does not occur before treatment of the hazardous constituents in the applied wastes is completed.

8.9.1.1 Surface Irrigation

Surface irrigation appears to be the easiest application technique for a liquid waste and requires the least capital outlay. This method is commonly used so all necessary equipment is readily obtainable. One method of surface irrigation involves laying out the area so that wastewater can be applied by a set of trenches, canals and ditches. Waste is pumped to the main canal where it flows by gravity through trenches and ditches to all areas of the field where it infiltrates into the soil. There are, however, some drawbacks to this system. Since the waste stands in the trenches until the water infiltrates, there is a potential for odor and insect problems. Another disadvantage to this system is nonuniform application since as the liquid flows through trenches and ditches, less of the waste is carried to the far end of the field. In addition, if the waste is especially dangerous, such as a strong corrosive agent, all persons and animals must be kept away from the active area.

Another common means of surface application involves using a truck or trailer mounted tank filled with waste to spread the material across the field. The liquid waste is released by gravity flow or pumped through a sprayer or manifold (Wooding and Shipp, 1979). Application rates with this system are easily controlled by varying the flow rate or travel speed. Difficulties encountered during periods of bad weather may require alternate application technologies or storage facilities. One possible modification is to construct all weather roads in a pattern that allows a truck or spray rig to discharge wastes from the sides onto the disposal area. This would make continued application during periods of inclement weather possible. Waste spread this way should be incorporated as soon as the soil conditions permit. One possible disadvantage of vehicular applications is the resulting compaction and deterioration of soil structure (Kelling et al., 1976). A listing of commercial equipment for land application of wastes is included in the Implement and Tractor Red Book (1979).

8.9.1.2 Sprinkler Irrigation

Spray application of wastewater has enjoyed much popularity (Powell et al., 1972), particularly for municipal wastewater effluents (Cassel et al., 1979). This is primarily due to the availability and reasonable cost of the equipment. Sprinkler systems for use in hazardous waste disposal need to be designed by a qualified specialist to conform to the American Society of Agricultural Engineers Standard 5376. Highest priority needs to be given to attaining a uniform application pattern (coefficient of uniformity). A completely uniform application pattern has a coefficient of uniformity of 100%. Average irrigation systems attain a coefficient of uniformity of approximately 60%. Information on uniformity, which is available from irrigation suppliers, should be considered before accepting a system. When trying to achieve a uniform waste distribution, a higher degree of uniformity is required than when disposing of runoff water or wetting down plots for dust control. All materials need to be tested for corrosivity with the waste to be disposed to ensure that premature equipment failure does not occur.

The basic sprinkler irrigation system consists of a pump to move waste from the source to the site, a pipe leading from the pump to the sprinkler heads, and the spray nozzles. When choosing a pump, it must be made of a material compatible with the proper capacity and pressure needed for the given situation. For sludge applications, 1 to 2 inch nozzles requiring 50-100 psi water pressure are recommended (White et al., 1975). Pumps for these nozzles generally cost more and require more energy to operate than those used for nonpressured systems such as surface irrigation.

Sprinkler systems, if properly designed, are applicable to flat, sloping and irregular terrain. A site can be vegetated at the time of waste application provided the vegetation will not interfere with the spray nozzle operation and waste interception by the vegetative cover will not present a hazard or inhibit waste treatment. Generally, sites are cleared of trees and brush and planted to a pasture grass. In some cases, however, it may be desirable to dispose of wastewater in a forested area with risers placed in a pattern that avoids interference by trees. Pipes can be either permanently buried below the frost line or cultivation depth, or laid on the surface as with a portable irrigation system.

Although numerous configurations have been developed for sprinkler irrigation systems, three variations are most widely used. The first of the three main techniques is the fixed, underground manifold with risers and rotating impact type sprinklers. This system is the most costly to install and is permanent for the life of the installation. A second approach is to use a traveling pipe and sprinkler. In this system, a sprinkler connected by a flexible hose to the wastewater supply is mounted on a self propelled trailer device which traverses a fixed route across the field. The third commonly used spray system is the center pivot irrigation system. Here a fixed central wastewater supply comes up from an underground main and a self propelled sprinkler system rotates around the supply. The coefficient of uniformity with this system is as high as 80%.

Of the three major systems, the trailer mounted sprinkler has the most versatility and can be easily moved from one location to another. Above ground detachable irrigation pipe, normally used for agricultural irrigation, is not commonly used because of the hazardous nature of the liquids being handled. In general, most spray systems require little land preparation and can operate under a wide range of soil moisture conditions. The major difficulties with spray irrigation of wastewater are odor control, power consumption by high pressure pumps, clogging of nozzles causing a nonuniform application, and aerosol drift of hazardous waste materials. Low angle impact sprinklers have been developed to reduce aerosol drift.

Terrain and weather conditions should also be considered when designing a sprinkler system. Spray irrigation on sodded or cropped fields should be done only on slopes of 0-15%. If the spray application area is forested, application can be done on slopes up to 30%. Slopes at HWLT units are generally less than 5%. Low lying, poorly drained areas need to be drained as described in Section 8.3.6. Designers of spray irrigation systems need to give particular attention to cold weather alternatives. Pipes will need to be drained and flushed to prevent freezing and clogging during down times. Provisions must be made to recycle the drained water back to the original source.

Two other irrigation systems less frequently used for waste application are the tow line and side wheel roll systems. These systems are generally limited to use with wastes having a very low solids since the small nozzles clog easily. A review of irrigation systems and their suitability for waste application is presented by Ness and Ballard (1979).

8.9.2   Semiliquids

Semiliquids, also called sludges, typically contain 5 to 15% solids by weight. Application of semiliquids is normally done either by surface spreading with subsequent incorporation or by subsurface injection. Each of these systems, with its inherent advantages and disadvantages, are discussed below. Some general factors to be considered when choosing and designing a system are vehicle traction and weight, power requirements, topography and spreading patterns.

8.9.2.1 Surface Spreading and Mixing

Surface spreading and subsequent mixing is the conventional application technique for farm manures. Sludge may be applied in a similar manner, by loading the waste material on a manure spreader which applies it uniformly over the area. The sludge is then mixed with the surface soil by means of discing, deep plowing or rototilling. The main advantage to this system is the low capital outlay required. Equipment is conventional, readily available and of reasonable cost. Since this technique requires traversing the land area twice, it is neither energy nor labor efficient. Commercial waste applicators using this system often use large vacuum tank trucks equipped with flotation tires and a rear manifold or gated pipe for spreading the waste. Another option for moving sludges is to use a hauler box or a truck equipped with a waterproof bed.

If the sludge is too thick to pump (over 15% solids), the only choice may be to bring the material to the site and dump it. Typically, a pile of sludge slumps to about twice the area of the truck bed. Additional equipment is then needed to spread the waste over the soil surface. The most efficient piece of equipment for uniform spreading appears to be a road grader with depth control skids mounted on the blade. A second choice for this job is a bulldozer similarly equipped with depth control skids on the blade. Dozer blades may require wings on the edges to avoid formation of windrows. Backblading with a floating blade helps to achieve a uniform distribution.

Uniformity of application must be stressed; excessive applications to small areas result in barren "hot spots" and may lead to other environmental problems. Underapplication is inefficient and requires more land for disposal than would otherwise be needed. Normal cultivation practices such as plowing and discing cannot be relied on to evenly distribute waste over a field. Windrows should be avoided in the spreading procedure. Consequently, there must be a definite planned procedure to evenly distribute the waste prior to incorporation.

There are several basic pieces of equipment that effectively mix waste material with topsoil. First, there is the moldboard plow which very effectively inverts the upper 15-30 cm of soil. Secondly, there are discs which accomplish more mixing and less turning of the soil material than a moldboard plow. Rotary tillers do an excellent job of thoroughly mixing the waste with the surface soil, but it is generally slow and requires large energy expenditures. It does, however, only require one pass to accomplish adequate mixing while other types of equipment require two passes. A tractor-like vehicle with a large auger mounted sideways is also a very effective method for incorporating wastes into the soil in one pass. A more extensive equipment review is provided in Section 8.9.4.

The surface spreading and mixing technique is not particularly well adapted for use in applying hazardous volatile wastes since the material lies directly on the soil surface and is exposed to the atmosphere. If waste fumes will endanger the operator or the general public, or are objectionable, this system will not be acceptable.

8.9.2.2 Subsurface Injection

Subsurface injection is the technique of placing a material beneath the soil surface. It was originally developed by the agricultural industry for applying anhydrous ammonia. Equipment has also been developed for subsurface injection of liquid manures and wastes. Basic equipment consists of a tool bar with two or more chisels attached to the rear of a truck or tractor. Adjustable sweeps are often mounted on or near the bottom of the chisels to open a wide but shallow cavity underground. A tube connected to the waste source leads down the back of the chisel, and as the sweep opens a cavity, the waste is injected. With proper adjustment and use, very little waste reaches the soil surface. If waste is forced back to the soil surface in the furrow created by the chisel, blades may be attached which fold the soil back into the furrow.

Subsurface horizontal spreading of the waste may be limited with this technique, but a horizontal subsoiler may be added to the chisel injector to increase the subsurface area of incorporation. The horizontal subsoiler moves through the soil prior to the injector. This also enhances the waste degradation rate due to the increased waste-soil contact.

Common depths of application vary from 10 to 20 cm below the soil surface (Wooding and Shipp, 1979). Application rates are usually about 375 liters/min/applicator with nominal loading rates of 22,000 to 66,000 kg of dry solids per hectare (Smith et al., 1977; Brisco Maphis, personal communication). An experienced operator can achieve a uniform application across the field.

Where subsurface applications are made repeatedly over long periods of time, an underground supply pipe is sometimes used to conduct the waste to different areas of the field. A long flexible hose is then used to connect from the supply pipe to the truck or tractor-mounted injectors. Sophisticated systems have radio controlled shut-off valves so the operator can turn the waste off when he needs to raise the injectors to make a turn.

8.9.3   Low Moisture Solids

Low moisture solids are characterized by moisture contents of less than 85%. Basically, they can be handled much as one would handle sand or soil. If the materials are dense and in large units, such as logs or railroad ties, it may be necessary to shred or chip the material before application. A dump truck is the conventional method of transporting and applying solids. Piles of solids are then spread over the field using either a roadgrader or bulldozer.

As is the case with surface spreading of sludge materials, the most important concern is to achieve an even distribution. Another common implement used for spreading solid wastes is the manure spreader, which is particularly useful for wastes having moisture contents causing them to be sticky or chunky. The main disadvantage of this system is the small capacity, resulting in a large number of trips required to spread the waste. If the material is granular and relatively free of large chunks, a sand spreader on the back of a dump truck may be useful. Such broadcasting methods are commonly used in northern states to spread sand and salt on icy roads. Regardless of the spreading system selected, the waste needs to be incorporated and mixed with the surface soil shortly after spreading. Generally, the sooner this is accomplished, the lower the potential for environmental damage. Waste incorporation can be done according to the options listed for semiliquids (Section 8.8.2.1.).

If the application of low moisture solids will cause a significant increase in the ground surface, special precautions may be required. Under proper operation, the treatment zone will be a fixed depth from the surface where aerobic conditions promote degradation. Excessive loading of wastes could prohibit proper degradation by isolating nondegraded material below the zone of aeration. Therefore, sufficient time must be allowed for degradation of the waste before applying of additional waste. This may be accomplished by using a multiple plot design and rotating waste applications between these plots to allow sufficient time for proper degradation to occur. Since this affects area and timing requirements it needs to be considered in the original design of the land treatment unit.

The main disadvantage of using a low moisture solid disposal system is the large energy requirements if wastes are initially wet. First, the material must be dried, then transported to the disposal site, spread, and finally incorporated. If the material is dry when initially generated, such as an ash residue, the system becomes much more economical.

8.9.4   Equipment

In general, most HWLT units use specialized industrial equipment or agricultural equipment adapted to satisfy to their needs. Care must be taken to obtain compatible implements; often an agricultural implement cannot be attached to an industrial tractor without special adaptors. Where power requirements are high, the use of crawler type and 4-wheel drive articulated tractors is common. As previously noted, a comprehensive summary of such equipment is available in the Implement and Tractor Red Book (1979).

The equipment used to incorporate waste materials into the soil vary according to the size and condition of the site. Discing is the most commonly used technique. Under adverse conditions, such as hard, dry soil, an agricultural disc may not penetrate the soil adequately to obtain satisfactory incorporation. In this case, industrial discs with weights may be used to obtain sufficient penetration. After discing a field, a spring tooth harrow is useful to further mix the waste into the soil. Moldboard plows are excellent for turning under surface applied waste. The disadvantages of the moldboard plow are the high power requirements, slow speed and poor mixing. Inadequate mixing may result in a layer of persistent waste. Chisel tooth plows may also be used for waste incorporation.

Tractor mounted rotary tillers may be used to create a thorough soil-waste mixture and to provide effective aeration, in a single pass. Compaction is kept to a minimum since only one pass is needed, while plows, spring tooth harrows, disc harrows, etc. generally require multiple passes. A rototiller also tends to be more maneuverable than many other types of equipment. The power requirement for this piece of equipment is quite high, however, these other considerations may be of greater importance and a single pass with a rototiller may take less time and energy than multiple passes with other equipment. A special tractor with an auger mounted on its side has been developed for use in spreading, turning and incorporating sludge. It has many of the same advantages of the rototiller.

Specialized equipment, such as tractors with low bearing pressure for use in wet soils, are readily obtainable. Farm equipment such as spreaders and tank wagons can often be purchased with flotation tires. Trucks designed for field use in spreading liquids can also be equipped with flotation tires, if necessary.

Equipment for hauling and spreading liquid and solid wastes are commonly available. Tank trucks, vacuum trucks and liquid manure spreaders are available for use with liquid wastes. Manure spreaders, broadcast type fertilizer spreaders, dump trucks, road graders and loaders may be used for working with dry solid wastes.

Subsurface injection equipment has been developed and there are a few specialized sources. Many use chisel tooth plows often with sweeps on the bottom. Other systems use discs to cut a trench followed with a tube that injects the waste into the ditch immediately behind the disc. Still others use a horizontal discharge pipe mounted on the side of a truck. The most efficient systems, however, use large diameter flexible pipe to feed the applicator, eliminating the need of nurse tanks and frequent stops for refilling. Illustrations of such equipment can be found in many publications (EPA, 1979; White et al., 1975; Overcash and Pal, 1979).

8.9.5 Uniformity of Waste Application

Efficient use of the land in an HWLT unit requires that maximum quantities of waste be applied while preventing microbial or plant toxicity and minimizing the potential for contaminated leachate or runoff. Thus, hazardous waste loading rates are selected that rapidly load the soil to a safe limit based on the concentration of the rate limiting constituent (RLC). The benefits of this method include a relatively small land area requirement, which minimizes the volume of runoff water to be collected and disposed, and low labor and energy costs for operation. When wastes are loaded to the maximum safe limit, uniformity of application is essential to prevent the occurrence of "hot spots." Hot spots are areas that receive excessive quantities of waste causing an increased probability of wastes being released to the environment and requiring special treatment or removal when closing the site.

8.9.5.1 Soil Sampling as an Indicator

Field sampling of soils, in the treatment zone may be used to determine if the hazardous wastes are being uniformly applied. Location of the samples should be selected after first visually inspecting a given plot for differences in color, structure, elevation or other characteristics that may be indicative of uneven application. When such differences are observed, samples of the treatment zone from these areas should be obtained and analyzed for elements or compounds that are characteristic of the waste. Often analysis of the RLC can be used to indicate hot spots.

8.9.5.2 Vegetation as an Indicator

Despite efforts to achieve uniform application of waste, excessive amounts of waste constituents may accumulate in relatively small areas of the waste plot. Nonuniformity in soil characteristics may contribute to the accumulation of certain constituents in isolated areas. For example, areas containing preexisting salts or areas with lower permeability may cause hot spots. Growing vegetation between applications of waste helps identify such hot spots so that they can be treated to correct the problem or so that future applications to these areas can be avoided. In areas where surface vegetation does poorly, it is also highly probable that microbial degradation of organic constituents is inhibited. Thus, vegetation serves as a visual indication of the differential application or degradation of the applied waste. Furthermore, if nonuniform application has resulted in areas where substances have accumulated to phytotoxic levels, these areas may also have an increased probability for waste constituents to leach to groundwater. The soils in and below the treatment zone should be sampled at vegetative hot spots to ascertain the cause of unsatisfactory growth and to determine if any hazardous constituents are leaching.

8.10   SITE INSPECTION

The site is required to be inspected weekly and following storm events (EPA, 1982); however, daily inspections of all active portions of the HWLT unit are desirable. These inspections should include observations to assure that wastes are being properly spread and incorporated. Furthermore, daily observations should be made to assure that adequate freeboard is available in the various retention structures at all times.

Weekly inspections are sufficient for all inactive portions and for dikes, terraces, berms and levees. Observations should include indentification of hot spots where vegetation is doing poorly. Dikes, terraces and levees should be inspected for seepage and for evidence of damage by burrowing animals or unauthorized traffic.

Operational, safety and emergency equipment should receive regular inspection for damage or deterioration. Special attention should be given to this equipment since it is used on an irregular basis. When this equipment is needed it must perform properly; therefore, it should undergo testing at appropriate intervals to ensure that it will be ready when needed.

8.11   RECORDS AND REPORTING

As mentioned previously, a land treatment unit must be a well planned and organized operation. Records and on-site log books must be maintained since they are essential components of an organized facility, and serve to aid the manager in assessing what has and has not been done and what precautions need to be taken. These records also serve as a permanent record of activities for new personnel and off-site personnel including company officials and government inspectors. Finally, records must be kept of monitoring activities and pertinent data should be maintained throughout the active life of the land treatment unit. Most of these records can be kept in a log book accompanied by a loose leaf file containing lab reports, inspection reports and similar items. A checklist of items to be included in the operating record is presented as Table 8.14. All reporting should conform to the requirements of 40 CFR Parts 264 and 122 and any applicable state regulations.

Records to be kept at the site should include a map showing the layout of the land treatment units indicating the application rates for the wastes disposed and the date and location where each waste was applied and results of waste analyses. In addition, records need to be kept on the date, location, and code number of all monitoring samples taken after waste application. These records will include analyses of waste, soil, groundwater, and leachate water from the unsaturated zone. This information may be needed in case questions arise about the operation of the unit. Efforts to revegetate the site may also be documented. This can be done by recording the date, rate and depth of planting, species and variety planted, and the type and date of fertilizer applications. Measurements of emergence and groundcover should be determined at appropriate intervals and recorded.

Although climatic records are not required by regulation, they are very useful for proper management. The amount of rainfall should be measured on-site and recorded daily. Additional climatic data recorded may include pan evaporation, air temperature, soil temperature and soil moisture. When water is present in the retention ponds, the depth of water should be recorded at least weekly during a wet season. These records are easiest to use if results are graphed. This allows visual interpretation of the data to determine important trends that influence management decisions.

In addition, all accidents involving personal injury or spills of hazardous wastes are to be recorded and remedial actions noted. Any violations of security (i.e., entry of unauthorized persons or animals) also need to be recorded. Notes should be kept on all inspections, violations and accidents. They should clearly indicate the problem and the remedial actions planned or taken.

Another helpful management tool is to keep a balance sheet for each section of the unit that receives waste applications indicating the maximum design loading rate of each of the rate limiting constituents and those within 25% of being limiting, as well as the maximum allowable cumulative load of the capacity limiting constituent. As waste applications are made, the amount of each constituent added is entered on the balance sheet and subtracted from the allowable application to indicate the amount that can be applied in future applications. A running account of the capacity of each plot receiving waste is a valuable guide to the optimum placement so

TABLE 8.14   CHECKLIST OF ITEMS NEEDED FOR A THOROUGH RECORD OF OPERATIONS
AT A LAND TREATMENT UNIT

1. Plot layout map

2. Inspections

    a. weekly observations on levees and berms*
    b. observations of odor, excessive moisture, need for maintenance,
       etc.*

3. Waste applications

    a. date
    b. amount and rate
    c. location

4. Waste analysis

    a. original
    b. quarterly waste analysis reports
    c. any changes in application rate needed due to change in waste

5. Fertilizer and lime applications*

    a. date
    b. amount
    c. location

6. Vegetation efforts*

    a. planting date
    b. species planted
    c. fertilizer applied
    d. emergence date
    e. groundcover

7. Monitoring sample analyses

    a. soil samples
    b. waste samples
    c. groundwater samples
    d. leachate samples
    e. runoff samples*
    f. plant tissue samples*

8. Climatic parameters*

    a. rainfall
    b. pan evaporation
    c. air temperature
    d. soil temperature
    e. soil moisture

--continued--

TABLE 8.14 (continued)

9.	Water depth in retention basins*
10.	Accidents
	a. personal injury
	b. amount and type of waste spilled
	c. location
11.	Breaches of security
12.	Breaches of runoff retention resulting in uncontrolled off-site transport
13.	Maintenance schedule
	a. levees and berms
	b. regrading of plots
	c. grassed waterways
	d. tilling activities
	e. roads

* Not required by regulation but important to successful management of an HWLT unit.

that the cumulative capacity of all of the available soil is used. Section
7.5 discusses how to determine the limiting constituents of the waste
streams to be land treated.

CHAPTER 8 REFERENCES

ASAE. Terracing Committee. 1980. ASAE standard S268.2. pp. 522-525. In ASAE Agricultural engineers yearbook, 1980-1981. Am. Soc. Agr. Engr.

Asmussen, L. E., A. W. White, Sr., E. W. Hauser, and J. M. Sheridan. 1977. Reduction of 2,4-D load in surface runoff down a grassed waterway. J. Environ. Qual. 6:159-162.

Austin, M. E. 1965. Land resource regions and major land resource areas of the United States (exclusive of Alaska and Hawaii). Agriculture handbook 296. S.C.S., U.S.D.A. 82 p.

Barber, S. A. 1967. Liming materials and practices. Agron. 12:125-160.

Beasley, R. P. 1958. A new method of terracing. Missouri Agr. Exp. Stat. Bull. 699.

Black, C. A. 1968. Soil-plant relationships. 2nd ed. John Wiley and Sons, Inc., New York. 792 p.

Bohn, H. L., B. L. McNeal, and G. A. O'Connor. 1979. Soil chemistry. John Wiley and Sons, Inc., New York. 329 p.

Bouwer, H. and J. van Schilfgaarde. 1963. Simplified method for predicting fall of water table in drained land. Trans. Am. Soc. Agr. Engr. 6:288-291.

Brady, N. C. 1974. The nature and properties of soils. 8th ed. MacMillan Publ. Co., Inc. New York. 639 p.

Brown, K. W. and L. J. Thompson. 1976. Feasibility study of general crust management as a technique for increasing capacity of dredged material containment areas. Contract Report D-77. Texas A&M University, Research Foundation. College Station, Texas.

Buckman, H. W. and N. C. Brady. 1960. The nature and properties of soils. 6th ed. MacMillan Publishing Co., Inc. New York.

Capp, J. P. 1978. Power plant fly ash utilization for land reclamation in the Eastern United States. pp. 339-354. In F. W. Schaller and P. Sutton (eds.) Reclamation of drastically disturbed lands. Am. Soc. Agron. Madison, Wisconsin.

Carlson, C. A., P. G. Hunt, and T. B. Delaney. 1974. Overland flow treatment of wastewater. Misc. Paper Y-74-3. U.S. Army Engineer Waterways Experiment Station. Vicksburg, Mississippi. 63 p.

Cassell, E. A., P. W. Meals, and J. R. Bouyoun. 1979. Spray application of wastewater effluent in West Dover, Vermont - an initial assessment. U.S. Army Corps of Engineers, Cold Regions Research and Engineering Laboratory, Hanover, New Hampshire. Special Report 76-6.

Chen, R. L. and W. H. Patrick. 1981. Efficiency of nitrogen removal in a simulated overland flow wastewater treatment systems. J. Environ. Qual., 10:98-103.

Cheremisinoff, N. P., P. N. Cheremisinoff, F. Ellerbusch, and A. J. Perna. 1979. Industrial and hazardous waste impoundment. Ann Arbor Sci. Publ. Inc. Ann Arbor, Michigan. pp. 257-280.

Chessmore, R. A. 1979. Profitable pasture management. Interstate Printers & Publ. Danville, Illinois. 424 p.

Christensen, P. D. and P. J. Lyerly. 1954. Fields of cotton and other crops as affected by applications of sulfuric acid in irrigation water. Soil Sci. Soc. Am. Proc. 15:523-526.

Coleman, N. T. and G. W. Thomas. 1967. The basic chemistry of soil acidity. Agron. 12:1-34.

Dickey, E. C. and D. H. Vanderholm. 1981. Vegetative filter treatment of livestock feedlot runoff. J. Environ. Qual. 10:279-284.

Dolon, W. P. 1975. Odor panels. pp. 75-82. In P. N. Cheremisinoff and R. A. Young (eds.) Industrial odor technology assessment. Ann Arbor Science Publ. Inc. Ann Arbor, Michigan.

Dravenieks, A. 1975. Threshold of smell measurement. pp. 27-44 In P. N. Cheremisinoff and R. A. Young (eds.) Industrial odor technology assessment. Ann Arbor Science Publ. Inc. Ann Arbor, Michigan.

Dummn, L. D. 1964. Transient flow concept in subsurface drainage: its validity and use. Trans. Am. Soc. Agr. Engr. 7:142-146.

EPA. 1977. Process design manual for land treatment of municipal waste water. EPA 625/1-27-008. PB 299-665/IBE.

EPA. 1979. Disposal to land. In Process design manual for sludge treatment and disposal. EPA 625-1-79-011.

EPA. 1980a. Lining of waste impoundment and disposal facilities. U.S. EPA SW-870.

EPA. 1980b. Identification and listing of hazardous waste. Federal Register Vol. 45, No. 98, pp. 33119-33133. May 19, 1980.

EPA. 1981. Hazardous waste management system; addition of general requirements for treatment, storage and disposal facilities. Federal Register Vol. 46, No. 7. pp. 2802-2897. January 12, 1981.

EPA. 1982. Hazardous waste management system; permitting requirements for land disposal facilities. Federal Register Vol. 47, No. 43. pp. 32274-32388. July 26, 1982.

Flemming, G. S. 1975. Computer simulation techniques in hydrology. Elsevier Sci. Publ., Amsterdam.

Glover, R. E. 1964. Ground-water movement. U.S. Dept. Interior, Bur. Reclam. Engr. Monogr. No. 31. 67 p.

Graham, E. H. 1941. Legumes for erosion control and wildlife. U.S. Dept of Agriculture. Misc. Pub. No. 412. 153 p.

Hafenrichter, A. L., J. L. Schwendiman, H. L. Harris, R. S. MacLauchlan, and H. W. Miller. 1968. Grasses and legumes for soil conservation in the Pacific Northwest and Great Basin States. U.S. Dept. of Agriculture. Agriculture Handbook No. 339. Washington, D.C. 69 p.

Hanson, A. A. 1972. Grass varieties in the United States. U.S. Dept. of Agriculture. Agricultural Handbook No. 170. Washington, D.C. 124 p.

Hatayama, H. K., J. J. Chen, E. R. de Vera, R. D. Stephens, and D. L. Storm. 1980. A method for determining the compatibility of hazardous wastes. EPA-600/2-80-076.

Heath, M. E., P. S. Metcalfe, and R. F. Barnes. 1973. Forages: the science of grassland agriculture. Iowa State University Press, Ames, Iowa. 755 p.

Herschfield, D. M. 1961. Rainfall frequency atlas of the United States. Engr. Division, Soil Conserv. Serv., USDA. Washington, D.C. Tech. Paper 40. 117 p.

Hillel, D. 1971. pp. 167-182 In Groundwater drainage. Academic Press, New York.

Hitchcock, A. S. 1950. Manual of the grasses of the United States. U.S. Dept. of Agriculture. Misc. Pub. No. 200. Washington, D.C. 1051 p.

Hoeppel, R. E., P. G. Hunt and T. B. Delaney. 1974. Wastewater treatment of soils of low permeability. U.S. Army Eng. Wasteways Expt. Sta. Misc. Paper Y-74-2. Vicksburg, Miss. 84 p.

Hooghaudt, S. B. 1937. Bijdregen tot de kennis van eenige natuurkundige grootheden van den grand, 6. Versl. Landb. Ond. 43:461-676.

Implement and Tractor Red Book. 1979. Implement and Tractor Publications, Inc., 1014 Wyandotte, St., Kansas City, Missouri.

Jenkins, T. F., D. C. Leggett, C. J. Martel, and H. E. Hare. 1981. Overland flow: removal of toxic volatile organics. U.S. Army Cold Regions Research and Engineering Laboratory. Special Report 81-1. Hanover, New Hampshire. 15 p.

Jenkins, T. F. and A. J. Palazzo. 1981. Wastewater treatment by a prototype slow rate land treatment system. U.S. Army Cold Regions Research and Engineering Laboratory. Report 81-14. Hanover, New Hampshire. 44 p.

Kelling, K. A., L. M. Walsh, and A. E. Peterson. 1976. Crop response to tank truck applications of liquid sludge. J. Water Pollution Control Federation 48(9):2190-2197.

Kirkham, D., S. Toksoz, and R. R. van der Ploeg. 1974. Steady flow to drains and wells. pp. 203-244. In J. van Schilfgaarde (ed.) Drainage for agriculture. Agron. Monogr. No. 17. Am. Soc. Agron., Madison, Wisconsin.

Leithead, H. L., L. L. Yarlett, and T. N. Shiflet. 1971. 100 Native forage grasses in 11 southern states. U.S. Dept. of Agriculture. Agricultural Handbook No. 389. Washington, D.C. 216 p.

Linsley, R. K., Jr., M. A. Kohler and J. L. H. Paulhus. 1975. Hydrology for engineers. McGraw-Hill, Inc. New York. 482 p.

Loehr, R. C., W. J. Jewell, J. D. Novak, W. W. Clarkson, and G. S. Friedman. 1979. Land application of wastes, Vol. 2. Van Nostrand Reinhold Co. New York. p. 305.

Luthin, J. N. 1957. Drainage of irrigated lands. pp. 344-371. In J. N. Luthin (ed.) Drainage of agricultural lands. Am. Soc. Agron., Madison, Wisconsin. 611 p.

Martel, C. J., T. F. Jenkins, C. J. Diener, and P. L. Butler. 1982. Development of a rational design procedure for overland flow systems. U.S. Army Cold Regions Research and Engineering Laboratory. Report 82-2. Hanover, New Hampshire. 29 p.

Miyamato, S., J. Ryan, and J. L. Stroehlein. 1975. Potentially beneficial uses of sulfuric acid in southwestern agriculture. J. Environ. Qual. 4:431-437.

Miyamato, S. and J. L. Stroehlein. 1974. Solubility of manganese, iron, and zinc as affected by application of sulfuric acid to calcareous soils. Plant and Soil 40:421-427.

Moody, W. T. 1966. Nonlinear differential equation of drain spacing. Proc. Am. Soc. Civil Engr., J. Irrig. Drain. Div. 92(IR2):1-9.

Myers, E. A. 1974. Sprinkler irrigation systems. Design and operation criteria. pp. 299-309. In Proc. Conf. on recycling treated municipal wastewater through forest and cropland. EPA 660/2-74-003. PB 236-313/3BA.

Ness, L. D. and R. J. Ballard. 1979. Land application distribution equipment alternatives. Paper presented at joint meeting of Am. Soc. of Agr. Engr. and Can. Soc. of Agr. Engr. Univer. of Manitoba, Winnipeg, Canada. June, 1979.

Nutter, W. L. and R. C. Schultz. 1975. Spray irrigation of sewage effluent on a steep forest slope. I. Nitrate renovation. Agron. Abst. Am. Soc. Agron. Madison, Wisconsin.

Overcash, M. R. and D. Pal. 1979. Design of land treatment systems for industrial wastes - theory and practices. Ann Arbor Science Pub. Inc. Ann Arbor, Michigan. 684 p.

Peech, M. 1965. Lime requirement. Agron. 9:927-932.

Perrier, E. R. and A. C. Gibson. 1980. Hydrologic simulation on solid waste disposal sites (HSSWDS). Prepared for the U.S. EPA Municipal Environmental Research Laboratory. SW-868.

Peters, R. E. and C. R. Lee. 1978. Field investigation of advanced treatment of municipal wastewater by overland flow. Vol. II. p. 45-50. In H. L. McKim (ed.) State of knowledge in land treatment of wastewater. U.S. Army Cold Regions Research and Eng. Lab. Hanover, N. H.

Phung, T., L. Barker, D. Ross, and D. Bauer. 1978. Land cultivation of industrial wastes and municipal solid wastes; state-of-the-art study, Vol. 1. Technical summary and literature review. EPA 600/2-78-140a. PB 287-080/AS.

Pohlman, G. G. 1966. Effect of liming different soil layers on yield of alfalfa and on root development and nodulation. Soil Sci. 34:145-160.

Pountney, P. J. and H. Turner. 1979. Hydrogen peroxide makes an excellent sludge deodorant. Water & Wastes Engineering, September 1979. p. 56-59.

Powell, G. M., M. E. Jensen, and L. G. King. 1972. Optimizing surface irrigation uniformity by nonuniform slopes. Paper presented at the 1972 Winter meeting of the ASAE. Chicago, Illinois.

Prather, R. J., J. O. Goertzen, J. D. Rhoades, and H. Frenkel. 1978. Efficient amendment use in sodic soil reclamation. Soil Sci. Soc. Am. J. 42:782-786.

Ruffner, J. D. 1978. Plant performance on surface coal mine spoil in Eastern United States. U. S. Dept. of Agriculture, SCS-TP-155. Washington, D.C. 76 p.

Ryan, J., S. Miyamoto, and J. L. Stroehlein. 1975. Preliminary evaluation on methods of acid precipitation. Plant and Soil 41:11-13.

Ryan, J. and J. L. Stroehlein. 1979. Sulfuric acid treatment of calcareous soils: effects on phosphorus solubility, inoranic phosphorus forms and plant growth. Soil Sci. Soc. Am. J. 43:731-735.

Safaya, N. M. and M. K. Wali. 1979. Growth and nutrient relations of a grass-legume mixture on sodic coal-mine spoil as affected by some amendments. Soil Sci. Soc. Am. J. 43:747-753.

Schwab, G. O., R. K. Frevert, K. K. Barnes, and T. W. Edminster. 1971. Elementary soil and water engineering. John Wiley and Sons, New York. 316 p.

Smith, J. L., D. B. McWhorter, and R. C. Ward. 1977. Continuous subsurface injection of liquid dairy manure. EPA-600/2-77-117. PB 272-350/OBE.

Soil Conservation Service. 1958. Engineering handbook for soil conservationists in the Corn Belt. Agr. Handbook #135. U.S. Government Printing Office, Washington, D.C.

Soil Conservation Service. 1972. National engineering handbook, Section 4, Hydrology. Chapter 10. U.S. Government Printing Office, Washington, D.C.

Sopper, W. E. and L. T. Kardos. 1973. Vegetation responses to irrigation with treated municipal wastewater. pp. 271-294. In Sopper, W. E. and L. T. Kardos (eds.) Recycling treated municipal wastewater and sewage through forest and cropland. Penn. State Univ. Press. University Park, Pennsylvania.

Strunk, W. G. 1979. Hydrogen peroxide treats diverse wastewaters. Industrial Wastes, January-February, 1979. p. 32-35.

Stubbendieck, J., S. L. Hatch, and K. J. Kjar. 1981. North American range plants. Natural Resources Enterprises, Lincoln, Nebraska. 468 p.

Thomas, R. E., B. Bledsoe and K. Jackson. 1976. Overland flow treatment of raw wastewater with enhanced phosphorus removal. EPA-600/2-76-131, U.S. EPA, Washington, D.C.

Thornburg, A. A. 1982. Plant materials for use on surface mined lands in arid and semi-arid regions. SCS-TP-157. EPA-600/7-79-134 .88 p.

Thornwaite, C. W. 1948. An approach toward a rational classification of climate. Geog. Rev. 38:55-94.

Tisdale, S. L. and W. L. Nelson. 1975. Soil fertility and fertilizers. 3rd ed. MacMillan Publ. Co., New York.

USDA. 1937. Range plant handbook. Washington, D.C.

USDA. 1948. Grass: the yearbook of agriculture. Washington, D.C. 892 p.

USDA. 1960. Plant hardiness zone map. Agricultural Research Service. USDA Misc. Pub. No. 814. Washington, D.C.

USDA. 1973. Kentucky guide for classification, use and vegetative treatment of surface mine spoil. Washington, D.C. 31 p.

Vallentine, J. F. 1971. Range development and improvements. Brigham Young Univ. Press. Provo, Utah. 516 p.

Van Arsdel, E. P. 1967. The nocturnal diffusion and transport of spores. Phytopathology, 57(11):1221-1229.

Van Arsdel, E. P., E. C. Tullis, and J. D. Panzer. 1958. Movement of air in a rice paddy as indicated by colored smoke. Plant Disease Reporter, 42(6):721-725.

van Schilfgaarde, J. 1963. Design of tile drainage for falling water tables. Proc. Am. Soc. Civil Engr., J. Irrig. Drain. Div. 89(IR2):1-12.

van Schilfgaarde, J. 1965. Transient design of drainage systems. Proc. Am. Soc. Civil Engr., J. Irrig. Drain. Div. 91(IR3):9-22.

van Schilfgaarde, J. 1974. Nonsteady flow to drains. p. 245-270. In J. van Schilfgaarde (ed.) Drainage for agriculture. Agron. Monogr. No. 17. Am. Soc. Agron., Madison, Wisconsin.

Wali, M. K. and F. M. Sandoval. 1975. Regional site factors and revegetation studies in western North Dakota. pp. 133-153. In M. K. Wali (ed.) Practices and problems of land reclamation in western North America. University of North Dakota Press. Grand Forks, North Dakota.

Warburton, D. J., J. N. Scarborough, D. L. Day, A. J. Muehling, S. E. Curtis, and A. H. Jensen. 1979. Evaluation of commercial products for odor control and solids reduction of liquid swine manure. Paper presented at the Illinois Livestock Waste Management Conference, March 6, 1979, Champaign, Illinois.

Welch and Haferkamp. 1982. Seeding rangeland. Texas Agr. Exp. Sta. B-1379. 10 p.

White, R. K., M. Y. Handy, and T. H. Short. 1975. Systems and equipment for disposal of organic wastes on soil. Ohio Agricultural Research and Development Center. Res. Cir. 197.

Whiting, D. M. 1976. Use of climatic data in estimating storage days for soils treatment systems. U.S. EPA, Ada, Oklahoma. EPA 600/2-76-250. PB 263-597/7BE.

Wooding, H. N. and R. F. Shipp. 1979. Agricultural use and disposal of septic tank sludge. In Pennsylvania information and recommendations for farmers, septage haulers, municipal officials and regulatory agencies. Pennsylvania State Univ. Coop. Ext. Serv. Spec. Circ. 257.

Yahia, T. A., S. Miyamota, and J. L. Stroehlein. 1975. Effect of surface applied sulfuric acid on water penetration into dry calcareous and sodic soils. Soil Sci. Soc. Amer. Proc. 39:1201-1204.

MONITORING 9

Gordon B. Evans, Jr.
James C. Thomas
K. W. Brown

A monitoring program is an essential component at any land treatment unit, and should be planned to provide assurance of appropriate facility design, act as a feedback loop to furnish guidance on improving unit management, and indicate the rate at which the treatment capacity is being approached. Since many assumptions must be made in the design of a land treatment unit, monitoring can be used to verify whether the initial data and assumptions were correct or if design or operational changes are needed. Monitoring cannot be substituted for careful design based on the fullest reasonable understanding of the effects of applying hazardous waste to the soil; however, for existing HWLT units (which must retrofit to comply with regulations), monitoring can provide much of the data base needed for demonstrating treatment.

Figure 9.1 shows the topics to be considered when developing a monitoring program. The program must be developed to provide the following assurances:

(1) that the waste being applied does not deviate significantly from the waste for which the unit was designed;

(2) that waste constituents are not leaching from the land treatment area in unacceptable concentrations;

(3) that groundwater is not being adversely affected by the migration of hazardous constituents of the waste(s); and

(4) that waste constituents will not create a food chain hazard if crops are harvested.

To accomplish these assurances the current regulations (EPA, 1982a) require the following types of monitoring.

565

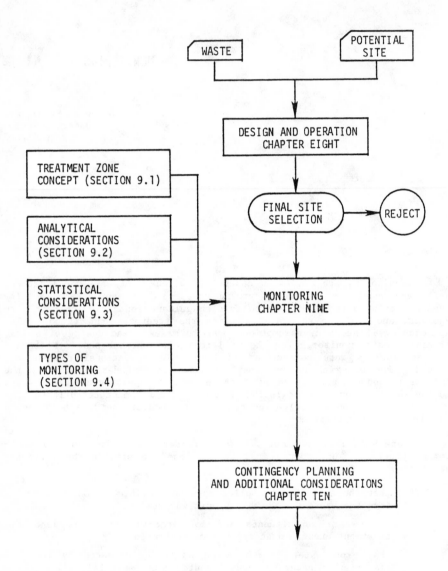

Figure 9.1 Topics to be considered in developing a monitoring program
for an HWLT unit.

(1) Groundwater detection monitoring to determine if a leachate plume has reached the edge of the waste management area (40 CFR 264.98).

(2) Groundwater compliance monitoring to determine if the facility is complying with groundwater protection standards for hazardous constituents (40 CFR 264.99).

(3) Soil pH and concentration of cadmium in the waste when certain food-chain crops are grown on HWLTs where cadmium is disposed (40 CFR 264.276).

(4) Unsaturated zone including soil cores and soil-pore liquid monitoring to determine if hazardous constituents are migrating out of the treatment zone (40 CFR 246.278).

(5) Waste analysis of all types of waste to be disposed at the HWLT (40 CFR 264.13).

In addition to these required types of monitoring, other types of monitoring may be needed in a thorough monitoring program (Fig. 9.2).

These secondary monitoring components, though not specifically regulated are important to successful land treatment. For instance, to complete the assurance that no unacceptable human health effect or environmental damage is occurring, air emissions, surface water discharge and worker exposure of hazardous constituents can be monitored. The treatment zone can be monitored to determine if degradation of waste organics is progressing as planned and whether adjustments in unit management (e.g., pH, nutrients, tillage) are needed to maintain the treatment process, and to gauge the rate at which the capacity limiting constituent (CLC) is accumulating in the land treatment unit and at what point closure should be initiated. Any of these components could be dropped from the proposed monitoring plan if treatment demonstrations show these types of monitoring are not needed to determine the proper performance of the HWLT unit.

9.1   TREATMENT ZONE CONCEPT

As is depicted in Fig. 9.2, the entire land treatment operation and monitoring program revolves about a central component, the treatment zone. Concentrating on the treatment zone is a useful approach to describing and monitoring a land treatment system. The treatment zone is the soil to which wastes are applied or incorporated; HWLT units are designed so that degradation, transformation and immobilization of hazardous constituents and their metabolites occurs within this zone. In practice, setting a boundary to the treatment zone is difficult. In choosing the boundaries of the treatment zone soil forming processes and the associated decrease in biological activity with depth should be considered. According to soil taxonomists, the lower limit of a soil must be set at the lower limit of biologic activity or rooting of native perennial plants, typically about 1 to 2 m (USDA, 1975). Since biological degradation of waste organics is

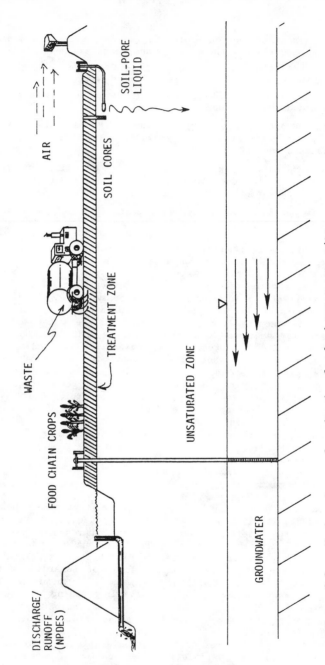

Figure 9.2.   Various types of monitoring for land treatment units.

often the primary objective in land treatment, the lower boundary of the treatment zone should not exceed the lower boundary of the soil. Current land treatment regulations place the lower limit of the treatment zone at 1.5 m (EPA, 1982a).

The choice of a lower boundary must be modified where shallow groundwater or perched water can encroach on this zone and thus increase the likelihood of contaminant leaching. A distance of 1 m is the required minimum separation between the bottom of the treatment zone and the seasonal high water table (EPA, 1982a). From soil physics considerations, this separation is necessary because the capillary fringe above the water table, resulting in elevated soil moisture content, is often observed to rise as much as 50 to 75 cm. A second reason for a 1 m separation is that the height of the seasonal high water table is generally an estimate based on limited observation and there may be periods when the saturated zone is at a higher elevation.

A final aspect of the treatment zone that should be considered is the rise in land surface elevation which may result from the accumulation of nondegradable waste solids. In some cases, this rise can be significant and the choice must be made whether to continually redefine the lower treatment zone boundary or define the lower boundary as a static value based on the original land surface elevation. The latter is the logical choice. If the lower boundary were continuously redefined, the waste material remaining below the redefined boundary would then be considered unacceptable since waste consituents must be degraded, transformed or immobilized within the treatment zone.

After considering the various aspects of the treatment zone, the generalized definition is the zone of waste and soil in which degradation, transformation and/or immobilization occurs, extending no more than 1.5 m below the original land surface and separated by at least 1 m from the seasonal high water table (EPA, 1982a). What constitutes "complete" treatment varies according to the specific hazardous constituent and the degree to which the constituent and its metabolites must be degraded or immobilized to prevent both short and long-term harm to human health or the environment. Where data are available, the required level of treatment may be relatively easy to designate; however, if data are lacking or inconclusive, the desired level of treatment must be resolved through laboratory, greenhouse, and/or field testing (Chapter 7).

9.2   ANALYTICAL CONSIDERATIONS

Certain nonhazardous waste constituents and/or their metabolites, either singly or in combination, are of concern when managing land treatment facilities because of their effect on treatment processes. A sound monitoring program should account for the potentially harmful effects of all waste constituents. Properly designed and conducted waste-site interaction studies should indicate the existence of environmental hazards. Nonhazardous inorganic constituents that are significant to the land treatment system should also be routinely included in the monitoring program. These unlisted constituents are often dealt with under the authority of State solid waste programs; therefore, facility permits should jointly address both hazardous and nonhazardous constituents. The permit officials and permit applicants should both recognize that in many cases a waste constituent, not regulated as hazardous, will be the limiting factor (ALC, RLC, or CLC) in facility design. Methods for determining the constituents that limit the amount of waste, the number of waste applications, and the cumulative capacity of a land treatment site are discussed in Section 7.5.

9.3   STATISTICAL CONSIDERATIONS

A monitoring plan can be judged by its ability to provide realistic, unbiased data from which valid comparisons between the values of monitored parameters and background quality can be made. The use of statistical principles in the monitoring design is therefore fundamental to providing the maximum amount of relevant information in the most efficient manner. In general, the most common monitoring approach compares the sample means of two populations assumed to be independent and normally distributed (i.e., parameter values from a uniform area or individual location compared with background, ambient values). It is suggested that the land treatment unit is designed and operated such that no significant movement of hazardous constituents occurs. Thus, the null hypothesis to be tested is that the population means are equal ($H:\mu_1 = \mu_2$; $A:\mu_1 \neq \mu_2$). The keys to valid comparison between these populations are the choice of sample size (number of replications) and the use of random sampling. Problems arise in planning monitoring systems when one must decide how best to meet the statistical requirements and what balance to establish between the needed data and economy of design. After defining the type of comparisons, the choice of test statistics can be made. The present problem is well suited to the "t" statistic, which is in fact generally suggested in EPA monitoring guidance and regulations (40 CFR 264 Subpart F in EPA, 1982a).

Often the difficulty of designing a monitoring plan is in choosing what is to be measured, how replicate samples are to be obtained, and how many replicates are needed. Basically, taking replicate samples is intended to provide a measure of the variability of the sampled medium. EPA (1982b) provides methods for developing a statistical approach for taking and analyzing monitoring samples. One must be careful to avoid interpreting analytical errors as actual differences in the sampled media. It is a good idea to obtain several samples in a random fashion and analyze these for the constituents of concern. For example, samples could be obtained from monitoring wells or soil-pore liquid samplers at random times over a period of several days, or soil core samples could be obtained from several random locations. The number of samples taken should depend on sampling variability and may be as few as three if variability is low. Sample variability must be established for the media to be sampled at the HWLT unit. A good starting point is to obtain and analyze five replicate samples; if the variance is low (e.g., 5-10% of the mean), then fewer samples would suffice while a high variance (e.g., >25% of the mean) indicates that more than five samples may be needed.

9.4   TYPES OF MONITORING

As discussed earlier the monitoring program centers around the treatment zone. The required types of monitoring for HWLT facilities are contained in the EPA (1982a) regulations and are also listed in Section 9.0. The frequency of sampling and the parameters to be analyzed depend on the characteristics of the waste being disposed, the physical layout of the unit, and the surface and subsurface characteristics of the site. Table 9.1 provides guidance for developing an operational monitoring program. Each of the types of monitoring are discussed below.

9.4.1   Waste Monitoring

Waste streams need to be routinely sampled and tested to check for changes in composition. A detailed description of appropriate waste sampling techniques, tools, procedures, and safety measures is presented in Section 5.3.2.1. These procedures should be followed during all waste sampling events. Analytical methods should follow established procedures for the given waste described in Section 5.3.2 which are based on standard protocols.

The frequency at which a waste needs to be sampled and the parameters to be analyzed depends greatly on the variables that influence the quantity and quality of the waste. When waste is generated in a batch, as would be expected from an annual or biannual cleanout of a lagoon or tank, the waste should be fully characterized prior to each application. When the waste is generated more nearly continuously, samples should be collected and composited based on a statistical design over a period of time to assure that that the waste is of a uniform quality. For example, wastes which are generated continuously could be sampled weekly or daily on a flow proportional basis and composited and analyzed quarterly or monthly. When no changes have been made in the operation of the plant or the treatment of the waste which could significantly alter concentration of waste constituents, the waste should, at a minimum, be analyzed for (1) the constituents that restrict the annual application rates (RLC) and the allowable cumulative applications (CLC), (2) the constituents that are within 25% of the level at which they would be limiting, and (3) all other hazardous constituents that have been shown to be present in the waste in the initial waste characterization. Since synergism and antagonism as well as unlisted waste metabolites can create hazards that cannot be described by chemical analysis alone, routine mutagenicity testing may be performed (Section 5.3.2.4) if the treatment demonstration has indicated a possible problem. In addition, waste should be analyzed as soon as possible after a change in operations that could affect the waste characteristics.

TABLE 9.1 GUIDANCE FOR AN OPERATIONAL MONITORING PROGRAM

Media to be Monitored	Purpose	Sampling Frequency	Number of Samples	Parameters to be Analyzed
Waste	Quality Change	Quarterly composites if continuous stream; each batch if intermittent generation.	One	At least rate and capacity limiting constituents, plus those within 25% of being limiting, principal hazardous constituents, pH and EC.
Soil cores (unsaturated zone)	Determine slow movement of hazardous constituents	Quarterly	One composited from two per 1.5 ha (4 ac); minimum of 3 composited from 6 per uniform area.	All hazardous constituents in the waste or the principal hazardous constituents, metabolites of hazardous constituents, and nonhazardous constituents of concern.
Soil-pore water (unsaturated zone)	Determine highly mobile constituents	Quarterly, preferably following leachate generating precipitation snowmelt.	Two per 1.5 ha (4 ac); minimum of 6 per uniform area.	All hazardous constituents in the waste or the principal hazardous constituents, mobile metabolites of hazardous constituents, and important mobile nonhazardous constituents.
Groundwater	Determine mobile constituents	Semiannually	Minimum of four suggested—one upgradient, three downgradient	Hazardous constituents and metabolites reasonably expected to be in waste or select indicators.
Vegetation (if grown for food chain use)	Phytotoxic and hazardous transmitted constituents (food chain hazards)	Annually or at harvests.	One per 1.5 ha (4 ac) or three of processed crop before sale.	Hazardous metals and organics and their metabolites.
Runoff water	Soluble or suspended constituents	As required for NPDES permit.	As permit requires, or one.	Discharge permit and background parameters plus hazardous organics.
Soil in the treatment zone	Determine degradation, pH, nutrients, and rate and capacity limiting constituents	Quarterly	7-10 composited to one per 1.5 ha (4 ac)	
Air	Personnel and population health hazards	Quarterly	Five	Particulates (adsorbed hazardous constituents) and hazardous volatiles.

The unsaturated zone as referred to in this document is described as the layer of soil or parent material separating the bottom of the treatment zone (defined earlier) and the seasonal high water table or groundwater table and is usually found to have a moisture content less than saturation. In this zone, the movement of moisture may often be relatively slow in response to soil properties and prevailing climatic conditions; however, in some locations, soils and waste management practices may lead to periods of heavy hydraulic loading which could cause rapid downward flux of moisture. An unsaturated zone monitoring plan should be developed for two purposes: 1) to detect any significant movement of hazardous constituents out of the system and 2) to furnish information for management decisions. In light of the variability in soil water flux and the mobility of hazardous waste constituents, the unsaturated zone monitoring plan should include sampling the soil to evaluate relatively slow moving waste constituents (soil core monitoring) and sampling the soil-pore liquid to evaluate rapidly moving waste constituents. Monitoring for hazardous constituents should be performed on a representative background plot(s) until background levels are established and immediately below the treatment zone (active portion). The number, location, and depth of soil core and soil-pore liquid samples taken must allow an accurate indication of the quality of soil-pore liquid and soil below the treatment zone and in the background area. The frequency and timing of soil-pore liquid sampling must be based on the frequency, time and rate of waste application, proximity of the treatment zone to groundwater, soil permeability, and amount of precipitation. The data from this program must be sufficient to determine if statistically significant increases in hazardous constituents, or selected indicator constituents, have occurred below the treatment zone. Location and depth of soil core and soil-pore liquid samples follow the same reasoning, but the number, frequency and timing of soil core sampling differs somewhat from that required for soil-pore liquid sampling. Thus, the unique aspects of these topics will be considered together with discussions of techniques for obtaining the two types of samples.

9.4.2.1 Locating Unsaturated Zone Samples

Soil characteristics, waste type, and waste application rate are all important factors in determining the environmental impact of a particular land treatment unit or part of a unit on the environment. Therefore, areas of the land treatment unit for which these characteristics are similar (i.e., uniform areas) should be sampled as a single monitoring unit. As will be used in further discussions, a uniform area is defined as an area of the active portion of a land treatment unit which is composed of soils of the same soil series (USDA, 1975) and to which similar wastes or waste mixtures are applied at similar application rates. If, however, the texture of the surface soil differs significantly among soils of the same series classification, the phase classification of the soil should be considered in defining "uniform areas." A certified profesional soil scientist should be consulted in designating uniform areas.

Based on the above definition, it is recommended that the location of soil core sampling or soil-pore liquid monitoring devices within a given uniform area be randomly selected. Random selection of samples ensures a more accurate representation of conditions within a given uniform area. It is convenient to spot the field location for soil-coring and soil-pore liquid devices by selecting random distances on a coordinate system and using the intersection of the two random distances as the location at which a soil core should be taken or a soil-pore liquid monitoring device installed. This system works well for fields of both regular and irregular shape, since the points outside the area of interest are merely discarded, and only the points inside the area are used in the sample.

The location, within a given uniform area of a land treatment unit (i.e., active portion monitoring), at which a soil core should be taken or a soil-pore liquid monitoring device installed should be determined using the following procedure:

(1) Divide the land treatment unit into uniform areas under the direction of a certified professional soil scientist.

(2) Set up coordinates for each uniform area by establishing two base lines at right angles to each other which intersect at an arbitrarily selected origin, for example, the southwest corner. Each baseline should extend far enough for all of the uniform area to fall within the quadrant.

(3) Establish a scale interval along each base line. The units of this scale may be feet, yards, meters, or other units depending on the size of the uniform area, but both base lines should have the same units.

(4) Draw two random numbers from a random numbers table (usually available in any basic statistics book). Use these numbers to locate one point along each of the base lines.

(5) Locate the intersection of two lines drawn perpendicular to the base lines through these points. This intersection represents one randomly selected location for collection of one soil core, or for installation of one soil-pore liquid device. If this location at the intersection is outside the uniform area, disregard and repeat the above procedure.

(6) For soil-core monitoring, repeat the above procedure as many times as necessary to obtain the desired number of locations within each uniform area of the land treatment unit. This procedure for randomly selecting locations must be repeated for each soil core sampling event but will be needed only once in locating soil pore liquid monitoring devices.

Locations for monitoring on background areas should also be randomly determined. Again, consult a certified professional soil scientist in determining an acceptable background area. The background area must have characteristics (i.e., at least soil series classification) similar to those present in the uniform area of the land treatment unit it is representing, but it should be free from possible contamination from past or present activities which could have contributed to the concentrations of the hazardous constituents of concern. Establish coordinates for an arbitrarily selected portion of the background area and use the above procedure for randomly choosing sampling locations.

9.4.2.2   Depth to be Sampled

Since unsaturated zone monitoring is intended to detect pollutant migration from the treatment zone, samples should logically be obtained from immediately below this zone. Care should be taken to assure that samples from active areas of the land treatment unit and background samples are monitoring similar horizons or layers of parent material. Noting that soils seldom consist of smooth, horizontal layers but are often undulating, sloped and sometimes discontinuous, it would be unwise to specify a single depth below the land surface to be used for comparative sampling. A convenient method for choosing sampling depths is to define the bottom of the treatment zone as the bottom of a chosen diagnostic soil horizon and not in terms of a rigid depth. Sampling depth would then be easily defined with respect to the bottom of the treatment zone. At a minimum, soil core and soil-pore liquid sampling should monitor within 30 cm (12 in) of the bottom of the treatment zone. Additional sampling depths may be desirable, for instance if analytical results are inconclusive or questionable. Core samples could include the entire 0 to 30 cm increment (or possibly only 0 to 15 cm) below the treatment zone while soil-pore liquid samplers should be placed so that they collect liquid from somewhere within this zone.

9.4.2.3   Soil Core Sampling Technique

Waste constituents may move slowly through the soil profile for a number of reasons, such as the lack of sufficient soil moisture to leach through the system, a natural or artificially occurring layer or horizon of low hydraulic conductivity, or waste constituents which exhibit only a low to moderate mobility relative to water in soil. Any one or a combination of these effects can be observed by soil core monitoring. Based on the treatment zone concept, only the portions of soil cores collected below the treatment zone need to be analyzed. The intent is to demonstrate whether significantly higher concentrations of hazardous constituents are present and moving in material below the treatment zone than in background soils or parent material.

Soil core sampling should proceed according to a definite plan with regard to number, frequency and technique. Previous discussions of statistical considerations should provide guidance in choosing the number of samples required. Background values for soil core monitoring should be established by collecting at least eight randomly selected soil cores for each soil series present in the treatment zone. These samples can be composited in pairs (from immediately adjacent locations) to form four samples for analysis. For each soil series a background arithmetic mean and variance should be calculated for each hazardous constituent. For monitoring the active portion of the HWLT, a minimum of six randomly selected soil cores should be obtained per uniform area and composited as before to yield three samples for analysis. If, however, a uniform area is greater than 5 ha (12 ac), at least two randomly selected soil cores should be taken per 1.5 ha (4 ac) and composited in pairs based on location. Data from the samples in a given uniform area should be averaged and statistically compared. If analyses reveal a large variance from samples within a given uniform area, more samples may be necessary. The frequency with which soil coring should be done is at least semiannually, except for background sampling which, after background values are established, may be performed only occassionally as needed to verify whether background levels are changing over time.

It is important to keep an accurate record of the locations from which soil core samples have been taken. Even where areas have been judged to be uniform, the best attempts at homogeneous waste application and management cannot achieve perfect uniformity. It is probable in many systems that small problem areas or "hot spots" may occur which cause localized real or apparent pollutant migration. Examples of "apparent" migration might include small areas where waste was applied too heavily or where the machinery on-site mixed waste too deeply. The sampling procedure itself is subject to error and so may indicate apparent pollutant migration. Therefore, anomalous data points can and should be resampled at the suspect location(s) to determine if a problem exists, even if the uniform area as a whole shows no statistically significant pollutant migration.

The methods used for soil sampling are variable and depend partially on the size and depth of the sample needed and the number and frequency of samples to be taken. Of the available equipment, oakfield augers are useful if small samples need to be taken by hand while bucket augers give larger samples. Powered coring or drilling equipment, if available, is the preferable choice since it can rapidly sample to the desired depths and provide a clean, minimally disturbed sample for analysis. Due to the time involved in coring to 1.5 m and sometimes farther, powered equipment can often be less costly than hand sampling. In any case, extreme care must be taken to prevent cross contamination of samples. Loose soil or waste should be scraped away from the surface to prevent it from contaminating samples collected from lower layers. The material removed from the treatment zone portion of the borehole can be analyzed if desired, to evaluate conditions in the treatment zone. It is advisable to record field observations of the treatment zone even if no analysis is done. Finally, bore holes absolutely must be backfilled carefully to prevent hazardous constituents from channelling down the hole. Native soil compacted to about field bulk density, clay slurry or other suitable plug material may be used.

Sample handling, preservation and shipment should follow a chain of custody procedure and a defined preservation method such as is found in EPA

(1982), Test Methods for Evaluating Solid Waste, or the analytical portion of this document (Section 5.3). If more sample is collected than is needed for analysis, the volume should be reduced by either the quartering or riffle technique. (A riffle is a sample splitting device designed for use with dried ground samples).

The analysis of soil cores must include all hazardous constituents which are reasonably expected to leach or the principal hazardous constituents (PHCs) which generally indicate hazardous constituent movement (EPA, 1982a).

9.4.2.4 Soil-Pore Liquid Sampling Technique

Percolating water added to the soil by precipitation, irrigation, or waste applications may pass through the treatment zone and may rapidly transport some mobile waste constituents or degradation products through the unsaturated zone to the groundwater. Soil-pore liquid monitoring is intended to detect these rapid pulses of contaminants, often immediately after heavy precipitation events, that are not likely to be observed through the regularly scheduled analysis of soil cores. Therefore, the timing of soil-pore liquid sampling is a key to the usefulness of this technique. Seasonality is the rule with soil-pore liquid sample timing (i.e., scheduled sampling cannot be on a preset date, but must be geared to precipitation events). Assuming that sampling is done soon after leachate-generating precipitation or snowmelt, the frequency also varies depending on site conditions. As a starting point, sampling should be done quarterly. More frequent sampling may be necessary, for example, at units located in areas with highly permeable soils or high rainfall, or at which wastes are applied very frequently. The timing of sampling should be geared to the waste application schedule as much as possible.

Land treatment units at which wastes are applied infrequently (i.e., only once or twice a year) or where leachate-generating precipitation is highly seasonal, quarterly sampling and analysis of soil-pore liquid may be unnecessary. Because soil-pore liquid is instituted primarily to detect fast-moving hazardous constituents, monitoring for these constituents many months after waste application may be useless. If fast-moving hazardous constituents are to migrate out of the treatment zone, they will usually migrate at least within 90 days following waste application, unless little precipitation or snowmelt has occurred. Therefore, where wastes are applied infrequently or leachate generation is seasonal, soil-pore liquid may be monitored less frequently (semi-annually or annually). A final note about timing is that samples should be obtained as soon as liquid is present. Following any significant rainfall, snowmelt or waste application, the owner or operator should check the monitoring devices for liquid within 24 hours.

The background concentrations of hazardous constituents in the soil-pore liquid should be established by installing two monitoring devices at random locations for each soil series present in the treatment zone.

Samples should be taken on at least a quarterly basis for at least one year and can be composited to give one sample per quarter. Analysis of these samples should be used to calculate an arithmetic mean and variance for each hazardous constituents. After background values are established, additional soil-pore liquid samples should occasionally be taken to determine if the background values are changing over time.

The number of soil-pore liquid samplers needed is a function of site factors that influence the variability of leachate quality. Active, uniform areas should receive, in the beginning, a minimum of six samplers per 5 ha (12 ac) or, for larger uniform areas, two samplers per 1.5 ha (4 ac). Samples may be composited in pairs based on location to give three samples for analysis. The number of devices may have to be adjusted up (or down) as a function of the variability of results.

To date, most leachate collection has been conducted by scientists and researchers and there is not an abundance of available field equipment and techniques. The EPA (1977) and Wilson (1980) have prepared reviews of pressure vacuum lysimeters and trench lysimeters. The pressure vacuum lysimeters are much better adapted to field use and have been used to monitor pollution from various sources (Manbeck, 1975); Nassau-Suffolk Research Task Group, 1969; The Resources Agency of California, 1963; James, 1974). These pressure vacuum samplers are readily available commercially and are the most widely used, both for agricultural and waste monitoring uses. A third type of leachate sampler is the vacuum extractor as used in the field by Smith et al. (1977). A comparison of in situ extractors was presented by Levin and Jackson (1977).

9.4.2.4.1 <u>Pressure-Vacuum Lysimeters</u>. Construction, installation, and sampling procedures for pressure-vacuum lysimeters are described by Grover and Lamborn (1970), Parizek and Lane (1970), Wagner (1962), Wengel and Griffen (1971) and Wood (1973). Some data indicate that the ceramic cups may contribute excessive amounts of Ca, Na, and K to the sample and may remove P from the sample (Grover and Lamborn, 1970); however, more recent work (Silkworth and Grigal, 1981) comparing ceramic samplers with inert fritted glass samplers showed no significant differences in Ca, Na, Mg, and K concentrations. No studies as yet have been done on the permeability of ceramic samplers to organic samplers. Recent data by Brown (1977) indicate that ceramics are permeable to some bacteria, while Dazzo and Rothwell (1974) found ceramic with a pore size of 3-8 m screened out bacteria. A special design (Wood, 1973) is needed if samples are to be collected at depths greater than 10 m below the soil surface. The basic construction of these devices is shown in Fig. 9.3 and consists of a porous ceramic cup with a bubbling pressure of 1 bar or greater attached to a short piece of PVC pipe of suitable diameter. Two tubes extend down into the device as illustrated. Data by Silkworth and Grigal (1981) indicate that, of the two commercially available sampler sizes (2.2 and 4.8 cm diameter), the larger ceramic cup sampler is more reliable, influences water quality less, and yields samples of suitable volume for analysis.

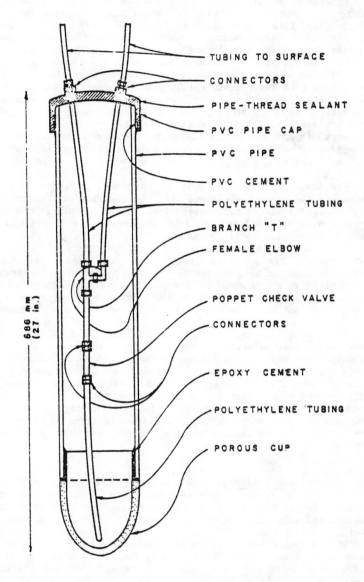

686 mm
(27 in.)

TUBING TO SURFACE

CONNECTORS

PIPE-THREAD SEALANT

PVC PIPE CAP

PVC PIPE

PVC CEMENT

POLYETHYLENE TUBING

BRANCH "T"

FEMALE ELBOW

POPPET CHECK VALVE

CONNECTORS

EPOXY CEMENT

POLYETHYLENE TUBING

POROUS CUP

Figure 9.3.   A modified pressure-vacuum lysimeter (Wood, 1973).
Reprinted by permission of the American Geophysical
Union.

Detailed installation instructions for pressure-vacuum lysimeters are given by Parizek and Lane (1970). Significant modification may be necessary to adapt these instruments to field use where heavy equipment is working. To prevent channelling of contaminated surface water directly to the sampling device, the sampler may be installed in the side wall of an access trench. Since random placement procedures may locate a sampler in the middle of an active area, the sample collection tube should be protected at the surface from heavy equipment by a manhole cover, brightly painted steel cage or other structure. Another problem associated with such sampler placement is that its presence may alter waste management activities (i.e., waste applications, tilling, etc. will avoid the location); therefore, the sampler would not yield representative leachate samples. This problem may be avoided by running the collection tube horizontally underground about 10 to 20 m before surfacing.

For sampling after the unit is in place, a vacuum is placed on the system and the tubes are clamped off. Surrounding soil water is drawn into the ceramic cup and up the polyethylene tube. To collect the water sample, the vacuum is released and one tube is placed in a sample container. Air pressure is applied to the other tube which forces the liquid up the tube and into the sample container. Preliminary testing should ensure that waste products can pass into the ceramic cup. An inert tubing such as Teflon may need to be substituted for the polyethylene to prevent organic contamination. Where sampling for possible volatiles in leachate, a purge trap such as suggested by Wood et al. (1981) or as described for volatiles in the waste analysis section (5.3.2.3.2.2) of this document may be used.

The major advantages of these sampling devices are that they are easily available, relatively inexpensive to purchase and install, and quite reliable. The major disadvantage is the potential for water quality alterations due to the ceramic cup, and this possible problem requires further testing. For a given installation, the device chosen should be specifically tested using solutions containing the soluble hazardous constituents of the waste to be land treated. Several testing programs to evaluate these devices are currently in progress, including programs sponsored by the U.S. Environmental Protection Agency and the American Petroleum Institute.

9.4.2.4.2 Vacuum Extractor. Vacuum extractors were developed by Duke and Haise (1973) to extract moisture from soils above the groundwater table. The basic device consists of a stainless steel trough that contains ceramic tubes packed in soil. The unit is sized not to interfere with ambient soil water potentials (Corey, 1974), and it is installed at a given depth in the soil with a slight slope toward the collection bottle which is in the bottom of an adjacent access hole. The system is evacuated and moisture moved from adjacent soil into the ceramic tubes and into the collection bottle from which it can be withdrawn as desired. The advantage of this system is that it yields a quantitative estimate of leachate flux as well as provides a water sample for analysis. The volume of collected leachate per unit area per unit time is an estimate of the downward movement of leachate water at that depth. The major disadvantages to this system are: it is delicate, requires a field vacuum source, is relatively difficult to install, requires a trained operator, estimates leachate quantity somewhat lower than actual field drainage, and disturbs the soil above the sampler. Further details about the use of the vacuum extractor are given by Trout et al. (1975). Performance of this type device is generally poor when installed in clay soils.

9.4.2.4.3 <u>Trench Lysimeters</u>. Trench lysimeters get their name from the large access trench or caisson necessary for operation. Basic installation as described by Parizek and Lane (1970) involves excavating a rather large trench and shoring up the side walls, taking care to leave open areas so that samplers can be placed in the side walls. Sample trays are imbedded in the side walls and connected by tubing to sample collection containers. The entire trench area is then covered to prevent flooding. One significant danger in using this system is the potential for accumulation of hazardous fumes in the trench which may endanger the health and safety of the person collecting the samples.

Trench lysimeters function by intercepting downward moving water and diverting it into a collection device located at a lower elevation. Thus, the intercepting agent may be an open ended pipe, sheet metal trough, pan, or other similar device. Pans 0.9 to 1.2 m in diameter have been successfully used in the field by Tyler and Thomas (1977). Since there is no vacuum applied to the system, only free water in excess of saturation is sampled. Consequently, samples are plentiful during rainy seasons but are nonexistent during the dry season.

Another variation of this system is to use a funnel filled with clean sand inserted into the sidewall of the trench. Freewater will drain into a collection chamber from which a sample is periodically removed by vacuum. A small sample collection device such as this may be preferable to the large trench since the necessary hole is smaller, thus making installation easier (Fig 9.4).

9.4.2.5 <u>Response to Detection of Pollutant Migration</u>

If significant concentrations of hazardous constituents (or PHCs) are observed below the treatment zone, the following modifications to unit operations should be considered to maximize treatment within the treatment zone:

(1) alter the waste characteristics;

(2) reduce waste application rate;

(3) alter the method or timing of waste applications;

(4) cease application of one or more particular wastes at the unit;

(5) revise cultivation or management practices; and

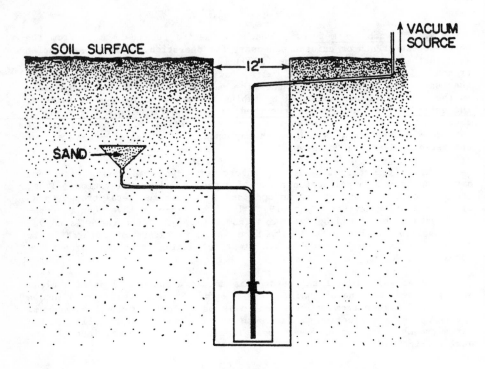

Figure 9.4.   Schematic diagram of a sand filled funnel used to collect
leachate from the unsaturated zone.

(6) alter the characteristics of the treatment zone, particularly soil pH or organic matter content.

Hazardous constituents movement below the treatment zone may result from
improper unit design, operation, or siting. Problems related to unit
design and operation can often be easily corrected, while serious problems
resulting from a poor choice of site are more difficult to rectify.
Certain locational "imperfections" may be compensated for through careful
unit design, construction, and operation.

If statistically significant increases of hazardous constituents are
detected below the treatment zone by the unsaturated zone monitoring program, the owner or operator should closely evaluate the operation, design
and location of the unit to determine the source of the problem. The characteristics of the waste should be evaluated for possible effects on treatment effectiveness. The rate, method, and timing of waste applications
should also be examined. Management of the treatment zone including maintaining the physical, chemical and biological characteristics necessary for
effective treatment, should also be reevaluated. Soil pH and organic
matter content of the treatment zone are two important parameters that
should be assessed. Finally, the owner or operator should determine if the
design or location of the unit is causing the hazardous constituents to migrate. Topographic, hydrogeologic, pedalogic, and climatic factors all
play a role in determining the success of the land treatment system.

In certain cases, the necessary unit modifications may be very minor,
while in other cases they may be major. Numerous unit-specific factors
must be considered to make this determination, and the exact elements of
the determination will vary on a case-by-case basis. Activities occurring
near the unit should be carefully investigated to confirm the source of the
contamination. The procedures used in the unsaturated zone monitoring program should also be closely examined. Resampling of the unit may be required to determine if errors occurred in sampling, analysis, or evaluation.

9.4.3   Groundwater Monitoring

To assure that irreparable groundwater damage does not occur as a
result of HWLT, it is necessary that the groundwater quality be monitored.
Groundwater monitoring supplements the unsaturated zone monitoring program,
but does not replace it. A contamination problem first detected in the
leachate water may indicate the need to alter the management program and
groundwater can then be observed for the same problem. It is through the
successful combination of these two systems that accurate monitoring of
vertically moving constituents can be achieved.

The complexity of groundwater monitoring is beyond the scope of this
document, and the reader is referred to a few of the numerous publications
which together cover much of what is to be known about the topic. These
sources of information include the following:

(1) Manual of Ground-Water Sample Procedures, (Scalf et al., 1981);

(2) Ground-Water Manual, (USDI, Bureau of Reclamation, 1977);

(3) Procedures Manual for Ground Water Monitoring at Solid Waste Disposal Facilities (EPA, 1977); and

(4) Ground-water Monitoring Systems, Technical Resource Document (EPA, in preparation); and

(5) Ground-water Monitoring Guidance for Owners and Operators of Interim Status Facilities, (EPA, 1982c).

In general, the success of a groundwater monitoring program is a function of many site-, soil- and waste-specific variables. The various aspects of planning and developing an appropriate groundwater monitoring program are interdependent and thus, design and development should be performed simultaneously. Mindful of these points, the following is a general outline of the major steps and considerations in establishing a groundwater monitoring program:

(1) develop an understanding of the potentially mobile constituents in the waste to be land treated and their possible reactions and behavior in groundwater, compatability with well casing and sampling equipment, and toxicity;

(2) perform a thorough hydrogeologic study of the land treatment site;

(3) choose well drilling, installation and sampling methods that are compatible with monitoring needs;

(4) locate wells based on hydrogeologic study results, but sample and analyze wells one by one as they are installed to help guide the placement of subsequent wells; and

(5) begin sampling and analytical program.

The wells should be placed to characterize background water quality and to detect any pollutant plume which leaves the site. The number of wells needed will vary from site to site based on local conditions. Wells should be sealed against tampering and protected from vehicular traffic. Finally, the frequency of sampling should be at least semi-annually for detection monitoring and at least quarterly for compliance monitoring (EPA, 1982a).

9.4.4   Vegetation Monitoring

Where food chain crops are to be grown, analysis of the vegetation at the HWLT unit will aid in assuring that harmful quantities of metals or other waste constituents are not being accumulated by, or adhering to surfaces of, the plants. Although a safety demonstration before planting is required (EPA, 1982a), operational monitoring is recommended to verify that crop contamination has not occurred. Vegetation monitoring is an important measurement during the post closure period where the area may possibly be used for food or forage production. Sampling should be done annually, or at each harvest. The concentrations of metals and other constituents in the vegetation will change with moisture content, stage of growth, and the part of the plant sampled, and thus results must be carefully interpreted. The number of samples to analyze is again based on a sliding scale similar to that used for sampling soils. Forage samples should include all aerial plant parts, and the edible parts of grain, fruit, or vegetation crops should be sampled separately.

9.4.5   Runoff Water Monitoring

If runoff water analyses are needed to satisfy NPDES permit conditions (EPA, 1981), a monitoring program should be instituted. This program would not be covered under RCRA hazardous waste land disposal requirements, but it would be an integral part of facility design. The sampling and monitoring approach will vary depending on whether the water is released as a continuous discharge or as a batch discharge following treatment to reduce the hazardous nature of the water. Constituents to be analyzed should be specified in the NPDES permit.

Where a relatively continuous flow is anticipated, sampling must be flow proportional. A means of flow measurement and an automated sampling device are a reasonable combination for this type of monitoring. Flow can be measured using a weir or flume (USDA, 1979) for overload flow water pretreatment systems and packaged water treatment plants while in-line flow measurement may be an additional option on the packaged treatment systems. The sampling device should be set up to obtain periodic grab samples as the water passes through the flow rate measuring device. A number of programmable, automated samplers which can take discreet or composite samples are on the market and readily available.

For batch treatment, such as mere gravity separation or mechanically aerated systems, flow is not so important as is the hazardous constituent content of each batch. Sampling before discharge would, in this case, involve manual pond sampling, using multiple grab samples. The samples would preferably represent the entire water column to be discharged in each batch rather than a single depth increment. Statistical procedures should again be used for either treatment and discharge approach.

9.4.6   Treatment Zone Monitoring

Treatment zone monitoring of land treatment units is needed for two purposes. One main purpose is to monitor the degradation rate of the organic fraction of the waste material and parameters significantly affecting waste treatment. Samples are needed at periodic intervals after application to be analyzed for residual waste or waste constituents. Such measurements need to be taken routinely as specified by a soil scientist. These intervals may vary from weekly to semi-annually depending on the nature of the waste, climatic conditions, and application scheduling. The second major function of treatment zone samples is to measure the rate of accumulation of conserved waste constituents as it relates to facility life.

9.4.6.1 Sampling Procedures

In order to monitor the treatment zone, a representative sample or set of soil samples must be collected. Since all further analysis, data, and interpretation are based on the sample(s) collected, the importance of obtaining a representative sample cannot be over-emphasized. Some of the needed samples may be obtained from soil cores taken from unsaturated zone monitoring, but additional samples are often desirable. The total area to be sampled should be first observed for its overall condition (i.e., waste application records, soil series, management techniques, soil color, moisture, vegetation type and vigor, etc.) and those areas having obvious differences need to be sampled separately. Where possible, sampling should most conveniently coincide with the "uniform areas" used in the unsaturated zone monitoring, but some deviation may be necessary. Uniform areas should be divided into 1.5 ha (4 ac) subsections. When sampling, care needs to be taken to avoid depressions, odd looking areas, wet spots, former fence rows, and edges of the field. Surface litter should not be included in the samples. Compositing of samples, when necessary, should be done in large inert containers, and subsampling of the mix should be done by the quartering technique or with a riffle subsampler.

Background soils should be sampled to the extent of the defined vertical treatment zone, while sampling an area that has had waste previously applied need extend only to about 15 cm below the depth of waste incorporation. If the waste is mixed poorly or not at all, the soil and waste should be mixed manually to the approximate expected depth of incorporation prior to sampling. Notes should be taken as to how well the waste is incorporated at the time of sampling. Plots that have had subsurface injections should be sampled by excavating a trench 10 to 20 cm wide and as long as the spacing between bands, perpendicular to the line of application and to a depth of 15 cm below the depth of incorporation. Useful equipment may include shovel, post hole digger, oakfield auger or bucket auger.

9.4.6.2 Scheduling and Number of Soil Samples

The sampling schedule and number of samples to be collected may depend on management factors, but a schedule may be conveniently chosen to coincide with unsaturated zone soil core sampling. For systems which will be loaded heavily in a short period, more (and more frequent) samples may be needed to assure that the waste is being applied uniformly, and that the system is not being overloaded. About seven to ten samples from each selected 1.5 ha (4 ac) area should be taken to represent the treatment zone, and these should be composited to obtain a single sample for analysis. In addition, if there are evidently anomalous "hot spots," these should be sampled and analyzed separately.

9.4.6.3 Analysis and Use of Results

Parameters to be measured include pH, soil fertility, residual concentrations of degradable rate limiting constituents (RLC), and the concentrations of residuals which limit the life of the disposal site (CLC), plus those which if increased in concentration by 25% would become limiting. Hazardous constituents of concern should also be monitored. Based on the data obtained, the facility management or design can be adjusted or actions taken as needed to maintain treatment efficiency. Projections regarding facility life can also be made and compared to original design projections. Since the treatment zone acts as an integrator of all effects, the data can be invaluable to the unit operator.

9.4.7 Air Monitoring

The need for air montitoring at a land treatment unit is not necessarily dictated only by the chemical characteristics of the waste. Wind dispersal of particulates can mobilize even the most immobile, nonvolatile hazardous constituents. Therefore, it is suggested that land treatment air emissions be monitored at frequent intervals to ensure the health and safety of workers and adjacent residents. This effort may be relaxed if the air emissions are positively identified as innocuous compounds or too low in concentration to have any effect. In any case, although air monitoring is not currently required, it is strongly suggested since this is a likely pathway for pollutant losses from a land treatment unit.

Sampling generally involves drawing air over a known surface area, at a known flow rate for a specified time interval. Low molecular weight volatiles may be trapped by solid sorbents, such as Tenax-GC. The high molecular weight compounds may be sampled by Florisil, glass fiber filters, or polyurethane foam.

CHAPTER 9 REFERENCES

Brown, K. W. 1977. Accumulation and passage of pollutants in domestic
septic tank disposal fields. Draft report to Robert S. Kerr, Environ.
Research Lab. EPA.

Corey, P. R. 1974. Soil water monitoring. Unpublished report to Dept. of
Agr. Eng. Colorado State Univ. Ft. Collins, Colorado.

Dazzo, F. B. and D. F. Rothwell. 1974. Evaluation of procelain cup water
samplers for bacteriological sampling. Applied Micro. 27:1172-1174.

Duke, H. R. and H. R. Haise. 1973. Vacuum Extractors to assess deep percolation losses and chemical constituents of soil water. Soil Sci. Soc. Am.
Proc. 37:963-4.

EPA. 1977. Procedures manual for groundwater monitoring at solid waste
disposal facilities. U.S. EPA Office of Solid Waste. SW-616.

EPA. 1980. Hazardous waste management systems; identification and listing
of hazardous waste. Federal Register Vol. 45, No. 98, pp.33084-33133. May
19, 1980.

EPA. 1981. Criteria and standards for the national pollutant discharge
elimination system. Title 40 Code of Federal Regulations Part 125. U.S.
Government Printing Office. Washington, D.C.

EPA. 1982a. Hazardous waste management system; permitting requirements for
land disposal facilities. Federal Register Vol. 47, No. 143. pp. 32274-
32388. July 26, 1982.

EPA. 1982b. Test methods for evaluating solid waste. U.S. EPA, Office of
Solid Waste. Washington, D.C. SW-846.

EPA. 1982c. Ground-water monitoring guidance for owners and operators of
interim status facilities. U.S. EPA, Office of Solid Waste and Emergency
Response. Washington, D.C. SW-963.

Grover, B. L. and R. E. Lamborn. 1970. Preparation of porous ceramic cups
to be used for extraction of soil water having low solute concentrations.
Soil Sci. Soc. Am. Proc. 34:706-708.

James, T. E. 1974. Colliery spoil heaps. pp. 252-255. In J. A. Coler (ed.)
Groundwater pollution in Europe. Water Information Center. Port Washington,
New York.

Levin M. J. and D. R. Jackson. 1977. A comparison of in situ extractors for
sampling soil water. Soil Sci. Soc. Amer. J. 41:535-536.

Manbeck. D. M. 1975. Presence of nitrates around home waste disposal sites.
Annual meeting preprint Paper No. 75-2066. Am. Soc. Agr. Engr.

Nassau-Suffolk Research Task Group. 1969. Final report of the Long Island groundwater pollution study. New York State Dept. of Health. Albany, New York.

Parizek, R. R. and B. E. Lane. 1970. Soil-water sampling using pan and deep pressure-vacuum lysimeters. J. Hydr. 11:1-21.

The Resources Agency of California. 1963. Annual report on dispersion and persistence of synthetic detergent in groundwater, San Bernadino and Riverside Counties. In a report to the State Water Quality Control Board. Dept. of Water Resources. Interagency Agreement No. 12-17.

Scalf, M. R., J. F. McNabb, W. J. Dunlap, R. L. Cosby, and J. Fryberger. 1981. Manual of ground-water sampling procedures. National Water Well Association, Worthington, Ohio. 93 p.

Silkworth, D. R. and D. F. Grigal. 1981. Field comparison of soil solution samplers. Soil Sci. Soc. Am. J. 45:440-442.

Smith, J. L., D. B. McWhorter, and R. C. Ward. 1977. Continuous subsurface injection of liquid dairy manure. EPA-600/2-77-117. PB 272-350/OBE.

Trout, T. J., J. L. Smith, and D. B. McWhorter. 1975. Environmental effects of land application of digested municipal sewage sludge. Report submitted to city of Boulder, Colorado. Dept. of Agr. Engr. Colorado State Univ., Ft. Collins, Colorado.

Tyler, D. D. and G. W. Thomas. 1977. Lysimeter measurements of nitrate and chloride losses and no-tillage corn. J. Environ. Qual. 6:63-66.

USDA. 1975. Soil taxonomy, a basic system of soil classification for making and interpreting soil surveys. Soil Conservation Service USDA Agriculture (Handbook No. 436. U.S. Government Printing Office, Washington, D.C.

USDA. 1979. Field manual for research in agricultural hydrology. USDA Agricultural Handbook No. 224. U.S. Government Printing Office, Washington, D.C.

USDI, Bureau of Reclamation. 1977. Groundwater manual. U.S. Government Printing, Washington, D.C.

Wagner, G. H. 1962. Use of porous ceramic cups to sample soil water within the profile. Soil Sci. 94:379-386.

Wengel, R. W. and G. F. Griffen. 1971. Remote soil-water sampling technique. Soil Sci. Soc. Am. Proc. 35:661-664.

Wilson, L. G. 1980. Monitoring in the vadose zone: a review of technical elements and methods. U.S. EPA. EPA-600/7-80-134.

Wood, W. W. 1973. A technique using porous cups for water sampling at any depth in the unsaturated zone. Water Resources Research. 9:486-488.

Wood, A. L., J. T. Wilson, R. L. Cosby, A. G. Hornsby, and L. B. Baskin. 1981. Apparatus and procedure for sampling soil profiles for volatile organic compounds. Soil Sci. Soc. Am. J. 45:442-444.

CONTINGENCY PLANNING AND OTHER CONSIDERATIONS

Beth D. Frentrup
K. W. Brown
David C. Anderson

Managers of all hazardous waste management facilities must take precautions to safeguard the health of both workers and nonworkers during normal facility operation and in the event of an environmental emergency. Routine health and safety considerations are discussed in Section 10.1. Preparedness and prevention measures and contingency plans appropriate for HWLT units are also discussed. Figure 10.1 indicates the key points considered by the permit evaluator. During the active life of an HWLT unit, changes in the management or operation of the unit may be made that require updating the closure plan. In some cases, changes in the waste stream being disposed may require modification of the permit as well as changes to management and closure plans. Changing waste streams are considered in Section 10.4. Requirements for contingency planning and other health and safety concerns are given in the EPA regulatons (EPA, 1980; EPA, 1981) and are discussed below.

10.1 ROUTINE HEALTH AND SAFETY

Although the management plans for HWLT units are designed to reduce the hazards associated with the particular waste being disposed (Chapter 8), there are some additional health and safety considerations that need to be specifically addressed. The type and amount of employee training necessary to safeguard human health and reduce environmental impacts from sudden or nonsudden releases of contaminants are based on the characteristics of the waste. Routine health and safety procedures must be developed and followed at all times. To protect the health of the nonworker population, access to the HWLT unit should be restricted.

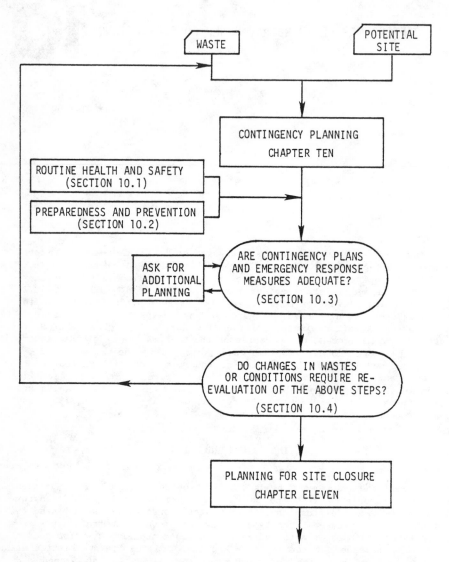

Figure 10.1. Contingency planning and additional considerations for HWLT units.

10.1.1 Site Security

The necessary site security measures vary with the location of the facility, the presence or absence of on-site storage, and the nature of the wastes being disposed. There are, however, certain minimum standards that apply to all HWLT units. For example, access to the site must be controlled at all times. At a minimum this may require fencing the entire HWLT site. When unknowing entry will not cause injury to people or livestock barbed wire fences are generally sufficient for the outer perimeter but fences intended to exclude people may be desirable around storage facilities, runoff retention ponds and office buildings. In heavily populated areas where the public can easily gain access, fences to exclude people may be needed around the entire perimeter to keep children and others off the site.

Appropriate warning signs designed to keep out unauthorized personnel should be posted at the main facility entrance, at all gates and at intervals along the site perimeter where access could be made by foot. Traffic control should be established to restrict unauthorized entry either through use of gates or a surveillance system. When the land treatment area is adjacent to the industrial plant where wastes are generated and where access can be gained only by passing through normal plant security, no further actions may be needed to restrict access.

10.1.2 Personnel Health and Safety

Events that endanger the health of workers at land treatment units include accidents while operating heavy equipment, fires and explosions. Exposure to toxic or carcinogenic wastes is also of concern since acute and chronic health effects may occur if proper precautions are not taken. The U.S. Occupational Safety and Health Administration (OSHA) has the primary responsibility for determining the adequacy of working conditions to ensure employee safety. This agency has developed specific operational criteria for most situations in the work place and may be consulted during the development of safety standards for a specific HWLT unit. Quick medical attention is often critical; an excellent guide to first-aid information is American Red Cross Standard First Aid and Personal Safety published by Doubleday and Company, Inc. It deals with such topics as heavy bleeding, stopped breathing, artificial respiration, shock, poisoning, burns, eye damage, heat stroke, and moving injured victims.

Accidents, fires and explosions often occur as a result of carelessness or vandalism and can therefore be reduced through proper training (Section 10.1.3) and controlled access (Section 10.1.1). Probably the most common cause of injury at land treatment units is operator error while handling heavy equipment; however, by following standard operating procedures, accidents such as these can be minimized. Fires are a continuous threat at facilities handling flammable wastes; waste storage areas may be set afire by vandalism, carelessness, sparks from vehicles or even spontaneous combustion. All sources of ignition including vehicles (where possible) and cigarette smoking should be prohibited near waste storage areas. Because the possibility of spontaneous combustion is greatly enhanced on very hot days, it may be advisable to keep certain storage tanks cool by continuous spraying with water or by a permanent cooling system. Waste storage areas and the actual land treatment area may be sources of explosive gases. Products of hazardous waste decomposition, oxidation, volatilization, sublimation or evaporation may include gases that are explosive. In sufficient concentrations, these low flash point gases might cause employee injury during tilling and waste spreading operations as well as during storage or handling operations. Fires, explosions or releases of toxic gases can also result from mixing incompatible wastes. Section 8.9 deals with this subject in detail and includes tables that can be used to determine incompatible waste combinations.

Acute or chronic exposure to toxic wastes may cause immediate sickness or long-term illness. Many wastes give off toxic vapors during storage or when they are applied to the soil. A simple respirator is often sufficient to eliminate the dangers associated with breathing these vapors. Long-term carcinogenic risks may be harder to protect against. If the hazardous waste being handled is known to be carcinogenic or acutely toxic, special protection is needed. Information on protective equipment may be obtained from the OSHA.

10.1.3   <u>Personnel Training</u>

As mentioned in the previous section, many sources of worker injury can be reduced through proper training. Training should be designed to ensure that facility personnel are able to respond effectively during an emergency and are able to implement contingency plans (Section 10.3). In addition to training sessions on standard operating procedures and use of equipment, two additional types of specialized training are appropriate for HWLT facility perosnnel, as follows:

(1) familiarization with the possible equipment or structure deterioration or malfunction scenarios that might lead to environmental or human health damages; and

(2) procedures for inspecting equipment and structures to determine the degree of deterioration or probability of malfunction.

10.2   PREPAREDNESS AND PREVENTION MEASURES

Preparedness and prevention measures are intended to minimize the possibility and effects of a contaminant release, fire or explosion which could threaten human health or the environment. Good management practices are the basis of preparedness; HWLT units should be operated to minimize the likelihood of spills, fires, explosions, or any other discharge or release of hazardous waste. Management concerns for HWLT are discussed in Chapter 8. Specific preparedness and prevention measures include adequate communications, arrangements with local authorities and regulatory agencies, and proper emergency equipment. Additionally, aisle space and roads should be clear and maintained to allow the unobstructed movement of emergency response personnel and equipment to any area of the facility at all times.

10.2.1 Communications

The following two types of communications systems may be needed at HWLT units (40 CFR 264.34; EPA, 1980):

(1) an internal communications or alarm system that is capable of providing immediate emergency instructions to facility employees; and

(2) a device capable of summoning external emergency assistance from local response agencies (e.g., telephone or 2-way radio).

Whenever hazardous waste is being mixed, poured, spread or otherwise handled, all personnel involved in the operation must have immediate access to an internal alarm or emergency communication device, either directly or through visual or voice contact with another employee. In addition, if there is ever only one employee on the premises while the facility is operating, he must have immediate access to a device, such as a telephone (immediately available at the scene of operation) or a hand held two-way radio, capable of summoning external emergency assistance.

10.2.2 Arrangements with Authorities

It is advisable to make arrangements to familiarize local and state emergency response authorities (such as police, fire, health, and civil defense officials) with the following:

(1) the layout of the unit;

(2) entrance to roads inside the unit that could be used as possible evacuation routes;

(3) places where personnel would normally be working; and

(4) the quantities and properties of the hazardous waste being handled at the unit along with any associated hazards.

When more than one police and fire department might respond to an emergency, an agreement should be made designating primary emergency authority to a specific department. This should be accompanied by agreements with other agencies to provide support to the primary emergency authority. Agreements should also be made with state emergency response teams, emergency response contractors, and equipment suppliers for their services or products if there is a potential need for these.

Arrangements should be made to familiarize local hospitals with the properties of the hazardous waste handled at the unit and the types of injuries or illnesses which could result from fires, explosions, waste releases, or other emergency related events.

All of the above arrangements agreed upon by local police departments, fire departments, hospitals, contractors, and state and local emergency reponse teams to coordinate emergency services should be included in the contingency plan for the HWLT unit (Section 10.3). In addition, a continuously updated list of names, addresses, and phone numbers (office and home) of all persons qualified to act as the emergency coordinator should be included in the contingency plan. Where there is more than one person listed, one must be named as the primary emergency coordinator and the home) of all persons qualified to act as the emergency coordinator should be included in the contingency plan. Where there is more than one person listed, one must be named as the primary emergency coordinator and the others must be listed in the order in which they will assume responsibility as alternates.

10.2.3   Equipment

To facilitate a quick response during an emergency, a continuously updated list of emergency equipment available at the unit should be kept. This list should include the location and physical description of each item and a brief outline of its capabilities.

10.2.3.1 Required Emergency Equipment

Federal regulations require certain types of emergency equipment to be maintained on-site (40 CFR 264.32; EPA, 1980). The types of communication equipment required are discussed in Section 10.2.1. The following equipment should also be maintained on-site:

(1) portable fire fighting equipment including special extinguishing equipment adapted to the type of waste handled at the facility;

(2) spill control equipment;

(3) decontamination equipment; and

(4) water in an adequate volume and pressure to deal with emergency situations.

10.2.3.2 Additional Equipment

In addition to the emergency equipment required by federal regulations, there are several other types of emergency equipment or material
that are specifically needed at HWLT sites. Materials that may be needed
on-site include the following:

(1) bales of hay and other materials that could be used as tem-
     porary barriers and as absorbents to soak up or slow the
     spread of spilled or accidentally discharged materials;

(2) sand bags and other materials that could be used for filling
     or blocking overflow channels in waste storage or water re-
     tention facilities;

(3) auxiliary pumps and pipelines to move or spray-irrigate ex-
     cess water to prevent overflow of retention facilities;

(4) appropriate boots, rain gear, gloves, goggles, and gas res-
     pirators for personnel;

(4) appropriate boots, rain gear, gloves, goggles, and gas res-
     pirators for personnel;

(5) basic hand tools to make "quick response" repairs to damaged
     or deteriorating equipment or structures; and

(6) lists of the closest emergency equipment suppliers or con-
     tractors (including sources of large vacuum trucks, and/or
     waterproofed dump trucks) to receive spill debris.

Plans and equipment should be available for removing, retaining, or
redistributing previously applied waste. This may become necessary where
waste has been accidentally applied at too high a rate or where waste which
has been applied is found to differ from that for which the application
rates were developed. Additionally, plans and equipment should be available to deal with the full variety of natural and man-made disasters which
may occur. Examples of these disasters include excessive rainfall, soil
overloads and surface water or groundwater contamination. When materials
are spilled in transit or in nontreatment areas of the facility, cleanup
will require the types of equipment described above.

10.2.3.3 Inspection and Maintenance

Development of and adherence to a written schedule for inspecting all
monitoring equipment, safety and emergency equipment, security devices, and
operating and structural equipment (such as dikes, waste storage or handling equipment, and sump pumps) that are important to preventing, detecting,
or responding to environmental or human health hazards is critical. The
frequency of these inspections is based on the rate of possible deterioration or malfunction of the equipment and the probability of an environmental or human health incident if the deterioration, malfunction, or an

operator error goes undetected between inspections. Areas subject to spills (such as waste loading, unloading and storage areas) should be inspected at least daily while they are in use. Any deterioration or malfunction of equipment or structures should be corrected to ensure that the problem does not lead to an environmental or human health hazard. Where a hazard is imminent or has already occurred, remedial action must be taken immediately.

10.3 CONTINGENCY PLANS AND EMERGENCY RESPONSE

Contingency plans and emergency responses are intended to minimize hazards to human health due to emergencies such as fire, explosions, or any unplanned sudden or nonsudden release of hazardous wastes to air, soil, groundwater or surface water. The plan must be carried out immediately whenever such an emergency occurs and should describe the actions that facility personnel must take. Copies of the contingency plan (and any revisions to the plan) should be maintained at the HWLT unit and supplied to all state and local emergency response authorities. At a minimum the plan should include the following (40 CFR 264.52; EPA, 1980):

(1) arrangements agreed upon with local and state emergency response authorities (Section 10.2.2);

(2) a continuously updated list with names and phone numbers of the people qualified to act as the emergency coordinator (Section 10.3.1);

(3) a continuously updated list of emergency equipment available on-site (Section 10.2.3); and

(4) an evacuation plan for personnel including signals to be used to begin evacuation, evacuation routes and alternate evacuation routes (in cases where the primary routes may be blocked as a result of the emergency situation).

The contingency plan and should be reviewed on a regular basis and amended as necessary. Examples of situations that would require amending the contingency plan include the following:

(1) the applicable regulations are revised;

(2) the plan fails in an emergency;

(3) the facility changes (in its design, operation, maintenance or in any way that would change the necessary response to an emergency);

(4) the list of emergency coordinator changes; and

(5) the list of emergency equipment changes.

At least one of the qualified emergency coordinators should be at the HWLT site or on call (i.e., available to respond to an emergency by reaching the site within a short period of time) at all times. The emergency coordinator has the responsibility for coordinating all emergency response measures. Specific responsibilities of the emergency coordinator are as follows:

(1) to be familiar with all aspects of the contingency plan, all operations and activities at the facility, the location and characteristics of the hazardous waste handled by the facility, the location of all records within the facility, and the facility layout;

(2) to have the authority and be able to commit the resources needed to carry out the contingency plan;

(3) to activate internal facility alarms or communication systems in case of emergency;

(4) to notify the appropriate emergency response authorities;

(5) to immediately identify character, exact source, amount, and extent of any released materials; and

(6) to immediately assess possible hazards to human health or the environment that may result from the emergency situation including both direct (fire, explosions, comtaninant releases) and indirect (generation of asphyxiating gas or contaminated runoff) effects of the emergency.

If, during an emergency response, the emergency coordinator determines that there may be a threat to human health or the environment outside the facility, he must report these findings. If his assessment indicates that evacuation is advisable, he must immediately notify the appropriate local authorities and be prepared to assist them in assessing whether local areas need to be evacuated. In addition, he must immediately notify either the government official designated as the on-scene coordinator for that geographical area or the National Response Center (using their 24-hour toll free number: 1-800-424-8802). His report should include the following:

(1) name and telephone number of reporter;

(2) name and address of the facility;

(3) time and type of accident;

(4) name and quantity of material involved;

(5) extent of injuries; and

(6) possible hazards to human health or the environment outside the facility.

During the emergency, he should take all reasonable measures so that fires, explosions, and waste releases do not occur, recur, or spread to other hazardous waste at the HWLT unit.

Immediately after an emergency, the emergency coordinator must provide for the treatment, storage or disposal of the recovered waste, contaminated soil or surface water, or any other material that results from the emergency (40 CFR 264.56; EPA, 1980). He must ensure that (in the affected areas of the facility) no wastes that may be incompatible with the released material are stored, disposed or otherwise handled until the released material is completely cleaned up. In addition, before operations resume, all emergency equipment listed in the contingency plan must be cleaned, refilled and made ready for its intended use. To prevent repetition of the emergency, the coordinator may need to do the following, where applicable:

(1) reject all future deliveries of incompatible waste;

(2) correct facility deficiencies;

(3) improve spill control structures;

(4) obtain proper first aid or other emergency equipment to address identified deficiencies; and

(5) retrain or dismiss responsible employees.

Before operations can resume, the owner or operator must notify the proper federal, state and local authorities that all cleanup procedures are complete and all emergency equipment is restored and ready for its intended use. The owner or operator must also record the time, date, and details of any incident that requires implementation of the contingency plan and, within fifteen days of the incident, he must submit a written report on the incident to the appropriate regulatory agency that includes the following:

(1) name, address, and telephone number of the owner or operator;

(2) name, address and telephone number of the facility;

(3) date, time, and type of incident;

(4) name and quantity of material(s) involved;

(5) the extent of injuries, if any;

(6) an assessment of actual or potential hazards to human health or the environment, where applicable; and

(7) estimated quantity and disposition of recovered material that resulted from the incident.

10.3.2   Specific Adaptations to Land Treatment

In addition to the general contingency plans discussed above that apply to all types of hazardous waste management facilities, some problems or emergency responses are uniquely characteristic of HWLT systems. Such contingences should be recognized and specifically addressed in an HWLT permit.

10.3.2.1 Soil Overloads

The capacity of the soil to treat and dispose of wastes may be overloaded despite the best of plans. There is always the possibility that occasional shipments of wastes will contain constituents which the facility was not designed to handle or in concentrations which exceed the designed application rates. In some cases it may be possible to see or smell that the waste is off-specification and, in such cases, it should be placed in a placed in a special holding basin or area. The waste should be sampled and analyzed before it is applied to the soil. In other cases, the differences may not be observed until the waste is applied to the land. In such instances, as much of the waste as possible should be picked up and placed in the off-specification holding area. In other instances, it may not be possible to pick up the waste and remedial treatment may be necessary.

Areas that need remedial treatment can often be identified because they have a different color or odor, remain wetter or drier, or do not support vegetation. On-site observations combined with reports from soil samples sent for analysis should be sufficient to determine the source of the problem. Several options for remedial measures to deal with waste "hot spots" are discussed below. One option is to physically remove the material and store the soil in an off-specification storage area until it can be analyzed to determine if the material can be respread over a larger area and degraded, or if it should be disposed elsewhere.

Certain remedial treatments and changes in HWLT management may be used to overcome the problem without removing the soil. Acids or bases may be used to neutralize areas which have become too basic or acidic. In most cases, it is advisable to use HCl or $CaCO_3$ or other neutralizing agents selected to avoid the accumulation of excess salts. If excessive sodium (Na) salts are causing the problem, it may be possible to overcome the problem by applying $CaSO_4$ or $CaCO_3$ to replace the Na with Ca. When excessive volatile organic materials cause a problem, it may be advisable to apply and incorporate powdered activated charcoal or other organic materials to adsorb and deactivate the chemicals until they can be degraded in the HWLT system. Where excessive amounts of oil have been applied, decomposition can often be enhanced by incorporating appropriate amounts of nutrients (particularly nitrogen) and hay or straw, which will help loosen the soil, absorb the oil, and allow oxygen to enter the system. In some instances where hot spots are small, it may be possible to solve the problem by spreading the treated soils over a larger area and subsequently regrading to eliminate any depressions.

In a few cases, however, a soil may become so overloaded with a toxic inorganic or nondegradable organic chemical that it is not economically feasible or environmentally sound to spread the soil over a larger area as a remedial measure. If there is no feasible on-site treatment that will alter the contaminated soil sufficiently to render it nonhazardous, the zone of contamination should be removed and disposed in a landfill authorized to accept hazardous waste. The zone of contamination will include the soil in the treatment area at least down to the depth of the waste incorporation (20 to 60 cm) and any additional underlying soil that is also contaminated.

10.3.2.2 Groundwater Contamination

The potential for migration of waste constituents to groundwater can be predicted from pilot studies (Sections 7.2.2 and 7.4) performed before land treatment of the waste begins. Thus, the facility can be designed to minimize this potential through waste pretreatment, in-plant process controls to reduce, eliminate, or alter the form of the waste constituents, or soil amendments. Groundwater contamination may occur at HWLT facilities when water percolates through soil if contaminants occur in leachable forms. Water enters contaminated soil in the treatment zone from direct precipitation, surface water run-on, applied wastes containing water, and from irrigation of the land treatment area to enhance waste biodegradation or cover crop growth. Where groundwater contamination occurs, remedial actions can be very extensive and costly. Hence, the key to minimizing the impact of the contamination incident and the resulting expenses is the early detection of contaminant migration. This can be accomplished through the proper use of unsaturated zone monitoring discussed in Chapter 9.

If the waste constituent that is leaching has not yet reached the groundwater, contingency plans may involve pressure-injecting a bowl-shaped grout bottom seal above the groundwater table and below the zone of contamination. The leachate contained by the bowl-shaped seal can then be pumped out and treated or land treated at rates that preclude water percolation. Further information is available in the publication, entitled Technical Information Summary: Soil Grouting, (Applied Nucleonics Company, Inc., 1976). Cost estimates for constructing portland cement bottom seals are given in Table 10.1. In some cases, it may be possible to remove the zone of waste incorporation to cut off the source of the leachate. Soil and waste in the zone of incorporation could then be disposed at another location.

TABLE 10.1 COSTS OF CONSTRUCTING A PORTLAND CEMENT BOTTOM SEAL UNDER AN
ENTIRE 10 ACRE (4.1 HECTARE) LAND TREATMENT FACILITY*

Thickness of injected grout layer		Voids in soil receiving grout	Cost of portland cement cement bottom liner
Meters	Feet	(%)	(Millions of 1978 dollars)
1.2	4	20	1.115 - 2.786
1.2	4	30	1.672 - 4.180
1.8	6	20	1.667 - 4.166
1.8	6	30	2.500 - 6.250

* Tolman et al. (1978).

If the leaching waste constituents have already reached the groundwater, the leachate may be recoverable downgradient from the land treatment
facility by using a well point interception system. This involves installation of short lengths of well screen on 5-8 cm diameter pipe that extend
into the water table. These well points should be spaced on 90 to 150 cm
centers (depending on the soil permeability) downgradient from the area of
leachate infiltration (Tolman et al., 1978). If suction extraction is
used, the depth of extraction is limited to 10 m. For extraction of
leachate from greater depths, air injection pumps may be required.

10.3.2.3 Surface Water Contamination

Surface water contamination may occur due to a break or leak in the
earthen wall of a water or waste retention facility or due to water runoff
from a treatment area. These problems can generally be avoided and
remedied with readily available earth moving or excavating equipment and
suitable fill material.

Prevention is the best approach to surface water pollution, as previously described in Section 8.3 and summarized below. To prevent surface
water from running onto active treatment areas, earthen berms or excavated
diversion ditches should be constructed upslope of active areas to direct
the water toward natural drainage ways downslope from the treatment area
(Tolman et al., 1978). These structures should be designed to control and
withstand water from the 25-year 24-hour storm. To prevent contaminated
water from leaving the land treatment unit, earthen berms or excavated
diversion ditches should be constructed to establish drainage patterns
which direct the water into the appropriate water retention facility. With
this in mind, water retention facilities should be constructed at the
lowest possible downslope position within the HWLT unit boundary while

leaving enough buffer area to permit access of emergency vehicles between the facility boundary and the retention pond.

Breaks or leaks in water diversion or storage facilities can be remedied by placing sandbags or fill material at the problem area. To prevent this problem from recurring, vegetation should be established on the sides of the diversion or storage structures. However, the vegetation may take a year to become fully established, so it may be necessary to use mulching and hay bales to maintain soil stability in the meantime.

Overflow of water or waste storage facilities usually can be overcome by sandbagging the low side wall. Unless the overflow is caused by an extraordinary event (i.e., one-time waste load, hurricane, or a 100-year storm), the owner or operator should immediately consider enlarging the existing water and/or waste capacity at the HWLT unit.

10.3.2.4 Waste Spills

Waste spills may affect soils, surface water and groundwater and, consequently, procedures developed in the sections dealing with soil overloads, surface water contamination, and groundwater contamination may all be important when dealing with spills. Spills of volatile wastes may also cause air quality problems. In the case of spills, rapid action is the key to limiting environmental damage.

If the spill occurs while the waste is being transported to the land treatment unit, the appropriate emergency equipment should immediately be dispatched to the scene. This equipment may include sandbags or fill dirt to check the spread of the spilled material, a vacuum truck to remove liquids from surface pools, and a backhoe or front-end loader and a waterproof dump truck to begin the excavation and removal of contaminated soil. If the waste was spilled at the land treatment unit, it may be a relatively simple matter to excavate the contaminated soil and respread it within the actual treatment area. If solid debris such as lumber pallets or trash are contaminated with the hazardous materials, they may also be disposed on-site after being ground.

Specialized equipment may be needed for some types of hazardous waste spills. The response time to spills of volatile wastes is particularly important to minimize air pollution. Techniques for handling spills of volatile hazardous substances have been reviewed (Brown et al., 1981). The use of dry ice or liquid nitrogen to cool the spill to reduce volatilization and the use of vapor containment methods were found to be most effective for dealing with volatile spills (Brown et al., 1981). If the spilled material is flammable, appropriate extinguishing equipment is needed at the accident site. If the material is toxic, breathing gear and protective clothing will be needed for all personnel active in the cleanup operations. If the spill involves explosive materials, an effort should be made to determine if there are deactivating procedures to reduce the chance of explosion. In any of these cases, area evacuation may be advisable. Where public health is threatened, the speed and appropriateness of the emergency response is of special importance.

For spills of oily liquids on soil, an approximation can be made for the volume of soil required to immobilize a known volume of the liquid (Davis, 1972), as follows:

$$V_s = \frac{0.20 \ (V_o)}{(P) \ (S_r)} \tag{10.1}$$

where

V_s = Volume of soil in cubic yards (1 yd^3 = 0.76 m^3);
V_o = Volume of liquid in barrels;
P = Porosity of the soil (percent); and
S_r = Residual oil saturation of the soil (percent).

Residual saturation (S_r) values which may be used in the equation are 0.10 for light oil or gasoline, 0.15 for light fuel oil or diesel, and 0.20 for heavy fuel oil or lube oil (Davis, 1972).

10.3.2.5 Fires and Explosions

Fires and explosions are ever present threats where hazardous materials are stored, disposed or otherwise handled. Safe handling of these wastes requires a knowledge of their physical and chemical properties. This information, as well as an understanding of any dangers associated with the waste, such as flammability, shock sensitivity and reactivity, should be obtained prior to transporting, storing, or disposing hazardous wastes. Where ignitable waste is to be land treated, subsurface injection is the suggested application technique. Subsurface injection reduces the rate of flammable vapor release and decreases the possibility that ignitable gases will accumulate to critical concentrations in the air at the HWLT unit. Timing applications to correspond with cooler weather will help to minimize the risks associated with treating ignitable wastes.

Flash point and ignition temperature are the most commonly used indicators of the hazards associated with ignitable materials. Although liquids do not burn, the flammable vapors given off by the stored or handled liquids can cause fires or explosions (Stalker, 1979). These low flash point vapors given off from hazardous wastes can travel long distances downwind or downhill to reach an ignition source and then flash back (NFPA Staff, 1979). Fires involving unconfined liquids resulting from a spill, leak, or storage vessel overflow may spread over a much greater area than is represented by the extent of the flammable liquid spill. During emergencies involving ignitable materials, immediate evacuation may be necessary to save lives.

Three types of explosions are possible at HWLT units. Combustion explosions involve the quick combination of flammable vapors with air where heat, light and an increase in pressure result. To explode, the flammable vapor and air must be within the explosive range and then ignited. Detonation explosions are similar to combustion explosions except the heat release is considerably higher for the detonation explosion and is accompanied by a shock wave that moves at approximately 1.5 to 8 km per second (Stalker, 1979). Boiling-liquid, expanding-vapor explosions (BLEVE) occur when sealed containers of flammable liquids are heated past their boiling points by an external heat source. The explosion occurs when released vapors are ignited by the external heat source. Explosions generally occur only in poorly ventilated areas where one of the following conditions exists (Stalker, 1979):

(1) the flash point of the liquid is less than −6.7°C;

(2) the flash point of the liquid is less than 43°C and the liquid is heated to greater than 16°C above its flash point; or

(3) the flash point of the liquid is less than 150°C, and the liquid is heated above its boiling point.

Sensitivity to shock is another important factor to consider when handling, storing or disposing explosives such as organic peroxides or wastes from the explosives industry. Another cause of explosions is the occurrence of a critical dust concentration in the presence of an ignition source.

The potential for an explosion can be minimized by the following:

(1) prevent a critical dust or vapor concentration from occurring;

(2) eliminate sources of ignition;

(3) keep all work areas well ventilated;

(4) train facility personnel about the dangers; and

(5) post warnings in critical areas.

Although fires and explosions are very similar processes, there is a difference in the speed of the reaction. With explosions, the event is almost instantaneous and hence cannot be controlled. This makes preventive measure even more important.

10.4   CHANGING WASTES

Since land treatment is a dynamic process, the demonstration of effective treatment considers the interaction of given waste applied to a particular treatment site. Not only is the waste altered by treatment, but the waste residuals continually change the character of the treatment medium. The characteristics of the waste and the specific waste-soil interactions form the basis for design and management decisions. Permits are also issued to HWLT units based on specific waste-soil combinations. Consequently, if waste stream characteristics change or if new wastes are substituted or added to the waste mixture being applied to the soil, changes may be necessary in both the design and management of the HWLT unit and permit modifications may also be required.

Assessing the capacity of an HWLT unit to accept a different waste often involves calculating a new application rate based on the new waste-soil combination (Chapter 7). In the case of a drastic change in waste characteristics, a complete facility redesign may be required. Waste characterization and pretreatment options should be reevaluated using the new waste mixture. To show that the goal of land treatment will be met, additional laboratory and/or field studies may be necessary to demonstrate that the wastes will be made less hazardous. If the soil is already in use for waste treatment, the demonstration must use the loaded soil and account for accumulated waste constituents. Modifications to the management, monitoring, contingency, and site closure plans may also be necessary.

CHAPTER 10 REFERENCES

Applied Nucleonics Company, Inc. 1976. Technical information summary: soil grouting. Prepared for U. S. Environmental Protection Agency. 15 p.

Brown, D., R. Craig, M. Edwards, N. Henderson, and T. J. Thomas. 1981. Techniques for handling landborne spills of volatile hazardous substances. EPA-600/S2-81-207. PB 82-105-230.

Davis, J. B. 1972. The migration of petroleum products in soil and groundwater: principles and counter measures. Am. Petr. Inst. Washington, D.C.

EPA. 1980. Hazardous waste and consolidated permit regulations. Federal Register Vol. 45, No. 98, pp. 33066-33258. May 19, 1980.

EPA. 1981. Hazardous waste management system; addition of general requirements for treatment, storage and disposal facilities. Federal Register Vol. 46, No. 7, pp. 2802-2897. January 12, 1981.

NFPA Staff. 1979. Industrial waste control. p. 901-918. In Gordon P. McKinnon (ed.) Industrial fire hazards handbook. National Fire Protection Assoc., Inc. Boston, Massachusetts.

Stalker, R. D. 1979. Flammable and combustible liquid handling and storage. p. 719-743. In Gordon P. McKinnon (ed.) Industrial fire hazards handbook. National Fire Protection Assoc., Inc. Boston, Massachusetts.

Tolman, A. L., A. P. Ballestero, Jr., W. W. Beck, Jr., G. H. Emrich. 1978. Guidance manual for minimizing pollution from waste disposal sites. EPA-600/2-78-142. PB 299-206/AS.

Gordon B. Evans, Jr.
K. W. Brown

CLOSURE AND POST-CLOSURE 11

The satisfactory completion of a land treatment operation depends on carefully planned closure activities and post-closure care. The necessary considerations in formulating closure and post-closure plans can be described, but the point of distinction between closure and post-closure is somewhat vague. This is because land treatment closure is a continuing process rather than a set of distinct engineering procedures. An exception would be the case where the treatment zone or the contaminated portion of the treatment zone is removed and disposed in another hazardous waste facility. Certification of the completion of closure and initiation of post-closure care would be based on the approved closure plan and such things as monitoring results, the degree of treatment achieved, changes in runoff water quality, and the condition of the final cover. Following the closure certification, the post-closure care period begins, this period is characterized by decreasing management and monitoring requirements over time. Figure 11.1 indicates the various aspects of closure and post-closure care discussed in this chapter.

11.1 SITE CLOSURE ACTIVITIES

After the last load of waste is accepted for treatment, the process of closing the land treatment unit begins. In practice, management and monitoring during closure differ very little from routine management during operation. The application of stored wastes continues along with cultivation to stimulate degradation. Cultivation, fertilization, liming to assure proper pH, and possibly irrigation continue until the organic constituents are sufficiently degraded. The required degree of degradation depends on the procedure to be used for final closure. Monitoring

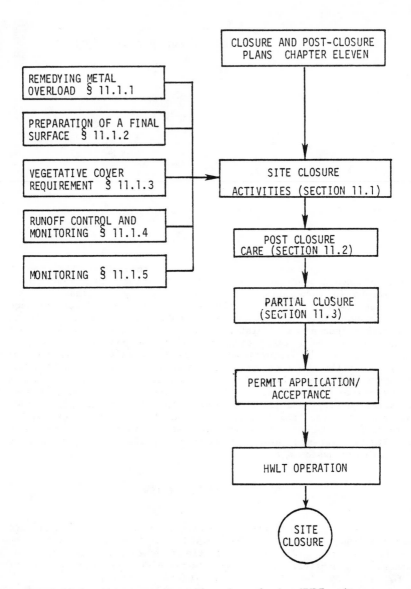

Figure 11.1. Factors to consider when closing HWLT units.

continues as before with some modification, as do run-on and runoff control. The time required for closure will vary considerably from site to site based on the rate at which waste organics are degraded and final cover is established.

11.1.1   Remedying Metal Overload

If immobile metallic elements have accumulated in the zone of waste incorporation to phytotoxic concentrations, consideration may be given to the use of deep plowing to mix the zone of incorporation with subsoil or addition of uncontaminated soil for mixing. Such a procedure will lower the concentrations of the phytotoxic elements to levels tolerated by plants. This option should be exercised only if there is sufficient field evidence that (1) the practice will not lead to mobilization of hazardous constituents, (2) deep plowing or dilution with clean materials will not disrupt a soil horizon which is instrumental in preventing migration, and (3) the organic components of the waste have degraded sufficiently to allow deeper incorporation without endangering groundwater. Furthermore, if the subsoil or the soil added has a pH below 6.5, sufficient lime to neutralize the mixed soil may need to be incorporated prior to plowing or soil additions. Greenhouse or field data should be used to determine if these actions will remedy the metal overload and allow the establishment of a permanent vegetative cover before deep plowing or dilution with uncontaminated soil is begun.

11.1.2   Preparation of a Final Surface

Closure generally requires that the treatment zone be revegetated (EPA, 1982). Planting can proceed as soon as the waste is sufficiently degraded, immobilized and detoxified to allow the establishment of a permanent vegetative cover. If the closure plan calls for the removal of the treatment zone, it will be advantageous to continue management until the last application of waste is sufficiently degraded to minimize the amount of material that needs to be removed. Whether or not material has been removed, the remaining surface should be terraced, fertilized, and limed as necessary and planted to establish vegetation. In the event the soil or subsoil exposed by removal of the treatment zone is not physically suitable to support vegetation, or if the desired contours cannot be achieved, it may be necessary to bring in additional suitable soil materials. Except for fairly level terrain, the final grade of any of the surfaces should be developed into a system of terraces and waterways to minimize erosion. The details of design procedures have been discussed in Section 8.5.

11.1.3   Vegetative Cover Requirement

Except where no significant concentrations of hazardous constituents remain in the treatment zone, the final surface must be covered with a permanent vegetative cover to prevent water and wind borne erosion and off-site transport of soil and/or waste materials (EPA, 1982). Where the soil in the treatment zone is removed or no hazardous constituents otherwise remain, a vegetative cover is not required by regulation; however, in the interest of soil erosion control, a vegetative or other cover (e.g., building construction) should be provided in any case. Following preparation of the final surface, the soil should be fertilized and limed again, if needed, and a seedbed should be prepared and planted. Depending on the season, it may be desirable or necessary to plant a temporary crop to provide a protective cover until the proper planting season for the permanent vegetation. If this is done, a clear plan must be provided for removing or destroying the temporary vegetation at the proper time in order to allow optimum conditions for establishing permanent vegetation. Guidance on the selection and establishment of permanent vegetation has been discussed in Section 8.7. Preferably, the permanent cover will consist of native, low maintenance plant species to eliminate the need for intensive long-term crop management.

11.1.4   Runoff Control and Monitoring

Along with the establishment of permanent vegetation, the collection, treatment, and on-site disposal or permitted discharge of runoff water must continue. As waste organics degrade and disturbances of the land surface decrease in frequency and effect, runoff water quality will gradually improve. This improvement is significant in two respects. First, better quality runoff means that less rigorous treatment may be needed to meet NPDES permit conditons. If a discharge permit had not been feasible before, improved runoff quality might make such a permit possible or economically more attractive. Second, when runoff monitoring reveals that water is practically free from hazardous and key nonhazardous constituents, this is one indication that closure is nearly complete and less management will be required at the HWLT unit.

11.1.4.1 Assessing Water Quality

Various criteria may be used to assess the quality of the runoff water. Certainly the runoff water should be analyzed for the hazardous constituents which were disposed at the site. Water quality criteria data should then be consulted to determine when concentrations are acceptable for direct discharge. Most states have developed discharge standards, but they often do not include guidelines on hazardous constituents and their metabolites. In general, water quality criteria depend on the type of receiving stream or the uses to be made of the receiving stream. Water

quality standards for drinking water, for irrigation, and for watering
cattle are given in Table 6.48. For organic constituents, data on the
specific biological activity should be consulted. For compounds which are
toxic to organisms present in the receiving streams, concentrations should
be less than 10% of the LD_{50}. Additional constraints will need to be
applied to compounds which are bioaccumulated or which are known to cause
genetic damage. A supplementary approach to chemical analysis of the
individual constituents and their metabolites is to use bioassay tests to
demonstrate the acceptability of runoff water quality (Section 5.3.2.4).

Classical indices of water quality, including BOD, COD, TOC, and oil
and grease, are valuable as indications of changes in the release of
organics from areas to which hazardous wastes have been applied. The
indices do not, however, adequately assess the degree of hazard, nor do
they provide assurance that the concentrations of hazardous waste constituents are decreasing, since many hazardous organic chemicals are biologically active at very low concentrations.

There is only scant information available on the concentrations of
hazardous chemicals or the biohazard in runoff water from soil which has
been treated with hazardous waste. However, there are data available for
selected pesticides which have been applied to lawns or agricultural
fields. The data have been summarized by Kaufman (1974).

Acceptability of runoff water quality for direct discharge should be
based on a series of samples taken over a period of time. Often there will
be only one or two parameters of concern. The impact of seasonal
variability on the release rate is likely to affect the data, but a general
trend should be evident. Runoff should be sampled at least quarterly on a
flow proportional basis from the entire hydrograph of a variety of
antecedent rainfall intensities and durations. Samples should be obtained
from channels leading from previously active plots to the retention ponds
rather than from grab sampling the ponds. The use of flumes or weirs along
with automated sample collectors is one possible approach. Runoff water
quality acceptability should be based on at least three consecutive
sampling events from representative storms.

11.1.4.2 Controlling the Transport Mechanisms

Chemicals applied to soil may be transported in the runoff waters
either in solution or in association with suspended particulate matter.
Water soluble organics are often rapidly degraded, so that it is anticipated that the major mode of transport will be in association with suspended particles. Thus, methods for decreasing runoff and erosion during
closure will probably decrease the amounts and concentrations of hazardous
chemicals which enter the runoff. Terracing and vegetative cover, both in
the treatment area and in adjacent buffer zones through which runoff water
will pass, may be effective in trapping suspended solids and thus decreasing transport.

The decreased concentration of organic constituents in runoff water with time is likely to depend on the mechanisms and rate of degradation. For materials which are photodegraded, the amount of material on the soil surface likely to be transported will decrease rapidly once cultivation ceases. For compounds which are metabolized by microorganisms, the decrease at the surface will depend on the impact of environmental parameters on the rate of decay. These factors and probable decay data are discussed in Section 7.2.1.

11.1.5   Monitoring

During the closure period soil core and groundwater monitoring must continue as in the operational plan. Soil-pore liquid sampling may be discontinued 90 days after the last application of waste. Runoff water monitoring (discussed above) and treatment zone monitoring are optional during closure. The treatment zone plan should be patterned after that described as optional during active land treatment unit operation (Section 9.4.6), particularly emphasizing analyses of the entire treatment zone by horizon or depth increments. Treatment zone monitoring allows the owner or operator to make a determination of the degree of degradation of hazardous constituents. This type of monitoring will also be needed to obtain a variance from certain post-closure requirements if the analyses show no significant increase over background of hazardous constituents. Even where a vegetative cover is not required, it may be important to establish vegetation to control soil erosion.

Cessation of soil-pore liquid monitoring is possible during closure due to the nature of the system and what it is intended to detect. Rapidly moving hazardous constituents are the targets for detection by the system, so movement of these constituents would logically occur very soon after the last waste application. Although soil-pore liquid monitoring may be terminated 90 days after the last waste application, it may be wise to continue monitoring these liquids until three consecutive samples are free of significant increases of hazardous constituents over background.

Monitoring of food chain crops if they are grown during closure, is also needed to provide assurance that residual materials in the soil are not being taken up by plants in concentrations that are phytotoxic or that could be bioaccumulated in animals. There is little information at this time, other than for selected pesticides and metals, on the uptake of hazardous materials by crops. If food chain crops are grown during closure, the pH must be maintained at a level sufficient to prevent significant crop uptake of hazardous constituents (e.g. pH 6.5 or greater) and all other food chain requirements must be met (EPA, 1982). Additionally, the harvested portion of the crop should be determined to be free of unacceptable concentrations of hazardous constituents.

During the post-closure period management activities are reduced. Present regulations call for continuation of post-closure activities for up to 30 years unless it can be demonstrated that a shorter period is acceptable (EPA, 1982). The intent of post-closure care at a land treatment unit is to complete waste treatment and stabilization of the remaining soil and waste residuals while checking for any unforseen long-term changes in the system. For example, if pH of a naturally acidic soil has been artificially raised to control metal mobility, gradual return to the native soil pH or some new equilibruim pH may mobilize metals.

An obvious advantage of land treatment is that wastes are degraded or otherwise made unavailable to the environment with time. Other land disposal techniques, especially landfills and surface impoundments, present long-term risks of contaminant leakage and lead to continued intensive monitoring liabilities. The post-closure monitoring schedule may be relaxed to include a decreasing number of samples over time. A land treatment unit that has been properly designed, managed, and closed should exhibit little potential for releasing undesirable constituents into the unsaturated zone or into the groundwater. A typical schedule for soil core and groundwater monitoring following the initiation of post-closure should include samples collected on a geometric progression at 1/2, 1, 2, 4, 8, 16 and 30 years. The parameters of interest should be plotted with time and additional samples should be taken, as needed, in the event unacceptable concentrations are found. Post-closure care should include activities for enhancing and sustaining treatment, and precautions for managing against unacceptable releases (e.g., run-on/runoff controls). Therefore, treatment may be completed during the post-closure care period without increased environmental risk. Soil pH, nutrient levels, and significant physical, chemical, or biological disturbances of the treatment zone may all play a major role in sustaining treatment and site stabilization. These factors should be examined and corrected periodically, if necessary, throughout the post-closure care period to ensure maintenance of treatment processes. Management should strive, however, for a system requiring only minimal attention since ultimately (after 30 years) all maintenance may cease and the system will then revert to an uncontrolled condition.

11.3   PARTIAL CLOSURE

Considerable management and expense may be involved in treatment or on-site disposal of runoff water from large areas; therefore, it may be desirable to design a land treatment unit with plots which will be carefully loaded to the CLC maximum in a few years or even one year, and then to proceed to close the area. In the meantime, waste would be applied to new plots which would be opened as needed. The system would need to be designed so that runoff water from the individual plots would be collected either in separate retention basins, or in a central retention basin. A more detailed description of this type of design is presented in Section

8.1.2. Once runoff water quality from a given plot is acceptable, its run-off can then be diverted and released under less restrictive permit conditions. Another advantage is that a portion of the unit can be released from long-term post-closure care sooner than remaining active plots. Finally, information learned through partial closure may be helpful in improving the management of active portions. The timetable for partial closure depends greatly on the rate at which the waste constituents of concern are degraded or sorbed by the soil.

CHAPTER 11 REFERENCES

EPA. 1982. Hazardous waste management system; permitting requirements for land disposal facilities. Federal Register Vol. 47, No. 143 pp. 32274-32388. July 26, 1982.

Kaufman, D. D. 1974. Degradation of pesticides by soil microogranisms. pp. 133-202. In W. D. Guenzie (ed.) Pesticides in soil and water. Soil Sci. Soc. Amer. Madison, Wisconsin.

Nash, R. G. 1974. Plant uptake of insecticides, fungicides and fumigants from soils. pp. 257-314. In W. D. Guenzi (ed.) Pesticides in soil and water. Soil Sci. Soc. Amer. Madison, Wisconsin.

APPENDIX **A**

LISTING OF HAZARDOUS WASTE LAND TREATMENT FACILITIES

The practice of land treatment for disposing of various types of wastes has been employed by industries for a considerable number of years. The petroleum refinery industry has historically been the primary industrial user, with records of organized landfarming operations dating to the early 1950's (Exxon Co. U.S.A., Personal Communication). Even predating what one would consider organized landfarming, it was recognized in a 1919 journal article that oil is degradable in soil. In the years hence, it became common practice to treat oily and leaded tank bottoms by first "weathering" them in soil to degrade the oil and oxidize the tetraethyl lead to less toxic form. However, it has not been until the last decade that land treatment was recognized as an environmentally sound and effective treatment and disposal technique which could be useful for many classes of industrial waste. Consequently, the data base for determining what constitutes a well-designed land treatment operation and which wastes are readily amenable to land treatment has been slow to develop. As the state of the art advances, some past practices have been found to be inadequate while important design and management considerations have begun to be understood.

However, many potentially land treatable wastes have not been tested, and many facilities at which land treatment is practiced have until recently lacked sufficient documentation as to their effectiveness and environmental safety. Therefore, this listing of land treatment facilities was prepared to aid the industrial user in identifying whether other companies in the given industry are using or have used land treatment. In addition, the geographical information can indicate the feasibility of land treating a waste based on climatic affects (Table A-1 through A-3). On the other hand, the failure of a given SIC code to appear on the list does not

619

Table A-1. Geographic distribution, by EPA region and state, of the 197
facilities identified in the survey.

EPA Region	Regional Office	Number of facilities
VI	Dallas, Texas	58
IV	Atlanta, Georgia	45
IX	San Francisco, California	19
VIII	Denver, Colorado	18
V	Chicago, Illinois	16
VII	Kansas City, Missouri	15
X	Seattle, Washington	12
II	New York City, New York	8
III	Philadelphia, Pennsylvania	7
I	Boston, Massachusetts	0

State or Territory	Number of facilities
Texas	29
California	18
Louisiana	13
Oklahoma	11
Ohio	9
Alabama	8
Kansas	8
Washington	8
Florida	7
Georgia	7
Mississippi	7
Montana	6
North Carolina	6
Wyoming	6
South Carolina	5
Missouri	4
Puerto Rico	4
Colorado	3
Illinois	3
Kentucky	3
New Mexico	3
Utah	3

Table A-1. (continued)

State or Territory	Number of facilities
Arkansas	2
Indiana	2
Iowa	2
New Jersey	2
Maryland	2
Minnesota	2
Pennsylvania	2
Tennessee	2
Virginia	2
Alaska	1
Delaware	1
Guam	1
Idaho	1
Michigan	1
Nebraska	1
New York	1
Oregon	1
Virgin Islands	1
American Samoa	0
Arizona	0
Commonwealth of the Northern Marianas	0
Connecticut	0
District of Columbia	0
Hawaii	0
Maine	0
Massachusetts	0
Nevada	0
New Hampshire	0
North Dakota	0
Rhode Island	0
South Dakota	0
Vermont	0
West Virginia	0
Wisconsin	0

Table A-2. Industrial classification of land treatment facilties.

SIC Code		EPA Region	State
025	Poultry Feed	IV	Tennessee
1321	Natural Gas Proc.	VI	Louisiana
1389	Oil & Gas Services	IX	California
203	Fruit Processing	IV	Florida
		IV	Florida
		IV	Florida
2067	Chewing Gum Manu.	IV	Georgia
222	Weaving Mills, Synthetics	III	Maryland
		IV	Georgia
229	Misc. Textile Goods	IV	North Carolina
		IV	South Carolina
249	Misc. Wood Products	IV	North Carolina
2491	Wood Preserving	IV	Alabama
		IV	Alabama
		IV	Mississippi
		IV	Mississippi
		VI	Texas
		VII	Missouri
2600	Paper & Allied Products	X	Washington
2611	Pulp Mills	V	Michigan
2621	Paper Mills	V	Mississippi
2819	Industrial Inorganic Chemicals	VI	Louisiana
		VI	Texas
2821	Plastics, Materials & Resins	VI	Louisiana
		VI	Texas
		VI	Texas
2834	Pharmaceutical Preparations	IV	Tennessee
2851	Paints & Allied Products	IV	Georgia
		VII	Iowa
		IX	California
2865	Cyclic Crudes & Intermediates	VI	Arkansas
2869	Industrial Organic Chemicals	VI	Arkansas
		VI	Louisiana
		VI	Louisiana
		VI	Oklahoma
		VI	Texas

Table A-2. (continued)

SIC Code		EPA Region	State
2869	Industrial Organic Chemicals (continued)	VI	Texas
		VI	Texas
		VII	Missouri
		IX	California
2873	Nitrogenous Fertilizers	VI	Texas
		VII	Iowa
		VII	Missouri
2874	Phosphatic Fertilizers	VII	Iowa
2875	Fertilizers, Mixing Only	IX	California
		X	Washington
2879	Agricultural Chemicals	IV	Georgia
289	Misc. Chemical Products	IV	South Carolina
		IV	South Carolina
2892	Explosives	IV	Alabama
		VII	Missouri
29	Petroleum Production	IV	Alabama
		IV	Mississippi
		VII	Nebraska
		IX	California
2911	Petroleum Refinery	II	New Jersey
		II	New Jersey
		II	Virgin Islands
		III	Delaware
		III	Maryland
		III	Pennsylvania
		III	Virginia
		III	Virginia
		IV	Alabama
		IV	Georgia
		IV	Mississippi
		IV	Mississippi
		V	Illinois
		V	Indiana
		V	Indiana
		V	Minnesota
		V	Ohio
		V	Ohio
		V	Ohio
		V	Ohio

Table A-2. (continued)

SIC Code	EPA Region	State
2911 Petroleum Refinery	V	Ohio
(continued)	V	Ohio
	V	Ohio
	VI	Arkansas
	VI	Louisiana
	VI	Louisiana
	VI	Louisiana
	VI	Louisiana
	VI	Louisiana
	VI	Louisiana
	VI	Louisiana
	VI	Louisiana
	VI	Louisiana
	VI	Louisiana
	VI	New Mexico
	VI	Oklahoma
	VI	Oklahoma
	VI	Oklahoma
	VI	Oklahoma
	VI	Oklahoma
	VI	Oklahoma
	VI	Oklahoma
	VI	Oklahoma
	VI	Oklahoma
	VI	Texas
	VI	Texas
	VI	Texas
	VI	Texas
	VI	Texas
	VI	Texas
	VI	Texas
	VI	Texas
	VI	Texas
	VI	Texas
	VI	Texas
	VI	Texas
	VI	Texas
	VI	Texas
	VI	Texas
	VI	Texas
	VI	Texas
	VI	Texas

Table A-2. (continued)

SIC Code	EPA Region	State
2911 Petroleum Refinery	VII	Kansas
(continued)	VII	Kansas
	VII	Kansas
	VII	Kansas
	VII	Kansas
	VII	Kansas
	VII	Kansas
	VII	Kansas
	VII	Missouri
	VIII	Colorado
	VIII	Montana
	VIII	Montana
	VIII	Montana
	VIII	Montana
	VIII	Montana
	VIII	Utah
	VIII	Utah
	VIII	Utah
	VIII	Wyoming
	VIII	Wyoming
	VIII	Wyoming
	VIII	Wyoming
	VIII	Wyoming
	IX	California
	IX	California
	IX	California
	IX	California
	IX	California
	IX	California
	IX	California
	IX	California
	IX	California
	X	Oregon
	X	Washington
	X	Washington
	X	Washington
	X	Washington
2969 Ind. Organic Chemicals	IX	California
3011 Pneumatic Tire Manu.	VI	Oklahoma
3317 Steel Pipe & Tubing Manu.	VI	Texas
3471 Plating & Polishing	IV	North Carolina
	VII	Iowa

Table A-2. (continued)

SIC Code		EPA Region	State
348	Ordnance & Accessories	IV IV IX X	Florida Kentucky Guam Idaho
3483	Ammunition	VI	Texas
349	Misc. Fabricated Metal Products	IV VI	Alabama New Mexico
3496	Misc. Fabricated Wire Products	IV VI	Georgia Texas
3498	Fabricated Pipe & Fittings	IV	Florida
3533	Oil Field Machinery	VI	Oklahoma
3589	Service Industry Machinery	IV IV	Georgia South Carolina
3621	Motors & Generators	IV	Mississippi
3641	Electric Lamps	IV	North Carolina
3662	Radio & TV Communication Equipment	IX	California
3679	Electronic Components	IV IX	Florida California
3743	Railroad Equipment	IV	Alabama
3999	Manufacturing Industries	II IV IV	New York Kentucky Kentucky
4441	Marine Terminal	VI	Louisiana
4463	Marine Cargo Handling	VI	Louisiana
49	Geothermal Energy Production	IX IX IX IX	California California California California
4953	Refuse Systems	III V VI VI VI VI	Pennsylvania Ohio Louisiana Louisiana Texas Texas

Table A-2. (continued)

SIC Code		EPA Region	State
4953	Refuse Systems	IX	California
	(continued)	IX	California
		IX	California
		IX	California
		IX	California
4990	Refuse Collection & Disposal	IX	California
5171	Petroleum Terminal	VI	Louisiana
7694	Armature Rewind Shop	VIII	Montana
7699	Repair & Related Services	VIII	Montana
8221	Colleges & Universities	VIII	Colorado
9711	National Security	IV	Alabama
		IV	Florida
		IV	North Carolina
		IV	North Carolina
		IV	South Carolina
		IV	Tennessee
		VI	New Mexico
		VIII	Colorado
		X	Washington

Table A-3. Land treatment usage by industry.*

SIC Code	Description	Number of facilities
2911	Petroleum Refinery	100
4953	Refuse Systems	11
2869	Industrial Organic Chemicals	9
9711	National Security	9
2491	Wood Preserving	6
49	Geothermal Energy Production	4
29	Petroleum Production	4
348	Ordnance & Accessories	4
203	Fruit Processing	3
2821	Plastics, Materials & Resins	3
2851	Paints & Allied Products	3
2873	Nitrogenous Fertilizers	3
3999	Manufacturing Industries	3
222	Weaving Mills, Synthetics	2
229	Misc. Textile Goods	2
2819	Industrial Inorganic Chemicals	2
2875	Fertilizers, Mixing Only	2
289	Misc. Chemical Products	2
2892	Explosives	2
3471	Plating & Polishing	2
349	Misc. Fabricated Metal Products	2
3496	Misc. Fabricated Wire Products	2
3589	Service Industry Machinery	2
3679	Electronic Components	2
025	Poultry Feed	1
1321	Natural Gas Proc.	1
1389	Oil & Gas Services	1
2067	Chewing Gum Manu.	1
249	Misc. Wood Products	1
2600	Paper & Allied Products	1
2611	Pulp Mills	1
2621	Paper Mills	1
2834	Pharmaceutical Preparations	1
2865	Cyclic Crudes & Intermediates	1
2874	Phosphatic Fertilizers	1
2879	Agricultural Chemicals	1
2969	Industrial Organic Chemicals	1
3011	Pneumatic Tire Manu.	1
3317	Steel Pipe & Tubing Manu.	1
3483	Ammunition	1

Table A-3. (continued)

SIC Code	Description	Number of facilities
3498	Fabricated Pipe & Fittings	1
3533	Oil Field Machinery	1
3621	Motors & Generators	1
3641	Electric Lamps	1
3662	Radio & TV Communication Equipment	1
3743	Railroad Equipment	1
4441	Marine Terminal	1
4463	Marine Cargo Handling	1
4990	Refuse Collection & Disposal	1
5171	Petroleum Terminal	1
7694	Armature Rewind Shop	1
7699	Repair & Related Services	1
8221	Colleges & Universities	1

* In some cases, the land treatment facility handled waste from more than industry.

necessarily indicate that land treatment is feasible for the associated wastes. Many industries have merely not attempted to use the technique. Although the bulk of the information was obtained from governmental agencies, several other sources proved useful in identifying or confirming facilities and in providing any missing data (Table A-4).

Table A-4. Sources consulted for information listed in the survey.

Category	Source
Governmental	EPA regional offices State environmental agencies Territorial environmental agencies
Industrial	Industrial associations Petroleum refiners Waste disposal companies Disposal equipment manufacturers Companies identified as operating land treatment facilities.
Other	Literature (e.g., journals, proceedings, and magazines) Environmental consultants

APPENDIX **B**

HAZARDOUS CONSTITUENTS REGULATED BY THE EPA

Acetaldehyde
(Acetato)phenylmercury
Acetonitrile
3-(alpha-Acetonylbenzyl)-4-
  hydroxycoumarin and salts
2-Acetylaminofluorene
Acetyl chloride
1-Acetyl-2-thiourea
Acrolein
Acrylamide
Acrylonitrile
Aflatoxins
Aldrin
Allyl alcohol
Aluminum phosphide
4-Aminobiphenyl
6-Amino-1,1a,2,8,8a,8b-hexahydro-
  8-[hydroxymethyl]-8a-methoxy-
  5-methylcarbamate azirino[2',3':
  3,4]pyrrolo[1,2-a]indole-4,7-dione
  [ester] [Mitomycin C]
5-[Aminomethyl]-3-isoxazolol
4-Aminopyridine
Amitrole
Antimony and compounds, N.O.S.*
Aramite
Arsenic and compounds, N.O.S.

Arsenic acid
Arsenic pentoxide
Arsenic trioxide
Auramine
Azaserine
Barium and compounds, N.O.S.
Barium cyanide
Benz[c]acridine
Benz[a]anthracene
Benzene
Benzenearsonic acid
Benzenethiol
Benzidine
Benzo[a]anthracene
Benzo[b]fluoranthene
Benzo[j]fluoranthene
Benzo[a]pyrene
Benzotrichloride
Benzyl chloride
Beryllium and compounds, N.O.S.
Bis[2-chloroethoxy]methane
Bis[2-chloroethyl]ether
N,N-Bis[2-chloroethyl]-2-naphthyl-
  amine
Bis[2-chloroisopropyl] ether
Bis[chloromethyl] ether
Bis[2-ethylhexyl] phthalate

Bromoacetone
Bromomethane
4-Bromophenyl phenyl ether
Brucine
2-Butanone peroxide
Butyl benzyl phthalate
2-sec-Butyl-4,6-dinitrophenol [DNBP]
Cadmium and compounds, N.O.S.
Calcium chromate
Calcium cyanide
Carbon disulfide
Chlorambucil
Chlordane [alpha and gamma isomers]
Chlorinated benzenes, N.O.S.
Chlorinated ethane, N.O.S.
Chlorinated naphthalene, N.O.S.
Chlorinated phenol, N.O.S.
Chloroacetaldehyde
Chloroalkyl ethers
p-Chloroaniline
Chlorobenzene
Chlorobenzilate
1-[p-Chlorobenzoyl]-5-methoxy-2-
  methylindole-3-acetic acid
p-Chloro-m-cresol
1-Chloro-2,3-epoxybutane
2-Chloroethyl vinyl ether
Chloroform
Chloromethane
Chloromethyl methyl ether
2-Chloronaphthalene
2-Chlorophenol
1-[o-Chlorophenyl]thiourea
3-Chloropropionitrile
alpha-Chlorotoluene
Chlorotoluene, N.O.S.
Chromium and compounds, N.O.S.
Chrysene
Citrus red No. 2
Copper cyanide
Creosote
Crotonaldehyde
Cyanides [soluble salts and
  complexes], N.O.S.
Cyanogen
Cyanogen bromide
Cyanogen chloride
Cycasin
2-Cyclohexyl-4,6-dinitrophenol
Cyclophosphamide
Daunomycin
DDD

DDE
DDT
Diallate
Dibenz[a,h]acridine
Dibenz[a,j]acridine
Dibenz[a,h]anthracene(Dibenzo[a,h]
  anthracene)
7H-Dibenzo[c,g]carbazole
Dibenzo[a,e]pyrene
Dibenzo[a,h]pyrene
Dibenzo[a,i]pyrene
1,2-Dibromo-3-chloropropane
1,2-Dibromomethane
Dibromomethane
Di-n-butyl phthalate
Dichlorobenzene, N.O.S.
3,3'-Dichlorobenzidine
1,1-Dichloroethane
1,2-Dichloroethane
trans-1,2-Dichloroethane
Dichloroethylene, N.O.S.
1,1-Dichloroethylene
Dichloromethane
2,4-Dichlorophenol
2,6-Dichlorophenol
2,4-Dichlorophenoxyacetic acid
  [2,4-D]
Dichloropropane
Dichlorophenylarsine
1,2-Dichloropropane
Dichloropropanol, N.O.S.
Dichloropropene, N.O.S.
1,3-Dichloropropene
Dieldrin
Diepoxybutane
Diethylarsine
0,0-Diethyl-S-(2-ethylthio)ethyl
  ester of phosphorothioic acid
1,2-Diethylhydrazine
0,0-Diethyl-S-methylester
  phosphorodithioic acid
0,0-Diethylphosphoric acid, 0-p-
  nitrophenyl ester
Diethyl phthalate
0-0-Diethyl-0-(2-pyrazinyl)
  phosphorothioate
Diethylstilbestrol
Dihydrosafrole
3,4-Dihydroxy-alpha-(methylamino)-
  methyl benzyl alcohol
Di-isopropylfluorophosphate (DFP)
Dimethoate

3,3'-Dimethoxybenzidine
p-Dimethylaminoazobenzene
7,12-Dimethylbenz[a]anthracene
3,3'-Dimethylbenzidine
Dimethylcarbamoyl chloride
1,1-Dimethylhydrazine
1,2-Dimethylhydrazine
3,3-Dimethyl-1-(methylthio)-2-
  butanone-0-[(methylamino)carbonyl]
  oxime
Dimethylnitrosoamine
alpha,alpha-Dimethylphenethylamine
2,4-Dimethylphenol
Dimethyl phthalate
Dimethyl sulfate
Dinitrobenzene, N.O.S.
4,6-Dinitro-o-cresol and salts
2,4-Dinitrophenol
2,4-Dinitrotoluene
2,6-Dinitrotoluene Di-n-octyl
  phthalate
1,4-Dioxane
1,2-Diphenylhydrazine
Di-n-propylnitrosamine
Disulfoton
2,4-Dithiobiuret
Endosulfan
Endrin and metabolites
Epichlorohydrin
Ethyl cyanide
Ethylene diamine
Ethylenebisdithiocarbamate (EBDC)
Ethyleneimine
Ethylene oxide
Ethylenethiourea
Ethyl methanesulfonate
Fluoranthene
Fluorine
2-Fluoroacetamide
Fluoroacetic acid, sodium salt
Formaldehyde
Glycidylaldehyde
Halomethane, N.O.S.
Heptachlor
Heptachlor epoxide (alpha, beta,
  and gamma isomers)
Hexachlorobenzene
Hexachlorobutadiene
Hexachlorocyclohexane (all isomers)
Hexachlorocyclopentadiene
Hexachloroethane

1,2,3,4,10,10-Hexachloro-1,4,4a,5,
  8,8a-hexahydro-1,4:5,8-endo,endo-
  dimethanonaphthalene
Hexachlorophene
Hexachloropropene
Hexaethyl tetraphosphate
Hydrazine
Hydrocyanic acid
Hydrogen sulfide
Indeno(1,2,3-c,d)pyrene
Iodomethane
Isocyanic acid, methyl ester
Isosafrole
Kepone
Lasiocarpine
Lead and compounds, N.O.S.
Lead acetate
Lead phosphate
Lead subacetate
Maleic anhydride
Malononitrile
Melphalan
Mercury and compounds, N.O.S.
Methapyrilene
Methomyl
2-Methylaziridine
3-Methylcholanthrene
4,4'-Methylene-bis-(2-chloro-
  aniline)
Methyl ethyl ketone (MEK)
Methyl hydrazine
2-Methyllactonitrile
Methyl methacrylate
Methyl methanesulfonate
2-Methyl-2-(methylthio)propional-
  dehyde-o-(methylcarbonyl) oxime
N-Methyl-N'-nitro-N-nitrosoguani-
  dine
Methyl parathion
Methylthiouracil
Mustard gas
Naphthalene
1,4-Naphthoquinone
1-Naphthylamine
2-Naphthylamine
1-Naphthyl-2-thiourea
Nickel and compounds, N.O.S.
Nickel carbonyl
Nickel cyanide
Nicotine and salts
Nitric oxide

p-Nitroaniline
Nitrobenzene
Nitrogen dioxide
Nitrogen mustard and hydrochloride
  salt
Nitrogen mustard N-oxide and
  hydrochloride salt
Nitrogen peroxide
Nitrogen tetroxide
Nitroglycerine
4-Nitrophenol
4-Nitroquinoline-1-oxide
Nitrosamine, N.O.S.
N-Nitrosodi-N-butylamine
N-Nitrosodiethanolamine
N-Nitrosodiethylamine
N-Nitrosodimethylamine
N-Nitrosodiphenylamine
N-Nitrosodi-N-propylamine
N-Nitroso-N-ethylurea
N-Nitrosomethylethylamine
N-Nitroso-N-methylurea
N-Nitroso-N-methylurethane
N-Nitrosomethylvinylamine
N-Nitrosomorpholine
N-Nitrosonornicotine
N-Nitrosopiperidine
N-Nitrosopyrrolidine
N-Nitrososarcosine
5-Nitro-o-toluidine
Octamethylpyrophosphoramide
Oleyl alcohol condensed with 2 moles
  ethylene oxide
Osmium tetroxide
7-Oxabicyclo[2.2.1]heptane-2,3-
  dicarboxylic acid
Parathion
Pentachlorobenzene
Pentachloroethane
Pentachloronitrobenzene (PCNB)
Pentacholorophenol
Phenacetin
Phenol
Phenyl dichloroarsine
Phenylmercury acetate
N-Phenylthiourea
Phosgene
Phosphine
Phosphorothioic acid, O,O-dimethyl
  ester, O-ester with N,N-dimethyl
  benzene sulfonamide
Phthalic acid esters, N.O.S.

Phthalic anhydride
Polychlorinated biphenyl, N.O.S.
Potassium cyanide
Potassium silver cyanide
Pronamide
1,2-Propanediol
1,3-Propane sultone
Propionitrile
Propylthiouracil
2-Propyn-1-ol
Pryidine
Reserpine
Saccharin
Safrole
Selenious acid
Selenium and compounds, N.O.S.
Selenium sulfide
Selenourea
Silver and compounds, N.O.S.
Silver cyanide
Sodium cyanide
Streptozotocin
Strontium sulfide
Strychnine and salts
1,2,4,5-Tetrachlorobenzene
2,3,7,8-Tetrachlorodibenzo-p-dioxin
  (TCDD)
Tetrachloroethane, N.O.S.
1,1,1,2-Tetrachloroethane
1,1,2,2-Tetrachloroethane
Tetrachloroethene (Tetrachloro-
  ethylene)
Tetrachloromethane
2,3,4,6-Tetrachlorophenol
Tetraethyldithiopyrophosphate
Tetraethyl lead
Tetraethylpyrophosphate
Thallium and compounds, N.O.S.
Thallic oxide
Thallium (I) acetate
Thallium (I) carbonate
Thallium (I) chloride
Thallium (I) nitrate
Thallium selenite
Thallium (I) sulfate
Thioacetamide
Thiosemicarbazide
Thiourea
Thiuram
Toluene
Toluene diamine
o-Toluidine hydrochloride

Tolylene diisocyanate
Toxaphene
Tribromomethane
1,2,4-Trichlorobenzene
1,1,1-Trichloroethane
1,1,2-Trichloroethane
Trichloroethene (Trichloroethylene)
Trichloromethanethiol
2,4,5-Trichlorophenol
2,4,6-Trichlorophenol
2,4,5-Trichlorophenoxyacetic acid
   (2,4,5-T)
2,4,5-Trichlorophenoxypropionic
   acid (2,4,5-TP) (Silvex)
Trichloropropane, N.O.S.
1,2,3-Trichloropropane
0,0,0-Triethyl phosphorothioate
Trinitrobenzene
Tris(1-azridinyl)phosphine sulfide
Tris(2,3-dibromopropyl)phosphate
Trypan blue
Uracil mustard
Urethane
Vanadic acid, ammonium salt
Vanadium pentoxide (dust)
Vinyl chloride
Vinylidene chloride
Zinc cyanide
Zinc phosphide

APPENDIX B REFERENCE

EPA. 1980. Identification and listing of hazardous waste. Part 261. Federal Register Vol. 45, No. 98. pp. 33132-33133. May 19, 1980.

C

SOIL HORIZONS AND LAYERS

Organic Horizons

O--Organic horizons of mineral soils. Horizons: (1) formed or forming in the upper part of mineral soils above the mineral part; (2) dominated by fresh or partly decomposed organic material; and (3) containing more than 30 percent organic matter if the mineral fraction is more than 50 percent clay, or more than 20 percent organic matter if the mineral fraction has no clay. Intermediate clay content requires proportional organic-matter content.

O1--Organic horizons in which essentially the original form of most vegetative matter is visible to the naked eye.

O2--Organic horizons in which the original form of most plant or animal matter cannot be recognized with the naked eye.

Mineral Horizons and Layers

Mineral horizons contain less than 30 percent organic matter if the mineral fraction contains more than 50 percent clay or less than 20 percent organic matter if the mineral fraction has no clay. Intermediate clay content requires proportional content of organic matter.

A--Mineral horizons consisting of: (1) horizons of organic-matter accumulation formed or forming at or adjacent to the surface; (2) horizons that have lost clay, iron, or aluminum with resultant concentration of quartz or other resistant minerals of sand or silt size; or (3) horizons dominated by 1 or 2 above but transitional to an underlying B or C.

A1--Mineral horizons, formed or forming at or adjacent to the surface, in which the feature emphasized is an accumulation of humified organic matter intimately associated with the mineral fraction.

A2--Mineral horizons in which the feature emphasized is loss of clay, iron, or aluminum, with resultant concentration of quartz or other resistant minerals in sand and silt sizes.

A3--A transitional horizon between A and B, and dominated by properties characteristic of an overlying A1 or A2 but having some subordinate properties of an underlying B.

AB--A horizon transitional between A and B, having an upper part dominated by properties of A and a lower part dominated by properties of B, and the two parts cannot conveniently be separated into A3 and B1.

A&B--Horizons that would qualify for A2 except for included parts constituting less than 50 percent of the volume that would qualify as B.

AC--A horizon transitional between A and C, having subordinate properties of both A and C, but not dominated by properties characteristic of either A or C.

B--Horizons in which the dominant feature or features is one or more of the following: (1) an illuvial concentration of silicate clay, iron, aluminum, or humus, alone or in combination; (2) a residual concentration of sesquioxides or silicate clays, alone or mixed, that has formed by means other than solution and removal of carbonates or more soluble salts; (3) coatings of sesquioxides adequate to give conpicuously darker, stronger, or redder colors than overlying and underlying horizons in the same sequum but without apparent illuviation of iron and not genetically related to B horizons that meet requirements of 1 or 2 in the same sequum; or (4) an alteration of material from its original condition in sequums lacking conditions defined in 1, 2, and 3 that obliterates original rock structure, that forms silicate clays, liberates oxides, or both, and that forms granular, blocky, or prismatic structure if textures are such that volume changes accompany changes in moisture.

B1--A transitional horizon between B and A1 or between B and A2 in which the horizon is dominated by properties of an underlying B2 but has some subordinate properties of an overlying A1 or A2.

B&A--Any horizon qualifying as B in more than 50 percent of its volume including parts that qualify as A2.

B2--That part of the B horizon where the properties on which the B is based are without clearly expressed subordinate characteristics indicating that the horizon is transitional in an adjacent overlying A or an adjacent underlying C or R.

B3--A transitional horizon between B and C or R in which the properties diagnostic of an overlying B2 are clearly expressed but are associated with clearly expressed properties characteristics of C or R.

C--A mineral horizon or layer, excluding bedrock, that is either like or unlike the material from which the solum is presumed to have formed, relatively little affected by pedogenic processes, and lacking properties diagnostic of A or B but including materials modified by: (1) weathering outside the zone of major biological activity; (2) reversible cementation, development of brittleness, development of high bulb density, and other properties characteristic of fragipans; (3) gleying; (4) accumulation of calcium or magnesium carbonate or more soluble salts; (5) cementation by such accumulations as calcium or magnesium carbonate or more soluble salts; of (6) cementation by alkali-soluble siliceous material or by iron and silica.

R--Underlying consolidated bedrock, such as granite, sandstone, or limestone. If presumed to be like the parent rock from which the adjacent overlying layer or horizon was formed, the symbol R is used alone. If alone. If presumed to be unlike the overlying material, the R is preceded by a Roman numeral denoting lithologic discontinuity as explained under the heading.

SYMBOLS USED TO INDICATE DEPARTURES SUBORDINATE TO THOSE INDICATED BY CAPITAL LETTERS

The following symbols are to be used in the manner indicated under the heading Conventions Governing Use of Symbols.

b--Buried soil horizon

ca--An accumulation of carbonates of alkaline earths, commonly of calcium.

es--An accumulation of calcium sulfate.

cn--Accumulations of concretions or hard nonconcretionary nodules enriched in sesquioxides with or without phosphorus.

f--Frozen soil

g--Strong gleying

h--Illuvial humus

ir--Illuvial iron

m--Strong cementation, induration

p--Plowing or other disturbance

sa--An accumulation of salts more soluble than calcium sulfate

si--Cementation by siliceous material, soluble in alkali. This symbol is
applied only to C.

t--Illuvial clay

APPENDIX C REFERENCE

USDA. 1975. Soil taxonomy: a basic system of soil classification for making
and interpreting soil surveys. Agricultural Handbook No. 436. 754 pp.

INDUSTRIAL LAND TREATMENT SYSTEMS CITED IN THE LITERATURE

A variety of experiences with land treatment of industrial wastes have been reported in the literature. No attempt was made to to verify whether the reported wastes were classified as hazardous, however, the list excludes references to wastes which were identified as likely to be non-hazardous.

Industry	References
Textile (SIC 22)	
Industrial Wastewater	Sayapin (1978)
Industrial Wastewater	Wallace (1976)
Wool Preserving	Wallace (1976)
Wool Scouring	Wadleigh (1968)
Lumber (SIC 24)	
Wood Distillation	Hickerson and McMahon (1960)
Pulp and Paper (SIC 26)	
Pulpmill	Wadleigh (1968)
Pulpmill	Hayman (1978)
Pulpmill	Watterson (1971)
Pulpmill	Blosser and Owens (1964)
Pulpmill	Kadamki (1971)
Pulpmill	Flower (1969)
Papermill	Vercher et al. (1965)
Papermill	Jorgenson (1965)

Industry	References
Papermill	Dolar et al. (1972)
Papermill	Das and Jena (1973)
Papermill	Aspitarte et al. (1973)
Papermill	Wallace (1976)
Papermill	Hayman (1978)
Hard Board	Parsons (1967)
Paper Board	Koch and Bloodgood (1959)
Straw Board	Meighan (1958)
Insulated Board	Phillip (1971)
Sulfite Pulp Mill	Crawford (1958)
Sulfite Pulp Mill	Wisniewski et al. (1955)
Sulfite Pulp Mill	Billings (1958)
Sulfite Pulp Mill	Blosser and Owens (1964)
Sulfite Pulp Mill	Gellman and Blosser (1959)
Sulfite Pulp Mill	Kolar (1965)
Sulfite Pulp Mill	Kolar and Mitiska (1965)
Sulfite Pulp Mill	Hashimoto (1966)
Sulfite Pulp Mill	Yokota and Hashimoto (1966)
Sulfite Pulp Mill	Pasak (1969)
Sulfite Pulp Mill	Yakushenko et al. (1971)
Sulfite Pulp Mill	Minami and Taniguchi (1971)
Sulfite Pulp Mill	Knowles et al. (1974)
Sulfite Pulp Mill	Flaig and Sochtig (1974)
Kraft (sulfate)	Blosser and Owens (1964)
Kraft (sulfate)	Crawford (1958)
Kraft (sulfate)	Wallace et al. (1975)
Semi-Chemical	Voights (1955)
Drinking	Flower (1969)
Not Specified (saline)	Hayman (1979)
Other Inorganic Chemicals (SIC 2819)	
Waste Sulfuric Acid	Wallace (1977)
Chemicals (SIC 282-289)	Shevstova et al. (1969)
Biological Chemical	Woodley (1968)
PCB	Griffin et al. (1978)
PCB	Tucker et al. (1975)
PCB	Griffin et al. (1977)
Pharmaceuticals (SIC 283)	
Mycelial Waste	Nelson (1977)
Fermentation	Colovos and Tinklenberg (1962)
Antibiotic Production	Uhliar and Bucko (1974)
High Nitrogen Industrial Wastewater	Brown (1976)
High Nitrogen Industrial Wastewater	Wallace (1976)
High Nitrogen Industrial Wastewater	Deroo (1975)
High Nitrogen Industrial Wastewater	Woodley (1968)

Industry	References
Explosives (SIC 2892)	Lever (1966)

Petroleum Refining (SIC 2911) and
Petroleum Refining (SIC 2992)

Refinery-Decomp. of Oily Waste in Soil	Jensen (1958)
Refinery-Decomp. of Oily Waste in Soil	Grove (1978)
Refinery-Decomp. of Oily Waste in Soil	Dhillon (1973)
Refinery-Decomp. of Oily Waste in Soil	Dotson et al. (1971)
Refinery-Decomp. of Oily Waste in Soil	Franke and Clark (1974)
Refinery-Decomp. of Oily Waste in Soil	Jobson et al. (1974)
Refinery-Decomp. of Oily Waste in Soil	Kincannon (1972)
Refinery-Decomp. of Oily Waste in Soil	Lewis (1977)
Refinery-Decomp. of Oily Waste in Soil	Maunder and Waid (1973)
Refinery-Decomp. of Oily Waste in Soil	Giddens (1974)
Refinery-Decomp. of Oily Waste in Soil	Nissen (1970)
Refinery-Decomp. of Oily Waste in Soil	Plice (1948)
Refinery-Decomp. of Oily Waste in Soil	Raymond et al. (1975)
Refinery-Decomp. of Oily Waste in Soil	Raymond et al. (1976)
Refinery-Decomp. of Oily Waste in Soil	Ongerth (1975)
Refinery-Decomp. of Oily Waste in Soil	Dibble and Bartha (1979)
Refinery-Decomp. of Oily Waste in Soil	Knowlton and Rucker (1978)
Refinery-Decomp. of Oily Waste in Soil	Baker (1978)
Tank Bottom	Cansfield and Racz (1978)
Refinery Wastes: Biosludge, Tank Bottoms, API Separator Sludge	Cresswell (1977)
Refinery Waste	Akoun (1978)
Refinery Waste	Huddleston (1979)
Refinery (1) Tank Bottom Crude	Lewis (1977)
(2) Slop Oil Immulsion	Ibid.
(3) API Separator Sludge	Ibid.
(4) Drilling Mud	Ibid.
(5) Cleaning Residue	Ibid.

Leather Tanning and Finishing (SIC 3111)

Leather Tanning and Finishing	Parker (1965)
Leather Tanning and Finishing	Parker (1967)
Leather Tanning and Finishing	Jansky (1961)
Leather Tanning and Finishing	
Leather Tanning and Finishing	Wallace (1976)
Leather Tanning and Finishing	S.C.S. Engineers (1976)

Blast Furnace Slag (SIC 3312) Steel	Volk et al. (1952)

Primary Aluminum Smelting (SIC 3334)

	Ongerth (1975)
Waste Oil from Aluminum Manufacturing	Neal et al. (1976)

Industry	References
Electricity Production (SIC 4911)	
Utility Waste	Page et al. (1977)
Fly Ash	Martens (1971)
Fly Ash	Plank and Martens (1974)
Fly Ash	Plank et al. (1975)
Fly Ash	Schnappinger et al. (1975)

Akoun, G. L. 1978. Hydrocarbon residues biodegradation in the soil. Study conducted by: Esso Saf, Shell Francaise, and Inra Rouen.

Aspitarte, T. R., A. S. Rosenfeld, B. C. Samle and H. R. Amberg. 1973. Pulp and paper mill sludge disposal and crop production. Tech. Assoc. Pulp Pap. Ind. 56:140-144.

Baker, D. A. 1978. Petroleum processing wastes. p. 1269-1270 In J. Water Poll. Control Fed. June 1978.

Billings, R. M. 1958. Stream improvement through spray disposal of sulfite liquor at the Kimberly-Clark Corp., Proc. of the 13th Industrial Waste Conference. Purdue Univ. 96:71.

Blosser, R. O., and E. L. Owens. 1964. Irrigation and land disposal of pulp mill effluent. Water and Sewage Works 3: 424-432.

Brown, G. E. 1976. Land application of high nitrogen industrial waste water. In Water - 1976 II biological waste water treatment. ALCHE symposium series. 73(167):227-232.

Cansfield, P. E., and G. J. Racz. 1978. Degradation of hydrocarbon sludges in the soil. Can. J. Soil Sci. 58:339-345.

Colovos, G. C., and N. Tinklenberg. 1962. Land disposal of pharmaceutical manufacturing wastes. Biotech. Bioengr. 4:153-160.

Crawford, S. C. 1958. Spray irrigation of certain sulfite pulp mill wastes. Sewage and Industrial Wastes 30(10):1266-1272.

Cresswell, L. W. 1977. The fate of petroleum in a soil environment. Proceedings of 1977 Oil Spill Conference. New Orleans, Louisiana. March 8-10. p. 479-482.

Das, R. C., and M. K. Jena. 1973. Studies on the effect of soil application of molybdenum, boron and paper mill sludge on the post harvest qualities of potato tuber (Solanum Tuberosum L.) Madras Agr. J. 60 (8):1026-1029.

Deroo, H. C. 1975. Agricultural and horticultural utilization of fermentation residues. Connecticut Agr. Exp. Sta. New Haven, Connecticut. Bull. 750.

Dhillon, G. S. 1973. Land disposal of refinery wastes. Environmental Development, Bethel Corp. San Francisco, California.

Dibble, J. T. and R. Bartha. 1979. Leaching aspects of oil sludge biodegradation in soil. Soil Science 0038-075 x/79 p. 365-370.

Dolar, S. G., J. R. Boyle, and D. R. Kenny. 1972. Paper mill sludge disposal on soils: effects on the yield and mineral nutrition of oats (Avena Satival). J. Environ. Qual. (4) 405-409.

Dotson, G. K., R. B. Dean, B. A. Kenner, and W. B. Cooke. 1971. Landspreading, a conserving and non-polluting method of disposing of oily wastes. Proc. of the 5th Int. Water Poll. Conference. Vol. 1 Sec. II. 36/1-36/15.

Flaig, W., and H. Sochtig. 1974. Utilization of sulphite waste of the cellulous industry as an organic nitrogen fertilizer. Netherlands J. of Agr. Sci. 22 (4):255-261.

Flower, W. A. 1969. Spray irrigation for the disposal of effluents containing deinking wastes." Tech. Assoc. Pulp Pap. Ind. 52:1267.

Francke, H. C., and F. E. Clark. 1974. Disposal of oily waste by microbial assimilation. Union Carbide, Doc Y-1934. May 16. Prepared by U. S. Atomic Energy Commission.

Gellman, I., and R. O. Blosser. 1959. Disposal of pulp and paper mill waste by land application and irrigation use. Proc. of the 14th Industrial Waste Conference. Purdue Univ. 104:479.

Giddens, P. H. 1974. The early petroleum industry. Porcupine Press, Philadelphia, 1950. 195 p.

Griffin, R., R. Clark, M. Lee and E. Chain. 1978. Disposal and removal of polychlorinated biphenyls in soil. Proc. of the Fourth Annual Research Symposium. August. EPA-600 9/78-016. p. 169-181.

Griffin, R. A., F. B. Bewalle, E.S.K. Chain, J. H. Kim, and A. K. Au. 1977. Attentuation of PCBs by soil materials and char wastes. p. 208-217. In S. K. Banerji (ed.) Management of gas and leachate in landfills. Ecological research series, EPA. Cincinnati, Ohio. EPA 600/9-77-026.

Grove, G. W. 1978. Use land farming for oily waste disposal. Hydrocarbon Processing. May. p. 138-140.

Hashimoto, T. Y. 1966. Edaphological studies on the utilization of wastepulp liquor. 2. Effects of the liquor on the absorption of nutrient elements by crops. J. Sci. Soil Manure Tokyo 37:223-225; J. Soil Sci. Pl. Nutr. 12 (3):39.

Hayman, J. P. 1978. Land disposal of mineralised effluent from a pulp and paper mill. In Progress in water technology. Vol. 9, No. 4.

Hayman, J. P., and L. Smith. 1979. Disposal of saline effluent by controlled-spray irrigation. J. Water Poll. Control Fed. 51(3):526-533.

Hickerson, R. D., and E. K. McMahon. 1960. Spray irrigation of wood distillation wastes. J. Water Poll. Control. Fed. 32(55).

Huddleston, R. L. 1979. Solid waste disposal: landfarming. Chemical Engineering. February 26. p. 119-124.

Jansky, K. 1961. Tannery-waste water disposal. Kozarstvi 2:327-329, 355-360.

Jensen, V. 1958. Decomposition of oily wastes in soil. Dept. of Microbiology and Microbial Ecology Royal Veterinary and Agricultural University. Copenhagen, Denmark. Rolishedsvej 21.

Jobson, A., McLaughlin, F. D. Cook, and D.W.S. Westlake. 1974. Effects of amendments on microbial Inc. utilisation of oil applied to soil. Appl. Microbiol. 27(1):166-171.

Jorgensen, J. R. 1965. Irrigation of slash pine with paper mill effluents. Bull. La. State Univ. Division of Engineering Res. No. 80. p. 92-99.

Kadamki, K. 1971. Accumulation of sodium in potted soil irrigated with pulpmill effluents. Consultant 16: 93-94.

Kincannon, C. B. 1972. Oily waste disposal by soil cultivation process. EPA. EPA-R2-72-110.

Knowles, R., R. Neufild and S. Simpson. 1974. Acetylene reduction nitrogen fixation by pulp and paper mill effluent and by klebsiella isolated from effluents and environmental situations. Appl. Microbiol. 28 (4):608-613.

Knowlton, H. E., and J. E. Rucker. 1978. Land farming shows promise for refinery waste disposal. Oil and Gas Journal, May 14. p. 108-116.

Koch, H. C., and D. E. Bloodgood. 1959. Experimental spray irrigation of paper board mill wastes. Sew. Ind. Wastes 31:827.

Kolar, L. 1965. Alteration of moisture properties of soil and technical soil constants by application of waste waters from cellulose works. Rada. 3 (5):91-95.

Kolar, L., and J. Mitiska. 1965. The influence of the presence of sulphide waste liquor in irrigation water on the biological, chemical and physico-chemical conditions of the soil. Budejovic. 3 (4):11-20.

Lever, N. A. 1966. Disposal of nitrogenous liquid effluent from modder fontein dynamite factory. Proc. of the 21st Industrial Waste Conference. Purdue University 121:902.

Lewis, R. S. 1977. Sludge farming of refinery wastes as practiced at Exxon's Bayway refinery and chemical plant. Proceedings of the National Conference on Disposal of Residues on Land. Information Transfer, Inc. Rockville, Maryland. p. 87-92.

Martens, D. C. 1971. Availability of plant nutrients in fly ash. Compost. Sci. 12:15-19.

Maunder, B. R., and J. S. Waid. 1973. Disposal of waste oil by land spreading. Proceedings of the Pollution Research Conference. Wairakei, New Zealand. June 20-21.

Meighan, A. D. 1958. Experimental spray irrigation of straw board wastes. Proc. of the 13th Industrial Waste Conference. Purdue University 96:456.

Minami, M., and T. Taniguchi. 1971. Effects of water pollution on the growth and yield of rice plants. 1. Effects of waste water from a pulp factory. Hokkaido Perfectual Agricultural Experiment Station. Bull. 24. p. 56-68.

Neal, D. M., R. L. Glover and P. G. Moe. 1976. Land disposal of oily waste water by means of spray irrigation. Proc. of the 1976 Cornell Agr. Waste Management Conference. Ann Arbor Sci. Publ. Inc. Ann Arbor, Michigan. p. 757-767.

Nelson, D. 1977. Laboratory and field scale investigation of mycelial waste decomposition in the soil environment. Presented to the Soc. for Industrial Microbiology meeting. Michigan State Univ. East Lansing, Michigan. Aug. 1977.

Nissen, T. V. 1970. Biological degradation of hydrocarbons with special reference to soil contamination. Tidssk 2 Pl. Arl. 74:391-405.

Ongerth, J. E. 1975. Feasibility studies from land disposal of a dilute oily waste water. To be presented at the 30th Industrial Waste Conference. Purdue Univ.

Page, A. L., F. T. Bingham, L. J. Lund, G. R. Bradford, and A. A. Elseewi. 1977. Consequences of trace element enrichment of soils and vegetation from the combustion of fuels used in power generation. SCE research and development series 77-RD-29. Rosemead, California.

Parker, R. R. 1965. Spray irrigation for industrial waste disposal. Canadian Municipal Utilities 103:7:28-32.

Parker, R. R. 1967. Disposal of tannery wastes. Proc. of the 22nd Industrial Waste Conference. Purdue Univ. 129:36.

Parsons, W. 1967. Spray irrigation from the manufacture of hard board. Proc. of the 22nd Industrial Waste Conference. Purdue Univ. p. 602-607.

Pasak, V. 1969. Sulphite waste liquor for protecting soil against wind erosion. Ved. Prace Vyskum. UST. Melior. 10:143-148.

Philipp, A. H. 1971. Disposal of insulation board mill effluent by land irrigation. J. Water Poll. Control Fed. 43:1749.

Plank, C. O., and D. C. Martens. 1974. Boron availability as influenced by application of fly ash to soil. Soil Sci. Soc. Proc. 38:974-977.

Plank, C. O., D. C. Martens, and D. L. Hallock. 1975. Effects of soil application of fly ash on chemical composition and yield of corn (Zea Mays L.) and on chemical composition of displaced soil solutions. Plant Soil 42:465-476.

Plice, M. J. 1948. Some effects of crude petroleum on soil fertility. Proc. of The Soil Sci. Soc. 13:412-416.

Raymond, R. L., J. O. Hudson, and V. W. Jamison. 1976. Oil degradation in soil. Appl. Environ. Microbiol. 31:522-535.

Raymond, R. L., J. O. Hudson, and V. W. Jamison. 1975. Assimilation of oil by soil bacteria, refinery solid waste proposal. Proc. 40th Mid-Year Meeting, API. May 14. p. 2.

Sayapin, V. P. 1978. Nutrition value of fodder harmlessness of plant output grown with textile industry waste water irrigation. Cold Regions Research and Engineering Laboratory. Draft Translation 671.

Schnappinger, M. G., Jr., D. C. Martens, and C. O. Plank. 1975. Zinc availability as influenced by application of fly ash to soil. Environ. Sci. Technol. 9:258-261.

S.C.S. Engineers. 1976. Assessment of industrial hazardous waste practices - leather tanning and finishing industry. NTIS PB 261018.

Shevtsova, I. I.,; V. K. Marinich; and S. M. Neigauz. 1969. Effects waste water from capron production on higher plants and soil micro-organisms. Biol. Nauk. (4):91-94. Chem. Abstr. 71.

Tucker, E. S., W. J. Litschg, and W. M. Mees. 1975. Migration of poly-chlorinated biphenyls in soil induced by percolating water. Bull. Environ. Contam. Toxicol. 13:86-93.

Uhliar, J., and M. Bucko. 1974. The use of industrial wastes for antibiotic production in crop production. Rostlinna Vyroba 20 (9):923-930.

Vercher, B. D., M. B. Sturgis, and O. O. Curtis. 1965. Paper mill waste water for crop irrigation and its effect on the soil. Bull. La. Agr. Exp. Stat. No. 604. p. 46.

Voights, D. 1955. Lagooning and spray disposal of neutral sulfite semi-chemical pulp mill liquors. Proc. of the 10th Industrial Waste Conference. Purdue Univ. 89:497.

Volk, G. W., R. B. Harding, and C. E. Evans. 1952. A comparison of blast furnace slag and limestone as a soil amendment. Res. Bull. 708, Ohio Agr. Exp. Sta., Columbus, Ohio.

Wadleigh, Cecil H. 1968. Wastes in relation to agriculture and forestry. USDA. Washington, D.C. Misc. Publication 1065.

Wallace, A. T. 1976. Land disposal of liquid industrial wastes. p. 147-162. In R. L. Sanks and Asano (ed.) Land treatment and disposal of municipal and industrial wastewater. Ann Arbor Science Publishers Inc. Ann Arbor, Michigan.

Wallace, A. T. 1977. Massive sulfur application to highly calcareous agricultural soil as a sink for waste sulfur. Res. 2:263-267.

Wallace, A. T., R. Luoma, and M. Olson. 1975. Studies of the feasibility of a rapid infiltration system for disposal of Kraft Mill effluent. To be presented at the 30th Industrial Waste Conference. Purdue Univ.

Watterson, K. G. 1971. Water quality in relation to fertilization and pulp mill effluent disposal. Consultant 16:91-92.

Wisniewski, T. K., A. J. Wiley, and B. J. Lueck. 1955. Ponding and soil infiltration for disposal of spent sulphite liquor in Wisconsin. Proc. of the 10th Industrial Waste Conference. Purdue Univ. 89:480.

Woodley, R. A. 1968. Spray irrigation of organic chemical wastes. Proc. of the 23rd Industrial Waste Conference. Purdue Univ. Lafayette, Indiana. 132:251-261.

Yakushenko, I. K., I. Y. Kazantsev, and V. G. Ovsyannikova. 1971. Waste sulphite liquors of the cellulous industry and their use for irrigation. Vest. Sel'- Khoz. Nauki. Mosk 1:87-92.

Yokota, H., and T. Hashimoto. 1966. Edaphological studies on the utilization of waste pulp liquor. 3. Effects of the liquor on phosphorus fixation. J. Sci. Soil Manure, Tokyo 37:294-297; J. Soil Sci. Pl. Nutr. 12 (4):40.

APPENDIX **E**

SAMPLE CALCULATIONS

   In order to illustrate the interpretation of data from the site assessment, waste analysis and pilot studies, sample calculations and design recommendations are given for a hypothetical land treatment unit and a given waste. The components of the waste are considered individually and compared to determine the application limiting constituents (ALC), rate limiting constituent (RLC) and capacity limiting constituent (CLC). The assumptions and calculations used in the design of the HWLT unit are discussed in detail in Section 7.5. The required treatment area size and the useful life of the HWLT unit are then calculated for the example waste (Appendix E-7). Additionally, an example of water balance determinations and runoff retention pond sizing is presented.

APPENDIX E-1

WATER BALANCE AND RETENTION POND SIZE CALCULATIONS

As discussed in Section 8.3.1.1 the water balance method can be used to evaluate hydraulic load and required storage for surface runoff. This is a very simplified approach to calculating the water balance and conservative values should be used to guard against any inaccuracy in parameter estimates used in the method. The value used in these calculations for discharge can be varied to account for the method of runoff water control, in this case the storage volume calculated includes the seasonal accumulation of water.

Initially, climatological data or estimates should be made for the parameters in the water balance. Precipitation values are derived from the long-term rainfall data collected at a nearby weather station, chosen according to the criteria given in Section 3.3. Estimates of the evapotranspiration can be obtained by using the class A pan evaporation value for each month (Figs. 8.9-8.20 show monthly pan evaporation data for the U.S.). This value is then multiplied by an appropriate annual pan evaporation coefficient. These coefficients are used to relate pan data to evaporation expected from lakes. An estimation of the amount of leachate may be calculated based on the hydraulic conductivity of the most restrictive layer as reported in the Soil Conservation Service (SCS) soil series description ("blue sheets") or, preferably, as measured for the soil. The actual leaching may be only 10-15% of that listed by SCS data yet to maintain a liberal estimate of runoff, leachate should be set at zero since waste application may affect the soil permeability. The depth of water applied monthly in the waste is calculated from water content of the waste, waste production rate, and total area of the land treatment unit watershed. In this example it is assumed that waste quality and quantity are relatively constant, but if it is known that these assumptions are false, monthly estimates will vary and can be ascertained from a more detailed accounting of the waste stream. For this example, water content of the waste is 70% and waste production rate (PR) is 20 metric tons or about 20,000 liters/day. The total watershed area of the HWLT unit is 6.6 hectares. Therefore, water application per month is calculated as

$$W(cm/mo) = \frac{PR \times \text{water content} \times 10^{-5} \times \text{\# of days in the month}}{\text{Watershed area (ha)}}$$

$$= \frac{2.0 \times 10^4 \text{ 1/day} \times 0.7 \times 10^{-5} \times \text{\# of days/month}}{6.6 \text{ ha}}$$

$$= 0.021 \text{ (days in the given month).}$$

Watershed area is generally larger than the unit area actively receiving wastes (A), to be determined later, but the watershed is a function of A. This is because for any unit area A, there are usually additional areas in the watershed made up of runoff ponds, waterways, roads, levees, etc.

Now using the water balance method from Section 8.3.1.1, first use the entire climatic record assuming zero discharge (Table E.1). The example shows only two years of record for illustrative purposes only. A much longer record is needed in practice (20 years if available). Since the last column in the table, cumulative storage, never drops to zero, some discharge or enhanced water loss will be necessary.

Next, one chooses a discharge rate (D) by taking the average annual increase in cumulative storage (CS) for the simulated period of record. In this example, CS is 9.66 and 8.76 for years one and two, respectively. The CS is thus 9.2 and D will assume a monthly value of 0.77 (9.2/12 = 0.77). Now rerun the simulated record, this time using the D term in the budget (Table E.2).

Based on the potential hazards of an uncontrolled release of water, a 0.10 probability is considered acceptable in this example. The storage value corresponding to this from the second run water budget is not readily apparent due to the short record.

If 20 years of data were available, then the highest annual value which is exceeded only in 10% of the years (i.e., in 2 years of the 20 years) would be chosen as the design value for normal seasonal storage. For convenience in this example, 15.62 cm storage is chosen.

In addition to this volume, capacity must be available to store the runoff from the 25-year, 24-hour storm. The 25-year, 24-hour rainfall for this site is 20.1 cm. Using the SCS curve number method described in Section 8.3.4, the runoff from the site would be 19.5 cm assuming antecedent moisture group III, fallow land use, and soil hydrologic group C.

Finally, management chooses to design an additional 10% volume for sludge and sediment buildup in the ponds. This would amount to 0.10(15.6 + 19.5) = 3.5 cm. Minimum freeboard (does not contribute to storage) of at least 60 cm should be provided above the 38.6 cm spillway level to guard against levee overtopping or failure. Since the HWLT unit area is 6.6 ha, this 38.6 cm storage translates into 254.75 ha-cm.

The assumption of zero leaching will be invalid in many circumstances, but it allows a sufficiently conservative water balance for safe retention pond design. Where leaching of waste constituents is of concern, however, better estimates of leaching are needed. In this case, use of the Perrier and Gibson (1980) computer model is suggested. Aside from computer techniques, a liberal leaching estimate can be estimated by assuming runoff and discharge are zero and setting leaching equal to the runoff values found in the first run of the water balance (Table E.1).

TABLE E.1   FIRST RUN WATER BALANCE, ASSUMING DISCHARGE RATE (D) EQUAL TO
ZERO

Month	Precip. (cm)	Water in Waste	Evaporation (cm)	Deep Percolation (cm)	Δ Storage (cm)	Cumulative Storage (cm)
S	6.4	0.63	6.0	0	1.03	1.03
O	6.0	0.65	5.4	0	1.25	2.28
N	6.5	0.63	4.6	0	2.53	4.81
D	8.1	0.65	3.8	0	4.95	9.76
J	8.2	0.65	4.0	0	4.85	14.61
F	7.2	0.59	4.9	0	2.89	17.50
M	6.7	0.65	6.1	0	1.25	18.75
A	8.3	0.63	6.9	0	2.03	20.78
M	7.8	0.65	7.6	0	−0.45	20.33
J	4.3	0.63	9.4	0	−4.47	15.86
J	5.4	0.65	10.6	0	−4.55	11.31
A	6.4	0.65	8.7	0	−1.65	9.66
S	5.2	0.63	6.3	0	−0.47	9.19
O	5.8	0.65	5.2	0	1.25	10.44
N	9.4	0.63	4.7	0	5.33	15.77
D	7.3	0.65	4.1	0	3.85	19.62
J	6.1	0.65	4.1	0	3.85	19.62
F	6.3	0.59	5.1	0	1.79	23.86
M	6.9	0.65	5.9	0	1.65	25.51
A	9.8	0.63	6.8	0	3.63	29.14
M	8.2	0.65	7.5	0	1.35	30.49
J	5.0	0.63	9.7	0	−4.07	26.42
J	4.1	0.65	10.8	0	−6.05	20.37
A	5.8	0.65	8.4	0	−1.95	18.42

TABLE E.2   SECOND RUN WATER BALANCE, ASSUMING CONSTANT DISCHARGE RATE (D)
OF 0.77 CM/MO

Month	Precip. (cm)	Water in Waste	Evapo-ration (cm)	Deep Perco-lation (cm)	Discharge (cm)	Δ Storage (cm)	Cumulative Storage (cm)
S	6.4	0.63	6.0	0	0.77	0.26	0.26
O	6.0	0.65	5.4	0	0.77	0.48	0.74
N	6.5	0.63	4.6	0	0.77	1.76	2.50
D	8.1	0.65	3.8	0	0.77	4.18	6.68
J	8.2	0.65	4.0	0	0.77	4.08	10.76
F	7.2	0.59	4.9	0	0.77	2.12	12.88
M	6.7	0.05	6.1	0	0.77	0.48	13.36
A	8.3	0.63	6.9	0	0.77	1.26	14.62
M	7.8	0.65	7.6	0	0.77	0.08	14.70
J	4.3	0.63	9.4	0	0.77	−5.24	9.46
J	5.4	0.65	10.6	0	0.77	−5.32	4.14
A	6.4	0.65	8.7	0	0.77	−2.42	1.72
S	5.2	0.63	6.3	0	0.77	−1.24	0.48
O	5.8	0.65	5.2	0	0.77	0.48	0.96
N	9.4	0.63	4.7	0	0.77	4.56	5.52
D	7.3	0.65	4.1	0	0.77	3.08	8.60
J	6.1	0.65	4.3	0	0.77	1.68	10.28
F	6.3	0.59	5.1	0	0.77	1.02	11.30
M	6.9	0.65	5.9	0	0.77	0.88	12.18
A	9.8	0.63	6.8	0	0.77	2.86	15.04
M	8.2	0.65	7.5	0	0.77	0.58	15.62
J	5.0	0.63	9.7	0	0.77	−4.84	10.78
J	4.1	0.65	10.8	0	0.77	−6.82	3.96
A	5.8	0.65	8.4	0	0.77	−2.72	1.24

LOADING RATE CALCULATIONS FOR MOBILE
NONDEGRADABLE CONSTITUENTS

Since mobile constituents are relatively free to migrate to the groundwater, some limits should be set for the acceptable leachate concentration of each species. The following concentrations in the leachate will be assumed to be the acceptable maxima (Table 6.48 contains a list of other elements). These values are the permissible water criteria for public drinking water supplies.

Constituent	Concentration in Water mg/1
N	100.0
Se	0.01
Cl	250.0

The values to be used in actual design may vary from site to site depending on the state regulations or the possible use of the groundwater. The leachate concentration limits may be used in conjunction with the composition of the waste and the depth of water leaching water (Appendix E-1) to compute the amount of a given waste that, if applied, will result in the maximum acceptable concentration in the leachate.

All soils will have some capacity to adsorb and retain limited amounts of mobile species. Additionally, plants may take up N, Se and Cl. If the adsorption capacity and plant uptake rates are known, they may be taken into account in the calculation. Once the adsorption capacities are satisfied, however, subsequent additions will likely leach to the groundwater. Since plant uptake is limited and sorption capacities will eventually be satisfied, it is best to calculate the required treatment area assuming that both are negligible.

For example, a waste containing 10 mg/kg Se and 580 mg/kg Cl is produced at a rate of 20 metric dry tons/day and is to be land treated on a site having an estimated leaching rate of 29 cm/yr. From the above information, the following can be computed.

Constituent	Concentration in Waste mg/kg	Annual Application (kg/yr)	Waste Loading Rate (kg/ha/yr)
Se	10	7.3	1.5×10^6
Cl	580	420	1.3×10^6

Chloride is the most limiting of the mobile constituents, with a maximum waste loading rate of 1.3×10^6 kg/ha/yr to maintain leachate concentrations at or below 250 mg/1.

APPENDIX E-3

CALCULATION OF WASTE APPLICATIONS BASED ON NITROGEN CONTENT

The fate of applied nitrogen (N) in soil has been extensively discussed in Section 6.1.2.1. There are many processes by which N may be lost from the system, but N transported in runoff and leachate water is of primary interest since it can have an adverse impact on the environment. Since direct discharge from HWLT units will be prevented, only the N concentration leaving the site in the leachate is generally of concern. Typically, 10 ppm nitrate-nitrogen is taken as the upper limit for drinking water and as the upper limit of acceptable leachate concentration. The equations used to calculate the acceptable load of nitrogen-containing waste are given in Section 7.5.3.4 and are shown below:

$$LR = 10^5 \; \frac{10(C + V + D) + (L_d)(L_c) - (P_d)(P_c)}{I + \sum\limits_{t=1}^{n} (M)(O)}$$

where

LR = waste loading rate (kg/ha/yr);
C = crop uptake of N (kg/ha/yr);
V = volatilization (kg/ha/yr);
D = denitrification (kg/ha/yr);
L_d = depth of leachate (cm/yr);
L_c = N concentration in leachate (mg/l);
P_d = depth of precipitation (cm/yr);
P_c = concentration of inorganic N in the waste (mg/l);
I = concentration of inorganic N in the waste (mg/l);
M = mineralization rate given in Table 6.4;
O = concentration of organic N in the waste (mg/l); and
t = years of waste application.

Example

A waste containing 30 mg/l inorganic N and produced at a rate of 20 metric tons/day, is to be land treated. From this information and that in Table E.3 loading calculations can be made and are shown in the following equation:

$$LR = 10^5 \; \frac{10(C + V + D) + (L_d)(L_c) - (P_d)(P_c)}{I + (M)(O)}$$

$$= 10^5 \; \frac{10(280 + 0 + 0) + (29)(10) - (63.5)(.5)}{30 + (.35)(260)}$$

$$= 2.53 \times 10^6 \; kg/ha/yr$$

TABLE E.3   WASTE CHARACTERISTICS USED IN EXAMPLE FOR NITROGEN
LOADING RATE CALCULATIONS

Parameter	Value
I (mg/1)	30
L_c (mg/1)	10
O (mg/1)	260
P_c (mg/1)	0.5
M	0.35, 0.1, 0.05
P_d (cm/yr)	63.5
C (kg/ha/yr)	280
D (kg/ha/yr)	0
V (kg/ha/yr)	0
L_d (cm/yr)	29
ρ (cm/gm^3)	1

APPENDIX E-4

EXAMPLES OF PHOSPHORUS LOADING CALCULATIONS

The equation presented in Section 7.5.3.5 is used to calculate the acceptable phosphorus application limit. Among the parameters that must be known are soil horizon depth (d_i), the P sorption capacity (b_{max}), P content of the waste, (P_{ex}), the rate of waste production, and the crop cover, if any. Using these values and the equation one can calculate the area needed for land treatment of a waste containing P.

A waste having wet weight P content of 2000 mg/kg is to be land treated on a soil having a 20 cm deep A horizon, 30 cm deep B horizon and 50 cm deep C horizon. The sorption capacities of the horizons are 54, 23 and 89 mgP/100 g, respectively.

Horizon	Depth (cm)	ρ g/cm^3	b_{max} mg/kg	P_{ex} mg/kg
A	20	1.3	540	2
B	30	1.35	230	1
C	50	1.45	890	3

The applicable equation $LC = (10) \sum_{t=1}^{n} d_i \, \rho (b_{max} - P_{ex})$

where

d_i = thickness of the ith horizon;
ρ = bulk density of the soil (g/cm^3);
b_{max} = P sorption capacity estimated from Langmiur isotherms (mg/kg);
P_{ex} = NaHCO$_3$ extractable P (mg/kg); and
LC = phosphorus loading capacity (kgP/ha).

Using the above data the P loading rate can be calculated as follows:

$$LC = 10 \sum_{t=1}^{n} (20)(1.3)(540 - 2) + 10 \sum_{t=1}^{n} (30)(1.35)(230-1)$$

$$+ 10 \sum_{t=1}^{n} (50)(1.45)(890 - 3)$$

$$= 139,800 + 92,745 + 643,075 = 875,700 \text{ kg P/ha.}$$

The phosphorus loading capacity (LC) of the soil is 875,700 kgP/ha, which for a waste containing 2000 mg P/kg is equivalent to a waste loading capacity of

$$\frac{875,700 \text{ kgP/ha}}{2000 \text{ kgP/10}^6 \text{ kg waste}} = 4.38 \times 10^8 \text{ kg waste/ha}$$

CHOICE OF THE CAPACITY LIMITING CONSITUTENT

The example contained in this section is designed to illustrate the appropriate approach to identifying the potential capacity limiting constituent from among the conserved species of a waste. Conserved refers to those constituents, usually only metals, which are practically immobile and nondegradable in the soil. It is important to be sure that the soil pH is at or adjusted to 6.5 or above before application. The soil CEC needs to be measured and if less than 5, the loading capacities should be reduced by 50%. For most purposes, the loading capacities presented in Table 6.47 are acceptable estimates.

A waste is to be land treated on a soil that has a pH of 7.0 and a CEC of 12.0 meq/100g. The choice of potential CLC is made easily using the ratio of each metal concentration in the waste residual solids fraction (RS) to its respective acceptable concentration in the soil as shown in the table below. The most limiting metal is Cr since it has the largest ratio, 4.1.

TABLE E.4   CHOICE OF CAPACITY LIMITING CONSERVED SPECIES BY THE RATIO METHOD

Metal	mg/kg in Waste Residual Solids	Metal Loading Capacity* (mg/kg)	Ratio
As	230	300	1.30
Cr	4,097	1,000	4.1
Cd	3.4	3	1.13
Cu	4.98	250	0.02
Pb	1,740	1,000	1.74
Ni	53	100	0.53
V	387	500	0.77
Zn	96	500	0.19

* Taken from 6.47.

ORGANIC LOADING RATE CALCULATIONS

This appendix includes examples of waste characteristics and the calculations which are used to determine the organic loading rate for each waste. The second example is the general example being used elsewhere in this appendix. The greenhouse and respirometer studies that can be used to generate data for these calculations are described in Sections 7.3 and 7.2.1, respectively. The first step in determining the organic loading rate is to determine the phytotoxicity or microbial toxicity limit. This limit is used as the maximum tolerable level of organic waste constituents from which the organic half-life is determined. There are two equations which are used in the determination of organic half-life. The first equation determines the fraction of the applied carbon evolved as CO_2.

$$D_t = \frac{(CO_2w - CO_2s)0.27}{C_a}$$

where

D_t = the portion of the applied carbon which is evolved as from the organic fraction after time t.
CO_2w = the cumulative CO_2 evolved by waste amended soil;
CO_2s = the cumulative CO_2 evolved by unamended soil;
t = time; and
C_a = carbon applied.

In addition to the fraction calculated from equation 1, the rate of degradation should be determined for the extractable organics and organic subfractions using the following equation:

$$D_{t_i} = \frac{C_{a_i} - C_{r_i} - C_{s_i}}{C_{ai}}$$

where

d_{t_i} = the portion of the carbon degraded from the organic fraction or fraction 1, 2 or 3;
c_{a_i} = the carbon applied in the organic fraction or fractions 1, 2 or 3;
c_{r_i} = the residual carbon in the organic fraction or fraction 1, 2 or 3; and
c_{s_i} = the background concentration in unamended soil of the organic fraction 1, 2 or 3.

The loading rate can be calculated for the bulk organic fraction or for any subfraction of interest which may better indicate the rate of degradation of the hazardous constituents.

The residual values given in the waste characteristics tables were calculated with the soil carbon content already subtracted. The lowest fraction of organics degraded (D_t) as calculated above is used to determine the halflife of the waste, as follows:

$$t_{1/2} = \frac{0.5t}{D_t}$$

The half-life is then used to calculate the organic loading rate in (C_{yr}) in kg/ha/yr.

$$C_{yr} = 1/2 \frac{1}{t_{1/2}} C_{crit}$$

where C_{crit} is the maximum tolerable limit (kg/ha) of organic waste constituents as determined by plant or microbial toxicity. This loading rate is based on laboratory data obtained under controlled conditions, and should be verified by field data. It is assumed that the waste has been demonstrated to be land treatable and will also be monitored in the field. The units are derived from laboratory data, an assumed plow or mixing depth, and the waste-soil mix bulk density.

The bulk waste loading rate (LR) based on organics applied is calculated as follows:

$$LR = (C_{yr})/C_w$$

where C_w is the fraction of the bulk waste constituted by degradable organics.

Example 1: An oil waste which is produced at a rate of 20 metric dry tons/day is to be land treated on a vegetated site. C_{crit} is determined to be 2.7% (1.2×10^5 kg/ha-15 cm) organics in soil. Waste characteristics are as follows (Data from Schwendinger (1968):

Waste characteristics:

Extractable organics (mg)	Total	F_1	F_2	F_3
Carbon applied (C_a)	2500mg	Data not given		
Carbon residual (C_r)		Data not given		
Respiration data – CO_2 (mg)	Day 14	28	49	
Waste + soil	620	1563	2104	
soil	20	63	104	

Calculations:

1) Residual Carbon:

   data not given

2) Evolved CO_2:

$$D_{49} = \frac{(2104-104).27}{2500} = 0.22$$

3) Half-life:

$$t_{1/2} = \frac{0.5t}{D_t} = 111 \text{ days} = .30 \text{ yr}$$

4) Organic loading rate:

$$C_{yr} = 1/2(1.2 \times 10^5 \text{ kg/ha}) \frac{1}{t_{1/2}} = 2 \times 10^5 \text{ kg/ha/hr}$$

5) $LR = \dfrac{2 \times 10^5}{0.10} = 2 \times 10^6$ kg/ha/yr

where the organic content of the waste C_w is 10% (0.10).

Greenhouse studies indicated that 2.7% oil in soil reduces the yield of rye grass by 25% compared to the yield of unamended soil, therefore C_{crit} is 2.78% or 1.2×10^5 kg/ha. A respiration study was conducted for 49 days and the cumulative CO_2 evolved determined for the entire time period. The percent of carbon evolved as CO_2 was calculated to be 22% over the 49 day period. The half-life of the carbon applied was then calculated to be 111 days, or 0.30 years. Using the half-life value, it was then determined that 2×10^5 kg/ha/yr oil or 2×10^6 kg waste/ha/yr could be applied to the soil at the land treatment facility while still retaining a vegetative cover. One limitation of this study is that no information is provided which describes the degradation of the organic subfractions.

Example 2: An API separator sludge from a petroleum refinery is
produced at a rate of 20 metric tons/day and is to be land
treated. The site will be vegetated with ryegrass. Waste
characteristics are as follows (Brown et al., 1980):

Waste characteristics:

Extractable organics (mg)	Total	F_1	F_2	F_3
Carbon applied (C_a)	550	396	121	33
Carbon residual (C_r)	220	153	52	14
Respiration data – CO_2 (mg)	Day 45	90	135	180
Waste + soil	675	954	1111	1241
soil	85	149	215	271

Calculations:

1) Residual Carbon:

$$D_{t\text{o}} = \frac{550-220}{550} = .6$$

$$D_{t1} = \frac{396-153}{396} = .61$$

$$D_{t2} = \frac{121-52}{121} = .57$$

$$D_{t3} = \frac{33-14}{33} = .58$$

2) Evolved CO_2:

$$D_{180} = \frac{(1241-271).27}{550} = .48$$

3) Half-life:

$$t_{1/2} = \frac{.50}{D_t}\, t = \frac{.50}{.48}(180) = 187 = .51\,\text{yr}$$

4) Organic loading rate:

$$C_{yr} = 1/2\ (C_{crit})\ \frac{1}{t_{1/2}} = 1/2(2.2 \times 10^5\ \frac{\text{kg}}{\text{ha}})\frac{1}{(.51\ \text{yr})} = 2.2 \times 10^5\ \text{kg/ha/yr}$$

5) $LR = \dfrac{2.2 \times 10^5}{0.10} = 2.2 \times 10^6\ \text{kg/ha/yr}$

It was determined in a greenhouse study that the yield of rye grass 100 days after application of 5% wt/wt (2.2×10^5 kg/ha) sludge was reduced 40% below control yields. After 180 days of incubation in a soil respirometer, the hydrocarbon was extracted and separated into subfractions. Data analysis indicated that the slowest rate of degradation was for carbon evolved as CO_2; the value 48% was used to calculate the half-life which was determined to be 187 days. This value was then used to determine the maximum loading rate with plant cover which was 2.2×10^5 kg organics/ha/yr. For this organics application rate, 2.2×10^6 kg/ha/yr of bulk sludge would be applied to the top 30 cm of soil.

CALCULATIONS OF FACILITY SIZE AND LIFE

The waste loading rate, unit size and the unit life are dependent on the waste and site characteristics. For the following calculations, the characteristics of the waste, the climate, and the soil used in the above examples (Appendices E-1 through E-6) will be assumed, and the resulting design conditions will be determined.

For the case under study, the RLC and the design waste loading rate are determined by a tabular comparison of values previously calculated for each waste constituent (Table E.5). By comparison, the RLC is found to be bulk organics degradation with a loading rate of 2.2×10^6 kg/ha/yr. For this example, no constituent was found to limit the size of individual applications (ALC).

Calculation of the required land treatment unit area is done using the equation from Section 7.5.4.

$$A = \frac{PR}{LR}$$

where

A = required treatment area (ha);
PR = waste production rate (kg/yr) on a wet weight basis; and
LR = waste loading rate (kg/ha/yr) on a wet weight basis.

Waste production is 20 metric tons/day, so the required area is as follows:

$$A = \frac{20 \text{ mt/day}(10^3 \text{ kg/mt})365 \text{ days/yr}}{2.2 \times 10^3 \text{ kg/ha/yr}} = 3.3 \text{ ha}$$

The capacity limiting constituent and unit life are determined by calculating unit lives for chromium (Cr) (the most limiting conserved species) and phosphorus and choosing the more restrictive value. For phosphorus, unit life is easily determined directly from the equation:

$$UL = \frac{LCAP}{LR_{RLC}}$$

where

UL = unit life in years;
$LCAP$ = maximum allowable waste load based on phosphorus (kg/ha); and
LR_{RLC} = loading rate based on RLC (kg/ha/yr).

This reduces to: $\dfrac{4.4 \times 10^8 \text{ kg/ha}}{2.2 \times 10^6 \text{ kg/ha/yr}}$

for the example and thus the facility will last 200 years based on phosphorus.

For chromium calculations, several choices or determinations must be made. In this case, assume a plow layer (Z_p) of 30 cm and a time between applications (t_a) of 1 for each plot. Given that the residual solids (RS) content of the waste is 0.2 and a bulk density (ρ_{BRS}) of the residual solids mix of 1.4 kg/l, the application depth (Z_a) is found as follows:

$$Z_a = \frac{LR_{RLC} \times RS}{\rho_{BRS}} \times 10^{-5}$$

$$= \frac{2.2 \times 10^6 (0.2)}{1.4} \times 10^{-5}$$

$$= 3.1 \text{ cm}$$

The background soil contains 100 mg/kg Cr (C_{po}), the application limit (C_{pn}) for Cr from Table 6.47 is 1000 mg/kg, and the given concentration of Cr in the waste residual solids (C_a is 4097 mg/kg. The number of applications of waste (n) may be made can thus be calculated:

$$N = \frac{Z_p}{Z_a} \ln \frac{C_{po} - C_a}{C_{pn} - C_a}$$

$$= \frac{30}{3.1} \ln \frac{100 - 4097}{1000 - 4097}$$

$$= 2.5$$

Unit life (UL) based on Cr is n t_a, and since t_a is one year, UL equals 2.5 yr. Comparing this with results for phosphorus, Cr is more limiting and is thus the CLC. In addition to hazardous constituents, the above results aid in the choice of monitoring parameters in the subsequent site monitoring program.

TABLE E.5 WASTE CONSTITUENTS TO BE COMPARED IN DETERMINING WASTE
APPLICATION (ALC) AND RATE (RLC) LIMITING CONSTITUENT

Constituent	Potential ALC (kg/ha/application)	Potential RLC (kg/ha/yr)
Organics	x	2.2×10^6
o volatization	x	
o leaching	x	
o degradation		2.2×10^6
Nitrogen	x	2.53×10^6
Inorganic acids, bases and salts		x
Halides		x

APPENDIX E REFERENCES

Brown, K. W., K. C. Donnelly, J. C. Thomas, and L. E. Deuel. 1980. Factors influencing biodegradation of an API separator sludge applied to soils.

Perrier, E. R. and A. C. Gibson. 1980. Hydrologic simulation on solid waste disposal sites (HSSWDS). Prepared for the U.S. EPA Municipal Environmental Research Laboratory. SW-868.

Schwendinger, R. B. 1968. Reclamation of soil contaminated with oil. J. Inst. Petroleum 54 (535):182-197.

GLOSSARY

acute toxicity: An adverse effect which occurs shortly after exposure to a substance.

adsorption: The attraction of ions or compounds to the surface of a solid. Soil colloids adsorb large amounts of ions and water.

aerosols: Microscopic droplets dispersed in the atmosphere.

ammonification: The biochemical process whereby ammoniacal nitrogen is released from nitrogen-containing organic compounds.

anaerobic: (i) The absence of molecular oxygen. (ii) Growing in the absence of molecular oxygen (such as anaerobic bacteria). (iii) Occurring in the absence of molecular oxygen (as a biochemical process).

annual crop: A crop which completes its entire life cycle and dies within 1 year or less; i.e., corn, beans.

application limiting constituent (ALC): A compound, element, or waste fraction in a hazardous waste which restricts the amount of waste which can be loaded onto soil per application (kg/ha/application).

aquifer: Stratum below the surface capable of holding water.

673

available water: The portion of water in a soil that can be readily absorbed by plant roots. Considered by most workers to be that water held in the soil against a pressure of up to approximately 15 bars.

base-pair mutation: Substitution mutation in which the wrong base is inserted into the DNA which then pairs with its natural partner during replication which results in a new pair of incorrect bases in the DNA.

base-saturation percentage: The extent to which the adsorption complex of a soil is saturated with exchangeable cations other than hydrogen. It is expressed as a percentage of the total cation-exchange capacity.

biodegradation: The breaking down of a chemical compound into simpler chemical components under naturally occurring biological processes.

bulk density: The mass of dry soil per unit bulk volume. The bulk volume is determined before drying to constant weight at 105°C.

calcareous soil: Soil containing sufficient calcium carbonate (often with magnesium carbonate) to effervesce visibly when treated with cold 0.1N hydrochloric acid.

capacity limiting constituent (CLC): A compound, element, or waste fraction in a hazardous waste which restricts the total amount of waste which can be loaded onto soil (kg/ha).

carbon cycle: The sequence of transformations whereby carbon dioxide is fixed in living organisms by photosynthesis or by chemosynthesis, liberated by respiration and by the death and decomposition of the fixing organism, used by heterotrophic species, and ultimately returned to its original state.

carbon-nitrogen ratio: The ratio of the weight of organic carbon to the weight of total nitrogen in a soil or in organic material. It is obtained by dividing the percentage of organic carbon (C) by the percentage of total nitrogen (N).

carcinogen: A chemical, physical, or biological agent which induces formation of cells that are no longer affected by normal regulations of growth; such formations are capable of spreading cells to other tissues resulting in the loss of the specific function of such tissues.

cation exchange: The reversible exchange between a cation in solution and another cation adsorbed onto any surface-active material such as clay or organic matter.

cation exchange capacity: The sum total of exchangeable cations that a soil can adsorb. Sometimes called "total-cation capacity," "base-exchange capacity," or "cation-adsorption capacity." Expressed in milliequivalents per 100 grams of soil (or of clay).

chelating properties: The property of certain chemical compounds in which a metalic ion is firmly combined with the compound by means of multiple chemical bonds.

chromosome aberration: Changes in the number, shape, or structure of chromosomes.

chronic toxicity: A prolonged health effect which may not become evident until many years after exposure.

clay: (i) Soil separate consisting of particles <0.002 mm in equivalent diameter. (ii) Soil material containing more than 40 percent clay, less than 45 percent sand and less than than 40 percent silt.

compost: Organic residues, or a mixture of organic residues and soil, that have been piled, moistened, and allowed to undergo biological decomposition. Often called "artifical manure" or "snythetic manure" if produced primarily from plant residues.

composite: To make up a sample of distinct portions so the sample is representative of the total material being sampled rather than any single portion.

denitrification: The biochemical reduction of nitrate or nitrite to gaseous nitrogen either as molecular nitrogen or as an oxide nitrogen.

diversion terrace: A terrace to divert runoff from the watershed above the land treatment area.

DNA repair: Repair of genetic material by cellular enzymes which can excise or recombine alterations in structure of DNA to restore original information.

drain tile: Concrete or ceramic pipe used to conduct water from the soil.

effluent: The liquid substance, predominately water, containing inorganic and organic molecules of those substances which do not precipitate by gravity.

electrical conductivity: An expression of the readiness with which an electrical impulse (generated by ionic activity) flows through a water or soil system.

erosion: (i) The wearing away of the land surface by running water, wind, ice, or other geological agents, including such processes as gravitational creep. (ii) Detachment and movement of soil or rock by water, wind ice, or gravity.

eutrophication: The reduction of dissolved oxygen in surface waters which leads to the deterioration of the aesthetic and life-supporting qualities.

evapotranspiration: The combined loss of water from a given area, and during a specified period of time, by evaporation from the soil surface and by transpiration from plants.

exchange acidity: The titratable hydrogen and aluminum that can be replaced from the adsorption complex by a neutral salt solution. Usually expressed as milliequivalents per 100 grams of soil.

fertility, soil: The status of a soil with respect to the amount and availability to plants of elements necessary for plant growth.

field capacity (field moisture capacity): The amount of water remaining in the soil after excess gravitational water has drained away and after downward movement of water has practically ceased (normally considered to be about 1/3 bar soil moisture tension).

forage crop: A crop such as hay, pasture grass, legumes, etc., which is grown primarily as forage or feed for livestock.

frameshift mutation: Mutation resulting from insertion or deletion of a base-pair from a triplet codon in the DNA; the insertion or deletion produces a scrambling of the DNA or a point mutation.

gene mutation: A stable change in a single gene.

genetic toxicity: An adverse event resulting in damage to genetic material; damage may occur in exposed individuals or may be expressed in subsequent generations.

groundwater: Water that fills all of the unblocked pores of materials underlying the water table, which is the upper limit of saturation.

heavy metals: Generally, those elements in the periodic table of elements which belong to the transition elements. They may include plant essential micronutrients and other nonessential elements. Examples are mercury, chromium, cadmium and lead.

heterotrophic organism: Requires preformed, organic nutrients as a source of carbon and energy.

hydraulic conductivity: The proportionality factor in Darcy's law as applied to the viscous flow of water in soil, i.e., the flux of water per unit gradient of hydraulic potential.

hydrologic cycle: The fate of water from the time of precipitation until the water has been returned to the atmosphere by evaporation and is again ready to be precipitated.

infiltration rate: A soil characteristic determining or describing the maximum rate at which water can enter the soil under specified conditions, including the presence of an excess of water. It has the dimensions of Velocity (i.e., cm^3 cm^{-2} sec^{-1} = cm sec^{-1}).

land treatment: The controlled application of hazardous wastes onto or into the aerobic surface soil horizon, accompanied by continued monitoring and management, to alter the physical, chemical, and biological state of the waste to render it less hazardous. The practice simultaneously constitutes treatment and final disposal.

leachate: Soil solution moving toward the groundwater under the pull of gravity.

lime requirement: The mass of agricultural limestone, or the equivalent of other specified liming material, required per acre to a soil depth of 15 cm to raise the pH of the soil to a desired value under field conditions.

loading rate: The mass or volume of waste applied to a unit area of land per unit time (kg/ha/yr).

lysimeter: (i) A container used to enclose a volume of soil and its contents and associated equipment used to measure the evaporative and/or drainage components of the hydrological balance. (ii) A device used to collect soil solution from the unsaturated zone.

metabolic activation: The use of extracts of plant or animal tissue to provide enzymes which can convert a promutagen into an active mutagen, or a procarcinogen into an active carcinogen.

metal toxicities: Toxicities arising from too great a content of metals in the soil.

micelle: A minute silicate clay colloidal particle that generally carries a negative charge.

microorganism: An organism so small it cannot be seen clearly without the use of a microscope.

moisture volume percentage: The ratio of the volume of water in a soil to the total bulk volume of the soil.

moisture weight percentage: The moisture content expressed as a percentage of the oven-dry weight of soil.

mulch: (i) Any material such as straw, sawdust, leaves, plastic film, loose soil, etc., that is spread upon the surface of the soil to protect the soil and plant roots from the effects of raindrops, soil crusting, freezing, evaporation, etc. (ii) To apply mulch to the soil surface.

mutagenic: Compounds with the ability to induce stable changes in genetic material (genes and chromosomes).

nitrification: The biochemical oxidation of ammonium to nitrate.

permeability, soil: (i) The ease with which gases, liquids, or plant roots penetrate or pass through a bulk mass of soil or a layer of soil. Since different soil horizons vary in permeability, the particular horizon under question should be designated. (ii) The property of a porous medium itself that relates to the ease with which gases, liquids, or other substances can pass through it.

pH, soil: The negative logarithm of the hydrogen-ion activity of a soil. The degree of acidity (or alkalinity) of a soil as determined by means of a glass, quinhydrone, or other suitable electrode or indicator at a specified moisture content or soil-water ratio, and expressed in terms of the pH scale.

primary degradation: Conversion of waste constituents into a form which no longer responds in the same manner to the analytical measurement used for detection.

rate limiting constituent (RLC): A compound, element, or waste fraction in a hazardous waste which restricts the amount of waste which can be loaded onto soil per year (kg/ha/yr).

respirometer: An apparatus which can be used to measure microbial activity and monitor waste decomposition under controlled environmental conditions.

retention basin: A basin or pond used to collect or store runoff water.

runoff: Any rainwater, leachate, or other liquid that drains over land from any part of a waste treatment or disposal facility. That which is lost without entering the soil is called surface runoff and that which enters the soil before reaching the stream is called groundwater runoff or seepage flow from groundwater. (In soil science, "runoff" usually refers to the water lost by surface flow; in geology and hydraulics, "runoff" usually includes both surface and subsurface flow.)

run-on: Any rainwater, leachate, or other liquid that drains onto any waste treatment area.

sand: (i) A soil particle between 0.05 and 2.0 mm in diameter. (ii) Any one of five soil separates, namely: very coarse sand, coarse sand, medium sand, fine sand, and very fine sand.

silt: A soil separate consisting of particles between 0.002 and 0.05 mm in equivalent diameter.

soil horizon: A layer of soil or soil material approximately parallel to the land surface and differing from adjacent genetically related layers in physical, chemical, and biological properties of characteristics such as color, structure, texture, consistency, kinds, and numbers of organisms present, degree of acidity or alkalinity, etc.

soil profile: A vertical section of the soil from the surface through all its horizons, including C horizons.

soil series: The basic unit of soil classification being a subdivision of a family and consisting of soils which are essentially alike in all major profile characteristics except the texture of the A horizon.

soil solution: The aqueous liquid phase of the soil and its solutes consisting of ions dissociated from the surfaces of the soil particles and of other soluble materials.

soil texture: The relative proportion of the various soil separates in a soil. The textural classes may be modified by the addition of suitable adjectives when coarse fragments are present in substantial amounts; for example, "stony silt loam," or "silt loam, stony phase."

sorption: See "adsorption."

subsurface injection: A method applying fertilizer and waste materials in a band below the soil surface.

suspended solids: Solid particles which do not precipitate out of solution or do not easily filter out. They may be colloidal in nature.

terrace: (i) A raised, more or less level or horizontal strip of earth usually constructed on or nearly on a contour and supported on the downslope side by rocks or other similar barrier to prevent accelerated erosion. (ii) An embankment with the uphill side sloping toward and into a channel for conducting water, and the downhill side having a relatively sharp decline, constructed across the direction of the slope to conduct water from the area above the terrace at a regulated rate of flow and to prevent the accumulation of large volumes of water on the downslope side.

toxicity: The ability of a material to produce injury or disease upon exposure, ingestion, inhalation, or assimilation by a living organism.

treatment zone: the area of a land treatment unit that is located wholly above the saturated zone and within which degradation, transformation, or immobilization of hazardous constituents occurs.

uniform area: Area of the active portion of an HWLT unit which is composed of soils of the same soil series and to which similar wastes are applied at similar application rates.

unsaturated flow: The movement of water in a soil which is not filled to capacity with water.

uptake: The process by which plants take elements from the soil. The uptake of a certain element by a plant is calculated by multiplying the dry weight by the concentration of the element.

volatilizaton – vaporization: The conversion of a liquid or solid into vapors.

waste: Any liquid, semiliquid, sludge, refuse, solid, or residue under consideration for disposal.

watershed: The total runoff from a region which supplies the water of a drainage channel.

water table: The upper surface of ground water or that level below which the soil is saturated with water; locus of points in soil water at which the hydraulic pressure is equal to atmospheric pressure.

USEFUL LAND TREATMENT CONVERSION FACTORS

1. a. 1 cubic yard (yd.^{3}) = 27 cu. ft. (ft^3)
   b. 1 gal. water = 8.34 lb.

2. a. 1 acre-inch of liquid = 27,150 gallons = 3,630 ft^3 = 102,800 liters = 0.01028 hectare-meters
   b. 1 hectare-cm of liquid = 100,000 liters = 100 m^3

3. 1 metric ton = 1,000 kg. = 2,205 lb.

4. cu. feet per second x 5.39 x mg./liter = lb./day

5. a. million gallons per day x 8.34 x mg./liter = lb./day
   b. (8.34 x 10^{-3}) x mg./liter = lb./1,000 gal.

6. 1 acre = 4,480 yards^2 = 43,560 feet^2 = 4,047 meters^2 = 0.4047 hectare

7. acre-inches x 0.266 x mg./liter = lb./acre

8. ha.-cm. x 0.1 x mg./liter = kg./hectare

9. English-metric conversions
   a. acre-inch x 102.8 = meter^3
   b. quart x 0.946 = liter
   c. English ton x 0.907 = metric ton
   d. English tons/acre x 2.242 = metric tons/hectare
   e. lb./acre x 1.121 = kg./hectare
   f. 1 lb. = 0.454 kg.

10. a. lbs. P x 2.3 = lbs. P_2O_5
    b. lbs. K x 1.2 = lbs. K_2O

11. Sludge conversions in English units
    a. wet tons sludge x % dry solids/100 = dry tons sludge
    b. wet tons/.85 = cubic yards sludge*
    c. wet tons sludge x 240 = gallons sludge*
    d. 1,700 lb. wet sludge = 1 yd^3 wet sludge*

12. Concentration conversions
    a. 10,000 ppm = 1%
    b. % x 20 = lb./ton
    c. (ppm/500) or (ppm x .002) = lb./ton

13. Wet weight conversions
    a. micrograms/milliliter (μg/ml) = milligrams/liter (mg/1) ppm (wet)
    b. ppm (wet x 100/% solids = ppm (dry)

14. Rate Conversions
    a. 1 lb/acre = 1.12 kg/ha
    b. 1 ton/acre = 2.24 ton (metric)/hectare

* Assumes a sludge density of about 1 g/cm^3.

CONVERSION FACTORS
U.S. Customary to SI (Metric)

U.S. Customary Unit Name	Abbreviation	Multiplier	Symbol	SI Name
acre	acre	0.405	ha	hectare
acre-foot	acre-ft	1.234	m³	cubic meter
cubic foot	ft³	28.32	1	liter
		0.0283	m³	cubic meter
cubic feet per second	ft³/s	28.32	1/s	liters per second
degrees Fahrenheit	°F	0.555(°F-32)	°C	degrees Celsius
feet per second	ft/s	0.305	m/s	meters per second
foot (feet)	ft	0.305	m	meter(s)
gallon(s)	gal	3.785	1	liter(s)
gallons per acre per day	gal/acre.d	9.353	1/ha.d	liters per hectare per day
gallons per day	gal/d	4.381x10⁻⁵	1/s	liters per second
gallons per minute	gal/min	0.0631	1/s	liters per second
horsepower	hp	0.746	kw	kilowatt
inch(es)	in.	2.54	cm	centimeter(s)
inches per hour	in./h	2.54	cm/h	centimeters per hour
mile	mi	1.609	km	kilometer
miles per hour	mi/h	0.45	m/s	meters per second
million gallons	Mgal	3.785	Ml	megaliters (liter x 10⁶)
million gallons per acre	Mgal/acre	8.353	m³/ha	cubic meters per hectare
million gallons per day	Mgal/d	43.8	1/s	liters per second
parts per million	ppm	1.0	mg/l	milligrams per liter
pound(s)	lb	0.454	kg	kilogram(s)
pounds per acre per day	lb/acre.d	1.12	kg/ha.d	kilograms per hectare per day
pounds per square inch	lb/in.²	0.069	kg/cm²	kilograms per square centimeter
square foot	ft²	0.0929	m²	Newtons per square centimeter
				square meter
square inch	in.²	6.452	cm²	square centimeter
square mile	mi²	2.590	km²	square kilometer
ton (short)	ton (short)	0.907	Mg (or t)	megagram (metric ton)
tons per acre	tons/acre	2.24	Mg/ha	megagrams per hectare

INDEX